Protective Coatings

Mei Wen • Karel Dušek
Editors

Protective Coatings

Film Formation and Properties

Springer

AXALTA®

Editors
Mei Wen
Axalta Coating Systems
Coatings Technology Center
Wilmington, DE, USA

Karel Dušek
Institute of Macromolecular Chemistry
Academy of Sciences of the Czech Republic
Prague, Czech Republic

ISBN 978-3-319-51625-7 ISBN 978-3-319-51627-1 (eBook)
DOI 10.1007/978-3-319-51627-1

Library of Congress Control Number: 2016962656

Printed on acid-free paper

This Springer imprint is published by Springer Nature
The registered company is Springer International Publishing AG
The registered company address is: Gewerbestrasse 11, 6330 Cham, Switzerland

Preface

The area of protective organic coatings is a field where many scientific disciplines meet. In order to synthesize designed binders and curatives, and apply them so as to get a coating of proper performance, one has to address disciplines like organic chemistry, gelation and network formation theory, rheology, surface science, physical chemistry, solid state physics, especially mechanics of homogeneous and inhomogeneous materials, and degradation and stability science. Because of this complexity, in the past the development of protective coatings was based mostly on empirical experience. One has to admit that on this empirical basis, coatings of excellent performance were developed, and a few "empirical" specialists could often solve a particular problem faster (and sometimes better) than a team of specialists in various disciplines. The methods to characterize coating film formation and properties were mostly empirical without understanding their physics, and several variants of characterization of a given property were available. Yet, gradually the situation has been changing: the coating systems have been becoming more complex, more exact and comparable description of properties has been required, and new methods have been needed to characterize additional properties for new applications. Predictions made on a theoretical basis have been helpful, at least in the form of "what happens, if…". Such development inspired us to collect articles on these topics and to publish them in the form of a book.

The book is composed of three parts: The first part "Network Formation and Modeling" is focused on the preparation of contemporary binders of complex composition characterized by the distributions of numbers and types of functional groups and the role of these distributions in network formation and properties development. In this context, modeling of the formation of other defects and the peculiarities of buildup of mechanical properties, when the cross-link density is high and stiff network chains develop, is important. In the second part "Coating Film Formation and Properties," characterization methods of film formation are addressed: rheology, cryogenic scanning electron microscopy methods, infrared spectroscopy, and methods characterizing volume shrinkage and stress development during film formation. This part also deals with diffusion technology as a method of designing drying of coating films. The third part of the book is "Coating Film Properties and Applications." The chapters of this part deal with methods of characterization of film properties mainly in the final state of a coating. Thermodynamic analysis of swelling, which is a common method for characterizing cross-link density and polymer solvent interactions, points to the dangers of misinterpretation of the results. Chapters on compositional depth profiles are based on slab microtomy and infrared spectroscopy, and confocal Raman microscopy. Several chapters deal with application properties of coatings including coating appearance and scratch and mar resistance. The reader will find information on the present state of the art of the methods. Also, the main factors causing defects and affecting degradation and durability of coatings are discussed. This part is concluded by comprehensive information on automotive paint application. As a whole, the book provides the reader a better understanding of the coating film formation process, coating properties, appearance, defect formation, and durability. The reader also finds information on contemporary trends of development in these areas.

The editors are grateful to the team of experts who helped review the chapters: Karlis Adamsons, David H. Alman, Robert J. Barsotti, Stuart G. Croll, C. Brent Douglas, Miroslava Dušková-Smrčková, Kevin Ellwood, Jason Ge, Eric C. Houze, Renee J. Kelly, Douglas M. Lamb, Herong Lei, Jun Lin, Robert Matheson, Alon V. McCormick, Mark E. Nichols, Kyle Price, Christine C. Roberts, Patricia M. Sormani, Shih-Wa Wang, Wenjun Wu, and Gann G. Xu. We would also like to thank Mary B. Hallberg for providing administrative support. Lastly, our special thanks go to Axalta leadership, especially Marc B. Goldfinger, Joanne R. Hardy, and Barry S. Snyder, for their encouragement and support during the preparation, and subsequent publication, of this book.

Wilmington, DE, USA — Mei Wen
Prague, Czech Republic — Karel Dušek

Contents

Part I
Network Formation and Modeling

Chapter 1
Role of Distributions in Binders and Curatives and Their Effect on Network Evolution and Structure

Karel Dušek, Jos Huybrechts, and Miroslava Dušková-Smrčková

Introduction

Distributions are the most typical attributes of macromolecular systems. Distributions of degrees of polymerization and molecular weights are typical for linear polymers, and distributions of molecular weights and composition are typical for copolymers. If a branching unit is present, the distributions are multiplied by the number and positional distributions of branch points. Distributions are always generated by cross-linking; at the gel point, the distributions of molecular weights get very wide (they diverge). One can say that distributions have turned *epitheton constans* of the polymeric world. Organic chemists hate distributions, physicists tolerate only those they like, but coating technologists have to live with them. They try to understand them and, possibly, to utilize them to develop more competitive products.

Where are the distributions encountered in practice? Already in raw materials purchased from producers—typical examples are several common polyisocyanates obtained by cyclotrimerization of diisocyanates or by addition of asymmetric diisocyanates to triols. Distributions in molecular weights and numbers of functional groups in telechelic polymers are usually narrower, but they should be accounted for when gelation and network formation are considered. Side reactions are another source of distributions—well-known examples are manifold reaction paths of isocyanate group leading to formation of allophanates, biurets, and isocyanurates, but important as well are transesterifications, transamidations,

K. Dušek (✉) • M. Dušková-Smrčková
Institute of Macromolecular Chemistry, Academy of Sciences of the Czech Republic, Prague, Czech Republic
e-mail: dusek@imc.cas.cz; mkdusek@yahoo.com

J. Huybrechts
Axalta Coating Systems Belgium B.V.B.A, Mechelen, Belgium

M. Wen, K. Dušek (eds.), *Protective Coatings*, DOI 10.1007/978-3-319-51627-1_1

ester–amide interchanges, various types of cleavage, and many other such reactions. Because the starting components are oligofunctional, there exists a variety of combinations in which the side (additional) reaction paths participate. Precursors of polymer networks, intentionally synthesized, are the main sources of distributions. Precursors are prepared in order to adjust group reactivity, processing properties like viscosity buildup, film-forming, and application properties, or specific chemical environment contained within the precursor can be introduced into the coating; not less important is a reduction of health hazards by lowering of volatility and penetration ability of toxic groups.

In this contribution, we will show examples of several ways by which distributions are generated in coatings binders or other thermosetting systems. For deeper understanding, we will refer to literature in which the generation of distributions and their treatment are described. The main aim of this contribution is to show the importance of distributions for cross-linking and for properties of the cross-linked systems, especially those of practical importance. Along with the search for simple model networks to verify or reject theories, the recent decades are characterized by growing interest in understanding and modeling systems of practical importance (for illustration, see examples in Refs. [1–13]). Since distributions are the central topic of this contribution, we first outline the ways the distributions can be described and transformed.

Distributions—Description and Transformation

The distributions can be described in several ways, for instance, by the number of objects, or their weights. The distribution can also be described by several moments of the distribution. For describing distributions and transforming distributions, we will be using the formalism of the generating functions. In the Appendix, we demonstrate the use of this formalism on the example of distributions of degrees of polymerization or molecular weights and reaction states of simple building units. No special mathematical knowledge is necessary, only the basic knowledge how to differentiate a function which is high-school mathematics. The types of units and bonds are tagged by a respective auxiliary (dummy) variable. The variable serves for the transform and is removed after the respective transformation is done (e.g., transfer of a number distribution into weight distribution). In this way, sometimes troublesome summations can be avoided. The great advantage of the use of the generating function tools is the simplicity of combination of various distributions, e.g., in blends of distributions, or for superimposed distributions like molecular weight and cross-link distributions. For further examples of the use of the probability generating functions (pgf) for description of distribution of complicated macromolecular systems, see Ref. [14]. Examples of other successful applications of pgf formalism in branching theory can be found in Refs. [1, 6–12, 14–22].

Distributions in Coatings Binders

In this contribution, we will show some cases in which the distributions in one or both components of a 2K cross-linking system either exist or are generated unintentionally or intentionally, before they enter the final cross-linking stage. The cross-linking stage is affected by the existence of these distributions; the distributions have an effect on gelation, the fraction of soluble materials, and cross-link density. The description of distributions occurring in real systems and an interpretation of their effects will be explained in the following sections:

- Distributions in commercial starting components
  - *Functionality distribution in polyisocyanates; effect on cross-linking of polyols*
  - *Functionality and molecular weight distribution in telechelics*
- Distributions generated by side reactions
- Two-stage reactive systems—preparation of a precursor in the first stage and cross-linking in the second stage
  - *Functional copolymers*
  - *Hyperbranched polymers*
  - *Off-stoichiometric (highly branched) copolyadducts*
  - *Systems with distributions of groups of different reactivities*
  - *Chain-extended systems*
- Multistage network formation processes

The methods used to determine the distributions are very important. The analysis of commercial materials is practically the only source of information on impurities or side products. For distributions in preprepared precursors, certain kinetic or equilibrium parameters are needed as input for theoretical generation of the distribution. Moreover, some of the theoretically generated distributions can be checked experimentally. Molecular ions mass spectrometry is very suitable for this purpose; unfortunately, while precise in molecular mass, the results are not so accurate in quantitative determination of components. Gas, size exclusion, and liquid chromatographies can be very helpful, but they are usually dependent on calibration, which may be not a simple task. Each case should be solved ad hoc by combinations of separation and spectroscopy techniques that help identify the composition of each separated fraction.

In this contribution, the effect of distributions on the cross-linking process, gelation, and network buildup is discussed only qualitatively or semi-quantitatively, but reference is made to work already published and to papers in which modeling of cross-linking is described. A large portion of relevant papers can be found in the references list.

Distributions in Commercial Starting Components

Functionality Distribution in Polyisocyanates

As an example, the functionality distribution in the commonly used triisocyanate obtained by trimerization of 1,6-diisocyanatohexane (HDI) is shown. It depends on the product type and also on the storage conditions. The NCO groups are first deactivated by reaction with a slight excess of an aliphatic alcohol (1-pentanol or benzyl alcohol) catalyzed by dibutyltin dilaurate (1000 ppm) or reacted with secondary amine (dibutylamine) and analyzed by ESI-TOF-MS or MALDI-TOF-MS [15, 25]. The samples of triisocyanate were of different origin obtained over a period of several years. As an example, the analysis of Desmodur® N3300 (Bayer) marked Desmodur® N3300_1 by ESI-TOF-MS is shown in Fig. 1.1 and Table 1.1.

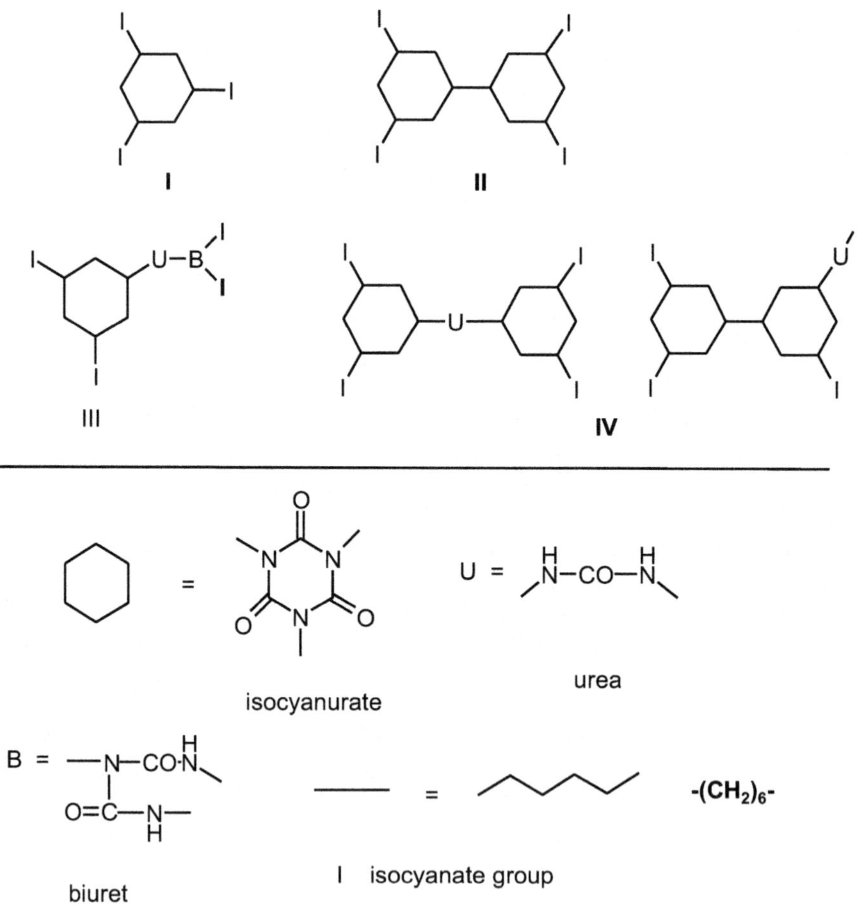

Fig. 1.1 Compounds found in a commercial sample of 1,6-hexane diisocyanate cyclo-trimer (cf., Table 1.1)

Table 1.1 Composition of trifunctional HDI-trimer (Desmodur® N3300_1) obtained by ESI-TOF-MS

Structure	Functionality	Content [mol %]
I	3	47
II	4	28
III	4	2
IV	4	23

Table 1.2 Composition of trifunctional HDI-trimer (Desmodur® N3600_1) obtained by ESI-TOF-MS

Structure	Functionality	Content [mol-%]
I	3	78
II	4	10
III	4	12
IV	4	0

Table 1.3 Average functionalities of trifunctional HDI-trimer (Desmodur®N) obtained by ESI-TOF-MS and MALDI-TOF-MS

Sample Desmodur®	$\langle f\rangle_1$	$\langle f\rangle_2$
N 3300_1	3.50	3.60
N 3300_2	3.42	3.57
N 3300_3	3.37	3.47
N 3600_1	3.23	3.28
N 3600_2	3.14	3.18

Apparently, the presence of urea and biuret groups in the sample suggests that the sample came in contact with water vapors during storage before it was used.

Desmodur® N3300 has higher functionality than Desmodur® N3600. Over several years, the functionality of Desmodur® N3300 has varied between 3.3 and 3.5; the distribution in Desmodur® N3600 is narrower (Table 1.2), and the number-average functionality varies around 3.2. The first-moment and second-moment averages are displayed in Table 1.3.

For modeling of cross-linking, the whole distribution is necessary, but for calculation of the number-average molecular weight and gel point the average values of f, $\langle f\rangle_1$ and $\langle f\rangle_2$, respectively, are sufficient:

$$\langle f\rangle_1 = \sum f n_f \text{ and } \langle f\rangle_2 = \sum f^2 n_f / \sum f n_f \tag{1.1}$$

where n_f is the molar fraction of polyisocyanate molecules of functionality f.

For instance, the gel point conversion of an ideal system A3 + B4 is described by the equation

$$\alpha_A \alpha_B = \left[\left(\langle f_A\rangle_2 - 1\right)\left(\langle f_B\rangle_2 - 1\right)\right]^{-1} \tag{1.2}$$

Thus, for a stoichiometric system, the critical value of α decreases from 0.408 for $\langle f_B\rangle_2 = 3$ to 0.358 for $\langle f_B\rangle_2 = 3.6$.

Functionality and Molecular Weight Distributions in Telechelics

Telechelic polyols also exhibit some functionality distribution, but it is usually narrower than for polyisocyanates. For polycaprolactone diol 1250, polycaprolactone triol 900, and polycaprolactone triol 300 (Aldrich), respectively, $\langle f \rangle_1$ values were found to be 1.87, 2.95, and 2.87, and $\langle f \rangle_2$ values were equal to 1.93, 2.97, and 2.93 (Ref. [17]). In an earlier work [16], for NIAX® LHT-240 polyoxypropylene triol (Union Carbide), $\langle f \rangle_1 = 2.89$ and $\langle f \rangle_2 = 2.92$. The lowering of average functionalities with respect to nominal values was generally small and was expressed as an admixture of monofunctional species to bifunctional or an admixture of bifunctional to trifunctional polyol.

Distributions Generated by Side Reactions

Side reactions, in which the cross-linkable group takes part or a new group is formed that takes part in the cross-linking reaction, are always important. Various interchange reactions such as transesterification, transamidation, ester–amide interchange, hydrolysis, or aminolysis rate among frequently occurring side reactions. In this contribution, we do not deal specifically with this subject. However, their occurrence, effect, and ways to treat them theoretically will be illustrated by some examples. Some side reactions may become the main reactions depending on reaction conditions and catalysts. The variety of reactions in which the isocyanate group takes part—formation of urethanes from alcohols, or urea from amines can be accompanied by cyclotrimerization, allophanate, and biuret formation—can serve as one of the examples [27]. The synthesis of linear polyesters by the reaction of dicarboxylic acids with diepoxides is another such example. The diepoxide unit, which is bifunctional in the addition reaction, can get tetrafunctional as a result of transesterification of the hydroxyester group which leads to formation of diester and glycol groups (Fig. 1.2).

Although the number of ester groups does not increase by this reaction, it is sufficient for gel formation, even if reaction conditions are such that neither the condensation reaction nor the etherification reaction takes place (Fig. 1.2). As shown in Ref. [1], the gel point condition reads

$$\alpha_C(\alpha_E + \alpha_T) = 1 \tag{1.3}$$

where α_C, α_E, and α_T are, respectively, conversions of carboxyl group, epoxy group, and transesterification conversion. Equation (1.3) shows that, if $\alpha_C = \alpha_E$, and $\alpha_T = 0$, this condition is fulfilled only at $\alpha_C = \alpha_E = 1$, i.e., 100% conversion of carboxyl and epoxide groups where an "infinitely" long linear chain is formed. If $\alpha_C = \alpha_E$, and $\alpha_T = 1$, according to Eq. (1.3) gelation occurs at conversion $\alpha_C = \alpha_E = (\sqrt{5} - 1)/2 \approx 0.618$. The gel fraction never reaches 100%, and about 20% of the material remains soluble.

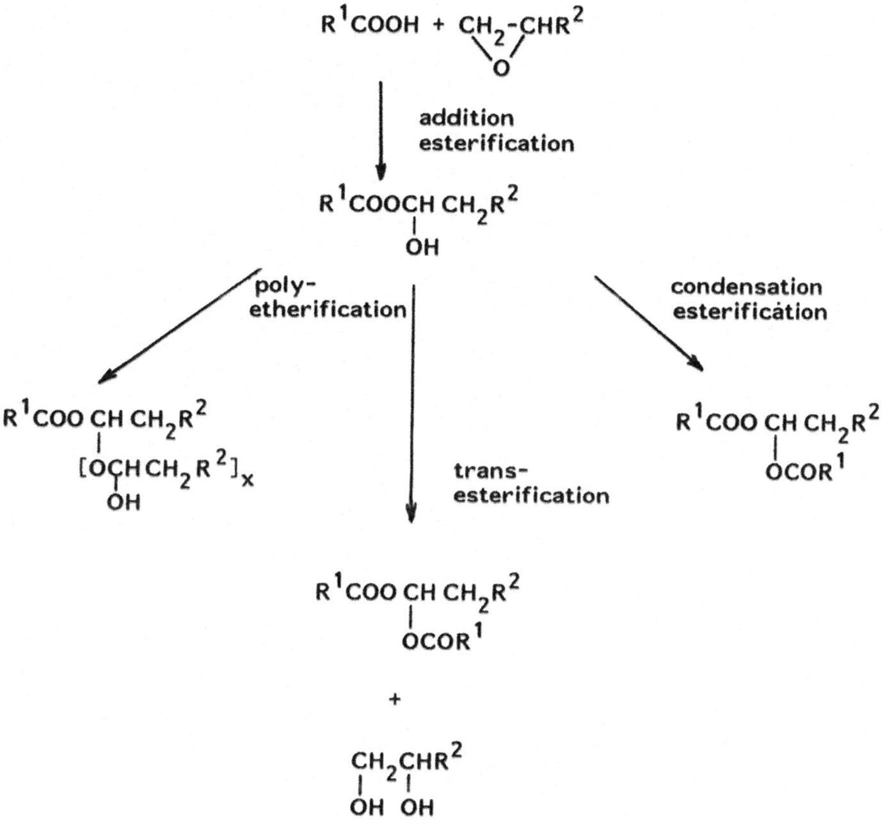

Fig. 1.2 Scheme showing possible transformations of hydroxyesters prepared by addition of a carboxylic acid to an epoxide compound

Two-Stage Reactive Systems—Preparation of a Precursor in the First Stage and Its Cross-Linking in the Second Stage

Two-stage systems are the most frequent and most important basis of modern organic coating technology. We will show a few examples and outline the ways by which such systems can be treated theoretically. The theoretical approach is ring-free, i.e., cyclization (formation of small loops) is first neglected. Then, the extent of cyclization is characterized experimentally, mostly by a shift of the gel point conversion as a function of dilution, and the parameters of the cross-linking system are rescaled against intermolecular conversion as a variable. A more detailed explanation and way of handling data can be found in Refs. [15, 17].

Functional Copolymers

Low-molecular-weight copolymers containing monomer units carrying a functional group are popular precursors commonly used. They can be prepared by various kinds of polymerizations. Most widely used is the free-radical copolymerization of a functional monomer, like hydroxyethyl or hydroxypropyl methacrylate or acrylate, glycidyl methacrylate, and methacrylic acid, with another or several nonfunctional monomers, like methyl methacrylate or acrylate, styrene, and isobornyl methacrylate [7], which adjust the precursor functionality and thermo-mechanical properties of the final coating. In some cases, a chain-transfer compound is added to achieve low molecular weight of the copolymers.

The distributions encountered in functional copolymers are of two kinds. If the copolymer is prepared batch-wise to high conversion, one has to deal with a compositional (and consequently degree of polymerization) distribution, because the comonomers are consumed at different rates. Such a distribution can be generated by copolymerization kinetics. The compositional heterogeneity due to conversion shift can be avoided—and this is usually the case in practice—by preparing the copolymers by the method of continuous monomer feeds, i.e., under starved conditions, by which a compositional distribution that is independent of reaction time is secured. Such a copolymer has a statistical distribution in composition and molecular weights. In Fig. 1.3, one species of the bivariate distribution (M and f) is shown.

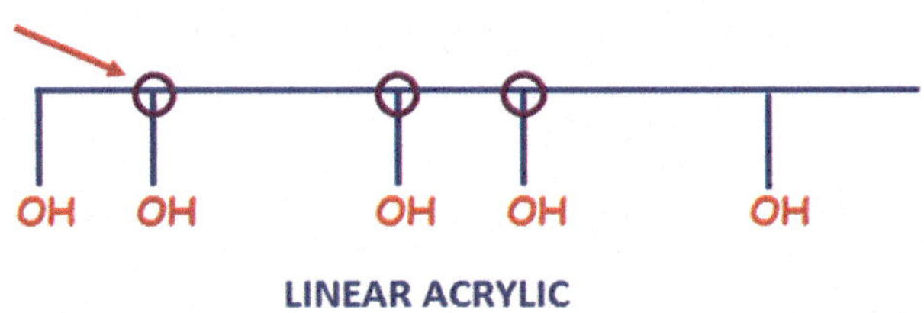

Fig. 1.3 Molecule of a hydroxyfunctional copolymer (linear acrylic copolymer), the *circles* show potential active branch points able to contribute to the number of EANCs

Each OH group can react, for instance, with an isocyanate group. The distribution in the number of reactive groups affects the molecular weight increase, gelation, the sol fraction, and cross-link density. The distribution in f and M can be expressed with the aid of the number fraction distribution function N_{Mc} in the form of the probability generating function (see also Appendix)

$$N_{\mathrm{M,c}}(Z_{\mathrm{A}}, Z_{\mathrm{B}}) = N_{\mathrm{M}}(N_{\mathrm{c}}(Z_{\mathrm{A}}, Z_{\mathrm{B}})) = \sum_{i=1}^{\infty} n_i \sum_{j=0}^{i} x_{j,i-j} Z_{\mathrm{A}}^{jM_{\mathrm{A}}}, Z_{\mathrm{B}}^{(i-j)M_{\mathrm{B}}} \tag{1.4}$$

where n_i is the number fraction of i-mer and $x_{j,i-j}$ is (within the i-mer, $\sum_{j=0}^{i} x_{j,i-j} = 1$) the fraction of i-mer molecules composed of j units A and i–j units B. For random distribution of monomer units A and B

$$
\begin{aligned}
N_{\mathrm{c}}(Z_{\mathrm{A}}, Z_{\mathrm{B}}) &= x_{\mathrm{A}} Z_{\mathrm{A}}^{M_{\mathrm{A}}} + x_{\mathrm{B}} Z_{\mathrm{B}}^{M_{\mathrm{B}}} \\
N_{\mathrm{M,c}}(Z_{\mathrm{A}}, Z_{\mathrm{B}}) &= \sum_{i=1}^{\infty} n_i \left(x_{\mathrm{A}} Z_{\mathrm{A}}^{M_{\mathrm{A}}} + x_{\mathrm{B}} Z_{\mathrm{B}}^{M_{\mathrm{B}}}\right)^i
\end{aligned}
\tag{1.5}
$$

where x_{A} and x_{B} are molar fractions of units A and B, respectively, in the copolymer. If we choose for the degree-of-polymerization distribution the Flory most probable distribution, Eq. (1.5) transforms into

$$
N_{\mathrm{c}}(Z_{\mathrm{A}}, Z_{\mathrm{B}}) = \frac{(1-q)\left(x_{\mathrm{A}} Z_{\mathrm{A}}^{M_{\mathrm{A}}} + x_{\mathrm{B}} Z_{\mathrm{B}}^{M_{\mathrm{B}}}\right)}{1 - q\left(x_{\mathrm{A}} Z_{\mathrm{A}}^{M_{\mathrm{A}}} + x_{\mathrm{B}} Z_{\mathrm{B}}^{M_{\mathrm{B}}}\right)}
\tag{1.6}
$$

where the parameter q is related to the number-average degree of polymerization P_{n} ($q = 1 - 1/P_{\mathrm{n}}$).

Choosing A for a nonfunctional monomer unit (or a mixture of them) and B for the functional monomer which subsequently reacts with the cross-linker C, the distribution function (Eq. (1.6)) is modified:

$$
\begin{aligned}
N_{\mathrm{c}}(Z_{\mathrm{A}}, Z_{\mathrm{B}}, z_{\mathrm{BC}}) &= \frac{(1-q)\left(x_{\mathrm{A}} Z_{\mathrm{A}}^{M_{\mathrm{A}}} + x_{\mathrm{B}} Z_{\mathrm{B}}^{M_{\mathrm{B}}} f_{\mathrm{B}}(z_{\mathrm{BC}})\right)}{1 - q\left(x_{\mathrm{A}} Z_{\mathrm{A}}^{M_{\mathrm{A}}} + x_{\mathrm{B}} Z_{\mathrm{B}}^{M_{\mathrm{B}}} f_{\mathrm{B}}(z_{\mathrm{BC}})\right)} \\
f_{\mathrm{B}}(z_{\mathrm{BC}}) &= 1 - \alpha_{\mathrm{B}} + \alpha_{\mathrm{B}} z_{\mathrm{BC}}
\end{aligned}
\tag{1.7}
$$

where α_{B} is the conversion of B groups in the reaction with C and z_{BC} is a variable denoting the bond $\mathrm{B} \rightarrow \mathrm{C}$; let us remind that x_{A} and x_{B}, respectively, are molar fractions of units A and units B in the copolymer.

For the trifunctional cross-linker designated, C_3, with three groups C of independent reactivity, the distribution of states is given by expansion of the function of Eq. (1.8) into power series of z_{CB}:

$$
F_{0\mathrm{C}}(z_{\mathrm{CB}}) = (1 - \alpha_{\mathrm{C}} + \alpha_{\mathrm{C}} z_{\mathrm{CB}})^3
\tag{1.8}
$$

where α_{C} is the conversion of C groups in the reaction with B and z_{CB} is assigned to the bond $\mathrm{C} \rightarrow \mathrm{B}$ ($[\mathrm{C} \rightarrow \mathrm{B}] = [\mathrm{B} \rightarrow \mathrm{C}]$). The coefficient at z_{CB}^k is equal to the molar concentration of C_3 participating in k bonds $\mathrm{C} \rightarrow \mathrm{B}$.

The distributions (Eqs. (1.7) and (1.8)) contain sufficient information to calculate number- and weight-average molecular weights (also of the sol beyond the gel point), the gel point conversion, sol fraction, molecular weight of sol, concentration of EANCs, and other quantities characterizing the cross-linking system. A sketch of such a system is shown in Fig. 1.4.

The gel point conversion calculated from Eqs. (1.6) and (1.7) is a simple equation independent of molecular weight of units which reads

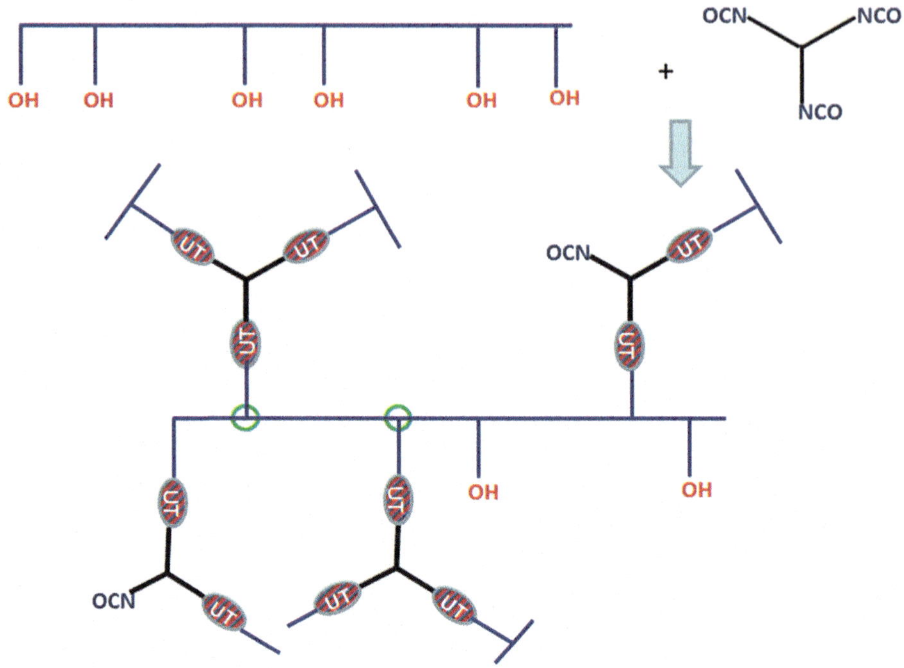

Fig. 1.4 Cross-linking of an OH-functional copolymer by a triisocyanate, *green circles*—elastically active trifunctional cross-links; *UT*—urethane bonds

$$\alpha_B\alpha_C = \frac{1}{2}\left(\frac{1-q}{2qx_B}\right) = \frac{1}{2}\frac{1}{2x_B(P_n-1)} = \frac{1}{2}\frac{1}{x_B(P_w-1)} \tag{1.9}$$

However, M_n, M_w, and sol fraction characteristics depend on the values of molecular weights of the monomer units. The transformations of Eq. (1.9) follow from the definition of $q = 1 - 1/P_n$ and $P_w - 1 = 2q/(1-q)$.

The expression in Eq. (1.9)

$$\alpha_B\alpha_C = \frac{1}{2}\frac{1}{x_B(P_w-1)}$$

is universally valid for a random distribution of cross-linked units, as shown in the Appendix. If the molecular weights of component units A and B are not equal, the weight-average molecular weight is not directly proportional to its second-moment-average degree of polymerization, and the gel point condition reads

$$\alpha_B\alpha_C = \frac{1}{2}\frac{M_{AB}^2}{x_B(M_w M_{AB} - \overline{M_{AB}^2})}$$
$$M_{AB} = x_A M_A + x_B M_B, \quad \overline{M_{AB}^2} = x_A M_A^2 + x_B M_B^2 \tag{1.10}$$

If one has more than one nonfunctional comonomer, the $x_A Z_A^{M_A}$ in Eq. (1.7) is replaced by $\sum_i x_{Ai} Z_A^{M_{Ai}}$.

Hyperbranched Polymers

Several tens of publications and even more patents deal with the application of functional hyperbranched polymers as precursors of organic coatings. They bear functional groups like OH, NH_2, NH, COOH, and vinyls. (cf., e.g., Refs. [28–30]). Hyperbranched polymers are thus functional precursors having a special type of degree of polymerization/molecular weight and functionality distributions. In contrast to most functional copolymers, they are nonlinear. Typically, they are formed from BA_f monomers by reactions $A + B \rightarrow AB$, and sometimes a polyfunctional core (A_g) is added to increase the degree of branching (Fig. 1.5).

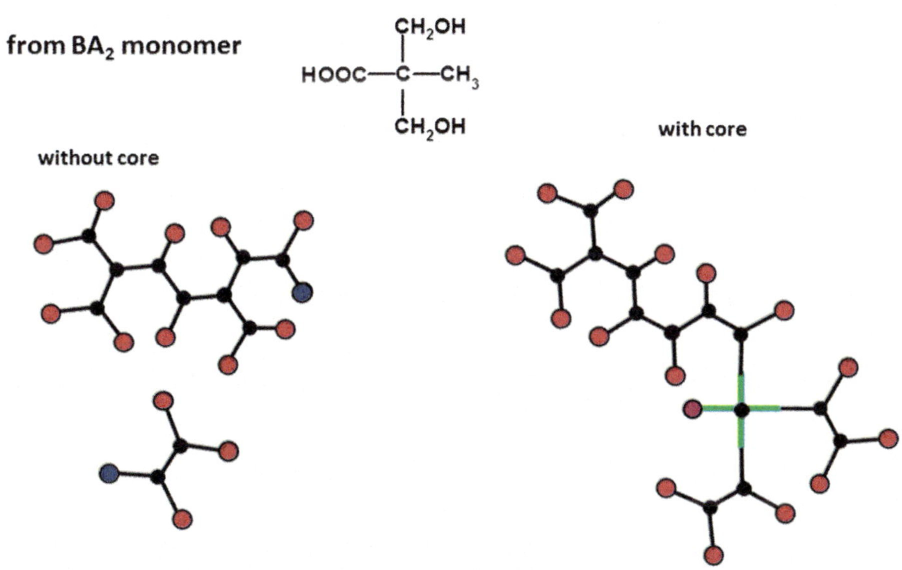

Fig. 1.5 Hyperbranched functional polymers; *red circles*—functional groups A (OH groups), *blue circle*—functional group B (COOH), *magenta circle*—A group of the core, *green cross*—tetrafunctional core, *black circles*—branch points, potential elastically active cross-links

The distribution of molecular weights and numbers of functional units in the case of random reaction without core is simple. The distribution of reaction states of a BA_f unit (expressed by probability generating function [31]) reads

$$F_0(\mathbf{Z},\mathbf{z}) = Z_{AB}^{M_{AB}}((1-\alpha_A)Z_{Af} + \alpha_A z_{AB})^{f_A}((1-\alpha_B)Z_{Bf} + \alpha_B z_{BA}) \tag{1.11}$$

A relation exists between conversions of A and B groups, $\alpha_A f_A = \alpha_B$. The variables Z_{Af} and Z_{Bf} denote, respectively, the free (unreacted) groups A and B in the hyperbranched polymer. One of them or both of them participate in the cross-linking reaction with a cross-linker C. The distributions and averages per molecule of the hyperbranched polymer (composed of units given by Eq. (1.11))—precursor of a polymer network—are obtained by substitution $\mathbf{z} \to \mathbf{u}$ making the equation recursive[1]

$$F_0(\mathbf{Z},\mathbf{u}) = Z^{M_0}((1-\alpha_A)Z_{Af} + \alpha_A u_{AB})^{f_A}((1-\alpha_B)Z_{Bf} + \alpha_B u_{BA}) \tag{1.12}$$

where

$$\begin{aligned} u_{AB} &= Z^{M_0}((1-\alpha_A)Z_{Af} + \alpha_A u_{AB})^{f_A} \\ u_{BA} &= Z^{M_0}((1-\alpha_A)Z_{Af} + \alpha_A u_{AB})^{f_A-1}(1-\alpha_B + \alpha_B u_{BA}) \end{aligned} \tag{1.13}$$

By differentiation with respect to Z one obtains the molecular weight averages, and by differentiation with respect to Z_{Af} the unreacted functionality averages:

$$M_n = M_0\left(\frac{1}{1-\alpha_B}\right); \quad M_w = M_0\left(\frac{1-\alpha_B/f_A}{(1-\alpha_B)^2}\right) \tag{1.14}$$

For the unreacted functional group A, the expressions for functionality averages read

$$\langle f\rangle_1 = \frac{f_A(1-\alpha_A)}{1-\alpha_B} \quad \langle f\rangle_2 = f_A\frac{(1-\alpha_A)^2}{(1-\alpha_B)^2} \quad \alpha_B = f_A\alpha_A \tag{1.15}$$

Figure 1.6 shows the functionality averages as a function of conversion of groups A (OH).

The average functionality of low-conversion hyperbranched polymers increases very slowly with increasing conversion because the monomeric molecules

[1]The recursive Eqs. (1.13) express the structure growth through repetition of progressive adding of building units. By solving these equations with respect to u_{XY} and inserting the solution into Eq. (1.12) one obtains the description, in the form of generating functions, of fractions of all possible molecules differing in the number of building units and unreacted functional groups. Description of these multiplicative processes by such recursive equations (also called cascade substitution) is the standard procedure employed in the statistical branching theories. It corresponds to the first-order Markov process controlled by transition probabilities.

Fig. 1.6 First-moment and second-moment functionality averages for A (OH) groups in hyperbranched polymers formed from BA_2 monomer (equal and independent reactivity of groups A, independent reactivity of groups B)

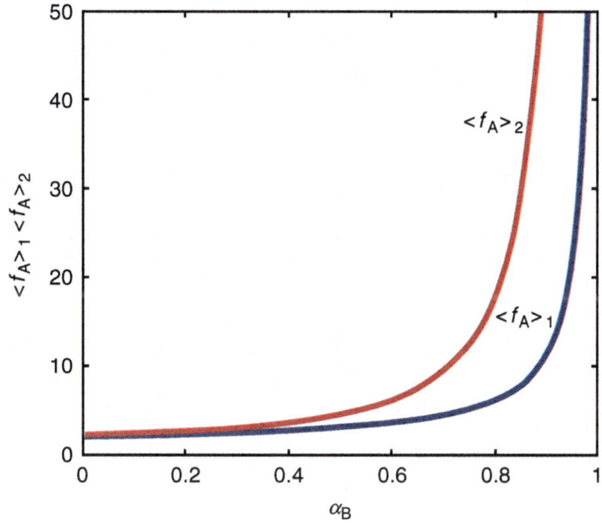

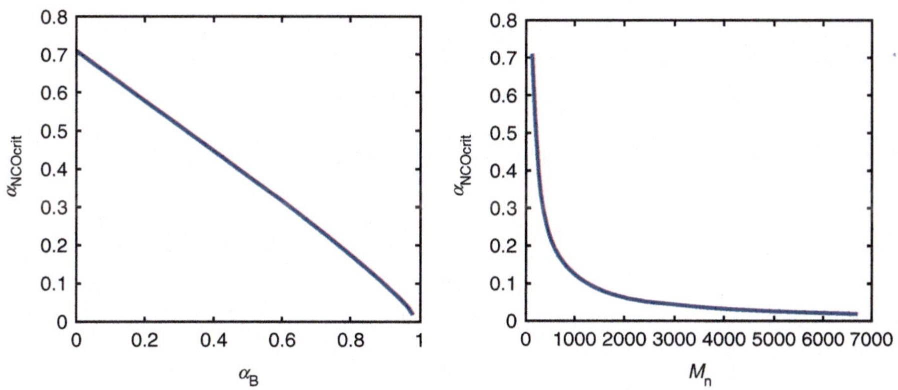

Fig. 1.7 Dependence of critical conversion of C (NCO) groups of a trifunctional cross-linker C_3 reacting with A (OH) groups of hyperbranched polymer of BA_2 monomer (stoichiometric ratio) as a function of final conversion of B groups (COOH) (**a**), or number-average molecular weight of the hyperbranched polymer (**b**); $M_0 = 134$ g/mol

predominate in the distribution. A sharp increase starts when α_B surpasses the value 0.8. Both averages diverge when $\alpha_B \rightarrow 1$, but the ratio $\langle f \rangle_2 / \langle f \rangle_1$ also diverges. The steep increase of $\langle f \rangle_2$ and $\langle f \rangle_2 / \langle f \rangle_1$ is unwanted because in the curing stage the critical conversion and pot life decrease sharply. An example of cross-linking a hyperbranched polymer BA_2 (having OH as A groups, B groups not reacting) with a trifunctional cross-linker C3 (triisocyanate) is shown in Fig. 1.7. The fall of the critical conversion with increasing molecular weight of the hyperbranched polymer is even sharper.

The polydispersity in molecular weights and functionality can be made narrower by addition of a core, i.e., a polyfunctional compound with A1 groups of $A1_g$; A1 groups may be of different reactivity than A groups. In this case, the probability generating function becomes

$$\begin{aligned} F_0(\mathbf{Z},\mathbf{z}) = \; & n_{AB} Z_{AB}^{M_{AB}} ((1-\alpha_A) Z_{Af} \\ & + \alpha_A (p_{AB} z_{AB} + p_{AB1} z_{AB1})^{f_A} ((1-\alpha_B) Z_{Bf} + \alpha_B z_{BA}) \\ & + n_{A1} Z_{A1}^{M_{A1}} ((1-\alpha_{A1}) Z_{A1f} + \alpha_{A1} z_{A1B})^g \end{aligned} \tag{1.16}$$

where n_{AB} and n_{A1} are molar fractions of the BA_f monomer and core, respectively. We have to distinguish between bonds of A to B group of BA_f monomer and B to A1 groups of core. For a kinetically controlled process, the respective conversions and weightings p_{AB} and p_{AB1} are found by solution of kinetic differential equations for population of dyads.

The calculation of molecular weight distributions and averages of hyperbranched polymers with and without a core and with and without a substitution effect has been the subject of a number of papers. The calculations are made by solving sets of differential equations for the concentrations of species or the set of equations for moments of distribution (cf., e.g., Refs. [32–34]). In the calculations, usually only intermolecular reactions are considered. However, intramolecular reactions also occur by which the molecule loses its B group, so that this group cannot actively contribute to structure growth [35]—it turns to a kind of core formed at a certain reaction conversion.

The existence of branch points—potential elastically active cross-links—is a special feature of this class of precursors when used as components of cross-linking systems. Close to the gel point, these branch points are mostly inactive, i.e., the paths of bonds issuing from them do not continue to infinity. As more new bonds are formed in the cross-linking stage, more branch points of the precursor get the elastically active status. How to deal with this transformation is described in Ref. [36]. The basic information about the states of units in a hyperbranched polymer (AB)–cross-linker (C) system is contained in the set of Eq. (1.17) in which the bonds extending from A to B are clearly distinguished from A to C bonds introduced during the cross-linking stage:

$$\begin{aligned} & F_0(\mathbf{Z},\mathbf{z}) = n_{AB} F_{0AB}(\mathbf{Z},\mathbf{z}) + n_C F_{0C}(\mathbf{Z},\mathbf{z}) \\ & F_{0AB}(\mathbf{Z},\mathbf{z}) = Z_{AB}^{M_{AB}} ((1-\alpha_{A1}) \xi_{A2} + \alpha_{A1} z_{AB1})^{f_A} ((1-\alpha_{B1}) Z_{Bf} + \alpha_{B1} z_{B1A1}) \\ & \xi_{A2}(Z,z) = (1-\alpha_{A2}) Z_{Af} + \alpha_{A2} z_{A2C2} \\ & F_{0C}(\mathbf{Z},\mathbf{z}) = Z_C^{M_C} ((1-\alpha_{C2}) Z_{Cf} + \alpha_{C2} (p_{C2A1} z_{C2A1} + p_{C2C2} z_{C2A2})^{f_C} \end{aligned} \tag{1.17}$$

Figure 1.8 shows the increase of the concentration of EANCs for hyperbranched polymers of BA_f of different degrees of polymerization cross-linked with tri- or bifunctional cross-linker C.

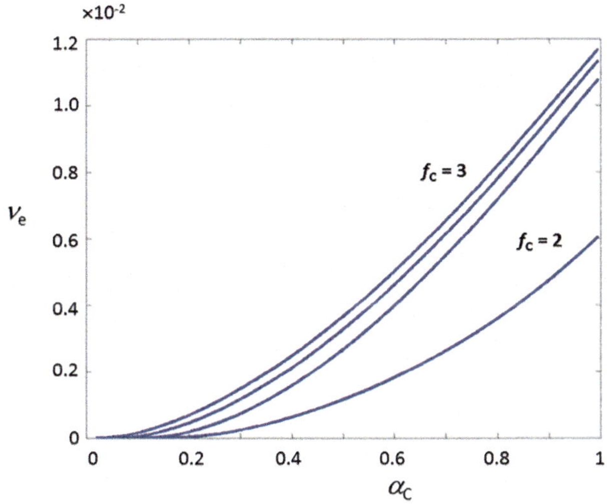

Fig. 1.8 Hyperbranched polymer carrying cross-linkable A groups and cross-linked with C3 or C2 cross-linker. Dependence of the concentration of EANCs (in mol/cm^3) on conversion of C groups of the cross-linker; the curves for $f_C = 3$ show dependence on the number-average degree of polymerization of hyperbranched polymer $P_n = 10, 20, 50$, respectively, displayed from bottom to top (based on results of Ref. [36]).The dependences curve upwards slightly as more branch points get activated. However, it seems that the cross-link density does not increase much with increasing P_n while the gel points are shifted to lower conversions

Off-Stoichiometric (Highly Branched) Copolyadducts

Polyaddition/polycondensation cross-linked systems are most frequently prepared by $A + B \rightarrow AB$ type of reactions. The lowest critical conversion and highest cross-link densities are achieved when the molar ratio for groups $[A]/[B] = 1$. There exist certain molar ratios (dotted curves in Fig. 1.9) outside of which the system does not gel at all and the final reaction products are composed only of branched but finite molecules (Fig. 1.9). This is called the *critical molar ratio*. There exist two such values for the A + B system corresponding to critical excess of groups A or B, r_{Acrit} and r_{Bcrit}. Between these critical values, the system can gel; beyond these values, only finite branched molecules containing just one of the functional groups, A or B, respectively, will be formed even when all of the minority groups have fully reacted. Such molecules are also called *highly branched* functional polymers.

The practical importance of preparing off-stoichiometric, highly branched, functional polymers and their use as precursors have been recognized by various industries several decades ago. For instance, in the epoxy resin adhesives and coatings areas, volatile low-viscosity di- and triamines are prereacted with epoxy resins to increase the viscosity and to eliminate the amine volatility and health hazards. In the polyurethane area, diols are prereacted with excess of diisocyanate to prepare isocyanate endcapped low molecular weight oligourethane diisocyanates. They are subsequently used for the preparation of hard block–soft block polyurethanes or for the

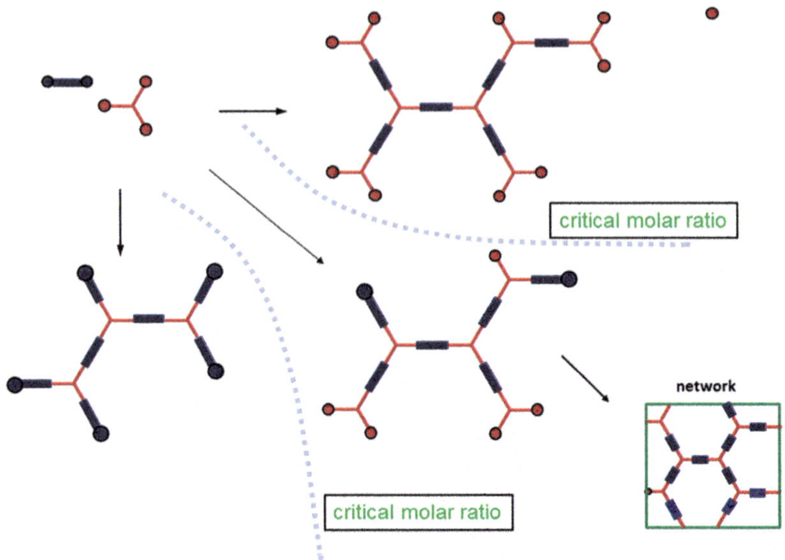

Fig. 1.9 Highly branched off-stoichiometric polymer obtained by polyaddition of A2 + B3 monomers and critical molar ratios delimiting the soluble polymers from the network

preparation of cross-linked systems (typical are mixtures of oligourethane diisocyanates with a triol). In all cases, the off-stoichiometric systems are composed of various molecules of varying molecular weight and functionality.

It is very useful and relatively simple to define the critical molar ratios—the boundaries between still soluble and cross-linked polymer. In other words, this condition says that the molar ratio of groups should be infinitesimally above the critical value to guarantee a discontinuation in bond sequence prior to infinity when one of the two types of functional groups has fully reacted in an A + B alternating reaction. For an ideal A + B system, this condition can be derived from Eq. 1.2:

$$\alpha_A \alpha_B = \left[\left(\langle f_A \rangle_2 - 1 \right) \left(\langle f_B \rangle_2 - 1 \right) \right]^{-1} \tag{1.18}$$

Introducing molar ratios defined as

$$r_A = [A]_0/[B]_0, \quad r_B = [B]_0/[A]_0 \tag{1.19a}$$

two critical molar ratios exist for each system.

(a) Excess of A ($\alpha_B = 1, \quad \alpha_A = 1/r_A = r_B$)

$$r_{Acrit} = [(\langle f_A \rangle_2 - 1)(\langle f_B \rangle_2 - 1)] \tag{1.19b}$$

(b) Excess of B ($\alpha_A = 1, \quad \alpha_B = 1/r_B = r_A$)

$$r_{Bcrit} = [(\langle f_A \rangle_2 - 1)(\langle f_B \rangle_2 - 1)] \tag{1.19c}$$

The question is how to describe branching, gelation, and network buildup when the network is prepared from such off-stoichiometric highly branched oligomers. In the prereaction and the cross-linking reaction, the same groups take part, but in two stages with different groups (e.g., A with B or with C) and under different conditions. Therefore, the bonds formed in Stage 1 (precursor formation) and Stage 2 (precursor reaction) must be distinguished. The basic distribution function for reaction states of building units reads

$$\begin{aligned}
F_0(\mathbf{Z},\mathbf{z}) &= n_A F_{0A}(\mathbf{Z},\mathbf{z}) + n_B F_{0B}(\mathbf{Z},\mathbf{z}) + n_C F_{0C}(\mathbf{Z},\mathbf{z})\\
F_{0A}(\mathbf{Z},\mathbf{z}) &= Z^{M_A}\Big((1-\alpha_{A1})\xi_{A2} + \alpha_{A1} z_{A1B1}\Big)^{f_A}\\
\xi_{A2}(Z,z) &= (1-\alpha_{A2})Z_{Af} + \alpha_{A2} z_{A2C2}\\
F_{0B}(\mathbf{Z},\mathbf{z}) &= Z^{M_B}\Big((1-\alpha_{B1})\xi_{B2} + \alpha_{B1} z_{B1A1}\Big)^{f_B}\\
\xi_{B2}(Z,z) &= (1-\alpha_{B2})Z_{Af} + \alpha_{B2} z_{A2C2}\\
F_{0C}(\mathbf{Z},\mathbf{z}) &= Z^{M_C}\Big((1-\alpha_{C2})Z_{Cf} + \alpha_{C2}(p_{C2A2} z_{C2A2} + p_{C2B2} z_{C2B2})\Big)^{f_C}
\end{aligned} \tag{1.20}$$

The conversions α and transition probabilities p are obtained by numerical solutions of sets of differential equations, which guarantees that α_2 for the second stage is equal to 0 until α_1 for the first stage has reached its final value (because $\alpha_{C2} = 0$ until that point).

The molecular weight and functionality distributions can be manipulated by employing differences in reactivities of functional groups in A-type and B-type monomers [19]. This approach was utilized for synthesizing precursors similar to hyperbranched polymers, by approaching the situation that a BA_2 monomer was formed in situ first and further polymerized [37] (see also Ref. [38]). Such a system is visualized in Fig. 1.10.

The distribution of reaction states of units in such system is given by the following pgf:

$$\begin{aligned}
F_{0n}(\mathbf{Z},\mathbf{z}) &= n_A Z_A^{M_A}[(1-\alpha_{A1})Z_{A1f} + \alpha_{A1}\xi_{1B}][(1-\alpha_{A2})Z_{A2f} + \alpha_{A2}\xi_{2B}]^2\\
&\quad + n_B Z_B^{M_B}[(1-\alpha_{B1})Z_{B1f} + \alpha_{B1}\xi_{1A}][(1-\alpha_{B2})Z_{B2f} + \alpha_{B2}\xi_{2A}]\\
\xi_{1A} &= p_{A1B1} z_{A1B1} + p_{A1B2} z_{A1B2} \quad \xi_{2A} = p_{A2B1} z_{A2B1} + p_{A2B2} z_{A2B2}\\
\xi_{1B} &= p_{B1A1} z_{B1A1} + p_{B1A2} z_{B1A2} \quad \xi_{2A} = p_{B2A1} z_{B2A1} + p_{B2A2} z_{B2A2}
\end{aligned} \tag{1.21}$$

In order to get correct result, four pairs of transition probabilities p must be distinguished that are based on reaction time (conversion)-dependent concentrations of dyads [A1B1], [A1B2], [A2B1], and [A2B2].

These concentrations are obtained by solution of the set of the respective differential equations. The effect of reactivity difference on the critical ratio of this A3 + B2 system is illustrated by Fig. 1.11.

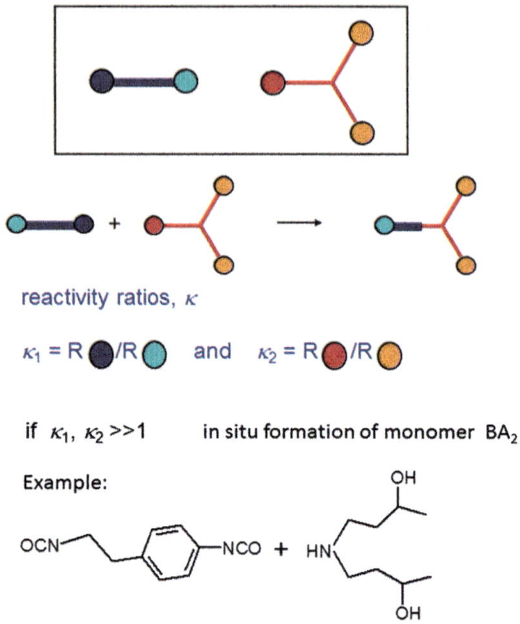

Fig. 1.10 Preparation of functional precursors from B2 + A3 monomers having groups of different reactivity (distinguished by depth of colors and reactivity ratios)

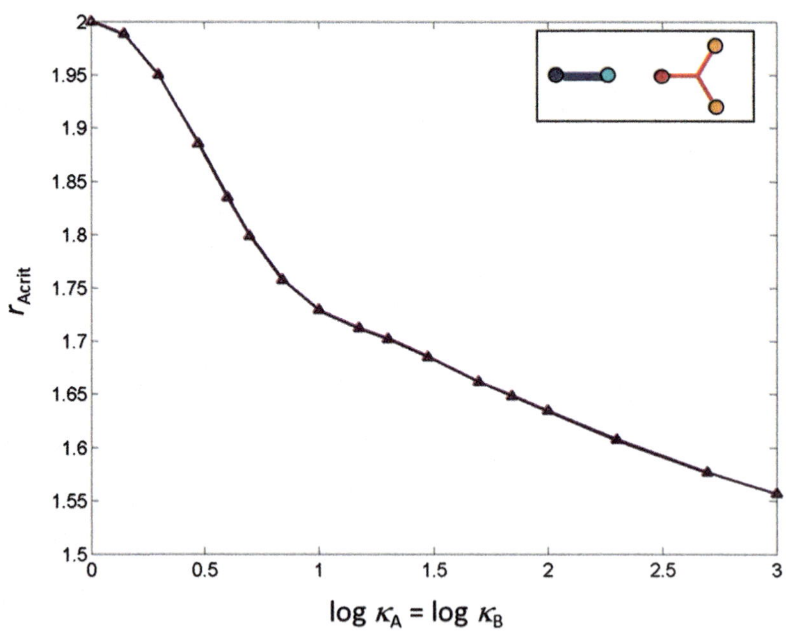

Fig. 1.11 Effect of reactivity ratios in monomer B_2 (B1B2) and A_3 ($A1(A2)_2$) (cf., Fig. 1.10) on the critical molar ratio for gelation (data of Ref. [19]); κ_A equals the ratio of rate constants for reaction of the *red group* of component A with respect to the reaction of each of the two orange groups; κ_B equals the same ratio for the reaction of the *dark blue group* of component B with respect to the *light blue group*. For simplicity of presentation, in this example both reactivity ratios were put equal

For equal reactivities of B1 and B2 groups and A1 and A2 groups, r_{Acrit} is equal to 2 (Eq. (1.19)), and it asymptotically decreases to the value 1.5 which is equal to the molar ratio of groups in the mixture of monomers B_2 and A_3. One can make a conclusion that for the preparation of functional branched precursors, a reactivity difference always helps in narrowing the functionality and molecular weight distributions. However, the main effect seems to be limited to the region of reactivity difference up to a factor of 10 (Fig. 1.11). Further synthetic efforts to find systems with reactivity differences larger than that value (i.e., 10) seem to be less productive. Also, theoretical analysis of this system has shown [19] that calculation of the transition probabilities just from conversions of respective groups ("equilibrium" approach) and not from kinetically derived concentrations of dyads would be incorrect. For small differences in reactivities (approximately, up to a factor of 5) the "equilibrium" approach offers a good approximation.

Systems with Distributions of Groups of Different Reactivities

Often, groups of the same type but different reactivity are distributed in a precursor more or less randomly, and both then participate in the curing reaction. A functional copolymer containing units with a primary OH group (e.g., 2-hydroxyethyl methacrylate) and with a secondary OH group (e.g., 2-hydroxylpropyl methacrylate) can serve as an example. Cross-linking of such copolymers has not yet been analyzed in-depth theoretically, but several options can be considered. Besides the neighbor effect, which can be operative also in single functionality copolymers, it is a question similar to that posed in the previous section, i.e., whether the kinetic dyad method would give results different from the conversion-based "equilibrium" approach.

A distribution of groups of different reactivities in a star oligomer has been analyzed theoretically and can serve as another example. A functional polyester star is synthesized from a polyol (tetrol) compound that reacts with a stoichiometric amount of cyclic anhydride by which an equivalent number of carboxyl groups are formed. In the second step, the carboxyls react with oxirane groups by which OH groups are formed: if the oxirane is ethylene oxide, the OH groups are primary, and if it is an epoxyester or an epoxyether, the OH groups are secondary. A relatively easy way to have a star oligomer with a distribution of primary and secondary OH groups is to use a less than stoichiometric amount of cyclic anhydride which leaves some of the primary OH groups of the tetrol unreacted, whereas the other will be converted to secondary OH group by this "chain extension." There exist practical reasons, why such distributions are prepared intentionally and used as coatings binders. The situation is illustrated by Fig. 1.12

If the tetrol groups are of equal and independent reactivities, the distribution of molecules differing in numbers of primary and secondary OH groups is binomial and can be described by the following number fraction generating function:

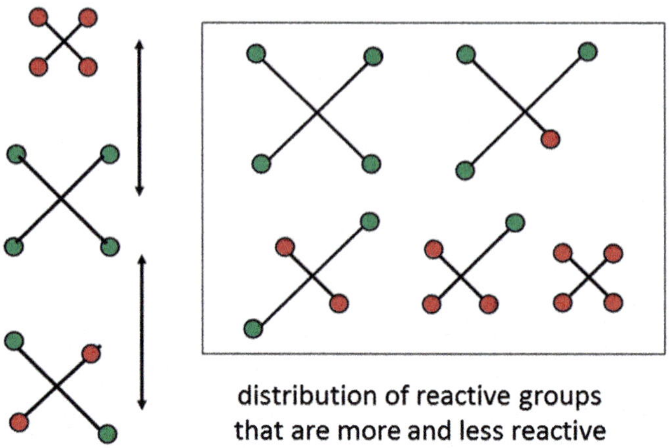

Fig. 1.12 A tetrafunctional oligomer with groups of different reactivities—*red circle* more reactive (primary OH) group, *green circle* less reactive (secondary OH) group; smaller star with all *red circles*—a tetrol with primary OH groups, larger star with all *green circles*—all arms were chain extended by converting primary OH groups to secondary ones; star with two red and two green circles—a single star with two primary and two secondary OH groups; enframed—distribution obtained by random chain extension of a part of tetrol arms

$$N(Z) = (x_P Z_P + x_S Z_S)^4 = \sum_{i=0}^{4} \frac{4!}{i!(4-i)!} x_P^i x_S^{4-i} Z_P^i Z_S^{4-i} \tag{1.22}$$

Here, x_P and x_S are molar fractions of primary and secondary OH groups, respectively; Z_P and Z_S are the pgf variables identifying these groups. Each of these groups can react with the cross-linker (polyisocyanate), but the reaction rates are different. For cross-linking with a triisocyanate, the basic pgf for reaction states of building units reads

$$\begin{aligned} F_0(\mathbf{Z}, z) = {} & n_{TS}\left[x_P Z_P^{M_P}(1-\alpha_P+\alpha_P z_{PI}) + x_S Z_S^{M_S}(1-\alpha_S+\alpha_s z_{SI})\right]^4 \\ & + n_I Z_I^{M_I}\left[1-\alpha_I+\alpha_I(p_{IP}z_{IP}+p_{IS}z_{IS})\right]^3 \end{aligned} \tag{1.23a}$$

Here, n_{TS} and n_I are molar fractions of tetrafunctional star molecules and triisocyanate, respectively, M_P and M_S are molecular weights of branches carrying primary and secondary OH groups, respectively, α_P and α_S are conversions of primary and secondary OH groups, respectively, and α_I is the conversion of isocyanate groups; p_{IP} and p_{IS} are transition probabilities calculated from concentrations of dyads [PI] and [SI] which are obtained from kinetic equations for transformation of primary and secondary OH groups, α_P and α_S, respectively.

If the gel point is calculated from Eq. 1.23a, one gets for the critical conversion, irrespective of the values of x_p and x_s, the equation

$$1 - 6\alpha_I(x_P\alpha_P + x_S\alpha_S) = 1 - 6\alpha_I\alpha_{OH} = 0 \tag{1.23b}$$

i.e., the same relation as for gelation of an ideal $A_4 + B_3$ system, with A and B groups in A_4 and B_3, respectively, of the same and independent reactivity. This somewhat surprising result is due to having a random (binomial) distribution of molecules differing in the number of primary and secondary OH groups. The gel times depend on x_p and x_s and increase with increasing x_s. When the primary and secondary OH groups are not randomly distributed, but their numbers are fixed, the gel point conversion is different [8]. In the latter case, Eq. 1.23a is changed to

$$\begin{aligned} F_0(\mathbf{Z}, z) &= n_{TS} Z_I^{M_j} [(1 - \alpha_P + \alpha_P z_{PI})]^j (1 - \alpha_S + \alpha_s z_{SI})]^{4-j} \\ &\quad + n_I Z_I^{M_I} [1 - \alpha_I + \alpha_I (p_{IP} z_{IP} + p_{IS} z_{IS})]^k \qquad (1.24) \\ j &= 0,\ 1,\ 2,\ 3,\ 4; \qquad M_j = jM_P + (4 - j) M_S \end{aligned}$$

In Table 1.4, the calculated conversions of isocyanate groups for tetrafunctional star and bifunctional isocyanate ($k = 2$) are shown. They depend on j, and the highest critical conversion is found for $j = 2$ where $\alpha_{I,crit}$ is equal to 0.611. For homogeneous star (all OH groups are primary or all secondary), $\alpha_{I,crit} = 0.577$. The same value, 0.577, is obtained for randomly distributed primary and secondary OH groups, irrespective of their average number.

Table 1.4 Critical conversion of isocyanate groups for tetrafunctional star and diisocyanate in dependence on the fixed number of branches with primary (j) and secondary ($4 - j$) OH groups. Reactivity ratio primary/secondary = 10 [8]

j	0	1	2	3	4
$\alpha_{I,crit}$	0.577	0.593	0.611	0.588	0.577

Chain-Extended Systems

In a number of systems of practical interest, the branches carrying functional groups are extended by a repeating reaction, for instance, by alternating addition of cyclic anhydride to epoxide, initiated by a carboxyl group by which a polyester chain is formed. By polyaddition of epoxide to an OH group, the formed branch is a polyether chain. In both cases, one obtains a distribution of arm lengths, i.e., distribution of their molecular weights. The precursor functionality remains constant. However, often one finds some functionality distribution due to the occurrence of side reactions. We will illustrate such a situation by analysis of a system, in which the branches of starlike polyols (diol to hexol) are chain extended with an epoxyester (glycidyl pivalate) [17]. The two-stage process is schematically shown in Fig. 1.13.

The primary OH groups react with the epoxide group by which a secondary OH group is formed. This group can also react with the epoxide group. The average

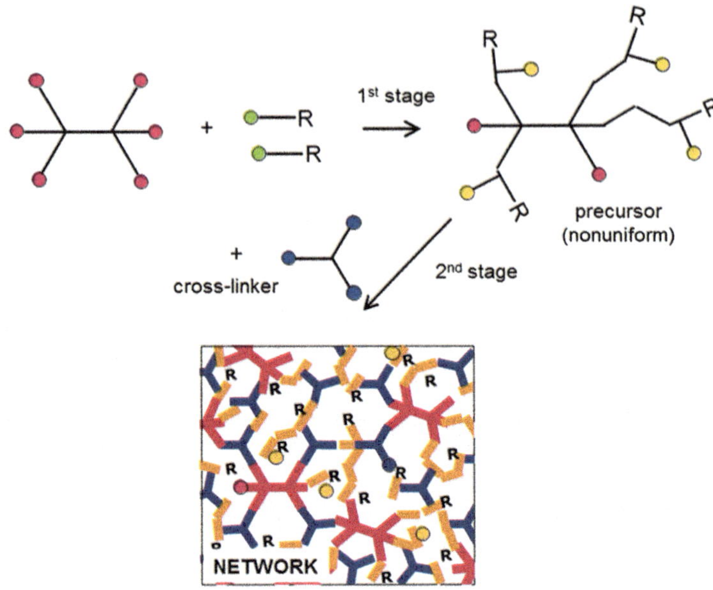

Fig. 1.13 The chain extension process (first-stage reaction) and cross-linking of the formed precursor (second-stage reaction)

length of the branches depends on the molar ratio of epoxy groups E to the primary OH groups. The number-average molecular weight depends on conversion of the etherification reaction and is independent of differences in the reactivity of OH groups. However, the weight-average molecular weight depends also on relative rates by which the primary and secondary groups react (Fig. 1.14).

The distribution of the lengths of the branches is described kinetically by a set of kinetic differential equations (for one branch):

$$
\begin{aligned}
&\frac{d[\mathrm{A1}]}{dt} = -k_1[\mathrm{A1}][\mathrm{E}] \\
&\frac{d[\mathrm{A1E_1}]}{dt} = k_1[\mathrm{A1}][\mathrm{E}] - k_2[\mathrm{A1E_1}][\mathrm{E}] \\
&\vdots \\
&\frac{d[\mathrm{A1E}_k]}{dt} = k_2[\mathrm{A1E}_{k-1}][\mathrm{E}] - k_2[\mathrm{A1E}_k][\mathrm{E}] \\
&\vdots
\end{aligned}
\tag{1.25}
$$

which can be easily transformed into equation for the generating function defined by

$$g_c(Z) \equiv \sum_{i=1}^{\infty} [AE_i]Z^i \tag{1.26}$$

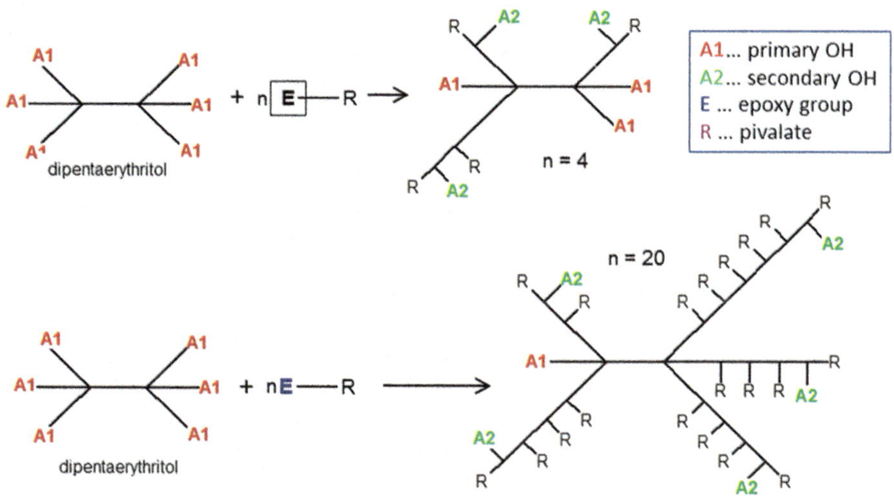

Fig. 1.14 Formation of secondary OH groups A2 of chain-extended dipentaerythritol by reaction with glycidyl pivalate

and solved [17]. As before, the variable Z is utilized for the cross-linking reaction. For a tetrol star (pentaerythritol), the dependence of polydispersity (nonuniformity) on the relative ratio of reactivities of groups A2 to A1 $\kappa_2 = k_2/k_1$ is shown in Fig. 1.15a. Figure 1.15b shows the polydispersity index as a function of the functionality of the precursors for $\kappa_2 = 0.3$.

These results show an interesting feature of certain chain-extended systems—the distributions are relatively narrow and become narrower with increasing conversion and functionality of the polyol core. These special features are due to the branches being a living type. This special feature—decrease of polydispersity with increasing functionality—seems to be unexpected, but a theoretical proof has been offered in Ref. [17].

How is this polydispersity reflected in curing? The distribution described by the solution of Eq. 1.25 was substituted into the statistical branching theory, and it was found that the polydispersity and reactivity differences do not affect the gel conversion and have a relatively small effect on the sol fraction and concentration of elastically active network chains. Figure 1.16 shows the dependence of the gel fraction, w_g, on conversion when precursors of various functionalities are cured with cross-linker B_3.

So far, all is theory, but the reality was somewhat different due to side reactions which were not negligible. The main side reaction was transesterification between the hydroxyl groups and the ester group of the glycidyl ester compound (glycidyl pivalate) similar to that shown in Fig. 1.2 where the reaction-releasing group was carboxyl. This side reaction produced more branching and more dead ends—terminated by a glycol unit. An example of the branching process is shown in the scheme below (Fig. 1.17).

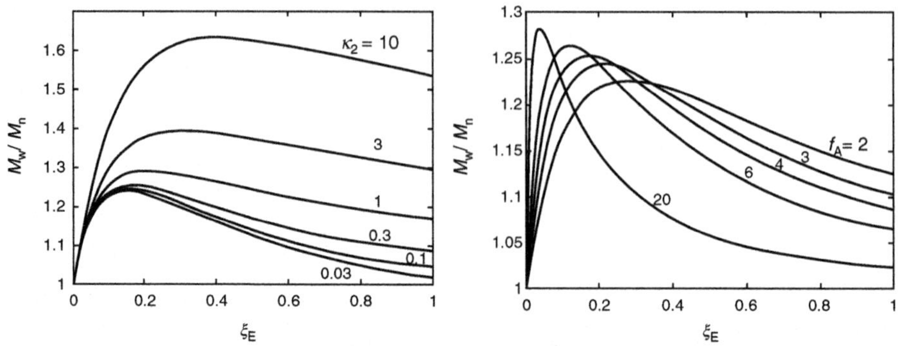

Fig. 1.15 Dependence of nonuniformity (polydispersity index) of a chain-extended $A1_4$ with R–E on conversion of epoxy groups for (**a**) reactivity ratio of groups A1 to A2, κ_2, for the initial molar ratio $[E]_0/[A1]_0 = 1$; (**b**) for $\kappa_2 = 0.3$ and $[E]_0/[A1]_0 = 1$ for various initial functionalities f_A indicated. Reprinted with permission from Ref. [17]. Copyright (2013) American Chemical Society

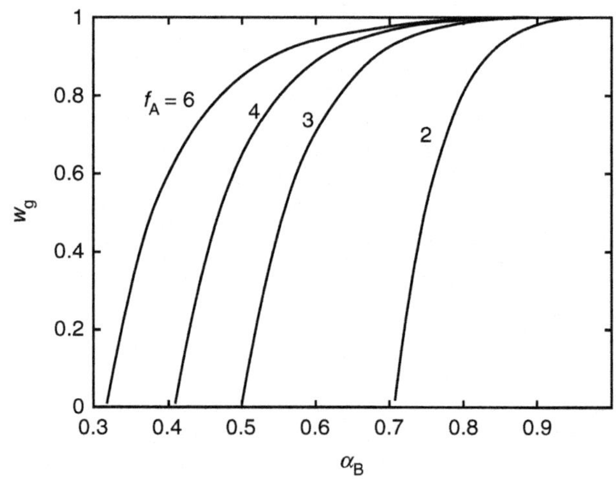

Fig. 1.16 Dependence of the gel fraction on the conversion of B groups of the cross-linker for a polyether precursor of different functionality (indicated), the same parameters as in Fig. 1.15b. Reprinted with permission from Ref. [17]. Copyright (2013) American Chemical Society

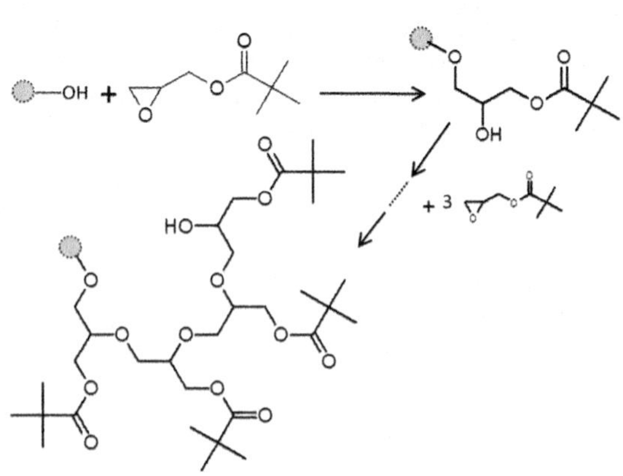

Fig. 1.17 Example of branching due to transesterification in the chain-extending reaction with a glycidyl ester

1. To generate the distribution of states of A- and B-units in the first stage marking the reacted as well as unreacted groups and to generate the w-distribution (second-moment distribution) in the standard way. This w-distribution function is converted into number fraction distribution by integration.
2. This number fraction distribution is used as the input state for the second stage—component AB and component C are added. The branching process gives the product ABC which is the resin component of the powder coating. Again, the distribution generated is a w-(second-moment) distribution. This distribution is then transformed into the number fraction distribution.
3. Both distributions are used in the cross-linking process with cross-linker D in which the gel threshold is surpassed. The gel connectivity determining the elastic response is controlled by probabilities that specific bonds have finite or infinite continuation.

In the second one, alternative to the output → input method just described in steps (1–3), one can generate the distributions by distinguishing bonds, even of the same type but formed in different stages, by assigning a different generating function variable to them. A similar approach was used for cross-linking of distributions of hyperbranched polymers.

Conclusions

Almost all precursors used for the preparation of contemporary cross-linked polymers of practical importance are not single chemical compounds, instead they contain many components. Many of them are series of compounds of increasing molecular weights or some other property. Their concentrations can be described by distribution functions. Distributions exist, for example, in molecular weight types and numbers of functional groups, or in some other property (e.g., the distribution of numbers of branch points). The origin of the distributions ranges from by-products in raw materials that actively participate in cross-linking to intentionally created distributions used for the purpose of controlling the processing and materials properties of the networks. For instance, distributions greatly affect the viscosity buildup and softening temperature, gelation, and amount and molecular weight distribution of the extractable material. As examples of the intentionally prepared distributions, functional copolymers, hyperbranched and highly branched functional polymers, and precursors formed in several stages were discussed. The important role of reactivity differences of functional groups participating in cross-linking was demonstrated.

Although distributions play such an important role in processing and materials properties, it is very difficult to characterize them experimentally. Also, modeling of the distributions and their effects is not easy. These obstacles were the main reason for empirical adjustment of the distributions by looking for the optimum of a certain property (e.g., viscosity buildup and gelation) but without understanding what is happening structurally or what the effect is on some other property (e.g., amount of extractable material). We have shown here that ways exist to describe and predict the evolutions of these distributions and how they affect certain properties. These methods are based on statistical branching theories employing the first-order (or, possibly, higher-order) Markov statistics. The use of the formalism of the probability generating function in expressing and transforming these distributions is of great help because it can easily identify populations relevant to either of these distributions. One can ask why not use other branching theories, such as the kinetic theories, or Monte Carlo simulation off-space or in-space (e.g., bond fluctuation model [42], or dynamic lattice liquid model [43], or molecular dynamics simulation [44]). In fact for certain cases, the use of kinetic theories would be manageable—we have shown here applications of combinations of kinetic and statistical theories. For in-space simulations working in finite spaces, the introduction of distributions would represent a serious complicating factor requiring an increase of the simulation space perhaps by several orders of magnitude. However, a combination of one of the in-space simulations with statistical formulation of the distribution could be possible.

Appendix

Definitions of Distributions and Their Transform Using the Formalism of Probability Generating Functions

In the following tables, the degree of polymerization, molecular weight, and functionality distributions and their averages are defined and expressed in the form of probability generating functions (pgf). The transform between various distributions and their averages by manipulation with the pgfs is also shown.

Probability Generating Functions

This communication deals with various distributions related to properties of the precursors, and often they are multivariate. To keep track of what property we are dealing with, the distributions are post-signed by labels and presented in the form of probability generating functions. They have been used for description of branching and cross-linking systems since a long time ago, and their formulation and use have been repeatedly explained (cf., e.g., Refs. [1, 14, 45]). Nevertheless, some explaining notes about their use are summarized below.

As examples, the following distributions in the form probability generating functions are discussed:

$$A(Z) = \sum_i a_i Z^i, \quad \sum_i a_i = 1$$
$$N(Z) = \sum_i n_i Z^i, \quad W(Z) = \sum_i w_i Z^i, \quad \sum_i n_i = \sum_i w_i = 1$$
$$Q(\mathbf{Z}, \mathbf{z}) = \sum_{i,j} q_{ij} Z_{\mathrm{A}}^{X_{\mathrm{A}}} Z_{\mathrm{B}}^{X_{\mathrm{B}}} z_{\mathrm{AB}}^{i} z_{\mathrm{BA}}^{j}, \quad Z = Z_{\mathrm{A}}, Z_{\mathrm{B}}, \quad z = z_{\mathrm{AB}} z_{\mathrm{BA}}, \quad \sum_{i,j} q_{ij} = 1$$

The function $A(Z)$ describes a set of probabilities $\boldsymbol{a} = a_1, a_2, \ldots, a_i, \ldots, a_n$ of finding the object A in all its possible states distinguished by the value i of a property. In the function $N(Z)$, the property is degree of polymerization, and its value is characterized by the exponent i, related is the set of number fractions **n**; for the function $W(Z)$, the property is also the degree of polymerization, and the related probability is the set of weight fractions **w**. Boldface quantities are vectors like **a** above. The function $Q(\mathbf{Z}, \mathbf{z})$ describes the probabilities of finding the building unit Q in states characterized by the properties X_{A} and X_{B} and by number of bonds i of AB bonds and j of BA bonds. **Z** and **z** are auxiliary variables of the pgfs; by manipulation with them, one can transform them from one to another and to get average values of the property, as will be shown below. The average value is obtained by differentiation and putting the auxiliary variable equal to 1; for example,

$$\frac{\partial N(Z)}{\partial Z} = \sum_i i n_i Z^i, \quad \left[\frac{\partial N(Z)}{\partial Z}\right]_{Z=1} = \sum_i i n_i = P_{\mathrm{n}}$$

Degree of Polymerization and Molecular Weight Distributions

Averages	Probability generating function formulation
Number-average degree of polymerization, P_{xn}, and molecular weight, M_{xn}, of homopolymer X $P_{\mathrm{Xn}} = \sum_i in_i$ $M_{\mathrm{Xn}} = \sum_i in_i M_{\mathrm{x}} = M_{\mathrm{X}} \sum_i in_i$	$N(Z) = \sum_i n_i Z^i$ $\frac{\partial N(Z)}{\partial Z} = \sum_i in_i Z^{i-1} \quad \left[\frac{\partial N(Z)}{\partial Z}\right]_{Z=1} \equiv N'(1) = \sum_i in_i = P_{\mathrm{Xn}}$ $N_{\mathrm{M}}(Z) = \sum_i n_i Z^{iM}$ $\frac{\partial N_{\mathrm{M}}(Z)}{\partial Z} = \sum_i in_i M Z^{iM-1}$ $\left[\frac{\partial N_{\mathrm{M}}(Z)}{\partial Z}\right]_{Z=1} \equiv N'_M(1) = \sum_i in_i M = P_{\mathrm{Xn}}$
w-Average (weight-average) degree of polymerization P_{Xw} and molecular weight M_{xw} of homopolymer X $P_{\mathrm{Xw}} = \sum_i iw_i = \sum_i i^2 n_i / \sum_i in_i$ $M_{\mathrm{Xw}} = \sum_i iw_i M_{\mathrm{x}} = M_{\mathrm{X}} \sum_i i^2 n_i / \sum_i in_i$	$W(Z) = \sum_i w_i Z' = Z \frac{\partial N(Z)}{\partial Z} / \left[\frac{\partial N(Z)}{\partial Z}\right]_{Z=1}$ $\frac{\partial W(Z)}{\partial Z} = \sum_i iw_i Z^{i-1}; \quad \left[\frac{\partial W(Z)}{\partial Z}\right]_{Z=1} \equiv W'(1) = \sum_i iw_i = P_{\mathrm{Xw}}$ $W_{\mathrm{M}}(Z) = \sum_i w_i Z^{iM} = Z \frac{\partial N_{\mathrm{M}}(Z)}{\partial Z} / \left[\frac{\partial N_{\mathrm{M}}(Z)}{\partial Z}\right]_{Z=1}$ $\frac{\partial W_{\mathrm{M}}(Z)}{\partial Z} = \sum_i iw_i M_{\mathrm{X}} Z^{iM-1}$ $\left[\frac{\partial N_{\mathrm{M}}(Z)}{\partial Z}\right]_{Z=1} \equiv N'_{\mathrm{M}}(1) = \sum_i in_i M_{\mathrm{x}} = M_{\mathrm{X}} P_{\mathrm{Xn}}$
Number-average degree of polymerization P_{ABn} and molecular weight M_{ABn} of copolymer AB $P_{\mathrm{ABn}} = \sum_i in_i = \sum_i \sum_{j=0}^{i} \left(jn_{\mathrm{A}j} + (i-j) n_{\mathrm{B}(i-j)}\right)$ $M_{\mathrm{ABn}} = \sum_i \sum_{j=0}^{i} \left(jn_{\mathrm{A}j} M_{\mathrm{A}} + (i-j) n_{\mathrm{B}(i-j)} M_{\mathrm{B}}\right)$	$N(Z_{\mathrm{A}}, Z_{\mathrm{B}}) = \sum_k n_k \sum_{i=0}^{k} x_{i,k-i} Z_{\mathrm{A}}^i Z_{\mathrm{B}}^{k-i} \overset{\text{random}}{=} \sum_k n_k (x_{\mathrm{A}} Z_{\mathrm{A}} + (1 - x_{\mathrm{A}}) Z_{\mathrm{B}})^k$ $\mathrm{DP}: Z_{\mathrm{A}} = Z_{\mathrm{B}} = Z, \quad \left[\frac{\partial N(Z)}{\partial Z}\right]_{Z=1} = \sum_k kn_k = P_{\mathrm{ABn}}$ $N_{\mathrm{M}}(Z_{\mathrm{A}}, Z_{\mathrm{B}}) = \sum_k n_k \sum_{i=0}^{k} n_{i,k-i} Z_{\mathrm{A}}^{iM_{\mathrm{A}}} Z_{\mathrm{B}}^{(k-i)M_{\mathrm{B}}} \overset{\text{random}}{=}$ $\sum_k n_k \left(x_{\mathrm{A}} Z_{\mathrm{A}}^{M_{\mathrm{A}}} + (1 - x_{\mathrm{A}}) Z_{\mathrm{B}}^{M_{\mathrm{B}}}\right)^k$ $\left[\frac{\partial N_{\mathrm{M}}(Z_{\mathrm{A}}, Z_{\mathrm{B}})}{\partial Z_{\mathrm{A}}}\right]_{Z_{\mathrm{A}}=Z_{\mathrm{B}}=1} + \left[\frac{\partial N_{\mathrm{M}}(Z_{\mathrm{A}}, Z_{\mathrm{B}})}{\partial Z_{\mathrm{B}}}\right]_{Z_{\mathrm{A}}=Z_{\mathrm{B}}=1} =$ $\sum_k n_k \sum_{i=0}^{k} (in_{i,k-i} M_{\mathrm{A}} + (k-i) n_{i,k-i} M_{\mathrm{B}}) \overset{\text{random}}{=}$ $\sum_k kn_k (x_{\mathrm{A}} M_{\mathrm{A}} + (1 - x_{\mathrm{A}}) M_{\mathrm{B}}) = M_{\mathrm{ABn}}$

w-Average (weight-average) degree of polymerization P_{Xw} and molecular weight M_{Xw} of copolymer AB	$W(Z_{\mathrm{A}},Z_{\mathrm{B}})=\sum_k w_k\sum_{i=0}^{k}x_{i,k-i}Z_{\mathrm{A}}^{i}Z_{\mathrm{B}}^{k-i}\overset{\text{random}}{=}\sum_k w_k\left(x_{\mathrm{A}}Z_{\mathrm{A}}+(1-x_{\mathrm{A}})Z_{\mathrm{B}}\right)^k$
$P_{\mathrm{ABw}}\equiv P_{\mathrm{AB2}}=\sum_i\sum_i i^2 n_i/\sum_i in_i$ $M_{\mathrm{ABw}}=\sum_i\sum_{j=0}^{i}\left(jw_{\mathrm{Aj}}M_{\mathrm{A}}+(i-j)w_{\mathrm{B}(i-j)}M_{\mathrm{B}}\right)$ $w_{\mathrm{Aj}}=\dfrac{jn_{\mathrm{Aj}}M_{\mathrm{A}}}{\sum_{j=0}^{i}\left(jn_{\mathrm{Aj}}M_{\mathrm{A}}+(i-j)n_{\mathrm{B}(i-j)}M_{\mathrm{B}}\right)}$	$\mathrm{DP}:Z_{\mathrm{A}}=Z_{\mathrm{B}}=Z$ as homopolymer $\left[\frac{\partial W(Z)}{\partial Z}\right]_{Z=1}=\sum_k kw_k=P_{\mathrm{ABw}}$ $W_{\mathrm{M}}(Z_{\mathrm{A}},Z_{\mathrm{B}})=\sum_k w_k\sum_{i=0}^{k}n_{i,k-i}Z_{\mathrm{A}}^{iM_{\mathrm{A}}}Z_{\mathrm{B}}^{(k-i)M_{\mathrm{B}}}\overset{\text{random}}{=}\sum_k w_k\left(x_{\mathrm{A}}Z_{\mathrm{A}}^{M_{\mathrm{A}}}+(1-x_{\mathrm{A}})Z_{\mathrm{B}}^{M_{\mathrm{B}}}\right)^k$ $W_{\mathrm{M}}(Z_{\mathrm{A}},Z_{\mathrm{B}})=\sum_k n_k\sum_{i=0}^{k}x_{i,k-i}Z_{\mathrm{A}}^{iM_{\mathrm{A}}}Z_{\mathrm{B}}^{(k-i)M_{\mathrm{B}}}\overset{\text{random}}{=}\sum_k n_k\left(x_{\mathrm{A}}Z_{\mathrm{A}}^{M_{\mathrm{A}}}+(1-x_{\mathrm{A}})Z_{\mathrm{B}}^{M_{\mathrm{B}}}\right)^k$ $W_{\mathrm{M}}(Z_{\mathrm{A}},Z_{\mathrm{B}})=(1/M_{\mathrm{ABn}})\left(Z_{\mathrm{A}}\frac{\partial N_{\mathrm{M}}}{\partial Z_{\mathrm{A}}}+Z_{\mathrm{B}}\frac{\partial N_{\mathrm{M}}}{\partial Z_{\mathrm{B}}}\right)=\sum_k n_k\sum_{i=0}^{k}x_{i,k-i}Z_{\mathrm{A}}^{iM_{\mathrm{A}}}Z_{\mathrm{B}}^{(k-i)M_{\mathrm{B}}}$ $=(1/M_{\mathrm{ABn}})\sum_k n_k\sum_{i=0}^{k}\left(iM_{\mathrm{A}}+(k-i)M_{\mathrm{B}}\right)x_{i,k-i}Z_{\mathrm{A}}^{iM_{\mathrm{A}}}Z_{\mathrm{B}}^{(k-i)M_{\mathrm{B}}}$ $M_{\mathrm{ABw}}=\left[\frac{\partial W_{\mathrm{M}}(Z_{\mathrm{A}},Z_{\mathrm{B}})}{\partial Z_{\mathrm{A}}}\right]_{Z_{\mathrm{A}}=Z_{\mathrm{B}}=1}+\left[\frac{\partial W_{\mathrm{M}}(Z_{\mathrm{A}},Z_{\mathrm{B}})}{\partial Z_{\mathrm{B}}}\right]_{Z_{\mathrm{A}}=Z_{\mathrm{B}}=1}$ $=(1/M_{\mathrm{ABn}})\sum_k n_k\sum_{i=0}^{k}\left(iM_{\mathrm{A}}+(k-i)M_{\mathrm{B}}\right)^2x_{i,k-i}\overset{\text{random}}{=}(1/M_{\mathrm{ABn}})$ $\sum_k k^2\left(x_{\mathrm{A}}M_{\mathrm{A}}+(1-x_{\mathrm{A}})M_{\mathrm{B}}\right)^2n_k=\left(x_{\mathrm{A}}M_{\mathrm{A}}+(1-x_{\mathrm{A}})M_{\mathrm{B}}\right)P_{\mathrm{ABw}}$

Functionality Distributions

Definition	Probability generating function formulation
Number-average and w-average (second moment) functionality average—single functional group $f_n = \sum_f f n_f$ $f_w \equiv f_2 = \dfrac{\sum_f f^2 n_f}{\sum_f f n_f}$	$F_{fn}(Z_f) = \sum_f n_f Z_f^f$ $f_n = \left[\frac{\partial F_{fn}(Z_f)}{\partial Z_f}\right]_{Z_f=1} \equiv F'_{fn}(1)$ $F_{f2}(Z_f) = \sum_f f n_f Z_f^f / f_n = Z_f \dfrac{\partial F_{fn}(Z_f)}{\partial Z_f} / f_n$ $f_w \equiv f_2 = \left[\frac{\partial F_{f2}(Z_f)}{\partial Z_f}\right]_{Z_f=1}$
Number-average and w-average (second moment) functionality average—two types of functional groups (R, S) $f_{An} = \sum_f f_A n_{fA}; \quad f_{Bn} = \sum_f f_B n_{fB}$ $f_{Aw} \equiv f_{A2} = \dfrac{\sum_f f_A^2 n_{fA}}{\sum_f f_A n_{fA}}$ Other second-moment averages possible (depends on reaction paths)	$F_{fn}(Z_{Af}, Z_{Bf}) = \sum_{i,j} n_{ij} Z_{Af}^i Z_{Bf}^j$ $f_{An} = \left[\frac{\partial F_{fn}(Z_{Af}, Z_{Bf})}{\partial Z_{Af}}\right]_{Z_{Af}=Z_{Bf}=1}$ $F_{Af2}(Z_{Af}, Z_{Bf}) = \sum_i i n_{ij} Z_{Af}^i Z_{Bf}^j / f_{An} =$ $Z_{Af} \dfrac{\partial F_{fn}(Z_{Af}, Z_{Bf})}{\partial Z_{Af}} / f_{An}$ $f_{Aw} \equiv f_{A2} = \left[\frac{\partial F_{Af2}(Z_{Af}, Z_{Bf})}{\partial Z_{Af}}\right]_{Z_{Af}=Z_{Bf}=1} = \sum_{i,j} i^2 n_{ij} / f_{An}$

Distribution of Reaction States of Building Units in the Branching Process

Cross-linking of f_A-functional precursor with f_B-functional cross-linker; groups of equal and independent reactivity	States in the form of a pgf: units, unreacted groups Af and Bf, and groups A reacted with B groups and B groups reacted with C groups of cross-linker marked; type of bond (A → B) indicated
$(1-\alpha_A+\alpha_A)^{f_A} =$ $\sum_{i=0}^{f} \dfrac{f_A!}{i!(f_A-i)!} \alpha_A^i (1-\alpha_A)^{f_A-i} =$ $(1-\alpha_A)^{f_A} + f(1-\alpha_A)^{f_A-1}\alpha_A + \ldots$ $+ f_A(1-\alpha_A)\alpha_A^{f_A-1} + \alpha_A^{f_A}$	n_A, n_B mole fractions of precursors A and B, respectively $F_0(Z_A, Z_{Af}, Z_B, Z_{Bf}, z_{AB}, z_{BA}) \equiv F_0(\mathbf{Z}, \mathbf{z}) =$ $n_A F_{0A}(Z_A, Z_{Af}, z_{AB}) + n_A F_{0B}(Z_B, Z_{Bf}, z_{BA})$; $F_{0A}(Z_A, Z_{Af}, z_{AB}) = Z_A((1-\alpha_A)Z_{Af} + \alpha_A z_{AB})^{f_A} =$ $Z_A(1-\alpha_A)^{f_A} Z_{Af}^{f_A} + Z_A(1-\alpha_A)^{f_A-1} Z_{Af}^{f_A-1} \alpha_A z_{AB} + \ldots +$ $Z_A f_A(1-\alpha) Z_{Af} \alpha_A^{f_A-1} + Z_A \alpha_A^{f_A} z_{AB}^{f_A}$
Cross-linking of functional copolymer with cross-linker C: Distribution of states of AB copolymer;	$F_0(\mathbf{Z}, \mathbf{z}) = n_{AB} F_{0AB} + n_C F_{0C}$ $F_{0AB}(\mathbf{Z}, \mathbf{z}) = N_{AB}(Z_A, F_{0B}(Z_B, z_{BC})) =$ $\sum_k n_k (x_A Z_A + (1-x_A) Z_B (1-\alpha_B+\alpha_B z_{BC}))^k$
B-unit carries a reactive group which reacts with C-group of C-cross-linker	$F_{0AB}(\mathbf{Z}, \mathbf{z}) = N_{AB}(\xi) = \sum_k n_k \xi^k$ $\xi(Z_A, Z_B, z_{BC}) = x_A Z_A + (1-x_A) Z_B (1-\alpha_B+\alpha_B z_{BC})$

References

1. Dušek, K.: Network formation in curing of epoxy resins. Adv. Polym. Sci. **78**, 1 (1986). doi:10.1007/BFb0051042
2. Bauer, D.R., Dickie, R.: Crosslinking chemistry and network structure in organic coatings. I. Cure of melamine formaldehyde/acrylic copolymer films. J. Polym. Sci. B Polym. Phys. **18**, 1997–2014 (1980). doi:10.1002/pol.1980.180181001
3. Bauer, D.R., Dickie, R.A.: Crosslinking chemistry and network structure in organic coatings. II. Effect of catalysts on cure of melamine formaldehyde/acrylic copolymer films. J. Polym. Sci. B Polym. Phys. **18**, 2015–2025 (1980). doi:10.1002/pol.1980.180181002
4. Computer Applications in Applied Polymer Science II. Automation, Modeling, Simulation. In: Provder, T. (ed.). ACS Symposium Series 404 (1989). ISBN: 13: 9780841216624
5. Dušek, K., Šomvársky, J.: Network formation and their application to systems of industrial importance. In: Aharoni, S., (ed.) Synthesis, Characterization and Theory of Polymeric Networks and Gels, pp. 283–301. Plenum Press, New York (1992). ISBN: 0306443066, 9780306443060
6. Dušek, K., Dušková-Smrčková, M., Lewin, L.A., Huybrechts, J., Barsotti, R.J.: Branching theories and thermodynamics used to help designing precursor architectures and binder systems. Surf. Coat. Int. B: Coat. Trans. **82**, 121–131 (2006). doi:10.1007/BF02699642
7. Huybrechts, J., Dušek, K.: Star oligomers for low VOC polyurethane coatings. Surf. Coat. Int. **81**, 117–127 (1998). doi:10.1007/BF02693852
8. Huybrechts, J., Dušek, K.: Star oligomers with different reactivity in low VOC polyurethane coatings: part II. Surf. Coat. Int. **82**, 172–177 (1998). doi:10.1007/BF02693857
9. Huybrechts, J., Dušek, K.: Star oligomers and hyperbranched polymers in low VOC polyurethane coatings: part III. Surf. Coat. Int. **82**, 234–239 (1998). doi:10.1007/BF02693865
10. Matějka, L., Dušek, K., Chabanne, P., Pascault, J.-P.: Cationic polymerization of diglycidyl ether of bisphenol A. 2. Theory. J. Polym. Sci. A Polym. Chem. **35**, 651–663 (1997). doi:10.1002/(SICI)1099-0518(199703)35:4<651::AID-POLA7>3.3.CO;2-N
11. Chivukula, P., Dušek, K., Wang, D., Dušková-Smrčková, M., Kopečková, P., Kopeček, J.: Synthesis and characterization of novel aromatic azo bond-containing, pH-sensitive and hydrolytically cleavable IPN hydrogels. Biomaterials. **27**, 1140–1151 (2006). doi:10.1016/j.biomaterials.2005.07.02
12. Wang, D., Dušek, K., Kopečková, P., Dušková-Smrčková, M., Kopeček, J.: Novel aromatic azo-containing pH-sensitive hydrogels: synthesis and characterization. Macromolecules. **35**, 7791–7803 (2002). doi:10.1021/ma020745k
13. Dušek, K., Dušková-Smrčková, M.: Modeling of polymer network formation from preformed precursors. Macromol. React. Eng. **6**, 426–445 (2012). doi:10.1002/mren.201200028
14. Dušek, K., Netopilík, M., Kratochvíl, P.: Nonuniformities of distributions of molecular weights of grafted polymers. Macromolecules. **45**, 3240–3246 (2012). doi:10.1021/ma3000187
15. Dušková-Smrčková, M., Valentová, H., Ďuračková, A., Dušek, K.: Effect of dilution on structure and properties of polyurethane networks. Pregel and postgel cyclization and phase separation. Macromolecules. **43**, 6450–6462 (2010). doi:10.1021/ma100626d
16. Ilavský, M., Dušek, K.: The structure and elasticity of polyurethane networks. I. Model networks from poly(oxypropylene) triols and diisocyanate. Polymer. **24**, 981–990 (1983). doi:10.1016/0032-3861(83)90148-9
17. Dušek, K., Dušková-Smrčková, M., Huybrechts, J., Ďuračková, A.: Polymer networks from preformed precursors having molecular weight and group reactivity distributions. Theory and application. Macromolecules. **46**, 2767–2784 (2013). doi:10.1021/ma302396u
18. Dušek, K., Dušková-Smrčková, M., Fedderly, J.J., Lee, G.F., Lee, J.D., Hartmann, B.: Polyurethane networks with controlled architecture of dangling chains. Macromol. Chem. Phys. **203**, 1936–1948 (2002). doi:10.1002/1521-3935(200209)203:13<1936::AID-MACP1936>3.0.CO;2-C

19. Dušek, K., Dušková-Smrčková, M., Voit, B.: Highly-branched off-stoichiometric functional polymers as polymer networks precursors. Polymer. **46**, 4265–4282 (2005). doi:10.1016/j.polymer.2005.02.037
20. Dušek, K., Scholtens, B.J.R., Tiemersma-Thoone, G.P.J.M.: Theoretical treatment of network formation by a multistage process. Polym. Bull. **17**, 239–245 (1987). doi:10.1007/BF00285356
21. Tiemersma-Thoone, G.P.J.M., Scholtens, B.J.R., Dušek, K., Gordon, M.: Theories for network formation in multi-stage processes. J. Polym. Sci. B Polym. Phys. **29**, 463–482 (1991). doi:10.1002/polb.1991.090290409
22. Dušek, K.: Applicability of statistical theories of network formation. Macromol. Symp. **256**, 18–27 (2007). doi:10.1002/masy.20075100
23. Dušek, K., Dušková-Smrčková, M.: Network structure formation during crosslinking of organic coating systems. Prog. Polym. Sci. **25**, 1215–1260 (2000). doi:10.1016/S0079-6700(00)00028-9
24. Dušek, K., Dušková-Smrčková, M.: Polymer networks. In: Matyjaszewski, K., Gnanou, P., Leibler, L., (eds.) Macromolecular Engineering. Precise Synthesis, Materials Properties, Applications, vol. 3, Structure–Property Correlation and Characterization Techniques, pp. 1687–1730. Wiley, Weinheim (2007). ISBN: 978-3-527-31446-1
25. Ni, H., Aaserud, D.J., Simonsick Jr., W.J., Soucek, M.D.: Preparation and characterization of alkoxysilane functionalized isocyanurates. Polymer. **41**, 57–71 (2000). doi:10.1016/S0032-3861(99)00160-3
26. Dušek, K., Dušková-Smrčková, M., Huybrechts, J.: Control of performance of nanostructured polymer network precursors by differences in reactivity of functional groups. J. Nanostruct. Polym. Nanocomposites. **1**, 45–53 (2005)
27. Dušek, K., Špírková, M., Havlíček, I.: Network formation in polyurethanes due to side reactions. Macromolecules. **23**, 1774 (1990). doi:10.1021/ma00208a036
28. Hult, A., Johansson, M., Malmström, E.: Hyperbranched polymers. Adv. Polym. Sci. **143**, 1–34 (1999). doi:10.1007/3-540-49780-3_1
29. Johansson, M., Glauser, T., Jansson, A., Hult, A., Malmström, E., Claesson, H.: Design of coating resins by changing the macromolecular architecture: solid and liquid coating systems. Prog. Org. Coat. **48**, 194–200 (2003). doi:10.1016/S0300-9440(03)00105-X
30. van Benthem, R.A.T.M.: Novel hyperbranched resins for coating applications. Prog. Org. Coat. **40**, 203–214 (2000). doi:10.1016/S0300-9440(00)00122-3
31. Dušek, K., Dušková-Smrčková, M.: Formation, structure and properties of the crosslinked state in relation to precursor architecture. In: Fréchet, J.M.J., Tomalia, D.A. (eds.) Dendrimers and Other Dendritic Polymers. Wiley (2011); Tomalia, D.A., Fréchet, J.M.J. (eds.) Dendritic Polymers, pp. 111–145. Wiley (2002). ISBN: 978-0-471-63850-6
32. Müller, A.H.E., Yan, D., Wulkow, M.: Molecular parameters of hyperbranched polymers made by self-condensing vinyl polymerization. 1. Molecular weight distribution. Macromolecules. **30**, 7015–1723 (1997). doi:10.1021/ma9619187
33. Zhou, Z., Yan, D.: Distribution function of hyperbranched polymers formed by AB2 type polycondensation with substitution effect. Polymer. **47**, 1473–1479 (2006). doi:10.1016/j.polymer.2005.12.035
34. Zhou, Z., Yan, D.: Kinetic theory of hyperbranched polymerization. In: Yan, D., Gao, C., Frey, H. (eds.) Hyperbranched Polymers: Synthesis, Properties, and Applications. Wiley, Hoboken (2001). doi:10.1002/9780470929001.ch13
35. Dušek, K., Šomvársky, J., Smrčková, M., Simonsick Jr., W.J., Wilczek, L.: Role of cyclization in the degree-of-polymerization distribution of hyperbranched polymers. Modelling and experiment. Polym. Bull. **42**, 489–496 (1999). doi:10.1007/s002890050493
36. Dušek, K., Dušková-Smrčková, M.: Polymer networks from precursors of defined architecture. Activation of pre-existing branch points. Macromolecules. **36**, 2915–2925 (2003). doi:10.1021/ma021666b

37. Bruchmann, B., Schrepp, W.: The AA* + B*B2 approach. A simple and convenient synthetic strategy towards hyperbranched polyurea-urethanes. e-Polymers. **3**(1), 191–200 (2003). doi:10.1515/epoly.2003.3.1.191
38. Voit, B.: Hyperbranched polymers—all problems solved after 15 years of research? J. Polym. Sci. A Polym. Chem. **43**, 2679–2699 (2005). doi:10.1002/pola.20821
39. Dušek, K.: Build-up of polymer network by initiated polyreactions. 2. Theoretical treatment of polyetherification released by polyamine|polyepoxide addition. Polym. Bull. **13**, 321–328 (1985). doi:10.1007/BF00262115
40. Dušek, K., Šomvársky, J., Ilavský, M.: Network build-up by initiated polyreactions. 4. Derivation of postgel parameters for postetherification in diamine-diepoxide curing. Polym. Bull. **18**, 209 (1987). doi:10.1007/BF00255112
41. Witte, F.M., Goemans, C.D., van der Linde, R., Stanssens, D.A.: The aliphatic oxirane powder coating system: development, crosslinking chemistry and coating properties. Prog. Org. Coat. **32**, 241–251 (1997). doi:10.1016/S0300-9440(97)00066-0
42. Carmesin, I., Kremer, K.: The bond fluctuation method: a new effective algorithm for the dynamics of polymers in all spatial dimensions. Macromolecules. **21**, 2819–2823 (1988). doi:10.1021/ma00187a030
43. Polanowski, P., Jeszka, J.K., Krysiak, K., Matyjaszewski, K.: Influence of intramolecular crosslinking on gelation in living copolymerization of monomer and divinyl cross-linker. Monte Carlo simulation studies. Polymer. **79**, 171–178 (2015). doi:10.1016/j.polymer.2015.10.018
44. Kroll, D.M., Croll, S.G.: Influence of crosslinking functionality, temperature and conversion on heterogeneities in polymer networks. Polymer. **79**, 82–90 (2015). doi:10.1016/j.polymer.2015.10.020
45. Fortunatti, C., Sarmoria, C., Brandolin, A. Asteasuain, M.: Modeling of RAFT polymerization using probability generating functions. Detailed prediction of full molecular weight distributions and sensitivity analysis. Macromol. React. Eng. 8, 781-795 (2014). doi: 10.1002/mren.201400020

Chapter 2
Heterogeneity in Crosslinked Polymer Networks: Molecular Dynamics Simulations

D.M. Kroll and S.G. Croll

Introduction

Crosslinked network polymers are a widely used class of chemicals and have been utilized since people used any form of paint, e.g., with proteins (casein, tempura) or unsaturated vegetable oils. Modern high-performance technology relies on different chemistry, such as epoxies, polyurethanes, and rubbers, but the continual search for improvements relies on increasing our understanding of how two or more reactive precursors form the desired network. When these modern technologies were first implemented, considerable advancements could be made rapidly; more recently however, as with all mature technologies, further progress is much more difficult and requires a greater degree of insight. Networks are formed from small molecules that react with other small molecules, and the resulting network topology depends on the specific interactions (chemistry) and the distribution and availability of these precursors, which is governed by their random location and motion. Thus, it is reasonable to expect that one can describe the resulting networks using statistical approaches which have been very successful at describing how representative molecular properties affect macroscopic network properties [1–4].

A simple calculation provides some perspective. We might consider a coating has failed when the density of visible features such as rust spots and blisters reaches a concentration of 1 per square centimeter. A polymer coating 1 cm^2 in area might have a typical thickness of 50 μm and thus a volume of $5 \times 10^{+21}$ Å^3. If we approximate the volume of most atoms as 1 Å^3, it is easy to appreciate that, while

D.M. Kroll
Department of Physics, North Dakota State University, Fargo, ND, USA

S.G. Croll (✉)
Department of Coatings and Polymeric Materials, North Dakota State University, Fargo, ND, USA
e-mail: Stuart.Croll@ndsu.edu

M. Wen, K. Dušek (eds.), *Protective Coatings*, DOI 10.1007/978-3-319-51627-1_2

most crosslinking reactions will proceed as anticipated, there is a chance that an appreciable number will not, at random, and will lead to a variety of heterogeneities. As long appreciated [1, 5, 6], as a crosslinking system proceeds toward its gel point, the various partially crosslinked components become much less able to move, and thus the reactive species cannot mix and react homogeneously. Heterogeneities can take the form of unreacted functionalities, dangling ends and chains that return to the same network junction and thus form loops. Since the numbers involved are so large, one can imagine several such occurrences in a neighborhood contributing to a more extended region of inconsistent crosslinking.

Although they are a natural consequence of the large numbers of entities involved in the crosslinking process, these heterogeneities inevitably detract from the performance of the network and will therefore be referred to here as imperfections, flaws, or defects. Polymer coatings are most often used to protect something or improve its appearance. Protective requirements can be diverse, but they often rely either on the mechanical properties of the crosslinked polymer or its ability to prevent aggressive chemicals from reaching the underlying engineering material, e.g., metals that may corrode. Network imperfections force the gelation point to occur at a higher value of chemical conversion than anticipated and may limit the toughness and mechanical strength or enable water, salts, or other chemicals to permeate too easily. Adhesive joints may fail, rust spots appear at unplanned places on a painted metal part, and caulks and sealants may spring a leak, seemingly at random places. Exposure to aggressive environments leads to chemical and physical degradation and thus failure. The quality of the starting material determines, at least in part, its success in service, so heterogeneities that arise during the network formation may be crucial.

Information on heterogeneities and imperfections, at the molecular level, is generally difficult [7], if not impossible, to obtain experimentally. In molecular dynamics simulations, however, molecular architecture and the crosslinking reactions can be specified, and a detailed picture of network structure and its heterogeneities can be determined. Modern computers and software enable us to investigate large system sizes and create crosslinked systems in which the location and connections of every molecule or bond are tabulated. For example, Duering et al. [8, 9] studied tetrafunctional networks with strand lengths ranging from 12 to 100 monomers and compared their results to the predictions of the rubber elasticity theories. Here, we also use coarse-grained molecular dynamics to investigate highly crosslinked networks typical of high-performance coatings, as distinct from rubbery materials which may have almost an order of magnitude lower crosslink density. The systems studied here prove to have heterogeneities that range from small, molecular size defects to features that extend across the whole simulated volume.

Background

For ideal networks, the statistical theories of Flory and others [1, 2, 10, 11] are very successful and provide algebraic equations that have been used for many decades to describe the rubbery elastic, mechanical properties in terms of a characteristic length of the network chains between crosslinks, their concentration, and the number of reactive functionalities at the network junctions. This approach was extended [12] to describe the ability of a solvent to swell a crosslinked polymer by including the Flory–Huggins theory for miscibility. A later approach recognized that gelation could be modeled as a process on a Bethe lattice [13, 14] and gave results supporting the findings of the statistical models. Others, notably Dušek [15–17] and Miller and Macosko [18, 19], used kinetic and recursive equations to calculate the likely crosslink chain length and density from the structure of the reactive precursors and thus enabled the prediction of macroscopic properties using the statistical models for rubber elasticity. Unless there is a specific effort to model the formation of loops [20], the recursive approach, like the Bethe lattice approach, is known to model dendritic structures, not networks [21]. However, these approaches are well known and successful for calculating the expected, average properties of a rubbery network, but they have no means of estimating the number and extent of heterogeneities, although one can compare predictions with experiment and, for example, deduce a ring-forming parameter [6] that accounts for the discrepancy.

Early in the development of these statistical models, it was realized that unreacted chain ends and other imperfections must be accounted for [1] and that spatial neighbors may not be joined by chemical bonds [22]. Reflecting the fact that only elastically active chains contribute, Flory proposed the cycle rank to calculate the mechanical properties of the network. In the so-called phantom network approach, the shear modulus of the rubbery material, G_{ph}, is proportional to the cycle rank, ξ, which is the number of complete loops in the network:

$$G_{\text{ph}} = \xi k_{\text{B}} T / V \tag{2.1}$$

where T = temperature, V is the volume of the system, and k_{B} is Boltzmann's constant.

The simpler "affine" model for rubber elasticity assumes that all network chains are elastically effective so that the modulus is

$$G_{\text{aff}} = \nu_{\text{e}} k_{\text{B}} T / V \tag{2.2}$$

The cycle rank, ξ, is given by the difference between the number of elastically active chains, ν_{e}, and the number of elastically active junctions, μ_{e}. In a perfect network without heterogeneities, the cycle rank is simply related to the network functionality [23]:

$$\xi = \nu_{\text{e}}\left(1 - \frac{2}{\phi}\right) \tag{2.3}$$

where ϕ is the functionality of the crosslink junctions (3 or 6 here). For a tetrafunctional network ξ is simply $\nu_{\text{e}}/2$ [24].

The Miller–Macosko (MM) approach can be used to predict values for the chain length between crosslinks, the cycle rank, etc., from the functionality, structure, and concentration of the reactive precursors. Like the statistical theories, and the Bethe lattice approach, MM calculations have no explicit recognition that networks form loops. Their calculations assume an ideal treelike growth of the network, and the criterion for an elastically effective link is that the path through the network continues indefinitely. There is also a considerable body of work by Dušek that has examined a variety of specific combinations of reactants to calculate the characteristics of the resultant networks [16]. Molecular dynamics simulations [25] have shown the variation possible in crosslink density and the effect on the mechanical properties of rubbers of long dangling chains and varying sizes of loops. The work here uses coarse-grained molecular dynamics and examines the heterogeneities in highly crosslinked networks characteristic of high-performance coatings and rigid composites and compares results with MM calculations, since they give very useful estimates of average properties of ideal networks.

Model and Simulation Method

Molecular dynamics is a very versatile approach that could be used to examine a myriad of networks. Here we study two simple, but distinctly different, coarse-grained models to model networks similar to those produced using a wide range of chemical options. The models consist of precursors composed of a two-bead "chain extender" and a crosslinker of functionality $f = 3$ or 6. These, and similar models, have been used before [26–29] to study fracture, thermomechanical properties, and cavitation in crosslinked networks. The 6-functional network is formed with one chain extender molecule already bonded to its 6-functional crosslinker bead prior to starting the simulations, but this pre-bonding has no noticeable effect on the resulting network connectivity [29]. The 3-functional model has a crosslinker with a central 3-functional node connected to three single functional beads. Diagrams of the network structures are shown in Fig. 2.1.

The interactions used to model bead behavior and interactions are completely conventional. All beads interact via van der Waals forces modeled by a Lennard-Jones (LJ) potential, $U(r)$, which is cut off at a radius, $r_c = 2.5\sigma$, and shifted upward so that the potential is equal to zero at r_c:

$$
\begin{aligned}
U(r) &= U_{LJ}(r) - U_{LJ}(r_c) \quad && \text{for } r \leq r_c \\
&= 0 && \text{for } r > r_c
\end{aligned}
\tag{2.4}
$$

where

$$U_{LJ}(r) = 4\varepsilon\left[\left(\frac{\sigma}{r}\right)^{12} - \left(\frac{\sigma}{r}\right)^{6}\right] \tag{2.5}$$

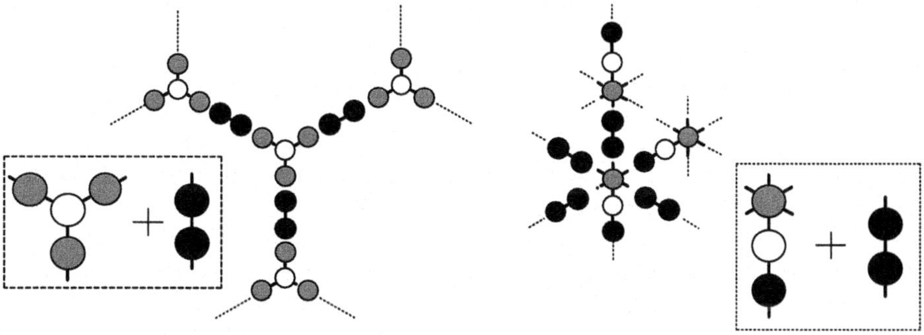

Fig. 2.1 Schematic diagrams showing the precursors, in *boxes*, and the connections for the 3-functional network (*left*) and the 6-functional network (*right*)

Here, σ is the characteristic length of the LJ potential, r is the distance between bead centers, and ε is the energy parameter for the strength of the interaction. This models van der Waals attractive forces very well and has a strong repulsive core that defines the size of the bead. Covalent bonds between beads, either preexisting or formed during crosslinking, are described by a bond potential that is the sum of the purely repulsive LJ interaction with a cutoff at $2^{1/6}\,\sigma$ (the minimum of the LJ potential) and a finite-extensible nonlinear elastic (FENE) attractive potential:

$$\begin{aligned} U_{\text{FENE}}(r) &= -0.5R_0^2 k \log_e\left[1-\left(\tfrac{r}{R_0}\right)^2\right] && r < R_0 \\ r &= \infty && r \geq R_0 \end{aligned} \tag{2.6}$$

where the force constant, $k = 30\varepsilon/\sigma^2$, and the maximum bond length, $R_0 = 1.5\sigma$. These are commonly used parameter values [29]. The FENE bond is ~10% shorter than LJ bonds and emulates the increased density in solid crosslinked polymers with respect to the liquid precursors; it also prevents chain crossing. The MD simulations were performed using the Large-Scale Atomic/Molecular Massively Parallel Simulator (LAMMPS) [30].

Results were obtained from simulations of stoichiometric mixtures, as before [29], of crosslinkers and dimers with a total system size of 11,424 beads. Averages were taken over five independent realizations for each set of parameters. A limited number of simulations were performed for larger systems. Before crosslinking, the precursor "molecules" are given random locations and equilibrated at a high temperature, $T = 1.0\varepsilon/k_B$, and at zero pressure using a Nosé–Hoover thermostat and barostat with a time step of 0.005τ, where $\tau = \sigma(m/\varepsilon)^{1/2}$ is the Lennard-Jones unit of time. Temperatures are given as multiples of the Lennard-Jones energy parameter with respect to Boltzmann's constant (henceforth, for convenience, the units of temperature will be omitted in the text). The networks are then dynamically crosslinked at constant temperature and zero load as before [26–29].

Two different approaches to crosslinking were taken. In the first, crosslinking was performed at several temperatures ranging from $T = 1.0$ down to $T = 0.1$ and $T =$ "0," since many coating and composite polymers are cured at a variety of temperatures, including many crosslinked systems that must be cured and used at ambient temperatures, e.g., epoxy adhesives and coatings, polyurethanes, and alkyd paints. Here, $T =$ "0" corresponds to the complete elimination of thermal fluctuations and therefore does not correspond to a situation attainable in the laboratory or applications. Nonetheless, it is an interesting limit because it provides insight into the influence of diffusion in network formation. In this case, crosslinking consisted of successive crosslinking and energy relaxation iterations until the same number of crosslinking attempts were performed as at finite temperature. The crosslinking procedure was the same as that applied at finite temperature [26–29]. The energy relaxation steps allowed the systems to relax to its lowest energy state as crosslinking proceeded. Results in this case were found to be very similar to those obtained when crosslinking at $T = 0.1$. Furthermore, percolating networks were also obtained in this case (without diffusion), indicating that the precursor solution was sufficiently randomized so that typical crosslinkers were surrounded by a sufficient number of active sites for crosslinking to proceed beyond the gel point. The pre-gel cluster growth process was found to be consistent with the predictions of percolation theory.

The 3-functional system was crosslinked for twelve million time steps and the 6-functional system for six million time steps. Networks crosslinked at high temperatures had properties consistent with what is observed with actual crosslinking polymers, which are frequently cured above their eventual glass transition temperature to ensure a high and stable degree of conversion in a reasonably short manufacturing process. Below T_g, crosslinking proceeded more slowly, resulting in lower conversions [31, 32]. Coating systems seldom cure completely; in these simulations, as in practice, conversion was continued until it was varying only very slowly and was characteristic of the conditions [31, 32].

Having determined these characteristic conversions, crosslinked networks with the same range of conversions were also made by crosslinking at the highest temperature, $T = 1.0$, at which all the components have a high degree of mobility. This is a very common and simple approach in computational simulations and provides another curing regime that can lead to insights on how network quality depends on the curing schedule.

A network search algorithm was developed [24, 25, 33] to determine the number of elastically active junctions, μ_e (a crosslink site that is connected by at least three paths to the gel), the number of elastically active chains, ν_e (an active chain is terminated by active junctions on both its ends), and the number of elastically active beads, as well as other quantities such as the number of dangling chains and primary loops. First, we identify the largest cluster (the backbone or gel) which forms the network, using a "depth first" search. Then, the backbone is searched using a "burning" algorithm [33] to examine the connectivity of junctions, i.e., if they connect to the gel, form a loop or end without connecting to the gel. Only those links which are connected by three paths to the gel are active junctions, and the number of bonds between them is the length of the active strands.

Prior work on these systems has shown that there are portions of the crosslinked network that are sufficiently heterogeneous to permit cavitation of regions much larger than the typical bond length in the network [29]. In order for this to occur, a sufficiently large number of nodes in close spatial proximity before cavitation must be distant neighbors along the network backbone, since the network bonds are essentially inextensional, and nodes near the surface of the cavitating region must be free to move apart to form the void. One of the advantages of molecular dynamics simulations is that the characterization of the network, topological defects, and pore space created under various conditions such as swelling can be studied in detail. We have used molecular dynamics simulations to analyze these topological defects and have found that their presence results in networks in which the number of elastically active chains, the cycle rank, and the number of elastically active junctions are smaller than predicted by the Miller–Macosko theory [34]. This effect is particularly pronounced in the high functionality, $f = 6$, model, where crosslinking at high temperatures leads to a large number of primary loops. Such defects adversely affect the mechanical properties, resistance to solvent swelling, and, possibly, the long-term protective properties of polymer networks.

Defects can be better seen if the network density is reduced. To that end, after the networks have been formed they are allowed to expand by removing the attractive part of the LJ interactions. While the bonding potentials along the network backbone provide network integrity, it is the attractive tails of the nonbonded Lennard-Jones interactions that maintain the system in the high-density condensed state. By replacing these interactions with a purely repulsive potential obtained by cutting of the LJ potential at its minimum and shifting it upward so that the potential is zero at the cutoff, thermal fluctuations will cause the networks to swell until the fluctuation-induced entropic forces are balanced by the network restoring forces. The strength of the entropic repulsion is proportional to T, so that its strength can be tuned if desired. Here, however, it is used simply to swell the network so that the resulting voids and network structure can be visualized and analyzed more readily.

Heterogeneities in Network Polymers

Although the interest here is on observing and quantifying the heterogeneities that occur in simulations of crosslinked network, it is useful to understand some basics of how they arise.

Figure 2.2 shows the temperature–density, T–ρ, phase diagram of pure LJ systems. Since the only attractive nonbonded forces in the system are provided by the LJ interactions, it is shown here to provide some perspective for interpreting the results. It is important to remember, however, that this is not the phase diagram of the precursor solutions or the bonded networks. The thick black curve (3-functional) and thick grey curve (6-functional) show the trajectory of the

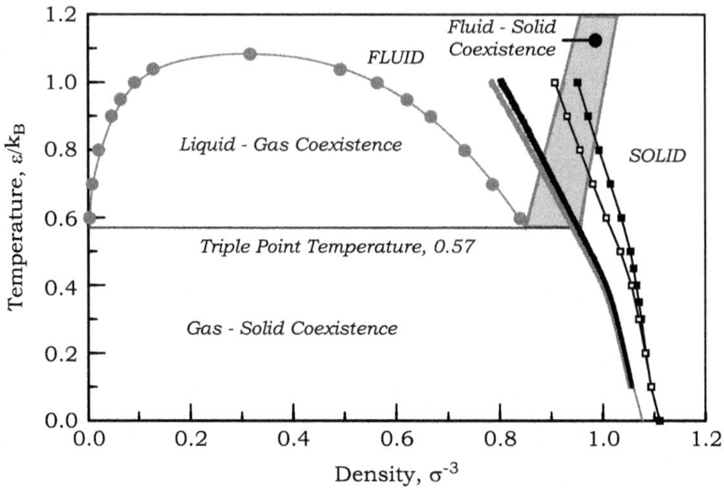

Fig. 2.2 Phase diagram of the cutoff, shifted Lennard-Jones fluid. The *grey bullets* mark the gas–liquid coexistence curve obtained from Gibbs ensemble simulations performed using the MCCCS Towhee simulation package [35]. The critical point estimate is taken from [36]. The *shaded region* is the liquid–solid coexistence region obtained using the procedure described in [37, 38]. The *horizontal grey line* is located at the triple-point temperature. The *grey* (6-functional) and *black* (3-functional) *lines* are the constant pressure ($P = 0$) trajectory of the precursor solution as it is cooled slowly from $T = 1$ to $T = 0.1$. *Squares* are data for the density after crosslinking at that temperature for the 3-functional system (*open squares*) and for the 6-functional system (*filled squares*)

precursor solutions as they are cooled at zero load at a rate of $4 \times 10^{-5}\tau^{-1}$ from $T = 1$ to $T = 0.1$. The high-temperature portion of the trajectories, for $T \geq 0.72$, is in the single phase, fluid region; at lower temperatures, they cross into the two-phase liquid–solid portion of the phase diagram before following the gas–solid coexistence line as the temperature diminishes toward zero.

Because of the comparatively short FENE bond length, the density of the system increases during crosslinking, as it does in practice when a solid polymer forms from the liquid reactants. The squares in Fig. 2.1 show the density of the crosslinked systems, and the change in slope indicates the location of the glass transition temperature, which is always in the solid part of the phase diagram. Rapid diffusion should play a large role during the initial crosslinking process for systems at $T \geq 0.8$ since the reacting system is fluid or in the solid–fluid coexistence region. Systems crosslinked at lower temperatures will almost immediately become solid, so large-scale diffusion will become difficult and one can expect that the structure of the resulting networks will depend on the crosslinking temperature. Many of the topological differences discussed later, if they occur early, must remain after gelation since subsequent changes happen only slowly and, in any case, crosslinking entails very strong bonds that inhibit significant movement away from the spatial arrangement at the point when the node is crosslinked.

Local-Scale Heterogeneities

Crosslinking reactions cannot occur uniformly because large molecules of reacted matter cannot diffuse rapidly. Sooner or later, molecules become much less mobile and can only react with their neighbors. Thus, an unreacted end might remain unreacted when opportunities around it are depleted or it reacts with the crosslink junction where its other end is attached, so a loop forms. In this way, the network can gain defects whose size is of the order of a coarse-grained bead or larger. Thus, depending on the defect concentration, one can appreciate that water or a corrosive ion would find it much easier to penetrate and cause harm in their neighborhood.

Figure 2.3 shows that a highly functional network, especially if it is cured rapidly at a high temperature, has many loops and an appreciable concentration of unreacted pendant network strands. A 6-functional network may have as many as 75% of the network junctions with a loop attached, or a 10% chance of a dangling end, and thus have a high probability of a defect in their neighborhood and a lower effective functionality. The redeeming feature of a 6-functional junction is that it probably has four other reactive sites for inclusion in a network even if two are occupied by a loop. The results for the 3-functional network are not so striking but, depending on the degree of conversion, may have almost 10% of its junctions with a loop. A 3-functional junction with a loop cannot be useful network junction; it is connected to the network through one functionality only and must form a pendant. The defect labelled here as a "dangling chain" is one with an unreacted site on the final bead, which is a different form of defect. All these defects are equivalent to a few beads in size and are local in extent, see Fig. 2.4. These defects diminish the number of elastically active loops in the network and lead to the use of cycle rank to quantify the strength of a network, rather than simply the crosslink chain density.

The illustrations in Fig. 2.4 visualize the results given in Fig. 2.3. They show that networks crosslinked at high temperature have a larger number of primary loops but comparatively few dangling or unreacted ends. The enhanced diffusion at high temperatures makes it possible for primary loops to close. In contrast, networks crosslinked at low temperatures have relatively few primary loops but many dangling and unreacted ends. Illustrations for the 6-functional network are very dense, but the differences are apparent. Loops and dangling ends reduce the effective functionality of the network. Figure 2.5 shows how that changes with the degree of conversion in both the 3-functional and 6-functional systems. Here, the effective functionality is calculated using

$$f_{\text{eff}} = [\text{bonds made}] - 2 \times [\text{primary loops}] - [\text{dangling chains}] \tag{2.7}$$

f_{eff} is the number of bonds of a crosslink junction leading to a neighboring crosslink site and is given as a fraction of the nominal functionality in each system.

These localized defects have consequences not only in the likely permeability but also in the macroscopic properties. Equations 2.1 and 2.2 are predictions of familiar theories that express the modulus of a rubbery network in terms of either

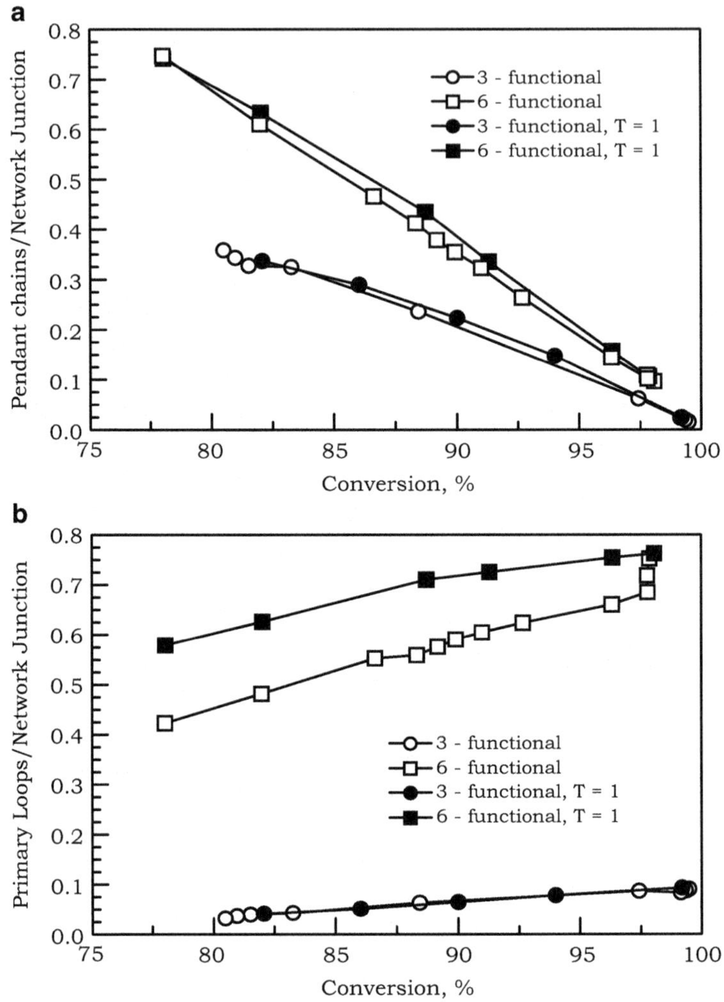

Fig. 2.3 The number of (**a**) dangling chains and (**b**) loops per network junction on the gel backbone as a function of the conversion of the 3- or 6-functional system. *Open symbols* are for systems crosslinked at different temperatures; *filled symbols* are results for crosslinking at $T = 1.0$ to a range of conversions

the cycle rank or the number of elastically active chains. Figure 2.6 provides a comparison of simulation results for the number of elastically active chains and nodes and the cycle rank with the predictions of the Miller–Macosko theory.

The MM approach makes the same ideal network assumptions as Flory, i.e., that all functional groups of the same type are equally reactive; all groups react independently of one another, and no intramolecular reactions occur in finite species.

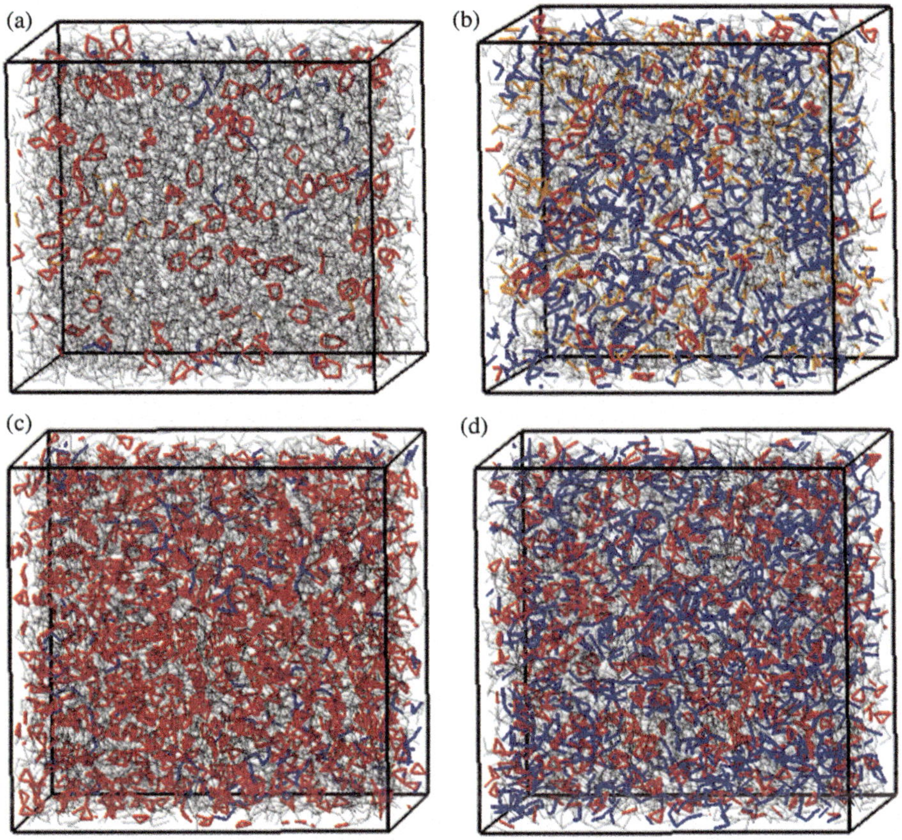

Fig. 2.4 Networks generated at $T = 1$ (*left*) and $T = 0.1$ (*right*). The network backbone bonds are shown in *black*; primary loops are *red*, dangling ends *blue*, and unreacted junction ends are *orange*. (**a**) 3-functional, $T = 1.0$, (**b**) 3-functional, $T = 0.1$, (**c**) 6-functional, $T = 1.0$, (**d**) 6-functional, $T = 0.1$

As can be seen in the figure, all quantities are diminished by approximately 20% for the 3-functional system regardless of conversion. However, for this system, the ratio between cycle rank and number of crosslink chains (Eq. 2.3) remains very close to that for the ideal network. Nevertheless, the modulus will be only approximately 80% of the expected value and, for analogous reasons, the network will swell more than an ideal network in the presence of a suitable solvent. The situation for the 6-functional system is more complex, and its performance relative to an ideal network will be worse. In this case, the number of elastically active crosslink chains is approximately 70% of the value predicted for an ideal network, and the cycle rank is 56% of that predicted for an ideal network, at high conversion. Even though cycle rank is used to describe the properties of a network in terms of the number of complete loops, and thus discount ineffective chains and junctions, any value predicted from statistical models is likely to be an overestimate.

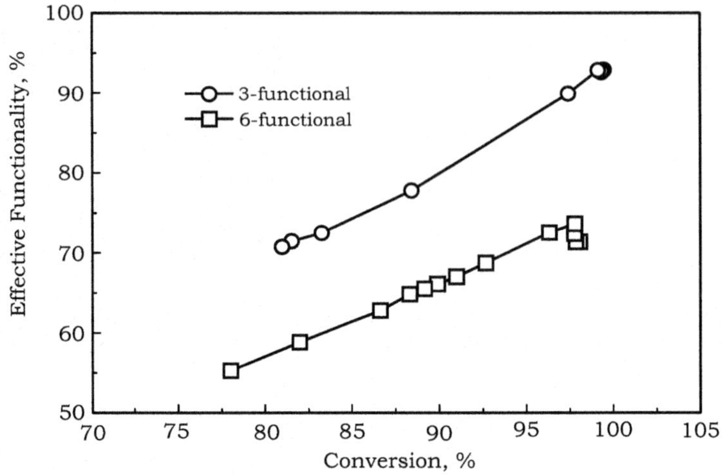

Fig. 2.5 Effective functionality of the 3- and 6-functional systems, quantifying how the functionality is reduced to a fraction of its nominal value by the imperfections

These localized defects, which are a natural consequence of the random processes of the curing of a huge number of molecules, in any system, provide a substantial number of places in the network where a penetrant molecule might more easily rest or pass through as well as reduce the stiffness and resistance to overall swelling.

So far the discussion has treated these defects as individual, perhaps isolated, features although there seems to be a high-enough concentration that they must provide some possibility of a defect pathway percolating through a network. The discussion now looks at how these defects might be organized into larger-scale features.

Extended Heterogeneities

Randomness intrinsic to the crosslinking process also leads to large-scale heterogeneities in network topology reflected by the fact that a significant number of nodes in close spatial proximity are only distant neighbors along the network backbone. This property, which is not shared by homogeneous, ideal, networks with uniform crosslinking, can be quantified in terms of the distribution of the length of the shortest path (along the network) between spatial nearest neighbors, determined by a Voronoi construction. Figure 2.7 shows the resulting distribution functions for networks crosslinked at $T = 1.0$ to conversions greater than 0.98. The number of paths is on a logarithmic axis, so one can see that the large majority of spatial neighbors are more or less directly connected, but there are significant tails

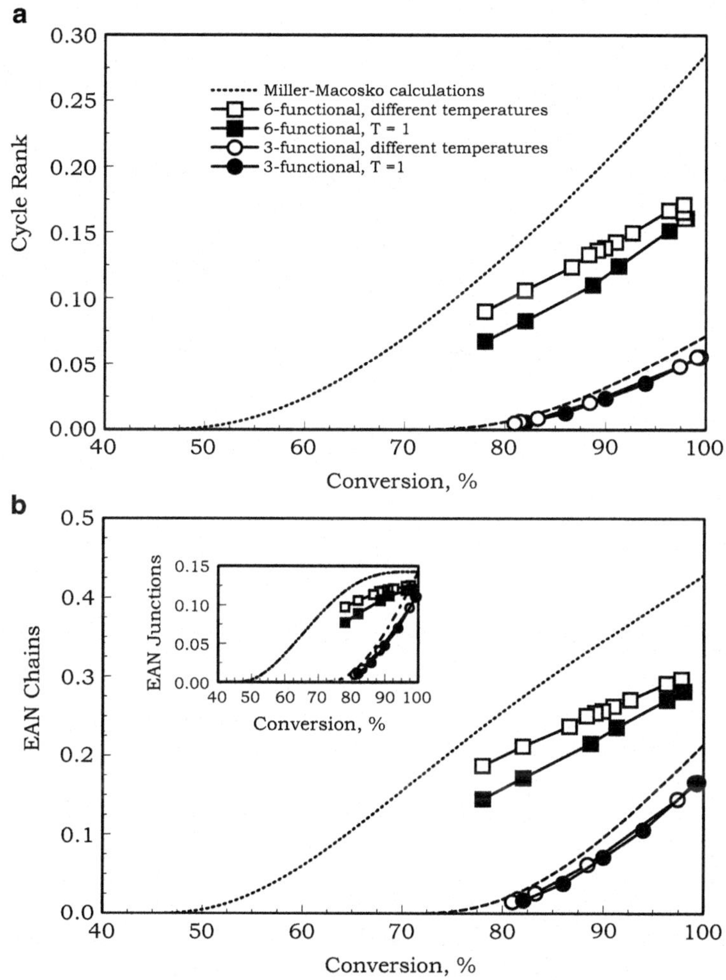

Fig. 2.6 Comparison of results from simulations with Miller–Macosko calculations (*dashed lines*) for the cycle rank (**a**), number of elastically active network chains and number of elastically active network junctions (**b**), for 3- and 6-functional systems. All network parameters are given as a fraction denominated by the total number of beads used in the simulation. The single legend is applicable to all the data

in the distributions. The much longer tail in the distribution for the 3-functional model is, in part, due to the fact that the shortest path between neighboring junctions is three bond lengths in the 6-functional model but five bond lengths in the 3-functional model.

There are more paths from a 6-functional junction that can be linked to the next, so if a crosslinked junction has a loop or pendant chain occupying one or more of

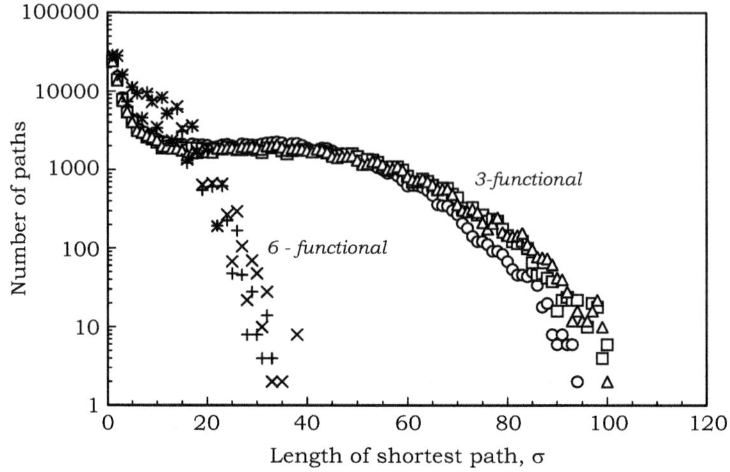

Fig. 2.7 Number distribution for the length of the shortest path, in terms of the number of bonds that must be travelled from one bead before arriving at a spatially neighboring bead. Networks crosslinked at $T = 1$ (conversion > 0.98). The *different symbols* indicate some of the (5) replicate simulations

the functionalities, there is still a good chance that it may be linked to the neighboring bead. On the other hand, a 3-functional junction has much reduced opportunities to be linked to the next bead if it also has a pendant or loop.

Some insight into the origin of these extended heterogeneities can be obtained by considering how the crosslinked clusters grow before gelation. Initially, each reactive group will find its co-reactant and form a larger molecule, but as the conversion progresses, some molecules will become larger than others, and when two growing molecules react together, they become a much larger molecule. The result is that the growing network forms clusters of reacted material in a way that is similar to diffusion-limited aggregation, DLA, [39] except that the molecules that accrete to an aggregate are often macromolecules rather than monomers, and so the process might be better labelled as cluster–cluster aggregation [40, 41]. This is visualized statically in Fig. 2.8. If the gel point is defined by the stage at which there is a bonded pathway that crosses the simulation space, gelation occurs when two (sometimes three) large clusters react together. These simulations have never observed gelation due to a single growing macromolecule. Macromolecular clusters are more or less reacted material whose surface has only limited remaining functionality with which to react to join a neighboring cluster. Thus, there are intercluster regions or fissures within the final crosslinked mass that are not spanned frequently by chemical bonds and where pendant chains may lie.

Supermolecular nodular structures in crosslinked networks created using a variety of chemistries have been observed and commented on since at least 1959 [42, 43]. Initially, it was natural to suspect that such heterogeneities, of some tens of nanometers in extent, were an experimental artifact; however, that does not appear

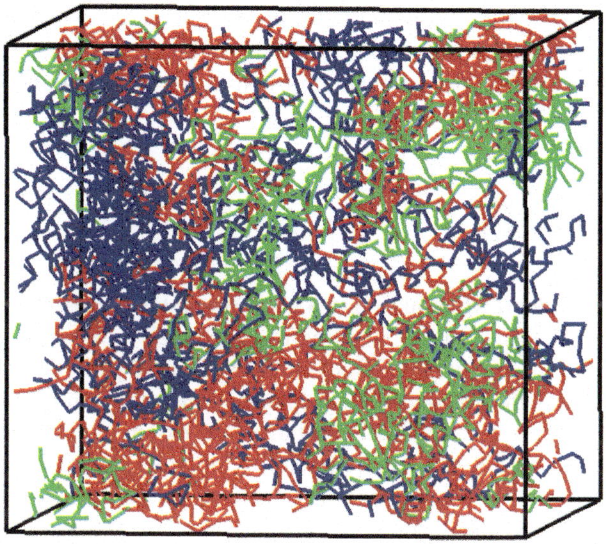

Fig. 2.8 The three largest clusters after crosslinking the 3-functional model for 13,600 time steps. The largest cluster (*red*) contains 2156 nodes, the second largest (*blue*) contains 1992 nodes, and the third largest cluster (*green*) contains 1154 nodes. The total system size is 11,424 nodes. Note the ramified structure of the clusters and the extent to which they are interdigitated. The fractal dimension of the clusters is close to that of percolation clusters

to be the case, since they have been observed a number of times using modern experimental techniques [44, 45]. Our simulation results are consistent with those observations. The center of clusters formed in our coarse-grained simulations would be the denser material seen in observations of such nodular structures. They arise from the growth of the network from a large number of starting points. This is consistent with experimental observations that the nodular heterogeneities form prior to gelation and do not change significantly thereafter [44, 46]. The network formation behavior via cluster aggregation exhibited in the simulations therefore provides a very simple, natural origin for supermolecular heterogeneities, without requiring any particular combination or variation in chemistry. Crosslinked networks do not, and cannot, form homogeneously [47].

Another consequence of the formation of clusters is that spatially neighboring beads may not be bonded together directly. A pendant is a chain of molecules connected to the elastically active network by a single junction. This type of structure is inevitable in randomly crosslinked systems and is caused by the highly ramified structure of pre-gel clusters during crosslinking (see Fig. 2.8). As might be seen in the figure, there are a significant number of linear chains; if these chains do not crosslink to adjacent chains before crosslinking is completed, they occupy a volume across which there are no bonds and which may open up to form a void if the network is subjected to swelling. If these pendants are a number of bonds in length, they will be coiled, at random, so the resulting volumes can be several bond lengths in size. As will be shown in the following, these voids occur throughout a crosslinked network, and regions of low crosslinking may be very significant in extent, as can be seen in the images published in the experimental investigations mentioned above [42–44, 46, 47].

Swelling and Voids

A crosslinked polymer, when placed in a good solvent, will absorb a portion of the solvent and swell. The extent of swelling is determined by a competition between the free energy of mixing, which will cause the solvent to penetrate and expand the network and an elastic reactive force which opposes this deformation. Since the steady-state swelling ratio is a direct function of the extent of crosslinking, swelling experiments are a simple way to characterize polymer networks. In addition, a molecular intruder will be better able to penetrate where the crosslinking level is low. As they penetrate, the intruding molecules cause the network to swell and thus open up pathways for further intrusion. Allowing a network to swell, after it has been cured, provides a useful and relevant way to visualize and quantify these extensive heterogeneities. These networks were also shown to cavitate if their boundaries were constrained against the shrinkage caused by the crosslinked bonds being shorter than the van der Waals bonds [29]. The tendency to form cavitation voids first brought attention to the possible presence of extended weaknesses in such networks.

The bond potentials along the network backbone provide network integrity, and the attractive tails of the nonbonded van der Waals interactions maintain the system in the high-density condensed state. Instead of swelling by a good solvent, a similar effect can be achieved by replacing the Lennard-Jones nonbonded interactions with a purely repulsive potential obtained by cutting off the LJ potential at its minimum and shifting it upward so that the potential is zero at the cutoff. Thermal fluctuations will then cause the networks to swell until the fluctuation-induced entropic forces are balanced by the network restoring forces. In order to better visualize the larger-scale heterogeneities, the network was allowed to expand in this way and expose any regions of low crosslink density, voids, or other network structures.

As can be seen in Fig. 2.9, which shows the number density of the networks as a function of the number of time steps, the networks swell very quickly at zero load when the attractive tails of the LJ interactions are removed, so this is a rapid, convenient approach.

We present here results only for networks crosslinked at one representative high temperature, $T = 1.0$, and one low temperature, $T = 0.1$. These two cases are sufficient to characterize the behavior in the highly crosslinked networks formed at high temperatures and the more loosely crosslinked systems, with a lower conversion, created at low temperatures. As anticipated, networks for the 3-functional model can expand significantly more than those for the 6-functional model because the length of the chains between junctions is longer (five bond lengths for the

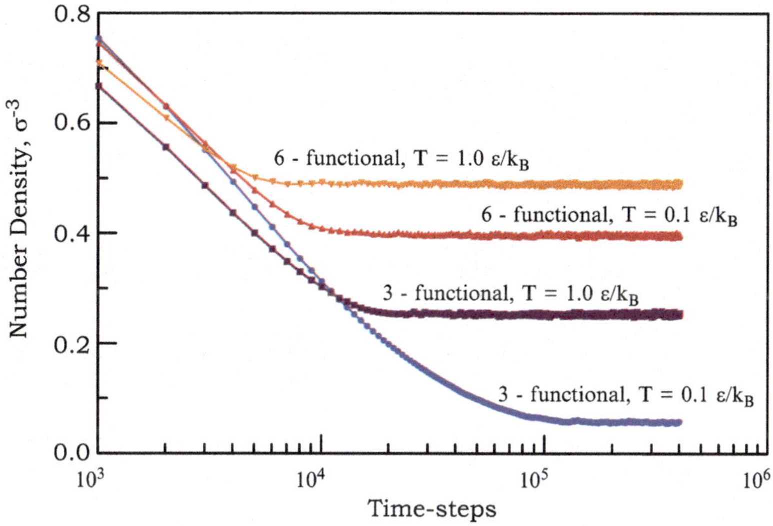

Fig. 2.9 Relaxation of the number densities as a function of the number of time steps for both the 6- and 3-functional networks at $T = 1$ and $T = 0.1$. Initial densities: 0.95 (6-functional, $T = 1$), 1.09 (6-functional, $T = 0.1$); 0.907 (3-functional, $T = 1$), 1.09 (3-functional, $T = 0.1$). The time step is 0.005τ, so the run is of duration 2000τ

3-functional as opposed to three bond lengths for the 6-functional model), and the more tightly crosslinked networks expand less than the networks crosslinked to a lower conversion at low temperatures.

As the network swells, voids open up, the sizes of which depend on the model and the conversion. In the following, we characterize the pore size by the distance, d, from any point to the closest network node on a uniformly spaced $80 \times 80 \times 80$ grid spanning the simulation cell. A distance d^* is chosen as a criterion to distinguish regions of low density from those of more normal density. The volume fraction of a void, such that a node is located a distance greater than d^* from another node, is therefore the number of grid points with separation $d > d^*$ (divided by the volume of the simulation space, 80^3). Isosurfaces are shown in Fig. 2.10 with d^* being the value defining these pore regions. An isosurface is the surface bounding a volume within which the distance to the nearest network node is greater than the chosen value of d^*. Thus, particles with a diameter less than $2d^*$ would be able to diffuse readily through the network inside these isosurfaces. Figure 2.10 shows the network backbone and void isosurfaces for $f = 3$ and $f = 6$ functionality model networks crosslinked at $T = 1$ (left figure) and $T = 0.1$ (right figure). For comparison, bonds in these networks are of approximately unit length. In these visualizations, one can see that the voids, which expand in regions not held together by crosslink bonded forces, are quite large indicating that there are extensive defected regions in these networks. This is true, even in the networks cured at a very high temperature to almost complete conversion at $T = 1$. Even when fully cured, the networks are not ideal and not homogeneous.

(a) (b)

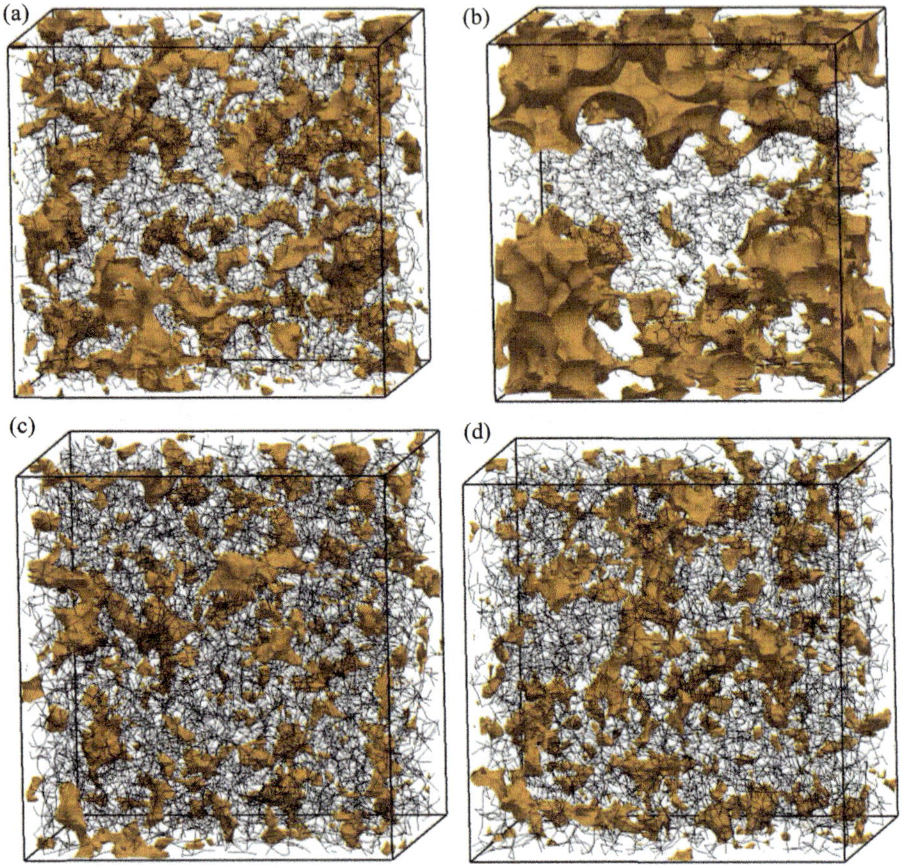

(c) (d)

Fig. 2.10 Pore space isosurfaces for expanded networks. The networks were formed first by crosslinking the 3-functional model for 12 million time steps and the 6-functional for 6 million time steps. In each case, the *left figure* was crosslinked and expanded at $T = 1.0$ with isosurfaces for d^* as labelled. The *right-hand side figure* is for $T = 0.1$. The *dark lines* show the network backbones. (**a**) 3-functional, $T = 1.0$, $d^* = 1.8$; (**b**) 3-functional, $T = 0.1$, $d^* = 4$; (**c**) 6-functional, $T = 1.0$, $d^* = 1.3$; (**d**) 6-functional, $T = 0.1$, $d^* = 1.4$

There are obvious and expected qualitative differences. The void extent is more pervasive in the 3-functional system at both temperatures. However, it is striking how many regions of low crosslinking there are even in the 6-functional system that has been cured at a high temperature. By no means can one consider the whole volume or thickness of the material as providing barrier properties or mechanical strength.

The Flory–Rehner calculation of swelling by a solvent [12] assumes that the crosslinked polymer forms an ideal homogeneous network, without significant open voids when swollen, which is not consistent with what we observe. Figure 2.11 gives the volume of the simulation cell of the crosslinked network as a function of

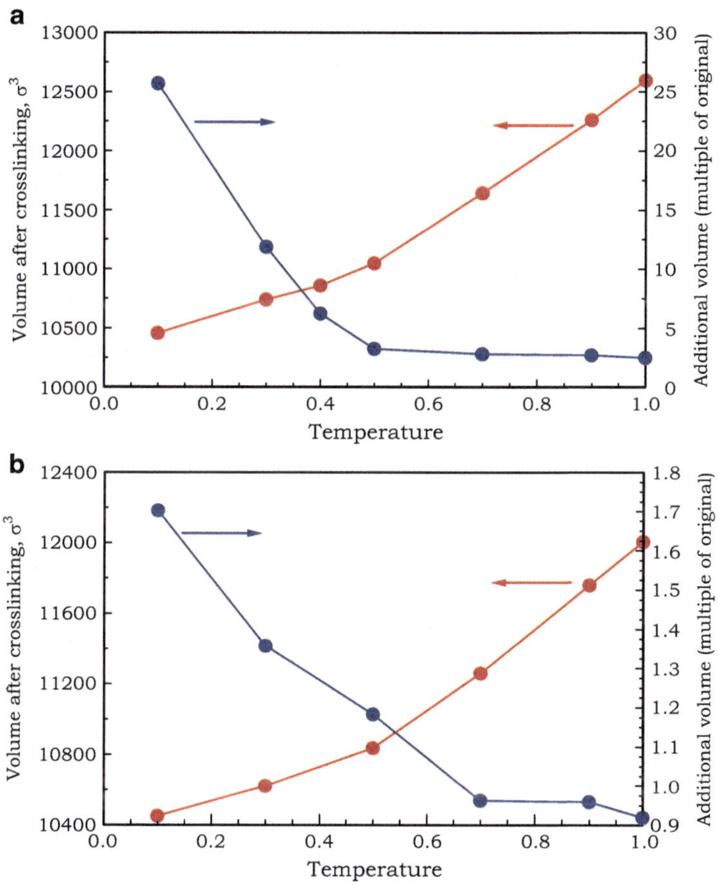

Fig. 2.11 Volume of the simulation cell of the crosslinked network before swelling, V_0, and the volume increase upon expansion, $(V - V_0)/V_0$, where V is the volume after expansion, for the 3-functional (**a**) and 6-functional (**b**) models, as a function of the crosslinking temperature, T

curing temperature, before swelling, V_0, and the reduced volume increase, $(V - V_0)/V_0$, where V is the volume after expansion, for the 3-functional and the 6-functional models. The networks were allowed to expand at the same temperature at which they were crosslinked. Figure 2.12 is a plot of the same data as a function of the conversion. For both models, the cured volume of the system increases with temperature, showing that the entropic contribution to the internal pressure is proportional to T (thermal expansion). On the other hand, the fractional volume change after swelling increases with decreasing temperature, particularly for $T < 0.5$ in the 3-functional model and $T < 0.7$ for the 6-functional model, where the conversion is lower and the networks have fewer saturated junctions. When

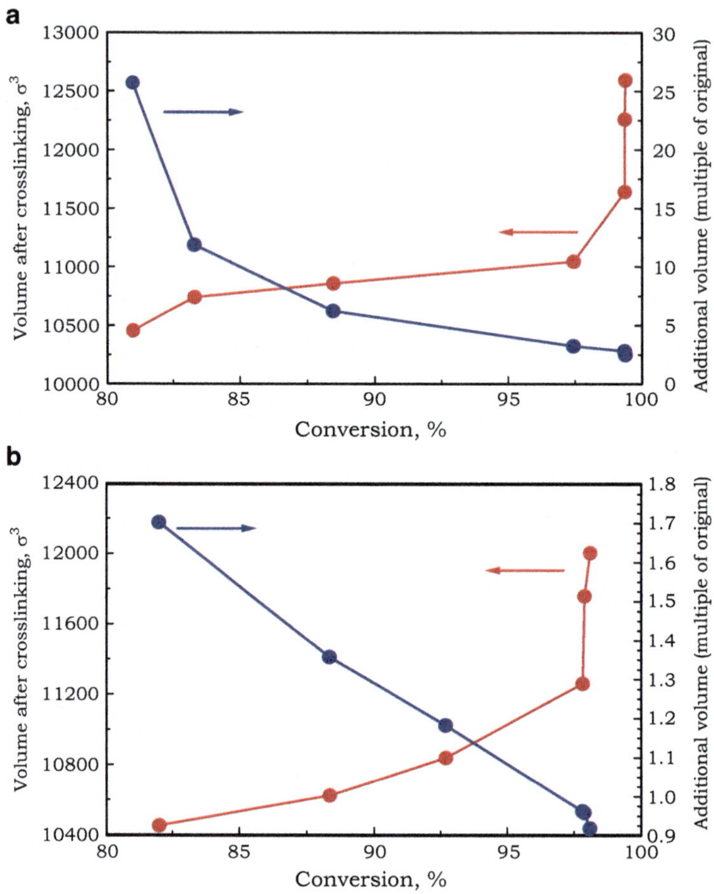

Fig. 2.12 Volume of the simulation cell of the crosslinked network before swelling, V_0, and the volume increase, $(V - V_0)/V_0$, where V is the volume after expansion, for the 3-functional (**a**) and 6-functional (**b**) models, as a function of the final conversion. The same data are used as in Fig. 2.11

expressed as a function of the conversion, the reduced volume change for the 6-functional model increases linearly as the conversion decreases. For the 3-functional model, the increase in the capacity to swell at low conversions is stronger, reflecting the fact that the low conversion networks in this case have conversions only slightly larger than at the gel point, i.e., a conversion of 0.77. The conversion at the gel point for the 6-functional system is 0.557 [29, 34].

Figure 2.13 shows the void volume, as a fraction of the overall expanded network, after the expansion procedure for networks crosslinked at temperatures ranging from $T = 0.1$ to $T = 1.0$, as a function of the choice of d^*. The left panel contains results for the 3-functional model and the right panel for the 6-functional model. Expanded void volumes are much larger for the 3-functional system than for

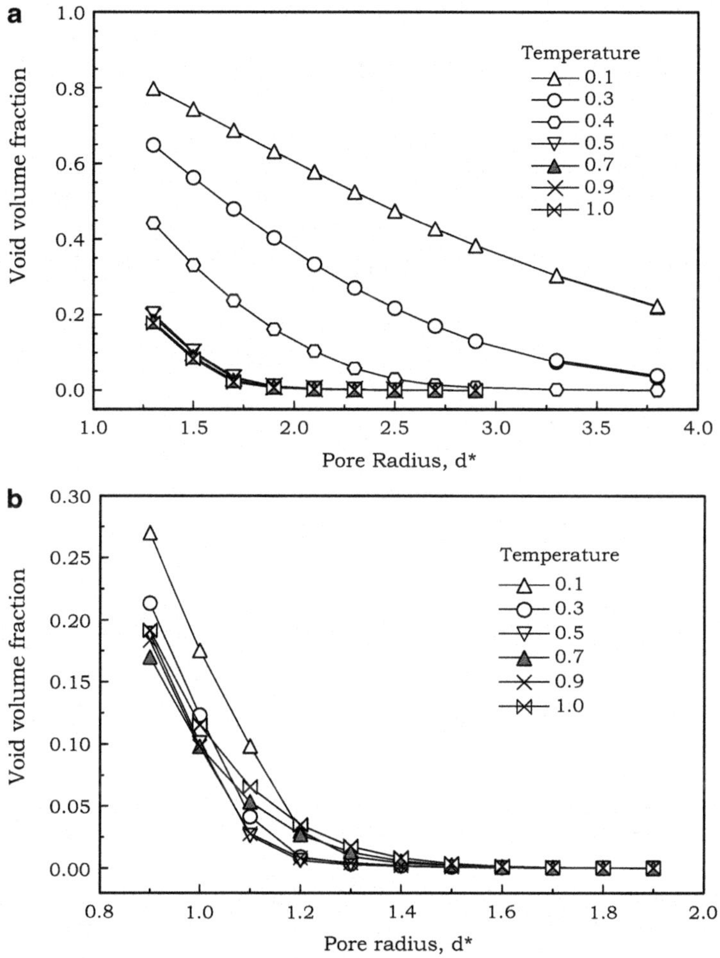

Fig. 2.13 Total void volume for (**a**) the 3-functional and (**b**) the 6-functional models, for different curing temperatures

the more tightly crosslinked 6-functional model, consistent with the longer tails for the 3-functional model in the topological distance distribution function shown in Fig. 2.6 and the larger interjunction separation in the model.

It can be seen in Fig. 2.10 that the voids are seldom spherical in nature but have an extended shape that supports the idea that their origin lies at the intersection of ramified molecular clusters, where pendant chains are more likely. In some cases, if there was some additional network degradation due to weathering, one can see that a percolating pathway might arise fairly readily. If we are interested in gauging whether an intrusive molecule might be able to penetrate the network, then

Fig. 2.13 demonstrates in a quantitative way how the intruder size will affect the outcome. For comparison, a bond is about one unit long in the dimensions of the expanded network. Small penetrating molecules will find many more weaknesses in the network to exploit than would a larger molecule.

Further investigation into the sizes of the nascent voids shows that when the expanded void volume is large, depending on the pore size parameter and the conversion, the vast majority of the pore volume is in the largest pore, with minor contributions from the second and third largest. There are comparatively few extensive voids in the volume simulated here, which is consistent with gelation being the result of two or possibly three molecular clusters dominating the gelation behavior in the simulation volumes here. It is also consistent with the number of cavitation voids formed in these systems under slightly different circumstances [29]. The contribution from the second largest void can be quantified by comparing results plotted in Fig. 2.13 with the corresponding data for the second largest void in Fig. 2.14.

As the pore size parameter, d^*, is increased, the size of the largest nascent pore decreases, and the size of the second (as well as third and higher) nascent pores increases to a peak before decreasing again for larger d^*. The position of this peak lies at slightly larger d^* than the value at which the largest pore would no longer percolate across the system. A similar relationship between the position of the peak in the size of the second largest cluster and the location of the gel point in the model systems used here has been noted before [48].

For the current models and system size used here, the largest pore percolates when its volume is on the order of 4% of the system size, a value which is attained in the 3-functional model for a pore radius in the range 1.5–1.6. For the 6-functional model, a smaller d^* is required, on the order of 1.0–1.1; even then, however, the diameter of the pore is significantly larger than the bead diameter in our models.

These results demonstrate that extensive heterogeneities, exposed as nascent voids, can occur throughout a crosslinked network and may occupy a significant fraction of the volume if it is swollen by fluctuations, as here, or in practice, by a solvent or internal stress. While normally closed because of the attractive van der Waals forces between network components, these regions can facilitate the passage of a penetrant molecule, or be the locus of a fracture surface, even before degradation. These weak regions are where an external molecule would have the best opportunity to penetrate and thus swell the network, opening it up further. The important point is that it was only the random fluctuations intrinsic to the crosslinking process that produced these regions not held together by bonded forces. There is no equivalent to a Maxwell demon that ensures that an ideal network is formed everywhere.

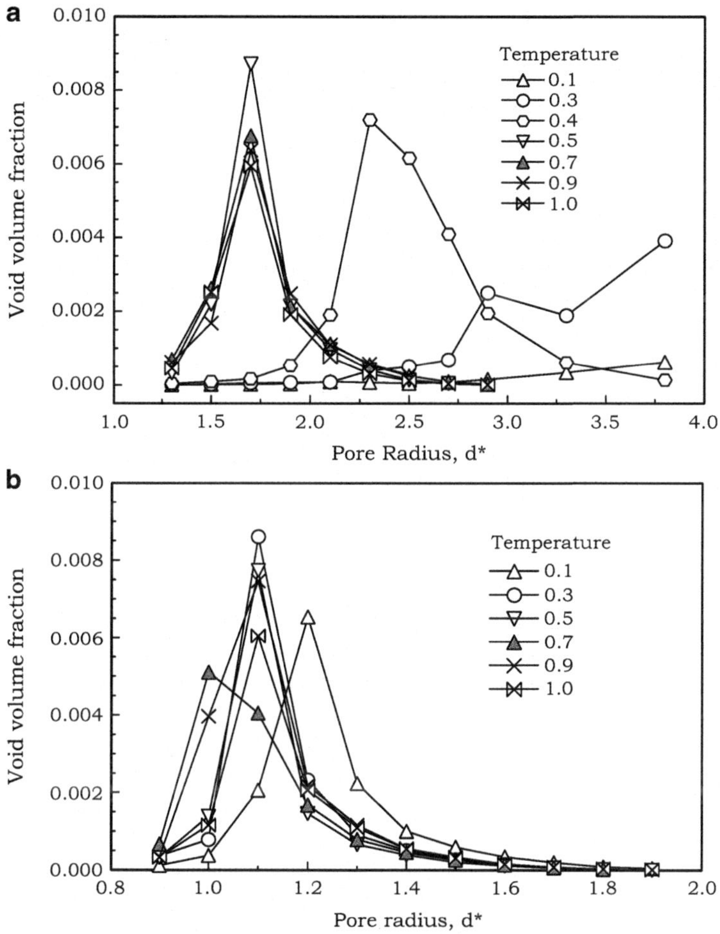

Fig. 2.14 Volume fraction of second largest pore (**a**) 3-functional system, (**b**) 6-functional system, for different curing temperatures

Summary

Modern computers provide the opportunity to analyze in detail the structure of networks and learn more about how crosslinked polymer networks form. Although it had been realized from the start that randomly crosslinked networks would contain dangling ends and loops, as well as other topological defects, which would diminish anticipated performance, predictive theories based on statistical approaches necessarily assumed that the eventual network was perfect and homogeneous in the sense that all reactive species would combine as expected. Molecular dynamic simulations show how the random nature of the crosslinking process involving huge numbers of small molecules inevitably produces nonideal, defected networks.

In terms of mechanical properties, the effect of those features that diminish the number of elastically effective network chains and junctions can be quantified by comparison between simulation results and the predictions for ideal networks given by the Miller–Macosko approach. In this case, the primary deviation from ideality is the formation of primary loops. In particular, network junctions in the 6-functional system have approximately a 75% chance of having a primary loop occupying two of its reactive sites at conversions close to 100%. At lower conversions, the number of primary loops decreases rapidly, and agreement with Miller–Macosko calculations is better. In contrast, simulations of the 3-functional system indicate that it is much less sensitive to the curing temperature. There is a smaller reduction in the absolute number of elastically effective crosslink chains and cycle rank, and the ratio of these parameters remains close to predictions for an ideal network. In this case, fewer loops can form after one functionality on a crosslinker has reacted with one end of a dimer because, for the other end of the dimer, there are only two sites on that crosslinker remaining to compete with unreacted sites on neighboring crosslinkers. In either case, a lower curing temperature might permit a more ideally formed network, but this may not be practical due to the very much longer curing time that would be necessary to achieve high conversion and thus good properties. Although cycle rank was introduced as a quantity for characterizing networks based on the number of elastically active, complete loops, any predictive calculation assumes that the network is ideal and thus must be an overestimate. These loops and unreacted groups are defects that are comparable in size to the initial precursor molecules.

Other imperfections occur that are more extensive in nature. Simulations show that polymer networks form by the aggregation of ramified cluster-like macromolecules, resulting in networks with a significant number of topological defects and heterogeneities. These clusters that form largely prior to gelation provide a natural explanation for the formation of nodular structures, or phase differences, seen experimentally, but where the explanation is derived more from the statistical nature of network formation rather than any combination of chemistry. Due to mobility restrictions and prior reactions, the surfaces of the erstwhile clusters are regions where there will be a reduced chance of a bond forming a link to a neighboring cluster or a long pendant chain remaining unreacted. These regions, where there are a reduced number of crosslinks to junctions in close spatial proximity, can extend over a significant fraction of the simulation cell and thus may join to form percolating pathways where external penetrating molecules could find a route to attack the substrate that the coating is protecting. The heterogeneous, nodular structure must diminish the mechanical properties compared to the potential in a perfectly homogeneous material. The number and type of defects depend on the crosslinker functionality, the length of the primary chains between crosslinkers, as well as the final conversion and crosslinking temperature. These features will have a scale comparable to the extent of the macromolecular clusters at the gelation point.

The extent of these intercluster fissures can be visualized by diminishing the LJ (van der Waals) forces between the beads after the simulation has formed the network. The resultant expansion of the network, and analysis of the void structure that appears where these defects occurred, provides information about the size and distribution of these extensive regions of low crosslinking. In any real polymer network, these fissures would be of the order of nanometers or tens of nanometers in size, so they are not large enough to initiate fracture themselves, but they could be the starting point for subsequent damage to grow and limit the ultimate strength of the material. It is also easy to appreciate the size of molecule that may be able to intrude and contribute to a variety of failure modes. The solvent swelling of a network is likely, in some cases, to be significantly more than that described by the Flory–Rehner and related models because of the potential for extensive voids that could form in the intercluster regions.

Regardless of the model details, networks contain defects of a small local nature and defects that are more extensive where the network is not held together by chemical bonds. These regions open up to form voids when the network is swelled or exposed to other stresses. While the systems examined here cannot represent all possible precursor combinations, we expect that the features we describe are typical of most types of resultant networks.

Acknowledgements The authors are glad to acknowledge computer access, financial, and administrative support from the North Dakota State University Center for Computationally Assisted Science and Technology and the U.S. Department of Energy through Grant No. DESC0001717.

References

1. Flory, P.J.: Network structure and the elastic properties of vulcanized rubber. Chem. Rev. **135**, 51–75 (1944)
2. Flory, P.J.: Principles of Polymer Chemistry. Cornell University Press, Ithaca (1953)
3. Stockmayer, W.H.: Theory of molecular size distribution and gel formation in branched-chain polymers. J. Chem. Phys. **11**(2), 45–55 (1943)
4. Stockmayer, W.H.: Theory of molecular size distribution and gel formation in branched polymers II. General cross linking. J. Chem. Phys. **12**(40), 125–131 (1944)
5. Dušek, K.: Crosslinking and networks. Makromol. Chem. Suppl. **2**, 35–49 (1979)
6. Cail, J.I., Stepto, R.F.T.: The gel point and network formation—theory and experiment. Polym. Bull. **58**(1), 15–25 (2007)
7. Zhou, H., Woo, J., Cok, A.M., Wang, M., Olsen, B.D., Johnson, J.A.: Counting primary loops in polymer gels. Proc. Natl. Acad. Sci. USA. **109**(47), 19119–19124 (2012)
8. Duering, E.R., Kremer, K., Grest, G.S.: Structure and relaxation of end-linked polymer networks. J. Chem. Phys. **101**, 8169–8192 (1994)
9. Duering, E.R., Kremer, K., Grest, G.S.: Structural properties of randomly crosslinked polymer networks. Progr. Colloid Polym. Sci. **90**, 13–15 (1992)
10. Gordon, M.: Good's theory of cascade processes applied to the statistics of polymer distributions. Proc. R. Soc. Lond. A Math. Phys. Sci. **268**, 240–256 (1962)
11. Treloar, L.R.G.: The Physics of Rubber Elasticity. Clarendon Press, Oxford (1958)
12. Flory, P.J., Rehner, J.: Statistical mechanics of swelling of crosslinked polymer networks. Chem. Phys. **11**, 521–526 (1943)

13. Coniglio, A., Stanley, H.E., Klein, W.: Site-bond correlated-percolation problem: a statistical mechanical model of polymer gelation. Phys. Rev. Lett. **42**(8), 513–522 (1979)
14. Gujrati, P.D.: Thermal and percolative transitions and the need for independent symmetry breakings in branched polymers on a Bethe lattice. J. Chem. Phys. **98**(2), 1613–1634 (1993)
15. Dušek, K.: My fifty years with polymer gels and networks and beyond. Polym. Bull. **58**, 321–338 (2007)
16. Dušek, K., Dušková-Smrčková, M.: Network structure formation during crosslinking of organic coating systems. Prog. Polym. Sci. **25**, 1215–1260 (2000)
17. Dušek, K., Dušková-Smrčková, M., Huybrechts, J., Ďuračková, A.: Polymer networks from preformed precursors having molecular weight and group reactivity distributions. Theory and application. Macromolecules. **46**, 2767–2784 (2013)
18. Miller, D.R., Macosko, C.W.: Molecular weight relations for crosslinking of chains with length and site distribution. J. Polym. Sci. B Polym. Phys. **25**, 2441–2469 (1987)
19. Miller, D.R., Macosko, C.W.: Network parameters for crosslinking of chains with length and site distribution. J. Polym. Sci. B Polym. Phys. **26**, 1–54 (1988)
20. Dušek, K., Spěváček, J.: Cyclization in vinyl–divinyl copolymerization. Polymer. **21**, 750–756 (1980)
21. Tiemersma-Thoone, G.P.J.M., Scholtens, B.J.R., Dušek, K., Gordon, M.: Theories for network formation in multistage processes. J. Polym. Sci. B Polym. Phys. **29**, 463–482 (1991)
22. Flory, P.J.: Statistical thermodynamics of random networks. Proc. R. Soc. Lond. A Math. Phys. Sci. **351**, 351–380 (1976)
23. Queslel, J.P., Mark, J.E.: Molecular interpretation of the moduli of elastomeric polymer networks of known structure. Adv. Polym. Sci. **85**, 135–176 (1984)
24. Flory, P.J.: Elastic activity of imperfect networks. Macromolecules. **15**, 99–100 (1982)
25. Grest, G.S., Kremer, K.: Statistical properties of random cross-linked rubbers. Macromolecules. **23**, 4994–5000 (1990)
26. Stevens, M.J.: Interfacial fracture between highly cross-linked polymer networks and a solid surface: effect of interfacial bond density. Macromolecules. **34**, 2710–2718 (2001)
27. Tsige, M., Stevens, M.J.: Effect of cross-linker functionality on the adhesion of highly crosslinked polymer networks: a molecular dynamics study of epoxies. Macromolecules. **37**, 630–637 (2004)
28. Tsige, M., Lorenz, C.D., Stevens, M.J.: Role of network connectivity on the mechanical properties of highly cross-linked polymers. Macromolecules. **37**, 8466–8472 (2004)
29. Zee, M., Feickert, A.J., Kroll, D.M., Croll, S.G.: Cavitation in crosslinked polymers: molecular dynamics simulations of network formation. Prog. Org. Coat. **83**, 55–63 (2015)
30. Plimpton, S.: Fast parallel algorithms for short-range molecular dynamics. J. Comput. Phys. **117**, 1–19 (1995)
31. Pascault, J.-P., Sautereau, H., Verdu, J., Williams, R.J.J.: Thermosetting Polymers. Marcel Dekker, New York (2002)
32. Gillham, J.K.: Formation and properties of thermosetting and high Tg polymeric materials. Polym. Eng. Sci. **26**(20), 1429–1433 (1986)
33. Herrmann, H.J., Hong, D.C., Stanley, H.E.: Backbone and elastic backbone of percolation clusters obtained by the new method of 'burning'. J. Phys. A Math. Gen. **17**, L261–L266 (1984)
34. Kroll, D.M., Croll, S.G.: Influence of crosslinking functionality, temperature and conversion on heterogeneities in polymer networks. Polymer. **79**, 82–90 (2015)
35. Martin, M.G.: MCCCS Towhee: a tool for Monte Carlo molecular simulation. Mol. Simul. **39** (14–15), 1212–1222 (2013)
36. Smit, B.: Phase diagrams of Lennard-Jones fluids. J. Chem. Phys. **96**, 8639–8640 (1992)
37. Ge, J., Wu, G.-W., Todd, B.D., Sadus, R.J.: Equilibrium and nonequilibrium molecular dynamics methods for determining solid–liquid phase coexistence at equilibrium. J. Chem. Phys. **119**, 11017–11023 (2003)

38. Ahmed, A., Sadus, R.J.: Effect of potential truncations and shifts on the solid liquid phase coexistence of Lennard-Jones fluids. J. Chem. Phys. **133**, 124515 (2010)
39. Witten, T.A., Sander, L.M.: Diffusion-limited aggregation. Phys. Rev. B. **27**(9), 5686–5697 (1983)
40. Meakin, P.: Diffusion-limited aggregation in three dimensions: results from a new cluster–cluster aggregation model. J. Colloid Interface Sci. **102**(2), 491–504 (1984)
41. Heinson, W.R., Chakrabarti, A., Sorensen, C.M.: Divine proportion shape invariance of diffusion limited cluster–cluster aggregates. Aerosol Sci. Technol. **49**(9), 786–792 (2015)
42. Erath, E.H., Spurr, R.A.: Occurrence of globular formations in thermosetting resins. J. Polym. Sci. **35**(129), 391–399 (1959)
43. Racich, J.L., Koutsky, J.A.: Nodular structure in epoxy resins. J. Appl. Polym. Sci. **20**(8), 2111–2129 (1976)
44. Morsch, S., Liu, Y., Lyon, S.B., Gibbon, S.R.: Insights into epoxy network nanostructural heterogeneity using AFM-IR. ACS Appl. Mater. Interfaces. **8**, 959–966 (2016)
45. Sahagun, C.M., Morgan, S.E.: Thermal control of nanostructure and molecular network development in epoxy-amine thermosets. ACS Appl. Mater. Interfaces. **4**, 564–572 (2012)
46. Cuthrell, R.E.: Epoxy polymers II. Macrostructure. J. Appl. Polym. Sci. **12**(6), 1263–1278 (1968)
47. Labana, S.S., Newman, S., Chompff, A.J.: Chemical effects on the ultimate properties of polymer networks in the glassy state. In: Chompff, A.J., Newman, S. (eds.) Polymer Networks, pp. 453–477. New York, Plenum Press (1971)
48. Zee, M.: Structures of crosslinked networks. Master's thesis, North Dakota State University. ProQuest Dissertations Publishing, 1589686 (2015)

Chapter 3
Rigidity Percolation Modeling of Modulus Development During Free-Radical Crosslinking Polymerization

Mei Wen, L.E. Scriven, and Alon V. McCormick

Introduction

Free-radical polymerization of multifunctional monomers, such as di-acrylates and di-methacrylates, produces highly crosslinked network polymers. The polymerization can take place rapidly at room temperature when it is initiated with ultraviolet light. Such network polymers find application in protective and decorative coatings, dental restoration, photoresists, optical fibers, and many other areas [1–4]. Modulus is not only an important mechanical property; it also determines the stress development of organic coatings caused by the volume shrinkage during polymerization. In a simple picture of elastic behavior, stress is the product of modulus and strain. When the stress developed is high, defects such as delamination, curling, and cracking can occur in a coating. Therefore, understanding modulus development during polymerization is important to make materials with a desired modulus and low stress.

Modulus development during crosslinking polymerization is determined by cluster and network growth. Before gelation, there generally are some locally crosslinked clusters or molecules—so-called microgels. These clusters are distinct and unconnected, so that there is no covalently bonded structure that spans the sample and, thus, no elastic modulus. After gelation, a load-bearing, sample-spanning covalently bonded structure, percolating network, forms, and so modulus starts to

† L.E. Scriven (Deceased)

M. Wen (✉)
Department of Chemical Engineering and Materials Science, University of Minnesota, Minneapolis, MN, USA

Axalta Coating Systems, Coatings Technology Center, Wilmington, DE, USA
e-mail: mei.n.wen@axaltacs.com

A.V. McCormick
Department of Chemical Engineering and Materials Science, University of Minnesota, Minneapolis, MN, USA

M. Wen, K. Dušek (eds.), *Protective Coatings*, DOI 10.1007/978-3-319-51627-1_3

develop. The higher the proportion of clusters that can be bonded into the percolating network, the faster the modulus develops.

In the polymerization of multifunctional monomers, cluster and network growth is complicated. First of all, crosslink density of formed networks can be much higher than the copolymerization of monofunctional monomers (e.g., monoacrylates) and a small amount of multifunctional monomers (e.g., divinylbenzene). Secondly, besides the high crosslink density, the reaction of pendant functional groups forms extensive primary and secondary cycles following the adopted definitions here [5]. A primary cycle forms when a free radical reacts with a pendant functional group on its own kinetic chain (the propagating path of a free radical). A crosslink forms when a free radical reacts with a pendant functional group on another kinetic chain. A secondary cycle forms when a free radical reacts with a pendant functional group on a kinetic chain that has already crosslinked with the radical's kinetic chain. Finally, as a result of the extensive cyclization, a heterogeneous structure, a network that consists of highly reacted regions and less highly reacted regions, forms. The structure complexity makes it hard to predict modulus as a function of conversion of functional groups.

The classical molecular theories of rubbery elasticity (i.e., *affine* network and phantom network) [6–9], which are based on statistical mechanics of freely joint chains in thermal motion, do not apply to the system any more. This is because these theories are valid in homogeneous networks at low crosslink density and without cycle formation. From a number of experimental studies through light scattering and small angle neutron scattering [10, 11], and confocal microscopy under shear deformation [12] analyses, it has been demonstrated that deformations in gels are indeed not affine. In other words, a macroscopic deformation applied on a sample cannot be translated uniformly to the microscopic level. Therefore, the cluster and network growth must be dictated in order to study modulus accurately.

Many attempts have been made to go beyond the affine deformation assumption to quantify gel or rubber elasticity. De Gennes proposed that the sol–gel transition is analogous to a percolation transition, and the shear modulus close to the gelation point behaves like the electric conductivity in conductor percolation problems [13], which is rigorous for a lattice model of gelation with an isotropic force between nearest neighbors [14]. Feng and Sen [15] later used a central force between neighboring sites, a spring element, to model network modulus. In addition, Kantor and Webman [16] have considered a model based on two contributions to the elastic energy, i.e., a central force term and a bending energy term through a beam element between neighboring sites. Therefore, a simple mechanical model can be used to calculate the deformation of dictated networks through replacing the classical entropic elasticity by spring or beam elements on a lattice to obtain modulus.

No theory exists that is satisfactory for predicting the growth of highly crosslinked, heterogeneous networks. Statistical theories such as the Flory–Stockmayer theory [17–19], the branching processes (cascade) theory [20, 21], and the recursive theory [22, 23] build up networks from monomeric units differing in reaction states. They can predict the details of the percolating network structure. However, they are not applicable to initiated network buildup because they disregard long-range time correlations like those determined by free-radical polymerization kinetics. Kinetic theories build up networks by solving a set of differential equations of the time evolution of concentrations of distinguishable molecules [24–27]. They provide the time correlations of network growth, but in the present state only limited information on the structure of the percolating network. The combined (kinetic-statistical) theory largely, but sometimes not fully, removes the deficiency of the statistical approach [28, 29]. Both statistical and kinetic theories are generally mean-field because they presume formation of homogeneous networks. In contrast, simulations in space, either lattice- or off-lattice-based, are non-mean-field and can take into account both time and space correlations. It should be remarked that the kinetic theory based on Smoluchowski generalized differential equations, although formally based on the mean-field method, can quite successfully approximate non-mean-field effects, both dynamic and static (cf., e.g., Refs. [29–31]).

One promising approach to simulate free-radical polymerization on lattice is kinetic gelation modeling [32–41]. It simulates free-radical polymerization on a lattice with rules of initiation, propagation, termination, and radical trapping. Reaction is restricted to occur between nearest neighbors, and in the earlier versions no physical movement or some simple motions of monomers and polymers [33, 34, 37, 39] are applied. It has been used to predict highly crosslinked, heterogeneous networks. Moreover, it can capture features of polymerization of multifunctional monomers, i.e., severe trapping of radicals and reaction-diffusion. The results are lattice-type dependent.

Another similar but more advanced approach is Monte Carlo simulation on lattice by incorporating bond fluctuations dynamics [42, 43], where monomers or repeat units on a polymer diffuse randomly along principal directions of the lattice by one lattice unit each time and bonds can form with a set of different lengths. This approach has been used to simulate mainly copolymerization of a monofunctional monomer with one or two types of crosslinkers. It provides information on kinetics, detailed structure development, and swelling of networks. Furthermore, off-lattice-based Monte Carlo simulation is available that can afford more realistic representation of monomer structure and dynamics than the lattice-based models, particularly at low conversions [44, 45].

The lattice-based kinetic gelation model is chosen here due to its increasing acceptance to model highly crosslinked, heterogeneous networks and the fact that it is easy to implement a rigidity percolation model to determine the modulus of the networks developed during polymerization. The kinetic gelation model used here is based on the rules developed in Ref. [40]. The model recasts polymerization kinetics as a Markov process through a stochastic approach. It simulates real reaction time by modifying and employing the probability density function and the associated Monte Carlo method to reactions on a lattice. Consequently, the rate of polymerization and cluster and network growth, which are strongly affected by how free radicals are generated, can be simulated quite accurately.

In statistical-mechanical models of elasticity of rubbery polymer networks, the stress–strain relation is obtained for various deformation geometries by a standard procedure from the Helmholtz energy. The stress σ is a function of deformation ε, $f(\varepsilon)$, with a prefactor, $\Phi(\nu_e, \alpha, \beta)$, which is usually directly proportional to the concentration of elastically active network chains (EANCs), ν_e, or a quantity related to it, the cycle rank. The prefactor is also a function of chain extension/compression in the undeformed state (factor α) and chain rigidity (factor β).

An EANC is defined as a sequence of units connected by two bonds with infinite continuation extending between junctions connected by three or more bonds with infinite continuation (for definition of states of units in a crosslinking system, see, e.g., Ref. [46]). Depending on the network model, the concentration of EANCs can be calculated analytically or numerically, or simulated. The cycle rank of a network is a quantity which can be obtained for any kind of (Monte Carlo) simulation in space. The cycle rank is equal to the number of cuts by which a network with cyclic paths can be converted into a spanning tree.

The relation $\sigma = \Phi(\nu_e, \alpha, \beta)\, f(\varepsilon)$ determines the "modulus," defined as σ/ε, or its small-strain limit $\lim(\sigma/\varepsilon)_{\varepsilon\to 0}$. Therefore, modulus is a strong function of the prefactor. Existing statistical-mechanical elasticity models based on the concept of EANCs can predict stress–strain behavior of networks ranging from very flexible to relatively stiff allowing only limited deformation before they break. Such models are, for instance, the finite-chain-extensibility models; their deformation limits in three-axial deformation are shown in Ref. [47]. Only the Gaussian chains allow for infinite extension irrespective of their degree of polymerization.

Another model of mechanics of network structures is that based on links and joints on a lattice. Depending on the type of mechanical element a bond or a link is treated, there are two basic approaches as introduced earlier. In one a bond is represented as a spring; in the other a bond is represented as a beam. In lattice simulation, both belong to rigidity percolation [48–50].

In the first approach, called central-force modeling, the bonds or links between nodes are springs that can be stretched, compressed, and freely rotate around their ends; they cannot transmit torque [15]. At the connectivity percolation threshold, where a sample-spanning cluster forms by connected bonds, the percolating cluster is, in general, unable to transmit any load across the system because along simply connected portions of the cluster the bonds are not aligned and ready to bear tension. In a random triangular net, the critical bond fraction where bulk modulus appears is 0.64, whereas the connectivity percolation threshold occurs at 0.35 [51]. The rigidity comes from sequences of fused triangles, so-called rigid structures [49]. Only after the rigid structures have formed across the system can a network become rigid (i.e., the bulk modulus of the network differs from zero). Termonia et al. [52–54] used the central-force model to predict the elastic properties of polymers, whose covalent bonds, hydrogen bonds, and van der Waals bonds, or segments between crosslinks and between entanglements, are modeled as springs in different ideal networks.

In addition to the central forces, bond-bending forces can be added when the angles between two bonds are changed during deformation [16, 51, 55]. Then the rigidity percolation threshold occurs at the connectivity percolation threshold. This approach is beam modeling, where bonds or links are beams that can be stretched, compressed, sheared, and bent. Nodes or joints are rigid joints that maintain constant angles between beams by exerting local torque. This is a simple kind of rigidity development. Rigidity percolation threshold appears at the connectivity percolation threshold. This model has been used to study the elastic properties of glass [56], rocks [57], and paper [50] and fracture properties of disordered media such as rocks and composite solids [51, 58]. It has also been applied to study the elastic properties of fibrous networks such as those forming cellular cytoskeleton in biomaterials [59] and fracture properties of polymers with network microstructure including amorphous and semicrystalline polymers, rubbers, thermosets, and carbon nanotubes [60].

The approach chosen is based on its simplicity and reflects the rigidity of some networks formed. In the free-radical polymerization of the most commonly used multifunctional monomers, i.e., difunctional monomers (each monomer has two functional groups such as a di-acrylate or a di-methacrylate monomer), the networks predicted by the kinetic gelation model [40, 41] on a two-dimensional triangular lattice cannot develop rigid structures (fused triangles spanning the system) even at the end of reaction. Therefore, if no secondary bonding (e.g., hydrogen bonding and van der Waals bonding) or bond-bending forces are added, the central-force modeling with free-rotating joints cannot be used to predict the rigidity of the networks formed. The beam modeling, however, is applicable and is used here. It should predict the modulus of the most rigid networks.

Bonds formed by the kinetic gelation model in a two-dimensional triangular lattice were represented as beams that meet at rigid joints. Networks formed were then stretched to calculate Young's moduli. The effects of initiation and primary cyclization rates on Young's modulus, the number of load-bearing bonds (elastically active bonds), and bond stress distributions were investigated.

Network Load–Displacement Relationship

The deformation of a network is determined by the deformation of each individual beam. Therefore, the deformation of a single beam is examined first. Figure 3.1 shows a general plane element loaded with forces and moments at its ends 1 and 2. The origin is chosen as end 1 at its undeformed state, and the x-direction is chosen as the undeformed axial direction from end 1 to end 2. The nodal loads and nodal displacements are both vectors as shown below:

$$\mathbf{p}_1 = \begin{bmatrix} p_{x1} \\ p_{y1} \\ m_1 \end{bmatrix} \quad \mathbf{p}_2 = \begin{bmatrix} p_{x2} \\ p_{y2} \\ m_2 \end{bmatrix} \quad \mathbf{d}_1 = \begin{bmatrix} \delta_{x1} \\ \delta_{y1} \\ \theta_1 \end{bmatrix} \quad \mathbf{d}_2 = \begin{bmatrix} \delta_{x2} \\ \delta_{y2} \\ \theta_2 \end{bmatrix}$$

where p_{x1}, p_{y1}, and m_1 are the axial forces in the x- and y-directions and bending moment at end 1, respectively, and p_{x2}, p_{y2}, and m_2 are the axial forces in the x- and

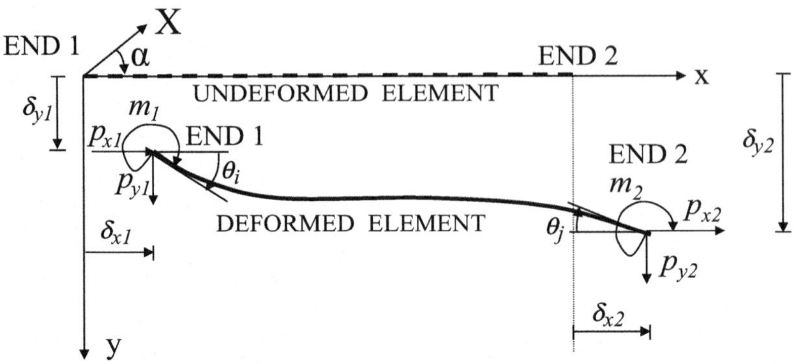

Fig. 3.1 Deformation of a beam from its undeformed state to its deformed state under nodal loads at end 1 and end 2. The x-direction is chosen as the undeformed axial direction from end 1 to end 2. The nodal loads at end 1 are p_{x1}, p_{y1}, and m_1, and the nodal loads at end 2 are p_{x2}, p_{y2}, and m_2. The displacements of end 1 and end 2 after the deformation are δ_{x1}, δ_{y1}, and θ_1 and δ_{x2}, δ_{y2}, and θ_2, respectively. The angle of rotation from the global frame to the element frame is α

y-directions and bending moment at end 2, respectively, δ_{x1}, δ_{y1}, and θ_1 are the displacements in the x- and y-directions and the angular change relative to the original beam position at end 1, and δ_{x2}, δ_{y2}, and θ_2 are the displacements in the x- and y-directions and the angular change at end 2.

According to the theory of deflection of a beam, when all the displacements are small compared to the dimensions of the element, the following linear equations describe the load–displacement relation [16]:

$$\begin{aligned}\mathbf{p}_1 &= \mathbf{K}_{11}\mathbf{d}_1 + \mathbf{K}_{12}\mathbf{d}_2 \\ \mathbf{p}_2 &= \mathbf{K}_{21}\mathbf{d}_1 + \mathbf{K}_{22}\mathbf{d}_2\end{aligned} \tag{3.1}$$

where

$$\begin{aligned}
\mathbf{K}_{11} &= \begin{bmatrix} \frac{EA}{L} & 0 & 0 \\ 0 & \frac{12EI}{L^3} & \frac{6EI}{L^2} \\ 0 & \frac{6EI}{L^2} & \frac{4EI}{L} \end{bmatrix} \quad
\mathbf{K}_{12} = \begin{bmatrix} -\frac{EA}{L} & 0 & 0 \\ 0 & -\frac{12EI}{L^3} & \frac{6EI}{L^2} \\ 0 & -\frac{6EI}{L^2} & \frac{2EI}{L} \end{bmatrix} \\
\mathbf{K}_{21} &= \begin{bmatrix} -\frac{EA}{L} & 0 & 0 \\ 0 & -\frac{12EI}{L^3} & -\frac{6EI}{L^2} \\ 0 & \frac{6EI}{L^2} & \frac{2EI}{L} \end{bmatrix} \quad
\mathbf{K}_{22} = \begin{bmatrix} \frac{EA}{L} & 0 & 0 \\ 0 & \frac{12EI}{L^3} & -\frac{6EI}{L^2} \\ 0 & -\frac{6EI}{L^2} & \frac{4EI}{L} \end{bmatrix}
\end{aligned} \tag{3.2}$$

Here E is Young's modulus of the beam, A is its cross-sectional area, L is its length, and I is the moment of inertia of the cross-sectional area with respect to the z-axis, i.e., the neutral axis. For a beam of rectangular cross section with width b and height h, the moment of inertia $I = bh^3/12$.

The linear relation between loads and displacements assumes that the only effect of an axial force (i.e., p_{x1} or p_{x2}) is to produce an axial strain. This is valid when the deflection in the y-direction is small enough to justify neglecting changes in lateral stiffness because of axial forces. When the effect of the axial forces on the bending moment distribution and hence on the components of the sub-stiffness matrices $\mathbf{K}_{ij}$ cannot be ignored, element stability functions ϕ_i can be used to keep the same format [61]. The sub-stiffness matrices then become

$$\mathbf{K}_{11} = \begin{bmatrix} \frac{EA}{L} & 0 & 0 \\ 0 & \frac{12EI\phi_1}{L^3} & \frac{6EI\phi_2}{L^2} \\ 0 & \frac{6EI\phi_2}{L^2} & \frac{4EI\phi_3}{L} \end{bmatrix} \quad \mathbf{K}_{12} = \begin{bmatrix} -\frac{EA}{L} & 0 & 0 \\ 0 & -\frac{12EI\phi_1}{L^3} & \frac{6EI\phi_2}{L^2} \\ 0 & -\frac{6EI\phi_2}{L^2} & \frac{2EI\phi_4}{L} \end{bmatrix}$$
$$\mathbf{K}_{21} = \begin{bmatrix} -\frac{EA}{L} & 0 & 0 \\ 0 & -\frac{12EI\phi_1}{L^3} & -\frac{6EI\phi_2}{L^2} \\ 0 & \frac{6EI\phi_2}{L^2} & \frac{2EI\phi_4}{L} \end{bmatrix} \quad \mathbf{K}_{22} = \begin{bmatrix} \frac{EA}{L} & 0 & 0 \\ 0 & \frac{12EI\phi_1}{L^3} & -\frac{6EI\phi_2}{L^2} \\ 0 & -\frac{6EI\phi_2}{L^2} & \frac{4EI\phi_3}{L} \end{bmatrix} \tag{3.3}$$

where the stability functions ϕ_1, ϕ_2, ϕ_3, and ϕ_4 are defined by

$$\begin{aligned} \phi_1 &= (\beta \cot \beta)\phi_2 \\ \phi_2 &= \frac{1}{3}\beta^2/(1 - \beta \cot \beta) \\ \phi_3 &= \frac{3}{4}\phi_2 + \frac{1}{4}\beta \cot \beta \\ \phi_4 &= \frac{3}{2}\phi_2 - \frac{1}{2}\beta \cot \beta \end{aligned} \tag{3.4}$$

where $\beta \equiv (L/2)\sqrt{|p_{x1}|/EI}$. Because the element stability functions are functions of p_{x1}, the load–displacement relationship becomes nonlinear.

Equation 3.1 is expressed in the coordinate basis defined by the element. To transform it into the form based on the global coordinates, a rotation matrix $\mathbf{R}$ is needed. It is determined by the angle of rotation from the global frame to the element frame, α, as shown in Fig. 3.1:

$$\mathbf{R} = \begin{bmatrix} \cos\alpha & -\sin\alpha & 0 \\ \sin\alpha & \cos\alpha & 0 \\ 0 & 0 & 1 \end{bmatrix} \tag{3.5}$$

At small displacements, α of each beam can be considered as constant when it is deformed. The matrix $\mathbf{R}$ is orthogonal, i.e., $\mathbf{R}^{-1}=\mathbf{R}^{\mathbf{T}}$, where $\mathbf{R}^{-1}$ is the inverse of $\mathbf{R}$ and $\mathbf{R}^{\mathbf{T}}$ is the transpose of $\mathbf{R}$. By left-multiplying Eq. 3.1 with matrix $\mathbf{R}$, the following relation is obtained:

$$\begin{aligned} \mathbf{p}_1{}' &= \mathbf{K}_{11}{}'\mathbf{d}_1{}' + \mathbf{K}_{12}{}'\mathbf{d}_2{}' \\ \mathbf{p}_2{}' &= \mathbf{K}_{21}{}'\mathbf{d}_1{}' + \mathbf{K}_{22}{}'\mathbf{d}_2{}' \end{aligned} \tag{3.6}$$

where $\mathbf{p}_i{}' = \mathbf{R}\mathbf{p}_i$, $\mathbf{d}_i{}' = (\mathbf{R}^{\mathbf{T}})^{-1}\mathbf{d}_i = \mathbf{R}\mathbf{d}_i$, and $\mathbf{K}_{ij}{}' = \mathbf{R}\mathbf{K}_{ij}\mathbf{R}^{\mathbf{T}}$ (for $i = 1, 2$; $j = 1, 2$). The prime designates variables in the two-dimensional global coordinates. Once the load–displacement equations have been found for all the elements, i.e., bonds simulated by the kinetic gelation modeling, the stiffness matrix of the whole network, which describes the load–displacement relation of all the sites connected by bonds, can be readily constructed by linear combination of contributions of bonds connected to each reacted site.

Constructing the stiffness matrix of the whole network from the element's sub-stiffness matrices depends solely on the way the bonds are connected. The reacted lattice sites are denoted by A, B, C, ..., and the end 1 of each bond corresponds to the "lower" and the end 2 to the "higher" lettered site. Equation 3.6 is used to calculate the contribution of a bond r between lattice sites I and J ($I < J$) to the final set of load–displacement equations. At joint mechanical equilibrium, the loads at lattice sites I and J are the sums of all the contributions from the bonds meeting at those two sites, respectively:

$$\begin{aligned} \mathbf{p}_I^g &= (\mathbf{p}_1{}')_r + \text{contributions from other bonds meeting at site } I \\ \mathbf{p}_J^g &= (\mathbf{p}_2{}')_r + \text{contributions from other bonds meeting at site } J \end{aligned} \tag{3.7}$$

where $\mathbf{p}_I^g$ is the load at lattice site I, $\mathbf{p}_J^g$ is the load at lattice site J, $(\mathbf{p}_1{}')_r$ is the contribution from bond r to lattice site I, $(\mathbf{p}_2{}')_r$ is the contribution from bond r to lattice site J, and the contributions from the other bonds are $\mathbf{p}_1{}'$ vectors from those bonds with end 1 at the lattice site considered and $\mathbf{p}_2{}'$ vectors from those bonds with end 2 at the lattice site considered. Substituting Eq. 3.6 into Eq. 3.7 leads to

$$\begin{aligned} \mathbf{p}_I^g &= (\mathbf{K}_{11}{}')_r\mathbf{d}_I{}' + (\mathbf{K}_{12}{}')_r\mathbf{d}_J{}' + \text{contributions from other bonds} \\ \mathbf{p}_J^g &= (\mathbf{K}_{21}{}')_r\mathbf{d}_I{}' + (\mathbf{K}_{22}{}')_r\mathbf{d}_J{}' + \text{contributions from other bonds} \end{aligned} \tag{3.8}$$

Because $\mathbf{K}_{21}{}'$ is the transpose of $\mathbf{K}_{12}{}'$, the contribution made by each individual bond to the stiffness matrix of the whole network is always symmetric.

In practice, only the leading diagonal coefficients and those above them, which form an "upper triangle" matrix, are needed in the stiffness matrix. The components of this upper triangle matrix follow the rules: (1) The leading diagonal sub-matrix in row I is the sum of the matrices $\mathbf{K}_{11}'$ or $\mathbf{K}_{22}'$ of all the bonds meeting at lattice site I, the matrix $\mathbf{K}_{11}'$ being selected if the member has its end 1 at site I, and the matrix $\mathbf{K}_{22}'$ being chosen if the member has its end 2 at site I. (2) The off-diagonal sub-matrices in row I, corresponding to columns J, K, L, $\ldots > I$, are the $\mathbf{K}_{12}'$ matrices of the members connecting site I to site J, K, L, ... If a reacted lattice site is not directly connected to I by a bond, then the associated sub-matrix is zero. The formula of the stiffness matrix of the whole network $\mathbf{K}^g$ is

$$\begin{aligned} \mathbf{K}^g{}_{II} &= \sum_{k=1}^{n_1} (\mathbf{K}_{11}{}')_k + \sum_{l=1}^{n_2} (\mathbf{K}_{22}{}')_l \\ \mathbf{K}^g{}_{IJ} &= \begin{cases} (\mathbf{K}_{12}{}')_r & \text{bond } r \text{ between } I \text{ and } J \\ 0 & \text{no bond between } I \text{ and } J \end{cases} \end{aligned} \tag{3.9}$$

where $\mathbf{K}^g{}_{II}$ and $\mathbf{K}^g{}_{IJ}$ are the leading diagonal sub-matrices at row I and column I and the upper triangle sub-matrices at row I and column J ($I < J$) of $\mathbf{K}^g$, respectively; n_1 and n_2 are the total numbers of bonds with end 1 and end 2 at site I, respectively; and r, k, and l are bond indexes. Once the complete matrix $\mathbf{K}^g$ is formed, the load–displacement equation of the whole network

$$\mathbf{p}^g = \mathbf{K}^g \mathbf{d}^g \tag{3.10}$$

is ready to solve. Equation 3.10 relates the deformation $\mathbf{d}^g$ ($=\mathbf{d}'$) of the whole network to load $\mathbf{p}^g$ imposed upon and developed in the network.

In this two-dimensional modeling, the degrees of freedom of each reacted site are the displacements in the x- and y-directions and an angular change. The angular change is the inclination of bonds at a reacted site from their original undeformed position. Because of the nature of rigid joints, the inclinations of all the bonds that meet at a lattice site must be equal. Therefore, the total number of unknowns in Eq. 3.10 is $3N$, where N is the total number of reacted sites.

Modeling Method

The kinetic gelation model developed in the reference [40] was used to generate polymer networks on a two-dimensional triangular lattice. Each monomer or initiator only occupied one lattice site. Reaction took place according to the local reaction probabilities of each radical, and the reaction time was calculated with a Monte Carlo technique. A propagation rate constant k_p and a termination rate constant k_t were used to specify the reaction probability per unit time of a certain combination of reactive entities, i.e., a radical and a functional group, and two radicals, respectively. Furthermore, a primary cyclization rate constant was used to specify the reaction probability per unit time for a radical reacting with a pendant functional group on its own kinetic chain to form a primary cycle. The rate constant was denoted as rk_p, where r is the primary cyclization factor, defined as the ratio of the primary cyclization rate constant to the propagation rate constant. The real-time calculation allowed initiation, which is a first-order reaction controlled by an initiation rate constant k_I, to be gauged more accurately. Details of this kinetic gelation model were described in the reference [40]. Periodic boundary conditions were applied to one direction of the lattice, called the transverse direction (relative to the other direction, the axial direction). In this direction, any molecule on one side of the lattice can react and form a bond with its counterpart on the opposite side. When a cluster first spanned from the bottom boundary to the top boundary in the axial direction, the connectivity or bond percolation threshold was reached, and modulus appeared. Thereafter, as more functional groups were converted and made fuller networks, their Young's moduli were calculated.

In this rigidity percolation modeling, the bonds between lattice sites were represented as beams that met in rigid joints. A single bond between two lattice sites represented a unit of –C–, presuming head-to-tail addition. Note that this bond or link is not a single covalent bond but two covalent bonds connected by a carbon. Comparing to other bond units on a lattice used in the literature, such as chains between crosslinks [13, 60], this is the shortest unit used due to the polymerization of di-acrylates or di-methacrylate directly. The single bond was considered as a beam with a square cross section. Its moment of inertia $I = w^4/12$, where w was the width of one square ($b = h = w$). Besides a single bond, multiple bonds, i.e., two or three bonds, could form between reacted sites through cyclization. When there was a single bond already formed between two sites, it could change into two bonds through primary or secondary cyclization; it could change into three bonds through two times of primary cyclization. Figure 3.2 gives an example of different bond

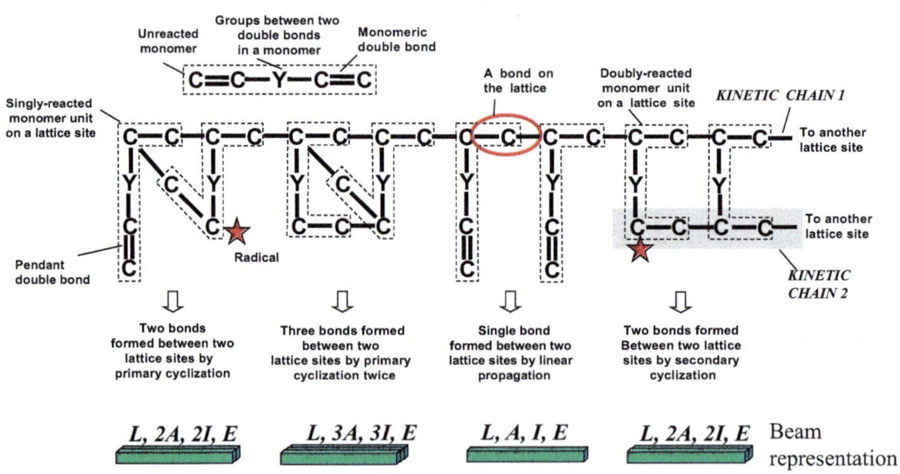

Fig. 3.2 Bond elements between lattice sites on part of a di-acrylate or di-methacrylate (methyl groups not shown) network molecule formed by two kinetic chains (chain 2 is *shaded*) in kinetic gelation modeling. Monomer units on lattice sites are shown enclosed in *dashed boxes*. Beam representation of the single bond, two bonds, and three bonds between two lattice sites are shown in the bottom part, where L is the beam length, A is the cross-sectional area of a single beam, E is the Young's modulus, and I is the moment of inertial of a single beam

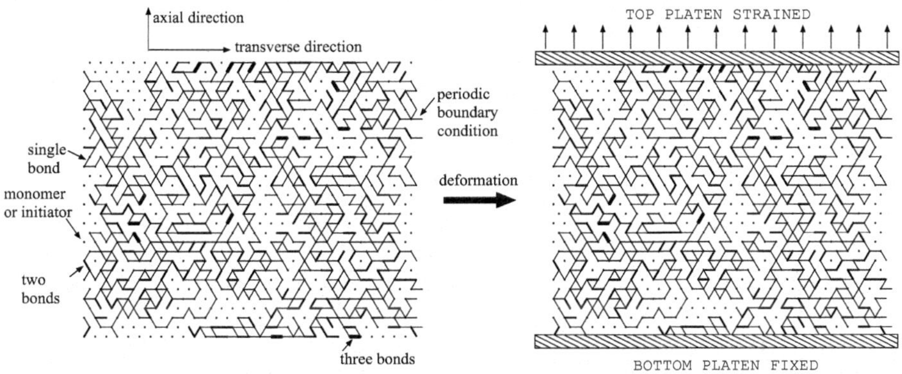

Fig. 3.3 Schematic diagram of the deformation of a network formed on a 30 × 30 triangular lattice at conversion 0.77. *Dots* without any connection are the monomer or initiator sites. A *thin line* between two sites represents a single bond, a *thicker line* represents two bonds, and the *thickest line* represents three bonds

elements between lattice sites on part of a di-acrylate network molecule formed by two kinetic chains in kinetic gelation modeling. The multiple bonds were considered as two or three beams put together in the width direction. In other words, multiple beams between a pair of lattice sites are equivalent to a new beam with the cross-sectional areas and the moment of inertia doubled and tripled, respectively ($b = 2w$ and $h = w$ for two bonds; and $b = 3w$ and $h = w$ for three bonds). This approximate treatment presumes the strongest connection possibly formed by these multiple bonds. It may be applicable to di-acrylate or di-methacrylate monomers with short chain length between their double bonds. When the chain length gets very long (such as in oligomers or telechelic polymers), bonds formed by cyclization are in loose loops, and thus multiple bonds between a pair of lattice sites can be treated as one single bond.

At a given conversion, the network formed was clamped in the axial direction of the lattice by two platens: the top and the bottom. Whereas the bottom platen was fixed, the top platen was then moved up to achieve a certain overall tensile strain of the system—this defines the top and bottom boundary conditions (illustrated in Fig. 3.3). With the top and bottom boundary sites (only those reacted) clamped under external forces, the force and moment balance (mechanical equilibrium) of the whole system became

$$\mathbf{K}^g \mathbf{d}^g + \mathbf{p}_\mathrm{b} = 0 \tag{3.11}$$

where $\mathbf{p}_\mathrm{b}$ denoted the external forces, whose components were all zero except on the top and bottom boundary sites. Equation 3.11 was solved by Newton's iteration method, in which the residual vector $\mathbf{r}$ ($\mathbf{r} \equiv \mathbf{K}^g \mathbf{d}^g + \mathbf{p}_\mathrm{b}$) was driven to an acceptably small magnitude by successively reevaluating the displacements:

$$\mathbf{d}^{g^{(k+1)}} = \mathbf{d}^{g^{(k)}} + [\mathbf{K}^{g^{(k)}}]^{-1}\mathbf{r}^{(k)} \tag{3.12}$$

Here $\mathbf{d}^{g^{(k+1)}}$ and $\mathbf{d}^{g^{(k)}}$ were the calculated displacements in the $(k + 1)$th and the kth iteration steps; $[\mathbf{d}^{g^{(k)}}]^{-1}$ was the inverse of $\mathbf{K}^{g^{(k)}}$, which was the stiffness matrix calculated by $\mathbf{d}^{g^{(k)}}$; and $\mathbf{r}^{(k)}$ was the residual in the kth step. Because the components of $\mathbf{d}^g$ at the top and bottom boundary sites were specified as boundary conditions, the residual vector $\mathbf{r}$ at those sites was set to zero and the forces on them $\mathbf{p}_b$ were not needed. For a given network, the initial values $(k = 0)$ of the displacements $\mathbf{d}^{g^{(0)}}$ were set as zero except for the top boundary sites, whose displacements were specified. Iteration would have converged faster had the displacements initially been set not to zero but to their values at the previous step of conversion; this less obvious procedure would have started the displacements at zero of the lattice sites only having new bonds. The Yale Sparse Matrix Package (YSMP), which solves large sparse systems of linear algebraic equations by Gaussian elimination without pivoting [62], was used to calculate $[\mathbf{K}^{g^{(k)}}]^{-1}$. The convergence criterion used was $\frac{|\mathbf{d}_i^{g^{(k+1)}} - \mathbf{d}_i^{g^{(k)}}|}{|\mathbf{d}_i^{g^{(k+1)}}|} \leq \xi$, $(i = 1, 3N)$, where ξ was a small number.

Young's modulus of each network was calculated as the ratio of the overall stress at the upper boundary in the axial direction to the overall tensile strain applied to the network. The modulus of a network was calculated at a small strain of 0.001. In the calculation of stress–strain relations, strain was increased in each step by an increment $(\Delta\varepsilon)$ of 0.001 over the calculated mechanical equilibrium state of the last step, and then the new equilibrium state was calculated. Both critical tensile strain ε_{ct} and critical shear strain ε_{cs} were assigned a priori. When either the tensile or the shear strain in a bond exceeded its critical value, the bond would break. The rate of convergence to the solution of the new problem with one fewer bonds was enhanced by breaking the bond in steps, i.e., solving a succession of problems with the modulus of the bond reduced stepwise to zero and starting the iterative solution of each with the solution of the last [57].

The generation of networks and calculation of modulus were continued until no more active radicals were present and only one initiator remained. At the end of reaction, the conversion of functional groups and the modulus were called asymptotic because it would have required infinitely long time to decompose the last initiator in a first-order kinetic reaction. Figure 3.4 is the flowchart of this modeling algorithm.

Base-case parameters used in the simulations are listed in Table 3.1. The initiation rate constant k_I, propagation rate constant k_p, termination rate constant k_t, and primary cyclization factor r were the same as those used in Ref. [41]. Young's modulus and shear modulus of a beam element, E and G, were estimated to be 300 GPa and 3 GPa according to the theoretical axial and shear moduli of crystalline polyethylene, respectively [63]. This brought the ratio of element's Young's modulus E to its shear modulus $G = 12EI/AL^2$, denoted as K $(K = (L/h)^2)$, to 100. The total width of the lattice was taken as one unit length. The average of 50 realizations at a given conversion was reported. In each realization, the seed of a random number generator,

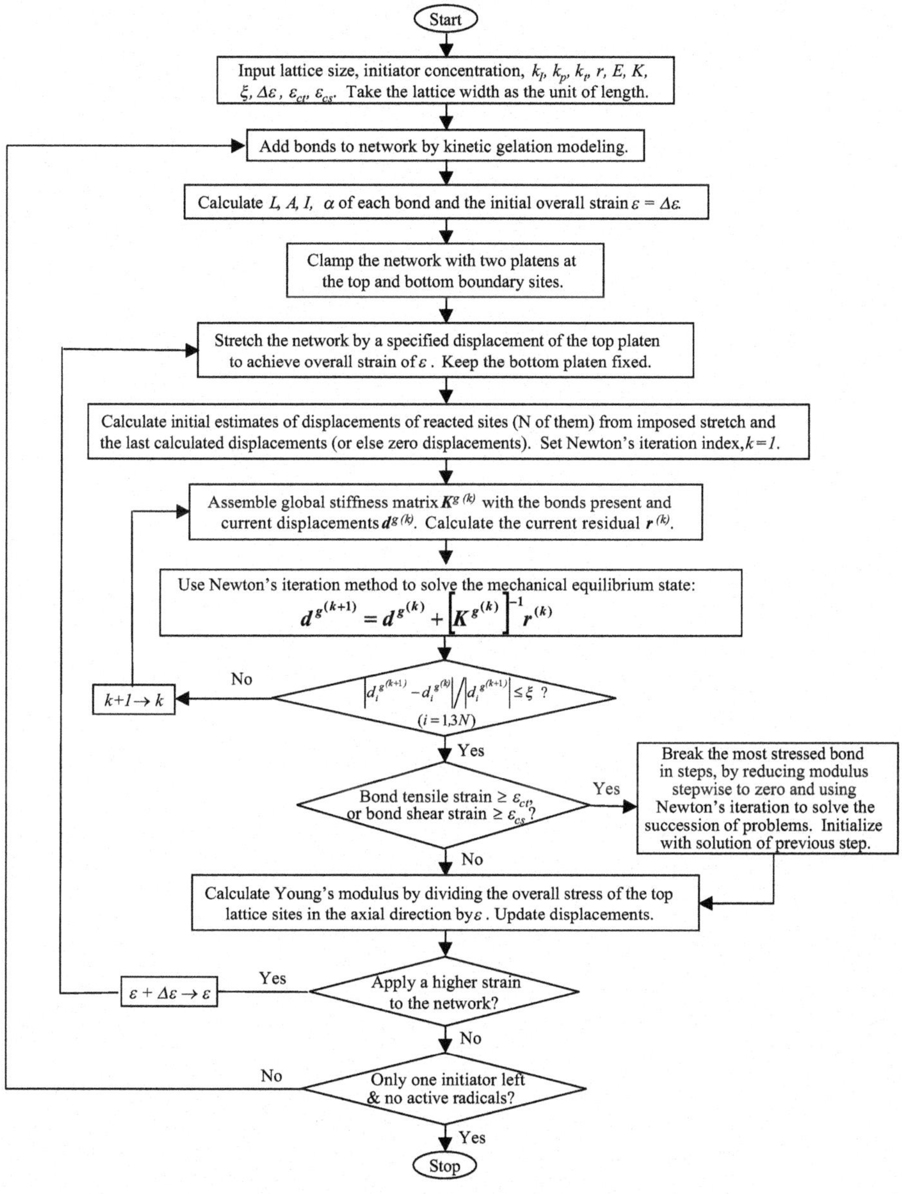

Fig. 3.4 Flowchart of rigidity percolation modeling algorithm

Table 3.1 Base-case parameters used in the simulations

Lattice Size: 200 × 200		
Kinetic gelation modeling parameters		
Initiator% = 5%	$k_{\mathrm{I}} = 1$ 1/s	$k_{\mathrm{p}} = 10^3$ 1/s
$k_{\mathrm{t}} = 10^7$ 1/s	$r = 1$	
Rigidity percolation modeling parameters		
$E = 300$ GPa	$K = 100$	$\xi = 0.001$
$\varepsilon_{\mathrm{ct}} = 0.04$	$\varepsilon_{\mathrm{cs}} = 0.04$	$\Delta\varepsilon = 0.001$

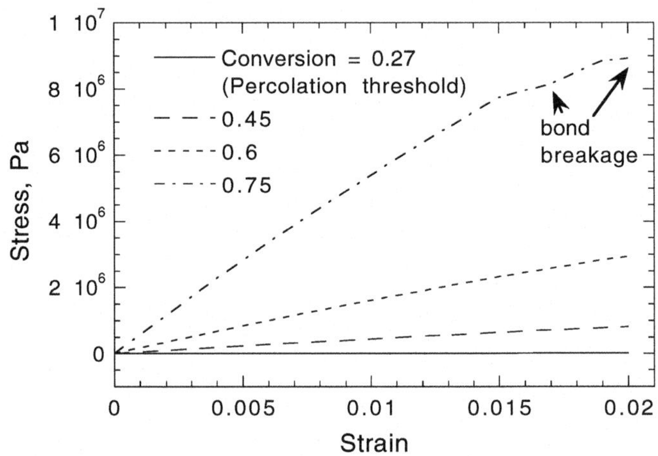

Fig. 3.5 Stress–strain relation at different conversions of functional groups: 0.27 (percolation threshold), 0.45, 0.6, and 0.75

Park and Miller's minimal standard generator [64], was changed in the kinetic gelation model.

Results and Discussion

Figure 3.5 shows the stress–strain relation up to 0.02 strain at different stages of reaction calculated by using the base-case parameters. This strain ensures that the deformations of bonds are small enough that the load–displacement relation described by Eqs. 3.1 and 3.3 is valid. Because the difficulty of convergence rises severely at high strains (e.g., higher than 0.01), a small lattice (80 × 80) was used here. Furthermore, only one sample (generated by one realization) at a given conversion was examined to illustrate the smooth change of stress (before cracking occurs) as strain was raised. The system starts to support stress when a rigid network percolates—spans the system—in the axial direction. The rigidity percolation threshold is of course the same as the connectivity or bond percolation threshold. Stress increases monotonically with strain. Bonds start to break at 0.75

conversion when strain reaches 0.017. This strain is much smaller than the critical tensile and shear strains at the element level, namely, 0.04. Plainly some bonds are under much higher strains than the applied one. The reason is that the irregular distribution of bonds induces stress concentrations. As polymerization proceeds, stress rises much faster with strain due to formation of more rigid networks.

Stress Distribution

Bond stress distributes nonuniformly when a network is deformed because the bonds are distributed irregularly. At mechanical equilibrium, there are two types of bonds: elastically active and inactive. The elastically active bonds bear axial or shear stresses, or moments, whereas the elastically inactive ones do not. The working criterion used here is that if a bond satisfies, (1) $|p_{x1}|/A \leq 300\,\text{Pa}$, (2) $|p_{y1}|/A \leq 300\,\text{Pa}$, and (3) $\max(|m_1|, |m_2|)/LA \leq 300\,\text{Pa}$, then it is elastically inactive. The stress level of 300 Pa is equivalent to a bond tensile strain of 10^{-9}, which is small enough to be neglected. Because only the elastically active bonds can bear loads, they are the only ones that contribute to the rigidity of a network. Furthermore, even though the whole network is under tension and the majority of bonds are under tension, some bonds are under compression. To illustrate the distributions of stresses in space, a network formed at 0.7 conversion of functional groups simulated with the base-case parameters was stretched to reach a strain level of 0.001, and then the modeled tensile, compressive, and shear stresses on bonds are marked based on their levels with four gray scales in Fig. 3.6. These four gray scales represent stress levels of around four orders of magnitude.

The bonds that bear tensile stress appear to be oriented along the direction of the applied strain ($\pm 60°$ from the transverse direction in the triangular lattice). The calculated orientations of bonds also show that among the bonds in tension, there are more bonds at $+60°$ or $-60°$ from the transverse direction than at $0°$ angle, whereas in the totality of bonds on the lattice, i.e., bonds bearing any type of loads, there is no such preference, i.e., around 1/3 of bonds are more or less aligned in each direction (see Table 3.2). Furthermore, though most bonds bear low tensile stress, there are some that bear high tensile stress. The latter form chains with high tensile stress, and these stress chains form a network that spans from the top boundary to the bottom boundary. These high tensile stress chains contribute most to the rigidity of the whole network. An analogous phenomenon was observed in the rigidity of colloidal suspension of hard particles, where arrays of particles form force chains to support the suspension under shear [65, 66]. Wool [60] also used the term *mechanical hot bonds* to describe the highly stressed bonds on a lattice in parallel to the *hot bonds* in conductivity percolation, where hot bonds arise from high current density and tend to overheat and break. These mechanical hot bonds are critical to the fracture of polymers in general and provide understanding why materials fracture under macroscopic stresses that are orders of magnitude less than molecular fracture stresses. Experimentally, it has been observed with Infrared and Raman

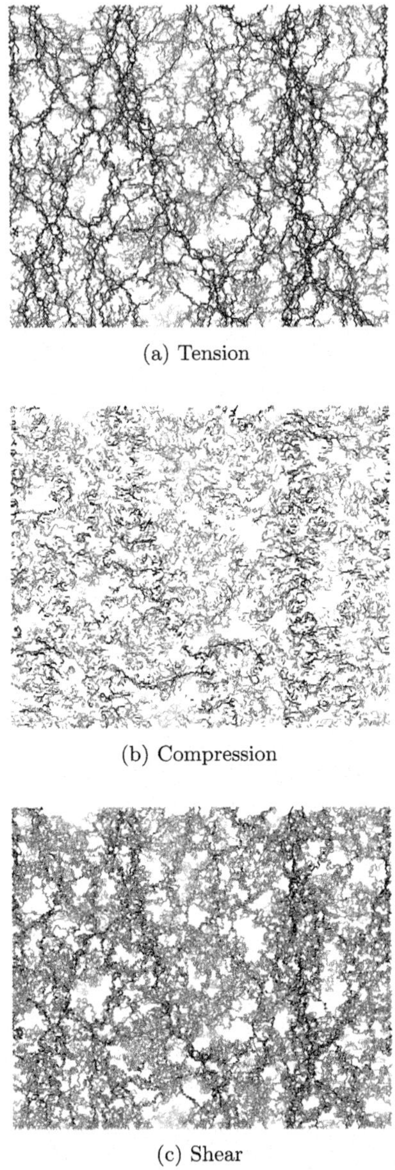

Fig. 3.6 Snapshots of (**a**) tensile, (**b**) compressive, and (**c**) shear stress distributions of a network formed at 0.7 conversion of functional groups in the base case at 0.001 strain. The lattice size is 80 × 80. Bonds are marked with four gray scales representing different levels of stress. Gray scale zero (*black*) represents stress bigger than 10^6 Pa, gray scale 0.4 (the *second darkest*) represents stress bigger than 10^5 Pa but smaller than 10^6 Pa, gray scale 0.6 (the *third darkest*) represents stress bigger than 10^4 Pa but smaller than 10^5 Pa, and gray scale 0.8 (the *lightest*) represents stress bigger than 300 Pa but smaller than 10^4 Pa. The *empty, white regions* do not have bonds formed yet, which are mainly filled with unreacted monomers

spectroscopy that the molecular stress distribution of polypropylene and polyethylene can be quite broad when the material is under a uniform tensile stress [67]. The bonds bearing high tensile stress tend to locate in the highly reacted regions and also near the monomer pocket-rich regions. The monomer pockets transmit no stress, but the axial force at each cross section of the network has to be equal; therefore chains near them should support more stress. Note that in this modeling, polymer–monomer interaction is not included. Further work should be

Table 3.2 Orientation of bonds bearing tension, compression, and shear, and bonds with any type of loads in the lattice at 0.7 conversion and 0.001 overall strain. The orientation was measured by the angle between each bond and the transverse direction. Each fractional bond count was calculated by the ratio of the number of corresponding bonds over the total number of bonds, and one standard deviation is listed

	Fractional bond count		
Stress	−60°	0°	60°
Tensile	0.193 ± 0.003	0.153 ± 0.003	0.193 ± 0.004
Compressive	0.107 ± 0.003	0.144 ± 0.003	0.107 ± 0.003
Shear	0.299 ± 0.002	0.297 ± 0.002	0.299 ± 0.002
Any type	0.333 ± 0.002	0.335 ± 0.002	0.333 ± 0.002

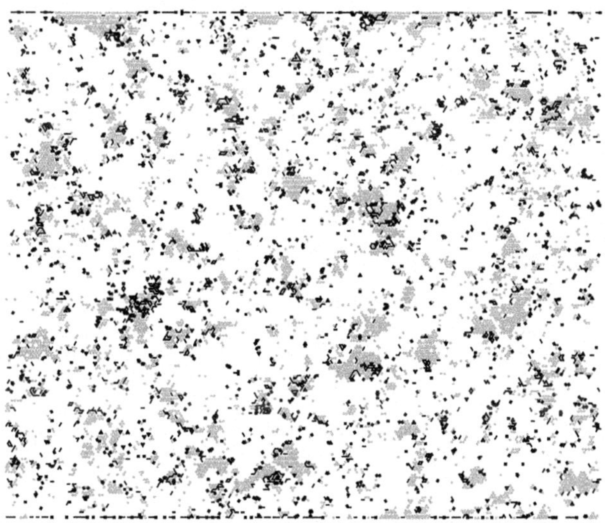

Fig. 3.7 Snapshot of elastically inactive bonds (in *black*) and monomers (in *gray*) at 0.7 conversion of functional groups in the base case at 0.001 strain. Each monomer site is indicated as a *gray dot*. The *white regions* are filled with elastically active bonds

considered to determine if the presence of monomer pockets is thermodynamically stable or not.

The bonds bearing compressive stress tend to be oriented perpendicular to the applied strain direction (0° angle from the transverse direction). The calculated orientations of bonds in compression also show that there are more of them at 0° angle from the transverse direction than at +60° or −60° (Table 3.2). This is because those bonds are squeezed when the whole network is stretched. There are more bonds bearing high tensile stress than bonds bearing high compressive stress. But most bonds bearing high compressive stress also locate in the regions where the high tensile stress develops.

The bonds bearing shear stress do not show any orientation preference. The calculated orientations of bonds in shear also show that there are about equal amount of bonds in each direction (Table 3.2). The bonds bearing high shear stress also locate in the regions where the high tensile stress develops.

Figure 3.7 shows a snapshot of elastically inactive bonds as well as monomers. Some elastically inactive bonds are scattered all over the space. These bonds mainly come from pendant short chains or small pendant loops attached to the big percolating network. Other elastically inactive bonds are near the monomer pocket regions. Those bonds form longer chains and even "clusters," both of which are either singly attached to the percolating network or isolated. As the *bond* or *connectivity* percolating network forms and becomes more widely spread and denser, formerly *site* percolated monomers gradually become isolated into pockets—a de-percolating process. In general, these isolated monomers are good swelling agent for the polymer network, leading to a clear, thermodynamically stable phase [68]. Near the monomer pockets, bonds are surrounded or partially surrounded by monomers; thus, they have no connection or loose connection with the percolating network.

In the next, the effects of initiation rate constant and primary cyclization rate constant on elastically active bonds, Young's modulus, and distributions of bond stresses are examined.

Effect of Initiation Rate Constant

Figure 3.8 shows the numbers of elastically active and inactive bonds and Young's modulus versus conversion of functional groups at different initiation rate constants k_I. Only modulus values greater than 10^5 Pa are shown because of big variance (e.g., one standard deviation) below that. Below the percolation threshold, there are no elastically active bonds—all bonds are elastically inactive, and their number is simply raised by further reaction due to formation of new bonds. The number of elastically inactive bonds peaks just one step of the reaction before the percolation threshold. At the threshold, some of the bonds in the newly formed percolating cluster suddenly become elastically active, and modulus appears. To appreciate the curves around the percolation threshold, it is necessary to recognize that because of the large variance of conversion at the percolation threshold in the small system studied, the average number of elastically active bonds and the average Young's modulus are not zero at the average conversion at the percolation threshold (i.e., 0.22, 0.26, and 0.32 conversion at $k_I = 0.1$, 1, and 10 1/s, respectively). Beyond the threshold, the number of elastically inactive bonds falls monotonically with conversion; the number of elastically active bonds and Young's modulus rise monotonically. This is because each functional group reacted builds a new bond between two lattice sites and further reaction connects more lattice sites together. The asymptotic moduli at which the reactions are stopped are 7.8×10^8, 7.6×10^8, and 5.3×10^8 Pa at $k_I = 0.1$, 1, and 10 1/s, respectively. These values are in the range of measured Young's modulus of polymers formed by difunctional monomers, which are typically in the range of 10^8 to 5×10^9 Pa [69–72].

Young's modulus increases with conversion of functional groups by orders of magnitude during the polymerization. This trend is overall consistent with experimental data. The experimental data of how elastic modulus develops as a function of conversion is yet to be obtained to compare with the modeling results here to

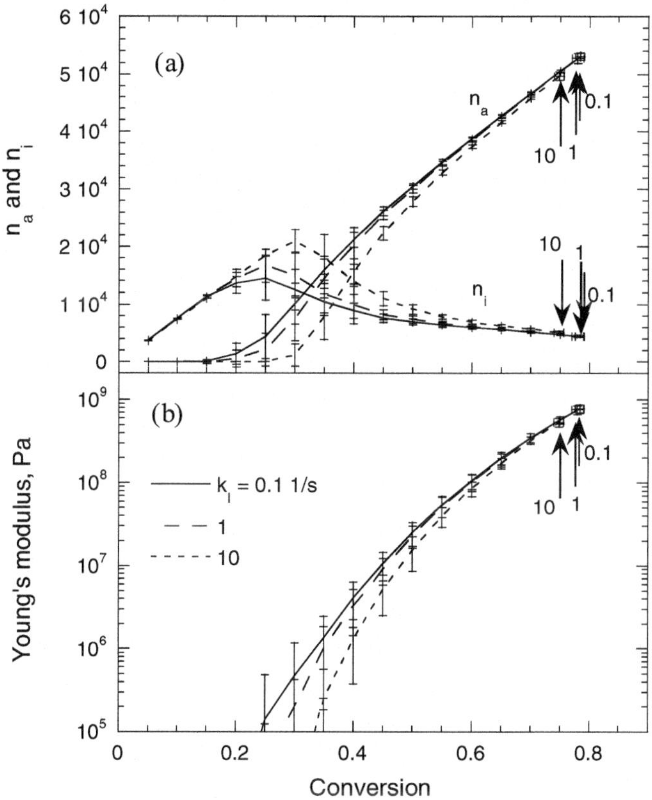

Fig. 3.8 (**a**) The numbers of elastically active bonds n_a and elastically inactive bonds n_i and (**b**) Young's modulus versus conversion of functional groups at three initiation rate constants: $k_I = 0.1$, 1, and 10 1/s. The *vertical range bars* indicate plus or minus one standard deviation of Young's modulus at a given conversion. At the end points (indicated by *arrows*), conversion is the mean asymptotic conversion, and the *horizontal range bars* there indicate plus or minus one standard deviation of the asymptotic conversion

further verify how well the rigidity percolation predicts modulus. It is possible that near the gelation point, the loosely connected chains with limited crosslinking points do not develop modulus that fast as what is predicted by the model here. Further modification of the model by replacing the joints that are non-crosslinking joints as free-rotating joints can be pursued in the future. In this way, the lattice will contain beams with rigid joints and with free-rotating joints, and a similar treatment has been proposed to model the cytoskeleton of living cells containing different types of crosslinkers [73]. Additional modification can also include changing Young's modulus and shear modulus of a beam element as a function of conversion or as a function of its environment (monomer or reacted site) considering the effect of polymer chain–monomer interaction. The simplified model used here provides the theoretically fastest development of modulus as a function of conversion, particularly close to the gelation point. At later stages of the reaction, when more and more joints become crosslinking points and truly rigid, the predicted results simulate the real system better. This is why a good agreement is obtained at the end of reaction as discussed above.

The greater the initiation rate constant, the higher the free-radical concentration in the early stages and thus the more frequent the formation of small clusters—and the smaller their sizes at a given conversion. Consequently, the percolation threshold and with it the appearance of elastically active bonds and nonzero Young's modulus are postponed to higher conversion. However, beyond the threshold, the elastically active bond count and the modulus rise more rapidly, the higher the initiation rate constant. This is presumably because each step of reaction is more likely to build a bond between a pair of previously formed clusters and thereby to strengthen the percolating network more rapidly. Interestingly, at any higher conversion, elastically active bond count and Young's modulus appear to be independent of the initiation rate constant. Yet the asymptotic conversion and with it the asymptotic elastically active bond count and Young's modulus are lower, the greater the initiation rate constant. The reason is that in the early stages the higher free-radical concentration produces more termination reactions, cutting short the mean life time of free radicals (kinetic chain length) and thereby lowering the asymptotic conversion.

What is indicated in Fig. 3.8 is that Young's modulus rises monotonically with the number of elastically active bonds. Moreover, at a given number of elastically active bonds, Young's moduli at different initiation rate constants all superpose (a different plot not shown here). Evidently, the bonding structure of the load-bearing, elastically active bonds is not sensitive to the initiation rate constant. That is, even though raising the initiation rate constant results in more and smaller clusters and in less connectivity between reacted sites, the spatial distribution of the elastically active bonds is not much affected.

In the percolation models on lattice based on random bonding rather than bond formation by reaction schemes (described by kinetic gelation models), in general, modulus is plotted against bonding probability [15, 74, 75]. In the previous kinetic gelation modeling [35–39], attention has been focused neither on the modulus of the networks formed nor on the elastically active bonds. Therefore, the rigidity percolation modeling based on the networks formed through the kinetic gelation modeling has enabled modeling modulus development of networks formed by reaction schemes. Still further work is needed to obtain more detailed information on the number of EANCs and their length distribution.

The information of elastically active bonds and elastically inactive bonds may be similar to the amounts of gel and sol formed, respectively, but two main differences exist. First, there are elastically inactive bonds (i.e., those in the pendant chains and elastically inactive cycles) in the gel. Secondly, in the calculation of the amount of gel, generally no distinction is made between the monomers with one and both functional groups reacted (apparently, monomers in the latter lead to denser networks). This perspective of elastically active bonds is more meaningful in terms of modulus development. As more bonds are built together to support stress, the system becomes stronger or more rigid.

Effect of Primary Cyclization Rate Constant

Figure 3.9 shows the numbers of elastically active and inactive bonds and Young's modulus versus conversion of functional groups as the primary cyclization rate constant rk_p is changed by the primary cyclization factor r. During the course of polymerization, r was treated as a single constant to simplify the effect of primary cyclization on network heterogeneity [40, 41]. As the primary cyclization factor is raised from 1 to 2, and to 3, more bonds are formed in primary cycles, where they

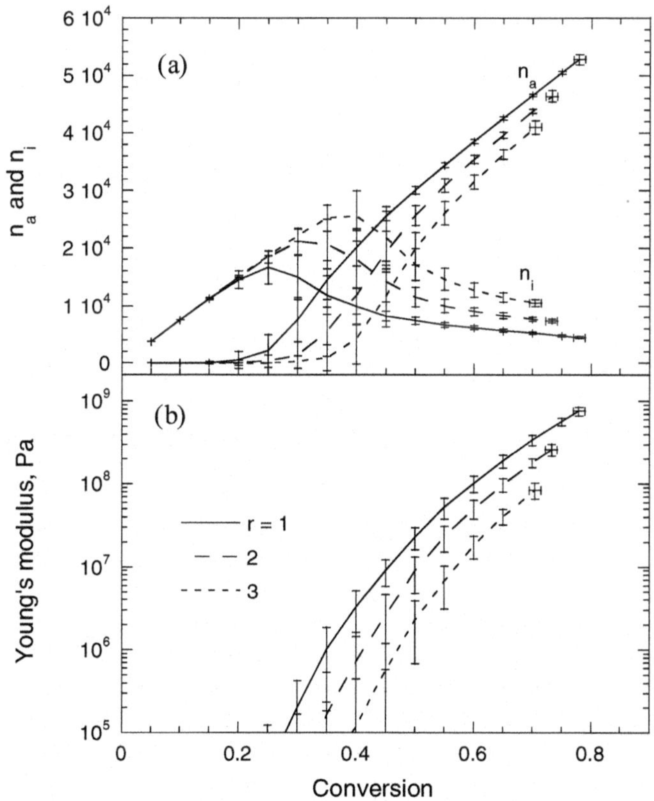

Fig. 3.9 (**a**) The numbers of elastically active bonds n_a and elastically inactive bonds n_i and (**b**) Young's modulus versus conversion of functional groups at different primary cyclization factors: $r = 1$, 2, and 3. The *vertical range bars* indicate plus or minus one standard deviation at a given conversion. At the end points of that conversion is the mean asymptotic conversion, and the *horizontal range bars* there indicate plus or minus one standard deviation of asymptotic conversion

contribute less to the overall connectivity than in backbones. Consequently, the percolation threshold, at which the elastically active bonds and Young's modulus appear, is shifted from conversion of 0.26 to 0.33 and to 0.39. Beyond the threshold at any given conversion, the elastically active bond count and Young's modulus are lower, because enhancing primary cyclization puts more of the bonds in denser and smaller clusters and leaves fewer that link up the clusters. This heterogeneity of bonding produced by the primary cyclization leads to a less rigid network at a given conversion or number of bonds. What is noteworthy is that this effect of primary cyclization is most significant at the percolation threshold regions. After that, the difference of Young's moduli with different levels of primary cyclization becomes smaller even though it is still appreciable. The decreased difference is likely due to the transformation of some of the cyclic structures originally elastically inactive into the active status (cycle activation [76, 77]). Due to more severe trapping of radicals when the primary cyclization is enhanced (higher r) [41], the final conversion is lower, leading to reduced Young's modulus at the end of the reaction. Furthermore, Fig. 3.9 indicates that responses of Young's modulus versus the number of elastically active bonds at different primary cyclization factors do not overlap. At a given number of elastically active bonds, the more severe the primary cyclization, the smaller the Young's modulus. This further indicates the heterogeneity of bonding brought on by the primary cyclization.

Conclusions

According to the rigidity percolation model, as a polymer network grows by emplacement of rigidly jointed bonds created by free-radical polymerization, load-bearing, elastically active bonds, and Young's modulus appear when the first network cluster spans the system. This is the connectivity percolation threshold and also the rigidity percolation threshold. One step of the reaction before the threshold, the elastically inactive bond count reaches a maximum. Beyond the threshold, the elastically active bond count and the modulus rise monotonically with further conversion of functional groups. More and more connected paths of bonds are formed that help resist imposed loading or deformation.

As the initiation rate constant is raised, the elastically active bond count and the Young's modulus appear at a higher conversion, but they grow faster with conversion beyond the percolation threshold, so their values attained at high conversions are not much affected by the delay of the threshold. As primary cyclization is enhanced, the elastically active bond count and Young's modulus appear at a higher conversion. The difference of Young's modulus at different levels of primary cyclization is most significant at the percolation threshold regions. After that the difference becomes smaller even though it is still appreciable. Changing the initiation rate does not affect the bonding structure of the elastically active bonds; but enhancing the primary cyclization forms less rigid bonding structure because of higher heterogeneity of network structures formed.

When the network is stretched, stress distributes nonuniformly because of the irregular distribution of bonds. Moreover, the tensile, compressive, and shear stresses that bonds bear vary by orders of magnitude. The bonds that bear high tensile, compressive, or shear stresses contribute most to the rigidity of the network. As polymerization proceeds, more of those bonds form, leading to a more rigid structure.

Rigidity percolation modeling in combination with kinetic gelation modeling turns out to be a useful tool to study how network structure affects its mechanical properties. This approach is particularly applicable to highly crosslinked, heterogeneous networks. This work builds up a framework of using this approach to study the modulus development of networks developed by free-radical homopolymerization of difunctional monomers such as di-acrylates and di-methacrylate. Further application of this modeling approach to examine other factors of free-radical crosslinking systems such as addition of solvent or plasticizer, addition of rigid, inorganic particles, and addition of higher functionality monomers should be considered.

Acknowledgments We would like to dedicate this chapter to Prof. L. E. Scriven, who was a coauthor for this chapter. We want to thank him for his dedicated and inspiring coaching of this thesis work. We would also acknowledge the support from the Center for Interfacial Engineering, an NSF Engineering Research Center at the University of Minnesota, through its Coating Process Fundamentals Program.

References

1. Kloosterboer, J.G.: Network formation by chain crosslinking photopolymerization and its applications in electronics. In: Electronic Applications. Advances in Polymer Science, vol. 84, pp. 1–61. Springer, Berlin (1988)
2. Pappas, S.P. (ed.): Radiation Curing: Science and Technology. Plenum, New York (1992)
3. Decker, C., Zahouily, K.: Light-fastness of UV absorbers in radiation-cured acrylic coatings. In: Polym. Mater. Sci. Eng. Preprints. 205 American Chemical Society national meeting, Denver, 28 March–2 April 1993
4. Fouassier, J.-P., Rabek, J.F. (eds.): Radiation Curing in Polymer Science and Technology. Practical Aspects and Applications, vol. 4. Elsevier Applied Science, London (1993)
5. Landin, D.T.: Formation-structure and structure-property relationships during vinyldivinyl copolymerization. Ph.D. thesis, University of Minnesota (1985)
6. Flory, P.J.: Principles of Polymer Chemistry. Cornell University Press, Ithaca (1953)
7. Flory, P.J.: Statistical thermodynamics of random networks. Proc. R. Soc. Lond. A. **351**, 351–380 (1976)
8. Flory, P.J.: The molecular theory of rubber elasticity. Polymer. **20**, 1317–1320 (1979)
9. Dusek, K.: Formation and structure of end-linked elastomer networks. Rubber Chem. Technol. **55**, 1–22 (1982)
10. Stein, R.S., Soni, V.K., Yang, H.E.: Optical studies of network topology. In: Lal, J., Mark, J.E. (eds.) Advances in Elastomers and Rubber Elasticity, pp. 379–392. Plenum, New York (1986)

11. Pyckhout-Hintzen, W., Westermann, S., Wischnewski, A., Monkenbusch, M., Richter, D., Straube, E., Farago, B., Lindner, P.: Direct observation of nonaffine tube deformation in strained polymer networks. Phys. Rev. Lett. **110**(19), 196002 (2013)
12. Basu, A., Wen, Q., Mao, X., Lubensky, T.C., Jaanmey, P.A., Yodh, A.G.: Nonaffine displacements in flexible polymer networks. Macromolecules. **44**(6), 1671–1679 (2011)
13. Gennes, D.: On a relation between percolation theory and the elasticity of gels. J. Physique Lett. **37**(1–2), (1976)
14. Sagis, L.M.C., Ramaekers, M., van der Linden, E.: Constitutive equations for an elastic material with anisotropic rigid particles. Phys. Rev. E. **63**(5-1), 051504/1–051505/8 (2001)
15. Feng, S., Sen, P.N.: Percolation on elastic networks: new exponent and threshold. Phys. Rev. Lett. **16**, 216–219 (1984)
16. Kantor, Y., Webman, I.: Elastic properties of random percolating systems. Phys. Rev. Lett. **52**, 1891–1894 (1984)
17. Flory, P.J.: Molecular size distribution in three dimensional polymers. I. Gelation. J. Am. Chem. Soc. **63**, 3083–3090 (1941)
18. Stockmayer, W.H.: Theory of molecular size distribution and gel formation in branched-chain polymers. J. Chem. Phys. **11**, 45–55 (1943)
19. Stockmayer, W.H.: Theory of molecular size distribution and gel formation in branched-chain polymers. II. General cross-linking. J. Chem. Phys. **12**, 125–131 (1944)
20. Gordon, M.: Good's theory of cascade processes applied to the statistics of polymer distributions. Proc. Roy. Soc. London. **A268**(240–256), (1962)
21. Dobson, G.R., Gordon, M.: Theory of branching processes and statistics of rubber elasticity. J. Chem. Phys. **43**, 705–713 (1965)
22. Macosko, C.W., Miller, D.R.: A new derivation of average molecular weights of nonlinear polymers. Macromolecules. **9**, 199–206 (1976)
23. Miller, D.R., Macosko, C.W.: A new derivation of post gel properties of network polymers. Macromolecules. **9**, 206–211 (1976)
24. Tobita, H., Hamielec, A.: Modeling of network formation in free radical polymerization. Macromolecules. **22**, 3098–3105 (1989)
25. Somvarsky, J., Dusek, K.: Kinetic Monte-Carlo simulation of network formation, I. Simulation method. Polym. Bull. **33**, 369–376 (1994)
26. Somvarsky, J., Dusek, K.: Kinetic Monte-Carlo simulation of network formation, II. Effect of system size. Polym. Bull. **33**, 377–384 (1994)
27. Elliott, J.E., Bowman, C.N.: Effect of primary cyclization on free radical polymerization kinetics: modeling approach. Macromolecules. **35**, 7125–7131 (2002)
28. Dušek, K., Šomvársky, J.: Modelling of ring-free chain crosslinking copolymerization. Polym. Int. **44**, 225–236 (1997)
29. Dušek, K., Dušková-Smrčková, M.: Modeling of polymer network formation from preformed precursors. Macromol. React. Eng. **6**, 426–445 (2012)
30. Family F., Landau, D.P.: Kinetics of aggregation and gelation. Proc. Int. Topical Conference on Kinetics of Aggregation and Gelation, April 2–4, 1984, Athens, Georgia, USA. Elsevier Science, Oxford (1984). ISBN: 9780444596581 0444596585
31. Šomvársky, J., Dušek, K., Smrčková, M.: Kinetic modelling of network formation: size-dependent static effect. Comput. Theor. Polym. Sci. **8**, 201–208 (1998)
32. Manneville, P., de Seze, L.: Percolation and gelation by additive polymerization. In: Dora, J. D., Demongeot, J., Lacolle, B. (eds.) Numerical Methods in the Study of Critical Phenomena. Springer Series in Synergetics, vol. 9, pp. 116–124. Springer, Berlin (1981)
33. Herrmann, H., Landau, D., Stauffer, D.: New university class for kinetic gelation. Phys. Rev. Lett. **49**, 412–415 (1982)
34. Bansil, R., Herrmann, H.J., Stauffer, D.: Computer simulation of kinetics of gelation by addition polymerization in a solvent. Macromolecules. **17**, 998–1004 (1984)
35. Boots, H., Pandey, R.B.: Kinetic gelation model: qualitative percolation study of free-radical cross-linking polymerization. Polym. Bull. **11**, 415–420 (1984)

36. Simon, G.P., Allen, P.E.E., Bennett, D.J., Williams, D.R.G., Williams, E.H.: Nature of residual unsaturation during cure of dimethacrylates examined by ^{13}C NMR and simulation using a kinetic gelation model. Macromolecules. **22**, 3555–3561 (1989)
37. Bowman, C.N., Peppas, N.A.: A kinetic gelation method for the simulation of free-radical polymerizations. Chem. Eng. Sci. **47**, 1411–1419 (1992)
38. Anseth, K.S., Bowman, C.N.: Kinetic gelation model predictions of crosslinked polymer network microstructure. Chem. Eng. Sci. **49**, 2207–2217 (1994)
39. Chiu, Y.Y., Lee, L.J.: Microgel formation in the free radical crosslinking polymerization of ethylene glycol dimethacrylate (EGDMA). II. Simulation. J. Polym. Sci. A Polym. Chem. **33**, 269–283 (1995)
40. Wen, M., Scriven, L.E., McCormick, A.V.: Kinetic gelation modeling: structural inhomogeneity during cross-linking polymerization. Macromolecules. **36**(11), 4140–4150 (2003)
41. Wen, M., Scriven, L.E., McCormick, A.V.: Kinetic gelation modeling: kinetics of cross-linking polymerization. Macromolecules. **36**(11), 4151–4159 (2003)
42. Lattuada, M., Del Gado, E., Abete, T., De Arcangelis, L., Lazzari, S., Diederich, V., Storti, G., Morbidelli, M.: Kinetics of free-radical cross-linking polymerization: comparative experimental and numerical study. Macromolecules. **46**(15), 5831–5841 (2013)
43. Lang, M., John, A., Sommer, J.-U.: Model simulations on network formation and swelling as obtained from cross-linking co-polymerization reactions. Polymer. **82**, 138–155 (2016)
44. Hutchison, J.B., Anseth, K.S.: Off-lattice simulation of multifunctional monomer polymerizations: effects of monomer mobility, structure, and functionality on structural evolution at low conversion. J. Phys. Chem. B. **108**(30), 11097–11104 (2004)
45. Hamzehlou, S., Reyes, Y., Leiza, J.R.: A new insight into the formation of polymer networks: a kinetic Monte Carlo simulation of the cross-linking polymerization of S/DVB. Macromolecules. **46**(22), 9064–9073 (2013)
46. Dušek, K., Dušková-Smrčková, M.: Network structure formation during crosslinking of organic coating systems. Prog. Polym. Sci. **25**, 1215–1260 (2000)
47. Dušek, K., Choukourov, A., Duškova-Smrčkova, M., Biedermann, H.: Constrained swelling of polymer networks: characterization of vapor-deposited cross-linked polymer thin films. Macromolecules. **47**, 4417–4427 (2014)
48. Thorpe, M.F.: Continuous deformations in random networks. J. Non-Cryst. Solids. **57**, 355–370 (1983)
49. Day, A.R., Tremblay, R.R., Tremblay, A.M.S.: Rigid backbones: a new geometry for percolation. Phys. Rev. Lett. **56**, 2501–2504 (1986)
50. Hamlen, R.C.: Paper structure, mechanics, and permeability: computer-aided modeling. Ph.D. thesis, University of Minnesota (1991)
51. Sahimi, M.: Applications of Percolation Theory. Taylor & Francis, Bristol (1994)
52. Termonia, Y., Meakin, P., Smith, P.: Theoretical study of the influence of the molecular weight on the maximum tensile strength of polymer fibers. Macromolecules. **18**, 2246–2252 (1985)
53. Termonia, Y.: Kinetic model for tensile deformation of polymers. 2. Effect of entanglement spacing. Macromolecules. **21**, 2184–2189 (1987)
54. Termonia, Y.: Molecular model for the mechanical properties of elastomers. 1. Network formation and role of entanglements. Macromolecules. **22**, 3633–3638 (1989)
55. Feng, S., Sahimi, M.: Position-space renormalization for elastic percolation networks with bond-bending forces. Phys. Rev. B. **31**, 1671–1673 (1985)
56. Thorpe, M.F.: Bulk and surface floppy modes. J. Non-Cryst. Solids. **182**(1,2), 135–142 (1995)
57. Pathak, P.: Porous media: structure, strength, and transport. Ph.D. thesis, University of Minnesota (1981)
58. Sahimi, M.: Brittle fracture in disordered media: from reservoir rocks to composite solids. Phys. A. **186**, 160–182 (1992)
59. Broedersz, C.P., Mao, X., Lubensky, T.C., MacKintosh, F.C.: Criticality and isostaticity in fibre networks. Nat. Phys. **7**, 983–988 (2011)

60. Wool, R.P.: Rigidity percolation model of polymer fracture. J. Polym. Sci. B Polym. Phys. **43**, 168–183 (2005)
61. Livesley, R.K.: Matrix Methods of Structural Analysis. Pergamon Press, New York (1975)
62. Eisenstat, S.C., Gursky, M.C., Schultz, M.H., Sherman, A.H.: Yale sparse matrix package I: the symmetric codes. Res. Rept. No. 112, Yale University, Dept. of Computer Science (1976)
63. Ward, I.M.: Mechanical Properties of Solid Polymers, 2nd edn. Wiley, New York (1983)
64. Press, W.H., Teukolsky, S.A., Vetterling, W.T., Flannery, B.P.: Numerical Recipes in Fortran: The Art of Scientific Computing, 2nd edn. Cambridge University Press, New York (1992)
65. Cates, M.E., Wittmer, J.P., Bouchaud, J.P., Claudin, P.: Jamming, force chains, and fragile matter. Phys. Rev. Lett. **81**, 1841–1844 (1998)
66. Cates, M.E., Wittmer, J.P., Bouchaud, J.P., Claudin, P.: Jamming and stress propagation in particulate matter. Phys. A. **263**, 354–361 (1999)
67. Wool, R.P., Bretzlaff, R.S., Li, B.Y., Wang, C.H., Boyd, R.H.: Infrared and Raman spectroscopy of stressed polyethylene. J. Polym. Sci. B Polym. Phys. **24**, 1039–1066 (1986)
68. Dusek, K. In: Haward, R.N. (ed.) Developments in Polymerisation-3, Chapter 4, pp. 143–206. Applied Science, London (1982)
69. Wilson, T.W., Turner, D.T.: Characterization of polydimethacrylates and their composites by dynamic mechanical analysis. J. Dent. Res. **66**, 1032–1035 (1987)
70. Kannurpatti, A.R., Anderson, K.J., Anseth, J.W., Bowman, C.N.: Use of "living" radical polymerizations to study the structural evolution and properties of highly crosslinked polymer networks. J. Polym. Sci. B Polym. Phys. **35**, 2297–2307 (1997)
71. Tobolsky, A.V., Katz, D., Takahashi, M., Schaffhauser, R.: Rubber elasticity in highly crosslinked systems: crosslinked styrene, methyl methacrylate, ethyl acrylate, and octyl acrylate. J. Polym. Sci. A Polym. Chem. **2**, 2749–2758 (1964)
72. Nielsen, L.E.: Cross-linking—effect on physical properties of polymers. J. Macromol. Sci. Polym. Rev. **C3**, 69–103 (1969)
73. Das, M., Quint, D.A., Schwarz, J.M.: Redundancy and cooperativity in the mechanics of compositely crosslinked filamentous networks. PLoS One **7**(5), e35939, 1–11 (2012)
74. Feng, S., Thorpe, M.F.: Effective-medium theory of percolation on central-force elastic networks. Phys. Rev. B. **31**, 276–280 (1985)
75. Thorpe, M.F., Garboczi, E.J.: Site percolation on central-force elastic networks. Phys. Rev. B. **35**, 8579–8586 (1987)
76. Dušek, K., Gordon, M., Ross-Murphy, S.B.: The graph-like state of matter 10. Cyclization and concentration of elastically active network chains in polymer networks. Macromolecules. **11**, 236–248 (1978)
77. Dušek, K., Vojta, V.: Concentration of elastically active network chains and cyclization in networks obtained by alternating stepwise polyaddition. Br. Polym. J. **9**, 164–171 (1977)

Part II
Coating Film Formation and Properties

Chapter 4
Rheology Measurement for Automotive Coatings

Michael R. Koerner

Introduction

Rheology control has long been considered a key enabler for coating application performance. Paint formulation design intent should include strategies to give the paint good stability, material handling, atomization, and coalescence during film formation. It is also understood that pigment, binder resins, and solvent all play key roles to define the morphology/rheology profile of a paint formulation. Selection of a rheology control agent (RCA) should not be an afterthought but rather designed at an early stage of formula development.

But which rheology profile should be used as an ideal target? Which aspect of rheology is most relevant to appearance or metallic flake control? What are the trade-offs when atomization is favored over flow and leveling control? How can we model and experimentally track the rheology of paint from application through film formation? In this chapter, we address these questions to outline methods currently used by coating formulators.

General reference books on coatings formulation [1, 2] make cursory summary of viscosity measurement and how it relates to coating performance. Historically, formulators limited viscosity measurement to a single point reading, such as the time for paint to drain from an efflux cup. This might have been acceptable for low solids polymer solutions where paint is reduced (i.e., diluted with solvent) to a target "spray viscosity." Ingredients in modern coatings have much stronger interactions which require consideration of viscosity under various conditions. Patton [3], Wicks et al. [4], and Schoff [5] provide rheology overviews with consideration to how viscosity changes at various conditions important to the painting process. More comprehensive reviews are provided by Eley [6, 7] and Mezger [8] to cover

M.R. Koerner (✉)
Axalta Coating Systems, Coatings Technology Center, Wilmington, DE, USA
e-mail: michael.r.koerner@axaltacs.com

M. Wen, K. Dušek (eds.), *Protective Coatings*, DOI 10.1007/978-3-319-51627-1_4

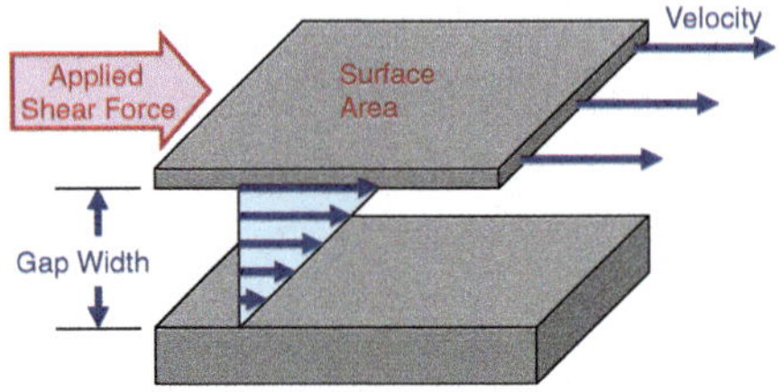

$$\text{Shear Stress} = \frac{\text{Applied Force}}{\text{Surface Area}} \qquad \tau\ [=]\ \frac{\text{N}}{\text{m}^2} = \text{Pa}$$

$$\text{Shear Rate} = \frac{\text{Velocity}}{\text{Gap Width}} \qquad \dot{\gamma}\ [=]\ \frac{\text{m/s}}{\text{m}} = \text{sec}^{-1}$$

$$\text{Viscosity} = \frac{\text{Shear Stress}}{\text{Shear Rate}} \qquad \mu\ [=]\ \frac{\text{Pa}}{\text{1/sec}} = \text{Pa·s}$$

$$1\ \text{cP} = 1\ \text{mPa·s}$$

Fig. 4.1 Geometry of simple shear flow where fluid is bound between parallel plates. The lower plate is held fixed while shear motion is applied to the upper plate. Relevant measures are given in relation to measurement units

typical coating processes and test methods. We start by summarizing this information with focus on methods most appropriate to automotive coating application. More specifically, the focus here is to spray applied liquid coatings (primer, colorcoat, or clearcoat) that are solvent-based or water-based in formula design.

Viscosity in Simple Shear

A fluid's viscosity describes its resistance to flow, and in turn that resistance should arise from the paint's morphology and association forces between ingredients. The study of paint rheology then should be concerned with experimentally applied flow in relation to commercially important processes. While in this controlled flow field, a rheologist should be concerned with the source of material resistance giving rise to viscosity.

It is most common to measure fluid behavior using material placed in simple shear geometry, namely, where force is applied in a direction transverse to the gradient of resulting shear. Figure 4.1 conceptually shows a fluid between parallel plates where the lower plate is stationary and an applied shear stress causes the upper plate to move at a measured velocity. This results in a simple shear flow field where fluid material elements adjacent to the lower plate remain fixed (bottom), move with the upper plate (top) or at some intermediate speed depending on the nature of the fluid. For rheology measure of pure fluids, we refer to a "no-slip" condition where material boundary conditions are set by the motion of the test geometry. In practice, however, we know that for colloidal dispersed paints, it is very hard to describe a compositionally uniform material element. Immediately adjacent to the geometry, we suspect that solvents will lubricate the geometry walls where dispersed particles have been depleted.

Increased shear rate can be achieved by increasing the upper plate velocity or by decreasing the gap width. On a material level, one should think of shear rate not as the fluid's speed but rather as the degree of tearing as one material element is forced past another.

If shear stress is the force needed to achieve a target rate scaled by the applied area, then viscosity (μ or η) can be defined as the material property that is the proportionality between shear stress (τ) and shear rate $\dot{\gamma}$:

$$\tau = \mu \dot{\gamma} \tag{4.1}$$

$$\tau = \eta(\dot{\gamma})\dot{\gamma} \tag{4.2}$$

Note that it's common practice to denote viscosity as μ only when it describes the constant proportionality between stress and shear rate. In this case, the material is said to be Newtonian. More generally for non-Newtonian fluids, we expect viscosity to depend on shear rate or shear history; we'll then denote viscosity with η. S.I. units for viscosity reflect this proportion, namely, Pascal (stress) divided by inverse seconds (shear rate) or Pascal-seconds. A more commonly used viscosity unit is Poise (from CGS units) or dyne·s/cm^2. Note the conversion rate that 1 cP (centipoise) is equivalent to 1 mPa·s.

Since viscosity is a local material property, we expect it to vary as a function of local conditions, i.e., temperature, shear rate, and recent history of shear exposure. For most coatings and other important fluids, we expect the viscosity to vary as a function of shear intensity. Since our intent is to describe the fluid as it applies to a certain process, we should consider a measurement that applies to the corresponding process.

Figure 4.2 provides an estimate of various processes important to coatings handling. When discussing the shear intensity of a certain process, one can expect

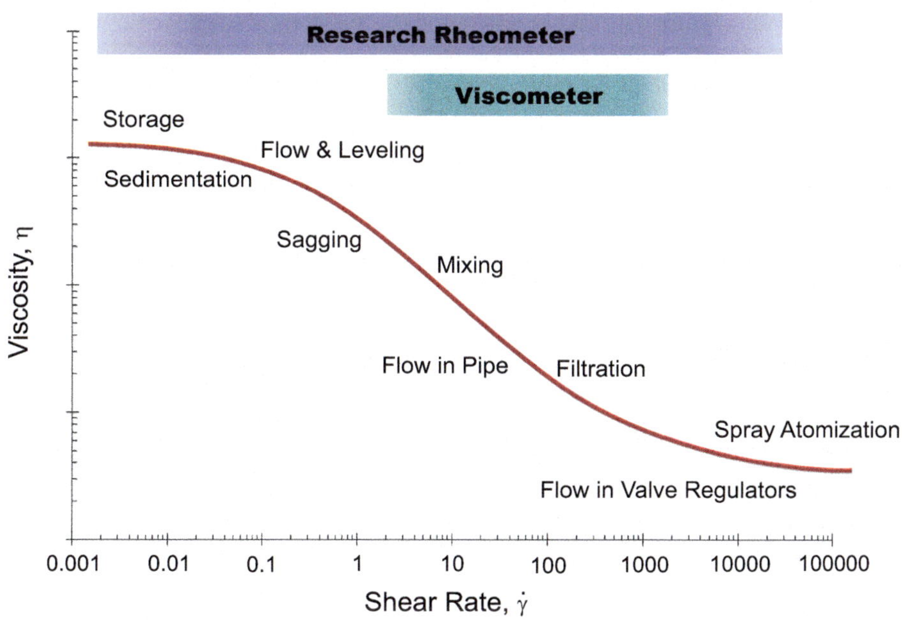

Fig. 4.2 Viscosity variation over a wide range of shear rates as related to common processes in coatings industry. Most liquid coatings exhibit a shear thinning profile with higher viscosity at low shear rates

variations over more than an order of magnitude in shear rate. For example, high flow rates through a back-pressure regulator might achieve 10^4 s^{-1}, while "low-flow" regulators could impart 10^3 s^{-1} of applied shear. Here we suggest that paint viscosity could also vary by orders of magnitude over such a broad range of shear rates. In general this is true, but certainly not in every case.

Someone interested in assessing the paint's tendency for sags should measure and control the low-shear viscosity (LSV), while viscosity for spray atomization should be determined at high shear rates (HSV).

Most commercial viscometers used for quality assurance testing have a limited range of applied shear and viscosity sensitivity. Various test geometries and shear rates are used to collect data in a range that is suitable to both the process of interest and the material being studied. To truly characterize the shear dependence of viscosity, one needs a research grade rheometer where shear stress and shear rate can be monitored over many orders of magnitude. Even then, there are measurement challenges to characterize very low or very high shear rate behavior.

Viscometers Used in Coatings Industry

Practical measurement of viscosity can be achieved by moving the simple shear geometry (Fig. 4.1) around a rotating axis. For concentric cylinder geometry, a paint sample is placed in the narrow wall gap between a stationary cup and a rotating cylindrical bob. The viscometer bob is driven to a known torque (shear stress), and the rotation rate can be converted to a resultant shear rate. Conversely if a fixed drive motor is set to a target rotation rate, a torque transducer can provide the fluid resistance. In either case, the viscosity is calculated as the proportionality between shear stress and rate.

Figure 4.3 shows some viscometers commonly used in the automotive paint industry. Figure 4.3a shows the concentric cylinder (cup and bob) configuration.

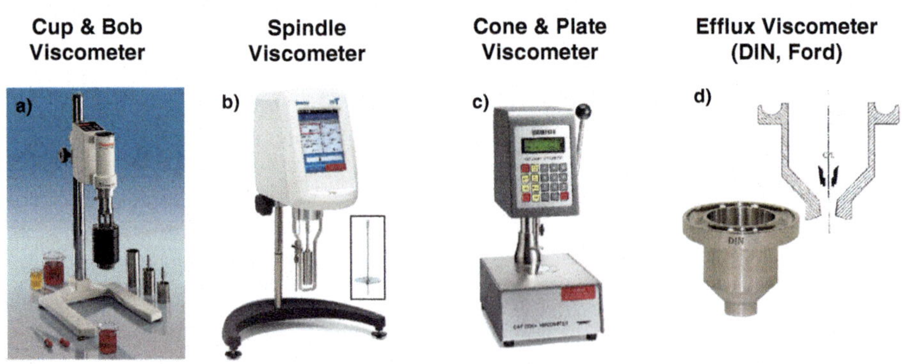

Fig. 4.3 Viscometers commonly used in automotive coatings industry. Photo credits: (**a**) Viscotester 550 from Thermo Fisher Scientific, (**b**) DV2T spindle viscometer and (**c**) Cap 2000+viscometer from Brookfield Engineering and (**d**) DIN efflux viscometer from Paul N. Gardner Company

A spinning cylindrical bob rotates in a fixed (temperature controlled) cup, where the test fluid is held between the concentric cylinders. Figure 4.3b uses a motor to support a rotating spindle. When the spindle is immersed in a sample, the paint's resistance is recorded as proportional to viscosity. Figure 4.3c shows another configuration where a few drops of paint are placed between a stationary plate and a spinning disk. Most commonly, the spinning upper plate is actually a shallow angle cone, so that all of the paint experiences a uniform shear rate. Finally, Fig. 4.3d shows an efflux cup viscometer. The cup is filled with paint and then timed as it drains under the force of gravity. Shear is concentrated in the small exit tube at the bottom of the cup, while the applied stress is a result of the liquid gravity head driving the flow.

Structures Causing Non-Newtonian Behavior

Figure 4.2 suggests that paints tend to be non-Newtonian or more specifically shear thinning. For most practical paints, we expect the fluid resistance to be dependent on the level of stress or the recent history of material deforming under applied stress. It is worth noting some morphology examples that support this idea. Figure 4.4 depicts typical paint ingredients as they might exist at rest vs. how they might deform under shear.

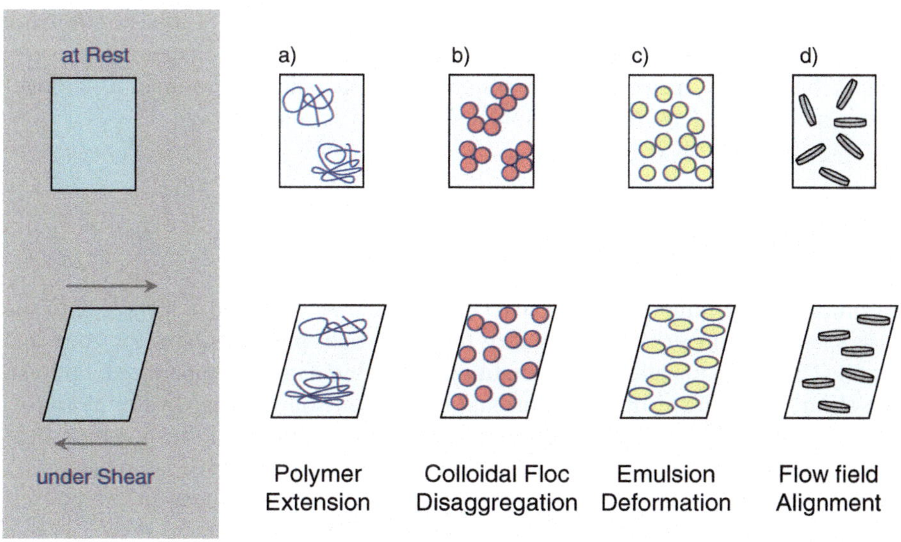

Fig. 4.4 Models of common paint ingredients that cause pseudoplastic or shear thinning behavior: (**a**) stretching of polymer chains dissolved in solution, (**b**) disaggregation of colloidal scale particles, (**c**) deformation of liquid–liquid emulsion, and (**d**) flow field alignment of microscopic flake particles

Depending on their solubility in surrounding solvent, polymer chains are configured in random coils that occupy a certain hydrodynamic volume. As shear is applied to the solution, the polymer chains must navigate around one another causing some resistance to flow. But under higher shear rates, we expect some level of affine deformation where chains align with the flow field. As deformed chains more easily pass one another, the overall resistance is reduced, and we measure a lower high-shear viscosity (HSV).

On a colloidal scale, in a liquid–liquid emulsion, we might also observe shear-induced deformation resulting in shear thinning or pseudoplastic behavior. A plausible mechanism is that under shear, there is a high droplet collision frequency creating more opportunities for coalescence or breakup of droplets. One doesn't need oil-in-water emulsions to observe this reorganization of phase boundaries in modern paint formulations.

Another common colloidal scale morphology deals with the interaction of primary particles to build weakly associated flocs or aggregates (Fig. 4.4b). Through a combination of steric and electrostatic interactions, conditions might favor some degree of flocculation. In the case where flocs form loose or open networks, we observe significant at-rest structure. Under shear conditions, we expect more disrupted structures resulting in smaller hydrodynamic volume and lower viscosity.

Even at larger macroscale, coating ingredients can give rise to non-Newtonian rheology. Rigid rod- or platelet-form materials can occupy a random configuration at rest. But under sustained shear, we expect these materials to align with flow. It's also noteworthy that under shear, say flow in a pipe, macroscale pigment will align with flow and become depleted immediately adjacent to the wall's non-flowing surface. Rheologists might observe apparent wall slip when doing careful measurements, but strictly speaking, in these conditions the paint is no longer a homogeneous material, at least not on colloidal scales and smaller.

Viscometry Test Practices

In order to determine the flow curve of a paint, we first load a sample into the viscometer/rheometer's test geometry (cup and bob, parallel plates, or cone and plate). The sample is allowed to condition by setting a target temperature, optional pre-shear, and some resting equilibrium time. The instrument then begins a ramp of increasing shear rates. At each level of applied shear rate, the corresponding shear stress is recorded. Alternatively for a stress-controlled rheometer, a ramp of stresses may be applied, where corresponding shear rates are recorded. In either case from Eq. 4.2, the viscosity is the proportionality between shear stress (τ) and shear rate ($\dot{\gamma}$).

Figure 4.5 shows that data may be graphed as shear stress vs. shear rate or in its more common format, viscosity vs. shear rate. Based on the relation of Eq. 4.2, we recognize that the viscosity for each shear rate is calculated as the proportionality of

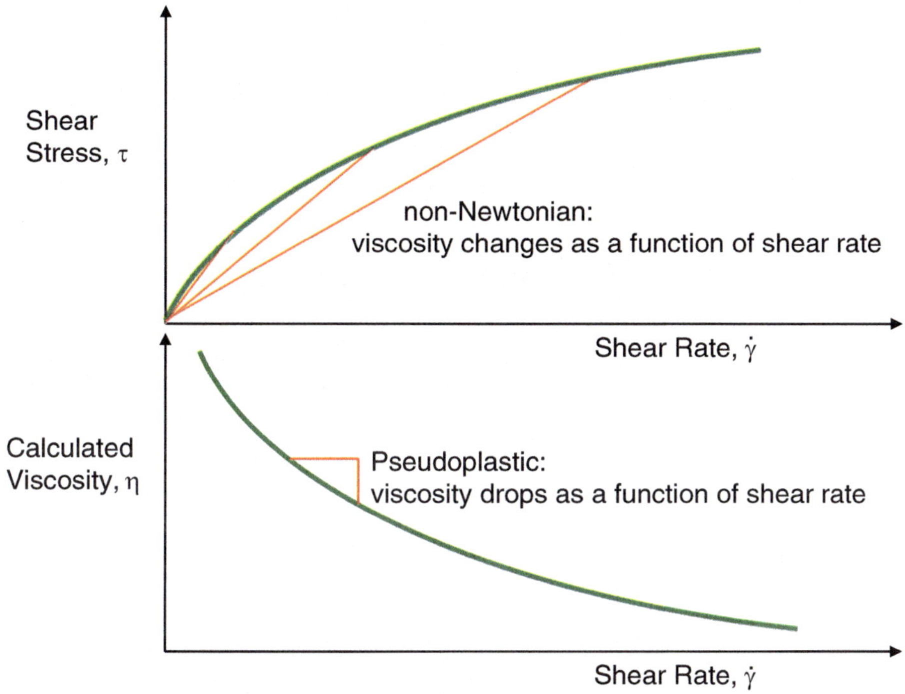

Fig. 4.5 Experiment scheme for running viscometry flow curve on a rate-controlled viscometer/ rheometer. The instrument records shear stress and shear rate over the desired testing range. Viscosity is derived from these collected results

shear stress to shear rate. It is typical for paints to exhibit a pseudoplastic or shear thinning flow curve characteristic. On a research rheometer, we are able to measure stresses and rates over several orders of magnitude. In this case, it is preferred to show flow curve results on a log–log scale.

More generally, we can think about the shear stress vs. shear rate response of various types of fluids. We've already described the difference between viscosity invariant (Newtonian) and shear thinning (pseudoplastic) materials. For many coatings, we notice a power law relationship between shear stress and shear rate:

$$\tau = K\dot{\gamma}^{n} \tag{4.3}$$

$$\eta = K\dot{\gamma}^{n-1} \tag{4.4}$$

where K is the fluid consistency and n is the power law (or shear thinning) index.

If the material thickens with additional shear, it is said to be dilatant (Fig. 4.6). This behavior might be the result of structures that build only when collision frequency is increased.

More common for coating samples is when a fluid at rest builds large-scale, weak structures. When very low stress is applied, the material deforms but does not flow. Like a pudding, the material exhibits a solid-like behavior. With added

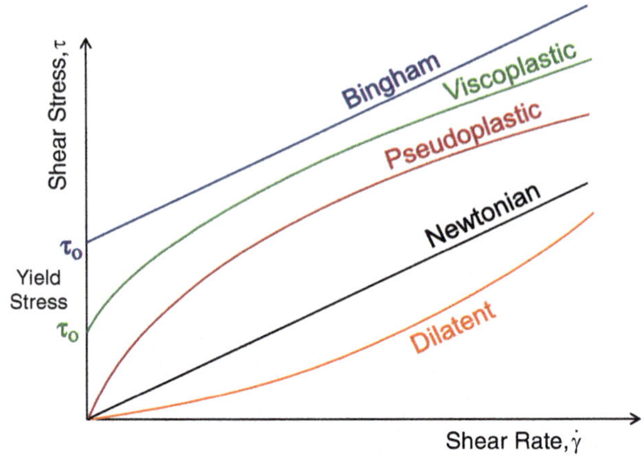

Fig. 4.6 Classification of common fluids based on time-invariant flow. Inclusion of material yield stress and/or rate-dependent variant can provide a variety of constitutive models

applied stress, the internal structure breaks, and the material can flow in liquid-like fashion. The yield point is defined by the critical stress above which we see flow. If the liquid-like behavior is Newtonian, then we call the material a Bingham fluid:

$$\tau - \tau_0 = \mu_0 \dot{\gamma} \tag{4.5}$$

$$\eta = \frac{\tau_0}{\dot{\gamma}} + \mu_0 \tag{4.6}$$

where τ_0 is the yield stress and μ_0 is the Newtonian viscosity after yield.

If the material has both a yield stress and exhibits shear thinning behavior, we call it viscoplastic. The Herschel–Bulkley model is an example of constitutive model that combines a power law fluid behavior with a yield stress. It has proven useful to describe the rheology of many coatings:

$$\tau - \tau_0 = K\dot{\gamma}^n \tag{4.7}$$

$$\eta = \frac{\tau_0}{\dot{\gamma}} + K\dot{\gamma}^{n-1} \tag{4.8}$$

There are many other constitutive models that have been used to model pseudoplastic and viscoplastic materials. Simple models mentioned here are useful over a limited range of shear rates. In order to build models to be applied over a broader shear range, more complex models use additional fitting parameters. It becomes challenging to relate these extra parameters to physical characteristics of the paint's structure and morphology.

With constitutive Eqs. (4.5) through (4.8), we discussed the potential for paint materials with a yield stress. This is most often observed with highly filled paints with strong rheological interactions. Especially after storage, the paint can have an elastic mass that appears like a pudding or ketchup. With some mild mixing, the gel structure breaks, and the material exhibits more fluid-like behavior. On a stress-controlled rheometer, it is rather straightforward to measure this yield stress, namely, the critical stress above which the material begins to flow (Fig. 4.7).

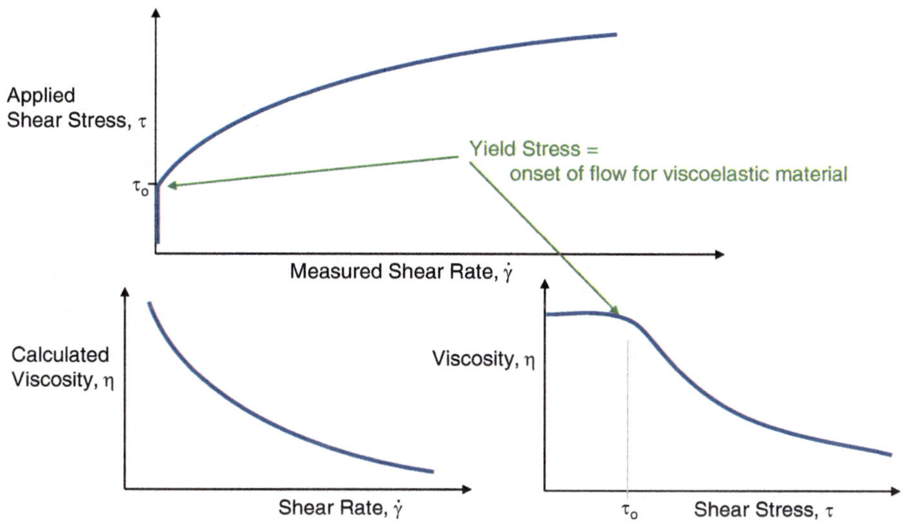

Fig. 4.7 Stress sweep experiment for static yield stress determination. The rheometer measures rates over a range of applied stresses. Calculated viscosity may be shown as a function of either shear rate or shear stress

In the stress sweep experiment, the sample is allowed some equilibrium time in the rheometer's test geometry. Then a series (usually a logarithmic progression) of shear stresses is applied. At very low stress, the material yields elastically as a soft solid but does not flow. Only after the yield stress is exceeded can we measure a shear rate and corresponding viscosity. If the rheometer reports a viscosity value prior to the yield stress, then it's usually an artifact of transient stress dissipation between applied stress steps.

In automotive coating applications, we associate paints with significant yield stress to be difficult to pump. In particular, start-up flow in dead-end legs of a fluid delivery system can be challenging. Conversely, some yield stress can be very beneficial to limit vertical sag and optimize appearance workability. Ideally, the paint's viscosity profile should be suitable for easy fluid delivery, provide good atomization application, and then on the work provide the right balance between sag resistance and good flow/leveling.

In Fig. 4.5, we considered taking viscosity readings for a series of rates from low- to high-shear intensity. As rheological associations are broken, overall structure is reduced, and we typically measure lower viscosity at these high shear rates. We might also consider extending this experiment by maintaining the high-shear treatment for some period, followed by another shear ramp from high to low shear rate. In Fig. 4.8, we might expect that the high-shear exposure causes some additional material structure breakdown. Furthermore when shear rates are reduced, we commonly observe that the measured viscosity is lower than the corresponding values on the curve from the earlier portion of the test. The material's thixotropy is

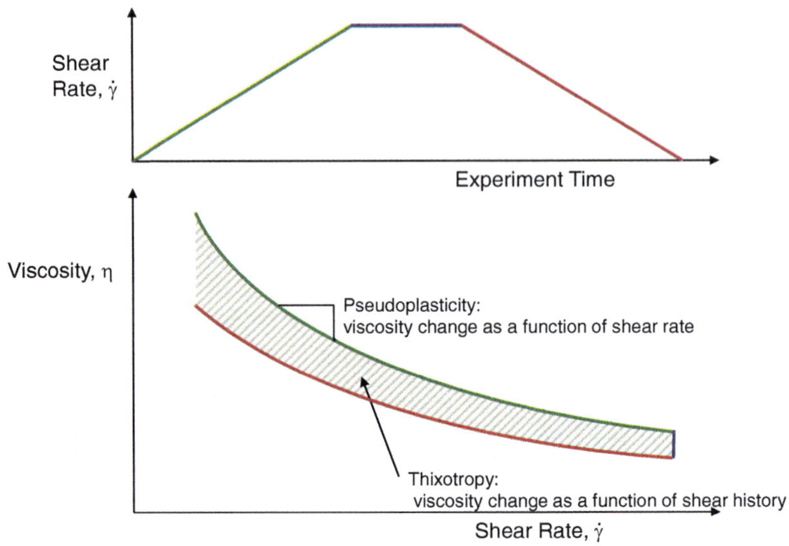

Fig. 4.8 Up-hold-down shear ramp to measure rheological thixotropic loop. The first graph shows target shear rate profile. The second graph shows expected viscosity profile of paints exhibiting both pseudoplastic and thixotropic behavior

the loss of structure observed as a result of the recent shear history of the sample. If the material is allowed to rest and rebuild its structure, we would expect to be able to remeasure the same flow curve behavior as from the initial shear ramp.

Many research rheometers provide software that will integrate the area between up and down flow curves to quantify the degree of thixotropy. One should, however, note that this result is strongly dependent on the specification of the flow curve parameters. But perhaps more to the point, if thixotropy describes the viscosity response to the paint's shear history, why not impose a shear history that is more directly related to paint application?

In a viscosity recovery test (Fig. 4.9), we typically measure a low-shear viscosity (LSV) at steady state, then impose a high-shear event to intentionally break down some of the paint's rheological structure, and then we switch back to measuring steady-state LSV. From this data we can ascertain the fraction of structure that is lost as a result of the high-shear event. We can also look at the time constant associated with the viscosity recovery. Both values have proven useful to predict paint viscosity performance in production environments.

Many important paint processes can be modeled using continuous viscometry with a simple shear geometry. In Fig. 4.2, we observe that viscosity measurement can be selected to elucidate behavior for a particular process. And in our most recent example, we can consider transient viscosity behavior when shear is removed from the system. For example, paint is sprayed at a high shear rate but then lands on an object where very low shear is experienced (i.e., flow-out and leveling). In this way, we are using the high-shear step for paint conditioning prior to tracking the LSV recovery.

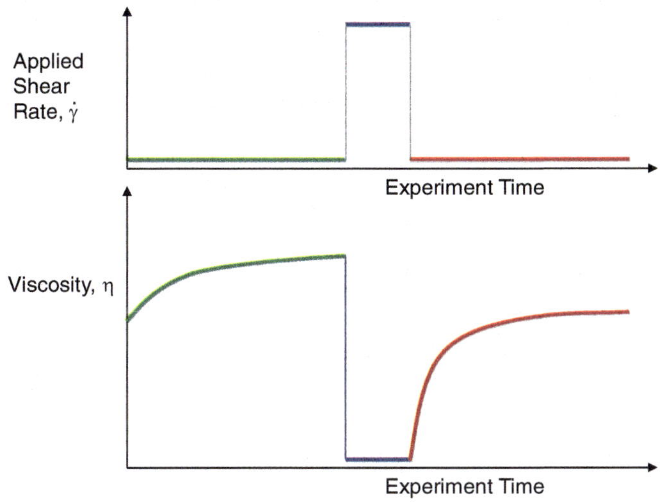

Fig. 4.9 Viscosity recovery testing at low shear, high shear, and low shear to study the effect of paint thixotropy. The first graph shows target shear rate profile. The second graph shows expected viscosity profile with focus on LSV recovery

An alternative to tracking the paint process is to consider fundamental rheological structures within the paint. We frequently see that rheological structures breakdown at a threshold stress and not necessarily at a particular shear rate. Since rheometers are intrinsically stress and rate measuring devices, it is quite arbitrary to show the flow curve as viscosity vs. shear rate. Figure 4.10 shows some test results of automotive waterborne basecoats.

The left-side graph is the traditional representation of the flow curve, namely, viscosity vs. shear rate. Near 1000 s^{-1}, these paints have similar viscosity. But we can also see that low-shear viscosity (LSV) is quite different. In fact, these paints contain various RCAs and resins with different particle–particle interactions. Figure 4.10b (viscosity vs. shear stress) provides additional information about threshold stress where these rheological structures break down. In the case of the Curve D, we observe a very distinct transition at about 7 Pa of applied stress.

Finally, we should consider the influence of temperature on viscosity. For pure materials, we expect viscosity thinning as temperature is increased. But for fully formulated paints and ingredient solutions, we often observe different results. Figure 4.11a shows a series of flow curves collected over a range of temperatures. In this example, we observe pseudoplastic (shear thinning) behavior at low temperatures and then more Newtonian behavior as temperature is increased. For selected shear rates, Fig. 4.11b shows the relationship of viscosity to temperature:

$$\eta(T) = A\, e^{\frac{-E_a}{RT}} \tag{4.9}$$

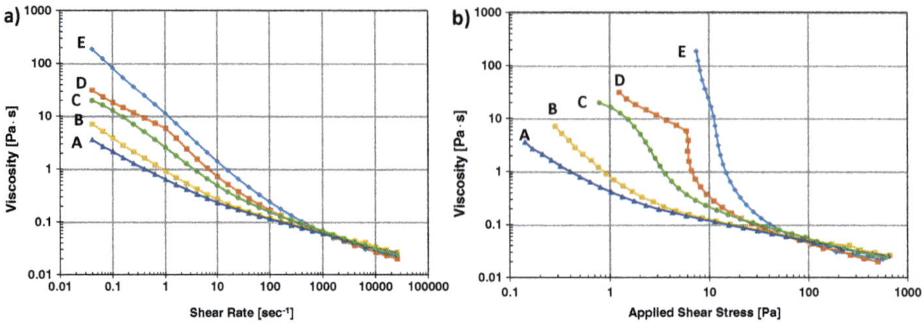

Fig. 4.10 Flow curve data for paints with similar high-shear behavior but substantially different low-shear rheology. Same data are plotted: (**a**) viscosity vs. shear rate and (**b**) viscosity vs. shear stress. For Curve D, note the sharp structure breakdown that occurs around 7 Pa stress

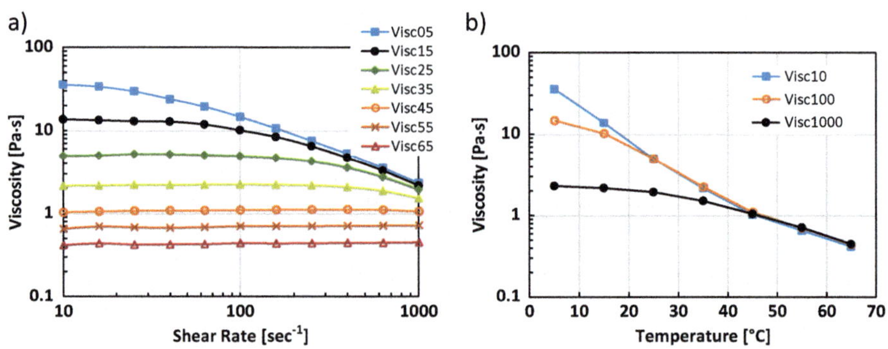

Fig. 4.11 Flow curve data of a resin solution collected on stress-controlled rheometer, in cone and plate geometry. (**a**) Traditional flow curve (viscosity vs. shear rate) at several measured temperatures from 5 °C to 65 °C. (**b**) Viscosity vs. temperature for shear rates 10, 100, and 1000 sec^{-1}

$$\ln\eta = \ln A + \left(\frac{-E_a}{R}\right)\frac{1}{T} \tag{4.10}$$

The Arrhenius equation (Eq. 4.9) relates viscosity to temperature using a prefactor (A) and activation energy (E_a) and gas constant (R). By plotting log viscosity vs. inverse (absolute) temperature, we can often linearize the data to model this dependency. This model works best for polymer solutions over temperature ranges without phase transitions. In the case of many aqueous-phase colloidal systems, we see very little temperature dependence. And in the case of self-reactive systems, we see cases where viscosity data is used to characterize safe handling limits of resin solutions.

Viscometry Applied to Paint Formulations

We've now introduced several viscometry methods to characterize the rheology of paint samples. We recognize that interactions between paint ingredients can impart a material yield stress and/or pseudoplastic structure. Furthermore that structure might be time (shear history) dependent giving rise to thixotropic behavior. We've also suggested that these ingredient interactions can occur on several length scales including microscopic (non-isotropic pigments), colloidal (flocculated emulsions or particles), or molecular (polymer associations or rearrangements). Paint formulators should use rheological measurements to guide the formulation process in a manner consistent with paint performance requirements. We should emphasize that an "ideal flow curve" is never a performance requirement but can be used to guide the formula development.

For improved storage stability, a paint should have a moderate to high low-shear viscosity. Certainly storage and transport of packaged material are a low-shear activity. To limit settling of high-density pigments, paints with some yield stress show improved shelf life.

Once the delivered paint is put in a circulation system, one should focus attention on medium- and high-shear processes. In a commercial application process, paint is typically loaded into a mix tank with some sort of impeller that maintains fluid motion. The material is then pumped through a filter and/or heat exchanger before delivery through a piping network to one or more paint applicators. In these medium-shear processes (1–1000 s^{-1} shear rate), paint viscosity should be sufficiently low that the pump can maintain design pressure at the applicator to assure reproducible fluid delivery. Very often circulation systems include a leg where unused paint is returned to the main circulation tank through a back-pressure regulator. This valve can impart relatively high shear rates onto paint passing through the valve. When we speak of circulation stability, the method described in Fig. 4.9 can be used to test recovery of viscosity following exposure to a shear event (i.e., filter or back-pressure regulator). Alternatively one can track the paint viscosity before and after conditioning for some period of heat and/or shear conditioning.

For paint delivery systems without a return line, or where there is a drop to a "dead" leg (i.e., one-way fluid delivery), one should consider the case where the system is shut down for some period of time. Paint should not phase separate or allow for pigment settling during the shutdown period, perhaps with some yield stress or higher low-shear viscosity. On the other hand, when the line is restarted, we expect normal fluid delivery where the pressure drop is sufficient to overcome yield stress and start-up viscosity.

Liquid spray atomization is a high-shear event where low paint viscosity is preferred in order to generate small, uniformly sized droplets. The previous chapter of this book takes up this topic in more detail, but here we can summarize by indicating that good atomization will transfer paint solids to the substrate after some portion of volatiles evaporates. For pigmented coatings that contain metallic or mica flakes, it becomes important to control the viscosity as volatiles are lost, as these droplets impact onto the target. In these cases, we often design paint rheology to include some thixotropy that allows for low viscosity for some seconds after the high-shear event of spray atomization.

Once on the substrate, one must recognize that the state of the paint is different from the in-can or as-delivered composition. As volatiles are removed during atomization and ambient flash period, we expect more concentrated interactions between solid paint ingredients. During this period, the paint surface should level under the influence of surface tension-driven flow but not so much to cause drips or sag under the influence of gravity. Of course, film thickness uniformity and part geometry have a significant impact on getting the right balance between optimal flow and sag resistance. The paint formulator should pay close attention to rheology as a function of increased nonvolatile content.

Dynamic Oscillatory Rheology

When we defined viscosity as the resistance to flow, we did not say anything about what happens to the energy that the flow field puts into the material. At a basic level, we can say that the material either stores that energy (think elastic band) or it dissipates the mechanical energy to create heat (mechanical energy is lost). The concept of oscillation rheology is to introduce a sinusoidal shear and then watch the response to determine both the magnitude and type of deformation. As was the case for rotational viscometry, rheometers have been designed to apply either a defined stress or strain, so that it can monitor the other. In either case, the material's modulus (G) is the proportionality of stress to strain.

Figure 4.12 schematically shows some limiting behaviors for materials we might consider. In this case, we apply a sinusoidal strain with known amplitude and frequency (ω). For a purely elastic soft solid in Fig. 4.12a, we note that the stress is perfectly inphase with the applied strain. For a purely dissipative viscous liquid in Fig. 4.12b, we note that the stress is 90° out of phase with the applied strain. And for a viscoelastic material in Fig. 4.12c, we observe something in between. We can define the phase angle (δ) as the offset between stress and strain:

$$\gamma(t) = \gamma_0 \sin(\omega t) \tag{4.11}$$

$$\tau(t) = G^* \gamma_0 \sin(\omega t + \delta) \tag{4.12}$$

$$\tau(t) = G' \gamma_0 \sin(\omega t) + G'' \gamma_0 \cos(\omega t) \tag{4.13}$$

Mathematically, we can define the time-dependent applied strain as $\gamma(t)$ and the corresponding stress as $\tau(t)$. The material constants can be expressed as complex modulus (G^*) and phase angle (δ) or alternatively as storage modulus (G') and loss modulus (G''). Figure 4.12d depicts the phase diagram of these material parameters, or they may be converted as follows:

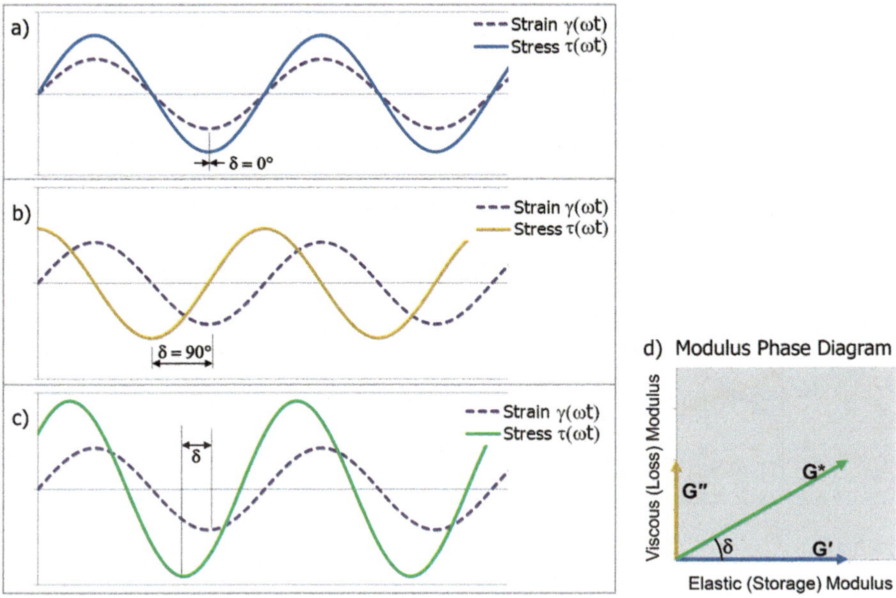

Fig. 4.12 Stress response to an applied sinusoidal strain for a: (**a**) purely elastic soft solid, (**b**) purely dissipative viscous liquid, and (**c**) viscoelastic material. (**d**) Moduli for each case represented as a phase diagram

$$|G^*| = \sqrt{G'^2 + G''^2} \tag{4.14}$$

$$\tan\delta = \frac{G''}{G'} \tag{4.15}$$

We should recognize that these property constants are defined for undisturbed material, where structure is neither created nor broken. It is good practice to start dynamic oscillatory rheology by first running an amplitude sweep on the material. At low oscillation amplitudes, we expect a constant elastic modulus. But once the limit of linear viscoelasticity (LLVE) is exceeded, we see a drop in elastic modulus as the microstructure of the sample begins to break down. Subsequent dynamic testing run at amplitudes below the LLVE is referred to as small amplitude oscillatory shear (SAOS).

One useful example of oscillation rheology might be an upgrade to the viscosity recovery test outlined in Fig. 4.9. There we conditioned the sample with a continuous shear and then dropped the rate to monitor the LSV recovery. If instead we started SAOS testing after a shear burst, we would be able to independently monitor the structure recovery in both elastic and loss modes.

Another application of SAOS testing is when the test material is too thick for continuous rotational viscometry. This is the case for paint materials at very high

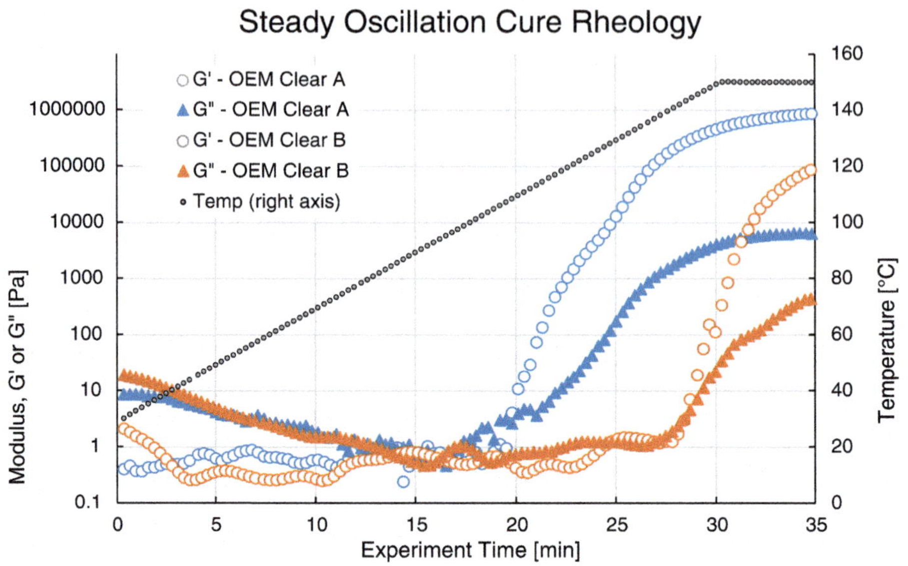

Fig. 4.13 SAOS test results following the cure profile of two automotive clearcoats with thermal responsive crosslinking. Clearcoat A exhibits much earlier cure response with gel point at about 110 °C

solids content or to monitor partially cured samples. Figure 4.13 provides some example data where SAOS is used to watch the speed and extent of cure of two automotive OEM clearcoats.

At the beginning of this experiment, both automotive clearcoats exhibit high viscosity liquid-like behavior (G'' is much larger than G'). When the oven is ramped from 30 to 140 °C, we observe that both moduli drop before crosslinking begins. After about 20 min (>100°C), we observe a rise in both moduli, but much faster in the elastic modulus, G'. This is directly related to the development of a crosslinked network. We can define the gel point as the crossover time (temperature) where the elastic modulus overtakes the loss modulus. When the cure is complete, we note a more solid-like material behavior, where G' is several orders of magnitude higher than G''.

One concern about this type of test is the management of volatiles in the paint sample. Liquid paints that are normally at low (sprayable) viscosity must have most of the volatiles removed before loading into the rheometer's test geometry. During a normal enamel bake cycle, the remaining volatiles and reaction by-products will diffuse to the coating surface and evaporate. However, between rheometer plates, these volatiles are trapped and often form bubbles within the sample.

Another application of SAOS rheology is the tracking of viscosity after spray, during ambient flash before drying/baking. Investigators are interested to monitor rheology interactions of paint ingredients as volatiles are removed. As free volume

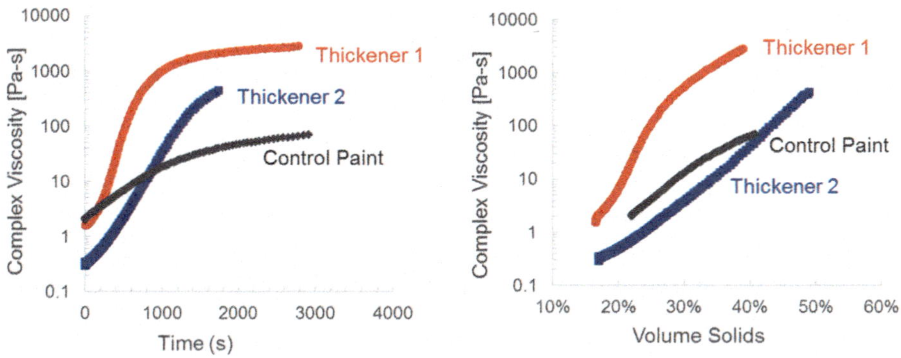

Fig. 4.14 Continuous SAOS testing of waterborne paints in an immobilization cell where volatiles are being removed during experiment. The left graph shows data as collected (vs. experiment time) while the right has been transformed to a volume solids basis to make more direct comparisons

is reduced and average solvent composition changes, we expect binders to coalesce around pigments and fixed particles. Historically, investigators can scrape partially evaporated films to transfer material into the rheometer's test geometry, but we are concerned about the effects of shear and air bubbles introduced by the transfer process. Another approach has been to replace the upper rheometer plate with a bar or ring where volatiles can escape around the geometry.

Yet another approach utilizes an immobilization cell [9]. Here the fixed lower plate is replaced with a vacuum chamber box where the top is fitted with a screen and filter paper. When wet paint is applied to the top of paper, the rheometer's upper plate can be used to start a steady SAOS measurement. The experiment is begun when the vacuum begins to pull volatiles out of the sample bottom through the filter paper.

In order to run the immobilization cell experiment, the rheometer must track normal force and correct upper plate position in real time to maintain contact with the sample, whose volume changes due to volatile loss. Success also depends on finding paper that retains pigments and resins but allows volatiles to pass. By retaining the solid components during measurement, the volume fraction of solids in the sample increases as the sample volume (gap height) decreases. Effectively, the immobilization cell permits a continuous measurement on a sample with changing solid content. Figure 4.14 provides some example data, where the control starts at 25 vol% solids and a 500 μm gap. After 2900 s, it is 42 vol% solids in a 270 μm gap. Addition of two different types of thickeners leads to different viscosity profiles during the drying process. From the left graph, one might conclude that both thickeners cause a more rapid viscosity increase. But recall that the volatiles removal process here might be quite different from what paint experiences during flash drying. Using the gap height data (i.e., sample volume vs. time), we can compute sample volume solids and calculate the right graph data that more correctly indicates in-process thickening behavior.

Nonlinear Dynamic Oscillatory Rheology

Material response is said to be linear if a change in input scale corresponds to a proportional change in the output. In SAOS testing, the strain amplitude is sufficiently small that material constants (moduli and phase angle) are independent of amplitude and the response stress wave remains sinusoidal. Many academic researchers have developed constitutive models to describe materials within the linear viscoelastic domain. It is an unfortunate reality that much of our interest in commercial products involves problems where we exceed the LLVE (dispersing, pumping, atomization, etc.)

Large amplitude oscillatory shear (LAOS) tests are made in conventional shear rheometers where amplitudes are beyond the LLVE. When a sinusoidal strain is applied, we see a distorted periodic wave as the response function. Ewoldt, McKinley et al. [10, 11] have provided several papers that review the mathematics and practice of fitting these output curves to higher harmonics of periodic functions. With these and other methods, we hope to use rheological measurements that utilize deformations that more closely resemble those occurring in commercial applications. If we can model these systems, we hope to advance our understanding of structure–property relations that control coatings behavior during the application process.

Summary/Outlook

A formulator's ability to control the rheology profile of paint enables many aspects of the product's performance. In this chapter, we've touched on processes within paint making, fluid delivery, spray atomization, leveling flow on the painted surface, and during baked curing operations. For less fluid materials, we provide rheology methods using dynamic oscillation in favor of continuous shear deformation. With SAOS methods, we can monitor rheology as a function of volatiles loss or baked cure. No one claims to describe "a perfect rheology flow profile," but we do recognize the importance to control pseudoplasticity, thixotropy, and yield stress as a bridge between microstructure morphology and end-use performance.

Certainly there is a very broad field of rheology which goes beyond the scope of this paint formulator's discussion. If oscillatory rheology provides insights to the material's relaxation time, then formulators should be interested in that performance not only in small amplitude limits (SAOS). Nonlinear methods are currently in development for analysis of LAOS rheology data. But to date, we have not yet seen application of these methods to end-use paint performance.

From an instrument development view, testing in simple shear geometry remains the correct starting point for liquid paint characterization. But rheology in extensional deformation geometries is another area of future development interest. In particular, process engineers are interested to understand extensional flow at high strain rates that should correlate to paint atomization. A reliable method would be a welcome addition to measurement tools for a paint formulator.

References

1. Streitberger, H.-J., Doessel, K.-F.: Automotive Paints and Coatings, 2nd edn. Weinheim, Wiley-VCH Verlag GmbH (2008)
2. Goldschmidt, A., Streitberger, H.-J.: BASF-Handbuch: Lackiertechnik. Vincentz Verlag, Hannover (2002)
3. Patton, T.C.: Paint Flow and Pigment Dispersion, 2nd edn. Wiley Interscience, New York (1979)
4. Wicks, Z.W., Jones, F.N., Pappas, S.P., Wicks, D.A.: Organic Coatings: Science and Technology, 3rd edn. Wiley Interscience, New York (2007)
5. Schoff, C.K.: Rheology. Federation of Societies for Coatings Technology, Blue Bell, PA (1991)
6. Eley, R.R.: Applied rheology in the protective and decorative coatings industry. Rheology Rev., Br. Soc. Rheol. **5**,173–240 (2005)
7. Eley, R.R.: Rheology and viscometry. In: Koleske, J.V. (ed.) Paint and Coating Testing Manual, 15th edn, pp. 415–451. ASTM International, West Chonshohocken, PA (2012)
8. Mezger, T.G.: The Rheology Handbook, 4th edn. Vincentz Network, Hannover (2014)
9. Cheng, V., Houze, E.C., Koerner, M.R.: Rheological characterization of the coating drying process with an immobilization cell. In: 89th ACS Coll. & Surf. Sci. Symp., Carnegie Mellon University, Pittsburgh, 15–17 June 2015
10. Hyun, K., Wilhelm, M., Klein, C.O., Cho, K.S., Nam, J.G., Ahn, K.H., Lee, S.J., Ewoldt, R.H., McKinley, G.H.: A review of nonlinear oscillatory shear tests: analysis and application of large amplitude oscillatory shear (LAOS). Prog. Polym. Sci. **36**(12), 1697–1753 (2011)
11. Ewoldt, R.H., Hosoi, A.E., McKinley, G.H.: New measures for characterizing nonlinear viscoelasticity in large amplitude oscillatory shear. J. Rheumatol. **52**(6), 1427–1458 (2008)

Chapter 5
Magnetic Microrheology for Characterization of Viscosity in Coatings

David J. Castro, Jin-Oh Song, Robert K. Lade Jr., and Lorraine F. Francis

Introduction

Coatings are used in a variety of applications from paints and varnishes for architectural coatings to advanced dispersions of functional materials and inks for biomedical, optical, and electronic devices. The estimated global coating industry in 2012 was approximately US$110 billion on 34 billion liters, and it could reach more than US$140 billion on more than 40 billion liters by 2017 [1]. The rheological characteristics of a coating liquid are of vital importance to its application onto a substrate. A low coating viscosity typically eases the application process and encourages leveling after deposition [2]. The viscosity of the as-deposited liquid layer increases during drying and/or chemical reaction (curing). Monitoring this viscosity rise is important for at least two reasons. First, viscosity provides information on the extent of the structural transformation that occurs as solvents depart during drying and reactions occur during curing. Second, the coating viscosity determines the rate of leveling and flow, both of which enter into the formation of surface defects [2–4].

Characterizing viscosity rise in a coating in situ during drying or curing is a challenge. Traditional rheological measurement techniques that employ rotating

D.J. Castro
Nalco Water—An Ecolab Company, Naperville, IL, USA

J.-O. Song
Department of Chemical Engineering and Materials Science, University of Minnesota, Minneapolis, MN, USA

LG Chem Research Park, Daejeon, South Korea

R.K. Lade Jr. • L.F. Francis (✉)
Department of Chemical Engineering and Materials Science, University of Minnesota, Minneapolis, MN, USA
e-mail: lfrancis@umn.edu

M. Wen, K. Dušek (eds.), *Protective Coatings*, DOI 10.1007/978-3-319-51627-1_5

plates or discs are not easily applied to the geometry of a coating and at best provide an average viscosity. To capture the viscosity gradients that can arise during drying or curing, a local measurement is required. Microrheological methods that characterize rheology on local length scales by monitoring the motion of small probe particles have promise for this application [5–10].

Microrheology is broadly divided into active methods, which drive particle motion using an external magnetic or electric field, and passive methods, which rely on Brownian motion for particle transport [6]. In both types, local rheological properties are determined from the motion of small tracer particle(s) in the liquid. Depending on the type of motion monitored, different properties can be obtained, such as the viscosity or viscoelastic properties of the fluid [6, 11, 12]. Particle motion is captured as a function of time using a microscope, and analysis is carried out with image analysis software.

The use of active methods like magnetic microrheology specifically has appeal for drying and curing coatings as the external magnetic force on the particle better accommodates a larger viscosity measurement range as compared with passive methods. The noncontact nature of the technique (all manipulation is done with a magnetic field) is also appealing in coating processes as it permits rheological measurements to be made in situ. Coupled with information on the position of the probe particle in the coating (e.g., distance from the substrate surface), microrheology is a powerful tool for measuring rheological properties not only as a function of time but also as a function of position.

Magnetic microrheometers vary in design and functionality, but all possess a magnet(s) capable of manipulating tracer particle(s) and a means of visualizing the resultant motion. Magnets can be permanent or electromagnetic, the latter permitting the measurement of dynamic properties such as creep and the storage and loss moduli [6]. Optical microscopy coupled with a high-speed camera is the most commonly employed method of capturing particle motion. The details of the measurement dictate the required sophistication of the camera and image analysis software; tracking complexity increases as particle speed or the number of particles increases.

Microrheology is also well suited to measuring the rheological properties of interfaces, and most of the active and passive techniques can be modified for such a measurement [13]. In interfacial magnetic microrheology, ferromagnetic rods [14–16] or nanowires [17–19] are typically used as probe particles. Their small size and large aspect ratio help maximize the rheological response from the interface, which can become obscured by the response of the bulk subphase below if appropriate precautions are not taken [17, 20]. The response of these magnetic rods to an applied magnetic field can be used to study the interfacial shear rheology. These types of measurements have important applications in characterizing foams and emulsions [21] and surfactant monolayers, especially at biological interfaces [19].

Characterization of spatial and temporal viscosity variations in drying and curing coatings permits formulation variables and processing conditions to be linked to structural development, defect formation, and coating performance. In this chapter, we review the design and features of a magnetic microrheometer specifically for coatings [22] and its application for the characterization of viscosity as a function of time and position in coatings. Then a more detailed case study on the application of the method to study viscosity development in a coating used in paper tissue manufacture is provided.

Magnetic Microrheometry for In Situ Characterization of Coating Viscosity

Magnetic Microrheometer Designed for Coatings

A schematic drawing and photo of a magnetic microrheometer designed to measure coating viscosity are shown in Fig. 5.1 [22]. Spherical, superparamagnetic probe particles are loaded into a sample liquid which is placed in a small container on the sample stage. This stage is situated between two neodymium-iron-boride (NdFeB)

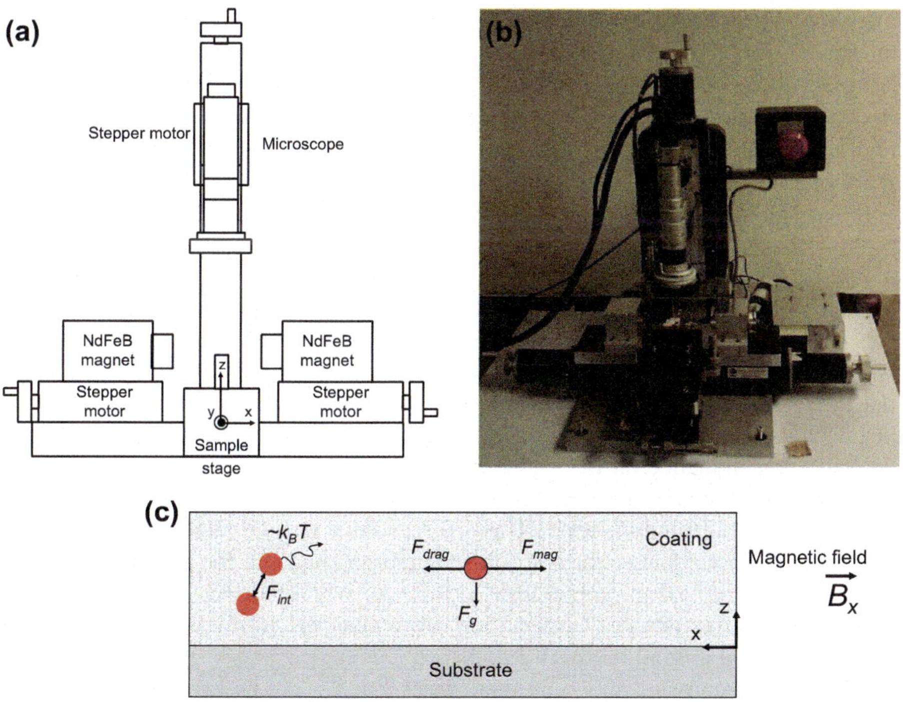

Fig. 5.1 Overview of magnetic microrheometer setup: (**a**) Schematic drawing of the apparatus and (**b**) picture of the apparatus. (Adapted from [22] with the permission of AIP Publishing.) (**c**) Schematic diagram showing forces acting on a magnetic probe particle

permanent magnets which are used to manipulate particle motion. The magnetic field gradient is controlled by adjusting the gap between the two magnets using stepper motors. A digital optical microscope, which is also mounted on a stepper motor for precise positioning of the focal plane throughout the depth of the coating, is used to capture particle motion. For control of drying and curing rate, a heating system with a temperature controller and adjustable air flow is connected to the sample stage. For ultraviolet (UV) curing studies, a UV lamp is positioned above the coating, and irradiance changed by adjusting the lamp-to-coating distance.

An equation for coating viscosity as a function of measurable parameters can be derived from a force balance on a single probe particle in the x direction (see Fig. 5.1c) [5]:

$$F_{\text{mag}} + F_{\text{drag}} + F_{\text{g}} + F_{\text{int}} + F_{\text{thermal}} = m\frac{dv}{dt} \tag{5.1}$$

where F_{mag}, F_{drag}, F_{g}, F_{int}, and F_{thermal} are the magnetic, drag, gravity, interaction, and thermal forces, respectively. These forces are balanced by particle inertia: the product of particle mass (m) and acceleration (dv/dt). Particles with diameters on the order of microns driven by a constant external force are expected to reach a terminal velocity within approximately 10^{-3} to 10^{-6} s of exposure to the force [23]. This time is predicted to be even shorter (~10^{-12} s) in highly viscous liquids like partially dried or cured coatings [24]. This is a result of the low Reynolds number flows that characterize magnetic microrheology experiments and permits inertia in Eq. (5.1) to be neglected.

Interaction forces are caused by hydrodynamic interactions between the particle of interest and other nearby particles or walls; they can be safely neglected at dilute particle concentrations ($<10^{-4}$ vol% [25]) and when particles are at least ~50 particle radii away from a wall [26]. The force on the particle due to gravity, F_{g}, can be minimized by matching the density of the particle to that of the liquid. Additionally, F_{g} acts perpendicular to the direction of the magnetic field, so it does not affect the motion of the particle along the x-axis. This permits F_{g} to be safely neglected from Eq. 5.1. However, it is important to realize that gravity can still have an important influence on the particle(s) via sedimentation (as discussed below). Thermal forces arise due to Brownian motion and tend to move the particle in a three-dimensional random walk. These forces have a magnitude on the order of kT/R (k is the Boltzmann constant, T is temperature, and R is the particle radius) or ~10^{-3} pN for typical microrheology setups. As will be discussed below, this is about three to five orders of magnitude lower than the magnetic force, allowing F_{thermal} to be safely omitted from Eq. 5.1.

Given the arguments above, in most cases Eq. 5.1 can be justifiably simplified such that the particle motion is governed only by the magnetic and drag forces:

$$F_{\text{mag}} + F_{\text{drag}} = 0 \tag{5.2}$$

The magnetic and drag forces exerted on a spherical particle are given by

$$F_{\text{mag}} = M\left(\frac{4\pi R^3}{3}\right)\frac{dB}{dx}, \qquad F_{\text{drag}} = -6\pi\eta R v \tag{5.3}$$

where M is the magnetization of the particle (in A/m), dB/dx is the applied magnetic field gradient along the x-axis, v is the particle velocity in the direction of the applied magnetic field, and η is the coating liquid viscosity. For a 1 μm particle, F_{mag} typically varies from 0.1 to 10 pN depending on the magnetization of the individual particle and the strength of the magnetic field gradient. When superparamagnetic particles are used, the magnetic field is chosen to be large enough for the particle magnetization to saturate and be nearly constant during the measurement. F_{drag} here applies to a spherical particle in a Newtonian fluid. The force balance equation allows the viscosity of coating liquids to be determined from the measured probe particle velocity under an applied magnetic field, assuming the coating liquid is Newtonian.

Combining Eqs. 5.2 and 5.3 allows the coating viscosity to be expressed as a function of experimentally measureable parameters:

$$\eta = \frac{2MR^2(dB/dx)}{9v} \tag{5.4}$$

While strictly applicable to Newtonian fluids, the shear rate (~$v/2R$) may be low enough for a Newtonian assumption to hold for liquids (such as some polymer solutions), which exhibit a Newtonian plateau at low shear rates. For non-Newtonian fluids, F_{drag} in Eq. 5.3 must be modified to account for variability in viscosity [27].

The relationship for the viscous drag force (Eq. 5.3) assumes that the particle is surrounded by liquid and is far from any other solid object, such as the wall of a container. In coatings, proximity to the substrate is a key issue. To account for the influence of a nearby wall, F_{drag} can be modified. To a first approximation [26]

$$F_{\text{drag, wall}} = -6\pi\eta R v\left(1 + \frac{9R}{16z}\right) \tag{5.5}$$

where z is the distance between the particle and the wall or substrate. This expression accounts for the increased drag on the particle caused by the wall [26, 28], and, as Eq. 5.5 shows, the effect is more pronounced for larger particles. Using this new expression for F_{drag}, Eq. 5.1 can be resolved to describe the local viscosity in the liquid based on the velocity of a particle moving near a wall:

$$\eta = \frac{2MR^2(dB/dx)}{9v\left(1 + \frac{9R}{16z}\right)} \approx \left(1 - \frac{9R}{16z}\right)\frac{2MR^2(dB/dx)}{9v} \tag{5.6}$$

where the second, simplified expression is valid (<1% error) when $z/R > 6$. For $z/R > 55$, the probe particle velocity reaches 99% of its uncorrected value.

The setup shown in Fig. 5.1 is limited to relatively low shear rates (~100 s^{-1}). Additionally, due to the use of permanent magnets, only shear viscosity can be measured. Practicality also limits coating systems to those that are transparent or translucent, as the method necessitates the visualization of particles. Opaque systems such as paints or concentrated suspensions are not compatible with this method. It is also important to consider the size of the particle with respect to the coating thickness. This is particularly important in drying coatings where thickness is a strong function of time and may shrink considerably over the course of a microrheology experiment. A coating that is initially thick enough to disregard particle–substrate interactions may not qualify for this simplification near the end of drying. Likewise, as the free surface moves down during drying, it is important to track its position relative to the position of probe particles. The free surface has been predicted to alter the drag force but only when the particle is within one diameter of the liquid–air interface [29]. When measurements are taken over long times, it is also important to consider sedimentation effects, which can move the tracer particle out of the field of view or move it closer than desired to the bottom of the sample container.

Coating Viscosity as a Function of Time and Position

The viscosity of a drying or curing coating rises as solidification proceeds. Tracking that rise is important to follow the structure development and to understand defect formation. When the composition of the coating and therefore its viscosity are assumed to be constant through the coating thickness, then the task of following the viscosity development with time can be accomplished by locating a magnetic particle significantly far from the substrate (see above) and tracking its position as a function of time. For example, the velocity of 1 μm diameter superparamagnetic iron oxide probe particles in an aqueous polyvinyl alcohol solution (12.1 wt% polymer) coating as a function of time at various drying temperatures is shown in Fig. 5.2 along with the calculated viscosities [24]. The data were gathered at 35 μm from the substrate in an initially 130-μm-thick coating to avoid wall effects. Calculations indicated that concentration gradients through the thickness should be minimal, but experiments were not done at different depths to test this assumption. The effect of temperature on the viscosity rise was significant, as expected, due to the rapid rise in viscosity with increasing polymer concentration during drying. Considering the coating dried at 27 °C, the liquid viscosity is shown to increase slightly over the course of the first 15 min followed by a steeper increase. At elevated temperature (40–60 °C), no discernible initial viscosity drop from the temperature effect was noted, indicating that within the time scale of the measurements (~10 s), the viscosity rise was dominated by drying. The time for viscosity rise decreases with increasing temperature, which follows the expectation based on faster drying at elevated temperature. Recently, Komoda et al. [9] developed a magnetic microrheology system that also characterizes viscosity of

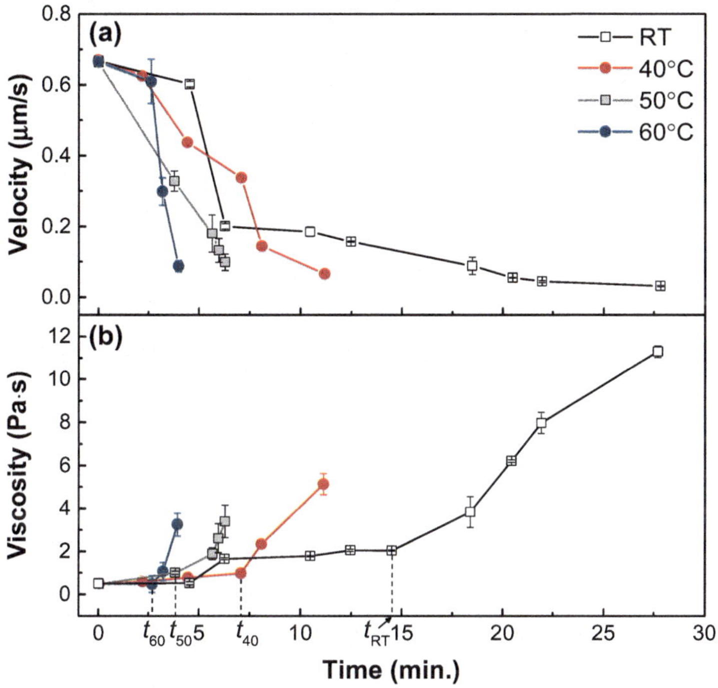

Fig. 5.2 (**a**) Magnetic probe particle velocities and (**b**) calculated viscosities as a function of drying time and temperature for coatings consisting of 12.1 wt% poly(vinyl alcohol) in water. Used by permission of the author from [24]

drying coatings; their apparatus uses an electromagnet to collect data from the oscillatory motion of a magnetic microdisk on the surface of the coating as it dries, which allows the determination of viscoelastic properties.

As mentioned, coating viscosity may vary as a function of position through the thickness of a coating during drying and curing [30, 31]. Drying, for example, removes solvent from the free surface, resulting in a higher local concentration of dissolved species, such as polymer, there. Under some conditions, diffusion is fast enough to keep the concentration fairly uniform during drying, but in others gradients are expected. In curing coatings, gradients in extent of crosslinking may come about in a variety of ways, including loss of volatile blocking agent for a catalyst, oxygen uptake, and absorption by a photoinitiator [31].

Kim and coworkers [32] used a magnetic microrheometer to study the top-down viscosity development and structuring in a drying block copolymer solution. The research employed vials of solution rather than coatings and demonstrated the utility as well as the challenges with tracking viscosity gradients during drying. Figure 5.3 shows the viscosity of a drying block copolymer solution at three different positions relative to the free surface as a function of drying time. To gather this data, the researchers were careful to track not only the magnetic particle motion but also the position of the magnetic particle relative to the free surface.

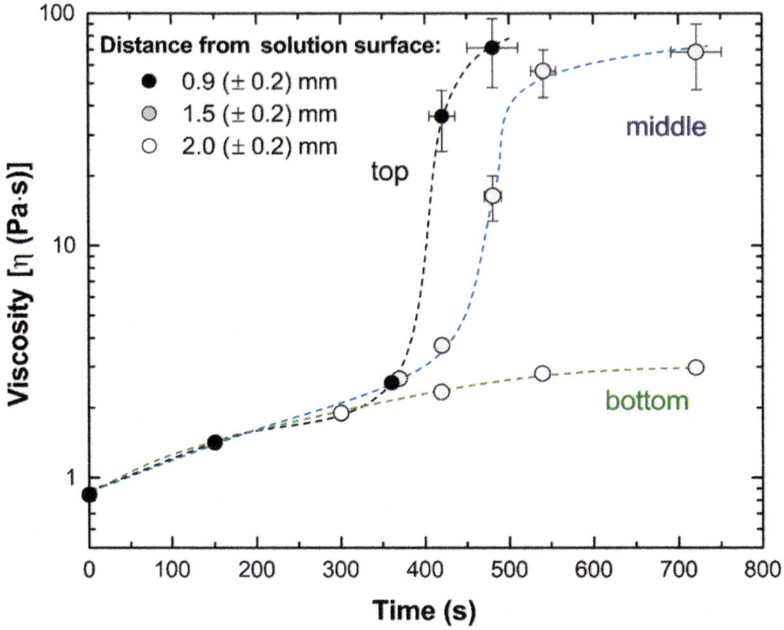

Fig. 5.3 (**a**) Viscosity of a 30 wt% solution of a block copolymer (polystyrene-polylactic acid, PS-PLA with $M_n = 75$ kg/mol, $f_{PS} = 0.72$) determined by magnetic microrheometer measurement during drying conditions at room temperature (298 K). Data were gathered at three locations relative to the surface of the drying coating, as noted. The *y*-axis error bars are the standard deviation from measurements of ten probe particles. The *dashed lines* show the trends in the data. Reprinted (adapted) with permission from [32]. Copyright 2013 American Chemical Society

Because the free surface of the coating moves downward with time, particles were tracked for a short period of time before the free surface was located once again by changing the focal plane of the microscope. Microphase separation in block copolymers leads to the formation of a nanostructure that boosts the viscosity relative to what one would expect in a homopolymer solution. The microrheology data in Fig. 5.3 along with small angle X-ray scattering measurements taken on block polymer solutions of different concentrations showed that the microphase separation starts at the surface of the drying solution and propagates downward as solvent evaporates and polymer concentration increases. This information is relevant to designing conditions for developing structure in coatings during drying.

For UV-curable coatings, viscosity variation through the coating thickness is easier to characterize because UV curing results in smaller shrinkages compared to drying, and therefore only minor changes in thickness during curing. Song and coworkers [33] explored the factors influencing viscosity gradients in UV-curable epoxy coatings. The experimental setup included a UV light source positioned above the sample stage. In parallel experiments, the magnetic particles at three different positions were imaged in the 80-μm-thick coatings: top (~10 μm from surface), middle (~25 μm from surface), and bottom (~45 μm from surface). The magnetic particle motion along the applied magnetic field gradient direction was

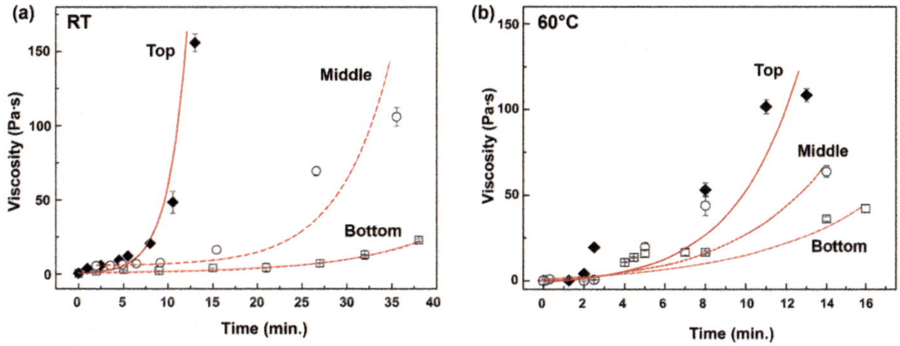

Fig. 5.4 Viscosity as a function of time for UV-cured epoxy coating at 10 μm (*top*), 25 μm (*middle*), and 45 μm (*bottom*) from the surface of an 80 μm thick coating for coatings cured at (**a**) room temperature (RT) and (**b**) 60 °C. Adapted with permission of John Wiley & Sons from [33]

monitored as a function of UV exposure time, and the results are shown in Fig. 5.4. The conditions for curing were such that a significant gradient in UV intensity through the thickness is found. The higher rate of conversion at the top of the coating results in a faster rise in viscosity relative to deeper into the coating. The depthwise gradient was less severe for coatings cured at 60 °C. The microrheology results were able to show that viscosity gradients through the thickness are linked to the formation of a wrinkle defect [31, 33].

Magnetic Microrheology of a Coating Used in Tissue Making

In tissue and towel paper manufacture, dilute coating formulations are continuously sprayed onto a section of the hot surface of a rotating metal drum [34, 35]. The wet paper web is applied to the coated drum at a short distance from the point of the coating application; the web dries as the drum rotates and then is removed as tissue or towel by a doctor blade at a point just before the coating station. This dryer is known in the paper industry as a Yankee dryer (Fig. 5.5). The coating in the Yankee dryer performs a number of functions, including adhering the wet sheet coming from the forming wire or transfer belt, facilitating drying of the sheet by heat transfer from the dryer to the sheet, holding the sheet in place for the "creping" action of the doctor blade, and protecting the dryer surface from wear and markings by the creping blade. Consequently, the Yankee coating undergoes considerable changes from the very moment it lands on the hot surface to the moment it is at least partially scraped off the surface of the metal. The steady-state condition of the film, especially in the latter stages of this sequence, is the subject of much interest in the industry because these conditions are expected to be similar to those of the coating when the creping occurs and many relevant properties of the tissue are developed.

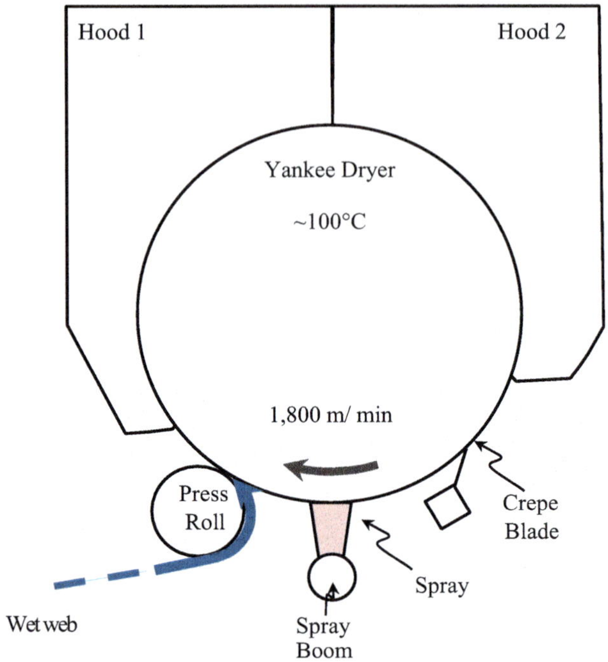

Fig. 5.5 Schematic drawing of a Yankee dryer showing coating application. To provide a sense of scale, a typical Yankee dryer diameter is about 5.4 m. The wet paper web (*blue arrow* coming from the bottom left) is pressed against the dryer by a press roll and dried by the heat of the Yankee and impinging hot air from the hoods. The paper is eventually "creped" out of the Yankee by the crepe blade on the bottom right

However, the earlier stages of the spray and initial film formation have received relatively little attention.

In this section, microrheology is used to explore the structure development of coatings used for these applications as a function of drying time. It is important to note that the experiments do not replicate the conditions on the dryer, but rather they were designed to explore trends in experimental variables such as temperature and coating composition. Additionally, to understand these trends over broad ranges, experiments were designed for straightforward data collection and not for the determination of the viscosity to high precision.

Experimental Methods

Magnetic Microrheometer Setup

The magnetic microrheometer described earlier in this chapter, and in more detail in [22], was used for this work. The goal of this study was to characterize coating viscosity as a function of time under different thermal conditions and with different coating solution chemistries. At the start of each experiment, a 100 μL volume of the coating solution containing a dilute concentration of superparamagnetic particles was pipetted onto a glass cover slip, which sat on a preheated copper stage (see Fig. 5.1). Drops approximately 2 cm in diameter formed on the cover slip, regardless of temperature or solution chemistry. A 100 μL spherical cap with a 1 cm radius has a height of ~400 μm, consistent with experimental range in which the probe particles were found. During the experiments, stage temperature was controlled using a heating element, thermocouple, and temperature controller (Omega CN 8500, ±1 °C resolution).

Two NdFeB permanent magnets (⌀ = 1.27 cm), aligned to a common x-axis, were used. The magnets were spaced approximately ~3 cm apart with the edge of the sample located almost adjacent to Magnet 2 (Fig. 5.6a). The magnetic field gradient was determined using a gaussmeter (F.W. Bell 5180). With the gaussmeter probe positioned next to the sample, magnetic field magnitudes were recorded after moving both magnets in steps of 50 μm in the same direction. After a total of 15 measurements, the magnets were brought to the starting position. The magnetic field measurements were plotted as a function of position (Fig. 5.6b). The absolute value of the slope of the linear fit to the data is the magnetic field gradient, $dB/dx = 46.6$ T/m, while the sign of the magnetic field is indicative of the direction of the field.

Coating Formulations

The compositions of the coating solutions used in this work are given in Table 5.1. The solutions are based on the two main polyaminoamide-epichlorohydrin (PAE) based chemistries for adhesives in Yankee coating applications. One of these is the high molecular weight, low crosslinking potential, NX PAE (Nalco number 64894; 15% solids), and the second is a low molecular weight, high crosslinking potential, X PAE (Nalco number 64800; 17% solids). The pH of the X PAE is intentionally maintained low at approximately 3.5 in order to minimize crosslinking. The variations in composition of solutions in Table 5.1 are designed to explore the effect of PAE concentration, the presence of a polyol-based plasticizer (Nalco DVP4V029), PAE crosslinking potential, and probe particle functionalization on viscosity development.

Probe Particles

Dynabeads M-270 Amine from Life Technologies, Inc. was used for this research. These probe particles have a nominal diameter of 2.7 μm and are functionalized at the surface with primary amines. A scanning electron microscopy (SEM) image is shown in Fig. 5.7; the measured diameter of the amine-functionalized particles was found by image analysis to be 2.41 ± 0.07 μm. This measured average was used for both viscosity and dilution calculations. To achieve particle concentration no greater than 4.5×10^{-6} vol%, ~6 μL of stock probe particle solution (with a concentration of 2×10^9 beads (30 mg/mL)) was added to the coating solution to produce a solution of 20 g in total. Additionally, carboxylic acid-functionalized particles (Dynabeads M-270 Carboxylic Acid) were also investigated. Preliminary

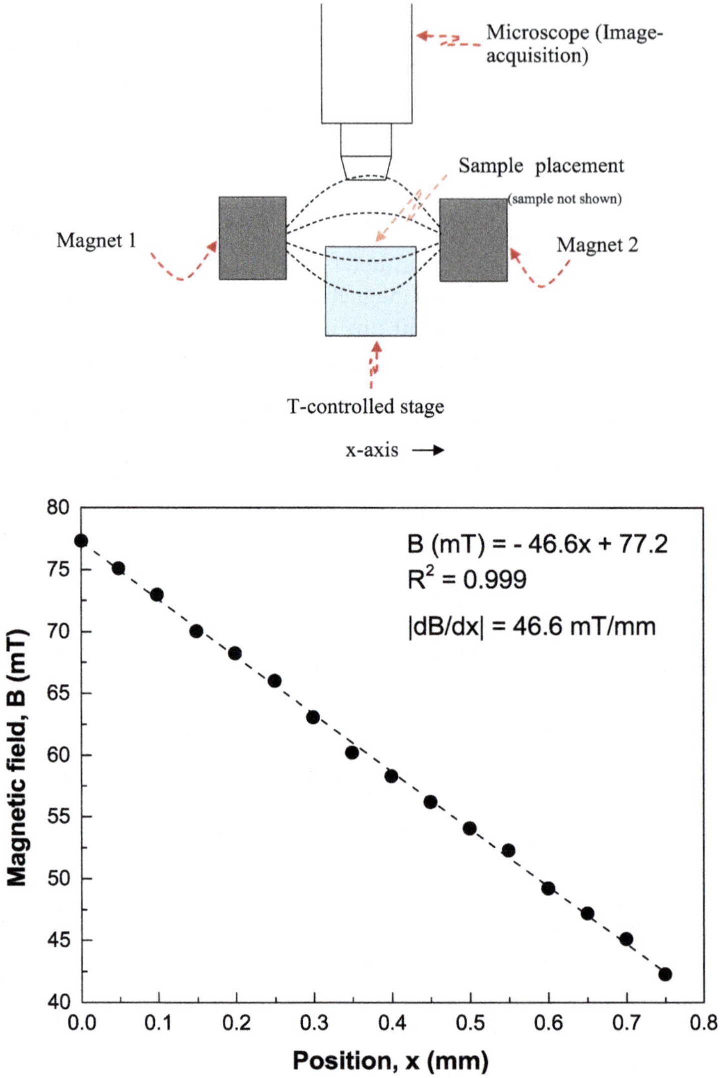

Fig. 5.6 (*Top*) Cartoon diagram of the laboratory microrheometer main components. The *grey dashed lines* in between the two magnets represent the magnetic field. Sample is not shown. (*Bottom*) Magnetic field as a function of gaussmeter tip position with respect to magnets in the *x*-axis. The line is a linear fit to the data

results indicated that the surface functionality may have an effect on the viscosity measurement but not on the qualitative trends that are reported here. Time did not permit an extensive study of the effect of particle functionality.

Particle magnetization was determined using a vibrating sample magnetometer (VSM). To account for an unequal distribution of the particles in the sample

Table 5.1 List of coating liquid compositions and viscosities at 21 °C

		Solids %				MMR[b]	Brookfield
	Concn. wt%[a]	NX PAE	X PAE	Plasticizer	pH	viscosity (cP)	viscosity (cP)
A	15	100	0	0	6.20	234	169
B	9	100	0	0	–	51	82
C	15	85	0	15	–	146	104
D	15	0	100	0	8.74	51	–
E	15	0	100	0	6.37	66	–
F	15	0	100	0	4.17	44	74

[a]"Concn. wt%" stands for the formulation total solids concentration
[b]*MMR* magnetic microrheometer

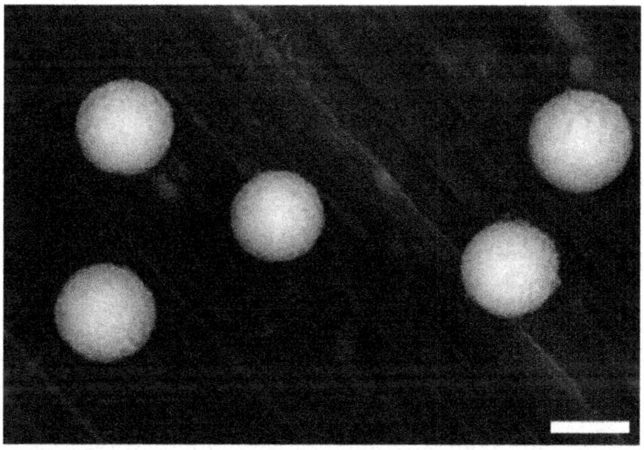

Fig. 5.7 Scanning electron microscopy (SEM) image of amine-functionalized probe particles. Bar = 2 μm

volume, the measurement is conducted a second time for each sample after rotating it 90° around the axis parallel to the VSM magnets. The measured suspension magnetization values were converted to particle magnetization using the sample volume (50 μL), the particle concentration (2×10^9 particles/mL), and particle volume (7.42 and 8.38 μm^3). Figure 5.8 shows the particle magnetization over a range of applied fields. The saturation magnetization, determined by averaging the final 30% of the magnetization measurements at high absolute magnetic field, was 31.4 and 24.2 kA/m for the amine- and carboxylic acid-functionalized particles, respectively. Based on Fig. 5.6, the magnitude of the magnetic fields used during the measurements was less than that needed to achieve saturation. Therefore, the value of magnetization that was used in the calculation of viscosity (Eq. 5.4) was determined from Fig. 5.8 based on the value of the field at the point of measurement.

For all results reported, the M-270 amine-functionalized particles were used.

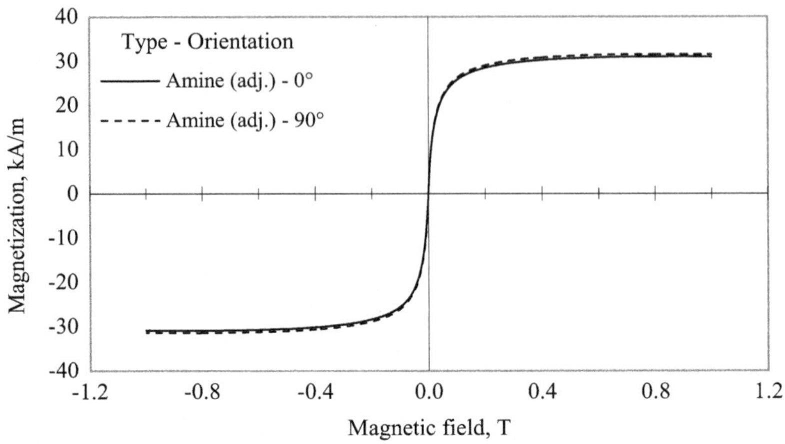

Fig. 5.8 Magnetization as a function of magnetic field for the amine-functionalized probe particles. To correct for unequally distributed beads in the sample volume, the measurement was conducted twice for each sample with a difference of 90° around the axis parallel to the vibrating sample magnetometer magnets

Imaging, Data Collection, and Analysis

Images were taken with a digital optical microscope (Hirox KH-7700), with an MXG-10C coaxial 10× zoom lens, and an OL-140 II objective (Hirox, USA). All experiments were conducted at a total magnification of 840×, resulting in a horizontal view of 369 μm and a resolution of 0.231 μm/pixel. This magnification accommodated a wide range of particle velocities and provided an optimal balance for viewing both slow- and fast-moving particles. Depending on the viscosity and temperature of the sample solution, particles moved anywhere from several microns to several hundred microns during the 10-s measurement period.

In a typical experiment, the microscope was positioned at the center of the pipetted drop of sample liquid; a particle was found and tracked for as long as possible, taking one image per second. Then, when the particle left the field of view, another particle was found, sometimes by adjusting the focal plane, and this particle was tracked. The process was then repeated until the particle velocity was too small (viscosity of the coating was too high). This procedure was convenient for data collection and allowed the study of a wide range of conditions; however, there were some drawbacks, as noted below. Care was taken to ensure that the particles were at least ~50 μm from the cover slip surface to lessen wall effects, and the particle velocities at the center of the specimen were monitored to minimize the influence of drying-induced convective flows. Images were analyzed to calculate particle velocity; ten frames (10 s) of data were used to find a velocity that was used to calculate viscosity.

Comparison with Traditional Rheology

Microrheology was used to determine the viscosity of all solutions in Table 5.1 at 21 °C. Measurements were conducted for at least 60 min. Averages of these measurements are listed in Table 5.1. Several identically prepared solutions were also analyzed with a Brookfield viscometer for comparison (see Table 5.1). Shear rates for the two instruments were similar (~5 s^{-1} for the Brookfield and a maximum of ~9 s^{-1} for the microrheometer). The magnetic microrheometer values are less than 40% different from their Brookfield measurement counterparts. The viscosity differences are likely connected to uncertainties related to the microrheometer procedure. Namely, the magnetic fields chosen were less than those needed for saturation, which added some uncertainty to the particle magnetization, and the process of tracking multiple particles at different focal planes during an experiment could lead to some variability based on the particle position relative to the substrate. Nevertheless, the procedure yielded data with sufficient accuracy for the comparisons made. Additionally, the five solutions for which both rheological methods were used had a reasonably high correlation coefficient ($r^2 = 0.93$).

Results and Discussion

Noncrosslinking PAE (NX PAE) Solutions

Figure 5.9 shows typical viscosity as a function of time profiles for a 15% NX PAE solution (A in Table 5.1) dried at 21, 60, and 100 °C. At 21 °C, the measured viscosity increases slowly with time, likely due to the natural water evaporation (and the related increase in polymer concentration) that occurs at room temperature. At higher temperatures, at least two other phenomena may be observed. First, a small drop in viscosity may occur during the first few minutes. This decrease in viscosity likely reflects the effect of the increase in the liquid temperature from close to room temperature to either 60 or 100 °C. The observation of this drop varied depending on the timing of the measurements. The second is the relatively steady viscosity (<500 cP) during the first few minutes of measurement, which was followed by a sudden increase in viscosity. The steady viscosities at early times are probably a combination of the increase in temperature (decreasing viscosity) and the increase in solids through water evaporation (increasing viscosity).

The sudden, steep nature of the viscosity increase at later times suggests that the polymer solution viscosity has a strong dependence on concentration. With this formulation, we do not expect appreciable crosslinking, and so the steep rise was somewhat unexpected. Since the microrheology technique requires data to be gathered over a specific length of time, capturing details of variations that occur rapidly with time is not easy. Further, we do not have concurrent data on weight loss, and so the rate at which the concentration increases may not be constant.

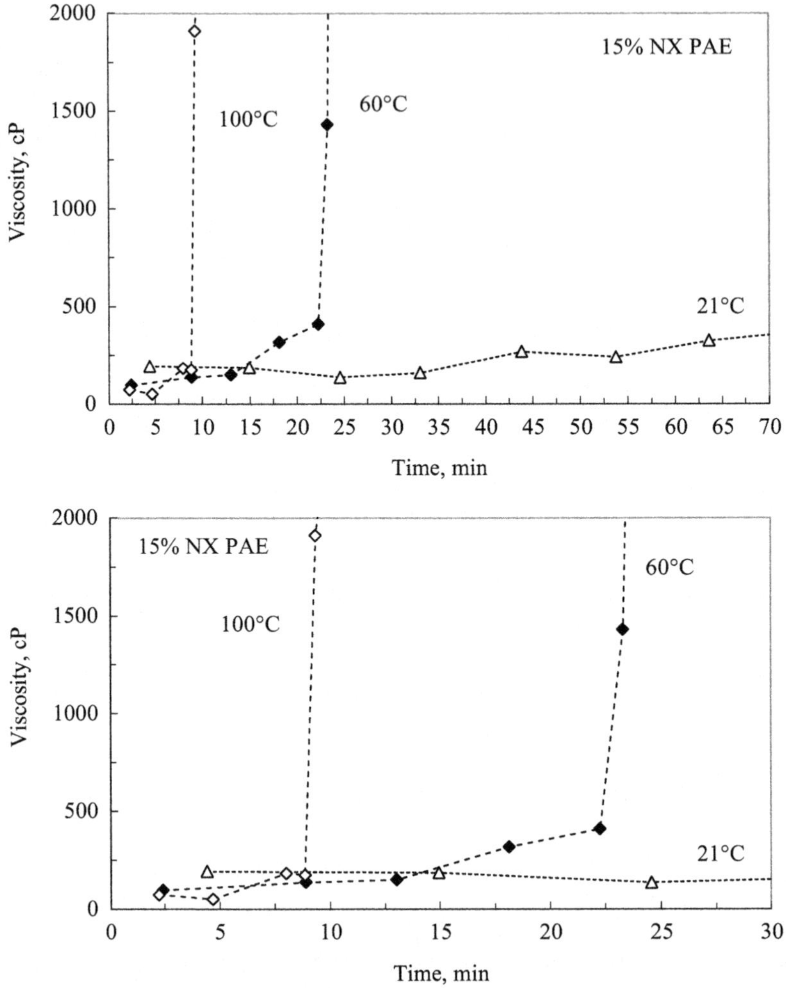

Fig. 5.9 Viscosity as a function of time for a 15% NX PAE film at 21, 60, and 100 °C. The lower plot magnifies the time range from 0 to 30 min (1 cP = 10^{-3} Pa·s)

Regardless, the coatings show a "setting" process in which the viscosity rises over a short period of time. Here, the term "setting time" is used to note the time when the viscosity increases sharply. In this case, the setting times are ~22 and 9 min for the 15% NX PAE at 60 and 100 °C, respectively. The fact that the higher temperature results in faster setting is consistent with faster drying.

In an effort to isolate changes in viscosity due to a changing polymer concentration, the setting curves for 15% NX PAE were compared with those of a 9% NX PAE solution (B in Table 5.1). Figure 5.10 shows the profiles for both the 15 and 9% solution concentrations. The more dilute solution sets at about 24 and 10 min at 60 and 100 °C, respectively, somewhat longer times than those measured for the

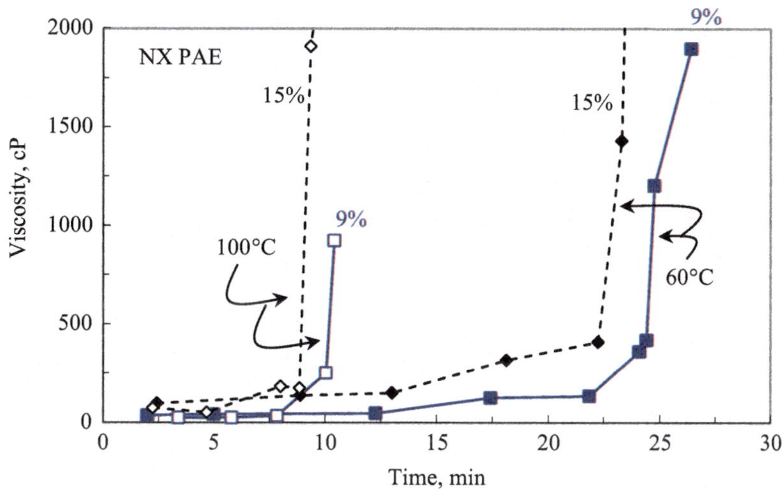

Fig. 5.10 Viscosity as a function of time for a 15% (*black diamonds*) and 9% (*blue squares*) NX PAE films at 60 and 100 °C (1 cP = 10^{-3} Pa·s)

15% NX PAE solution. The smaller observable difference at the higher temperature (1 min vs. 2 min) is related to faster evaporation, which presents some challenges for the technique. Hence, we focus mainly on observations at the lower temperature, 60 °C, and assume that the effect is present at 100 °C, although more difficult to capture.

Plasticizer Concentration in NX PAE

Polyol-based chemistries are known plasticizers of PAE films in Yankee coating applications [36, 37], and the effect of the latter in NX PAE films has been recently measured in our laboratories (Fig. 5.11). This plasticizing effect may have multiple advantages, whether on the creping blade or on the ability of the coating to adhere to the wet sheet at the press roll. To develop a better understanding of the latter, the viscosity development of drying films composed of a 15/85 plasticizer/NX PAE blend (C in Table 5.1) was compared to that of films consisting only of NX PAE.

Figure 5.12 shows the viscosity development profiles of the NX PAE containing the polyol-based plasticizer. At both 60 and 100 °C, the presence of the plasticizer lessens the steepness of the viscosity rise at setting and may have an effect on the setting time. While the results show a reasonable trend based on the expected changes with drying, the setting times are very similar, and more work is needed to verify that the times are significantly different. Like the 9% NX PAE curve, a net reduction in NX PAE content occurs when the formulation containing plasticizer is used (from 15 to 12.75% NX PAE). But, unlike the 9% NX PAE, the NX PAE taken out has been replaced by the polyol-based plasticizer rather than water.

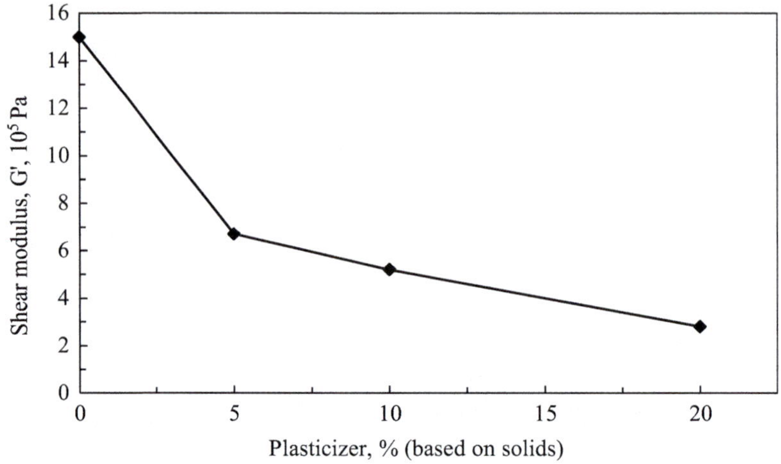

Fig. 5.11 Shear modulus of NX PAE as a function of polyol-based plasticizer concentration

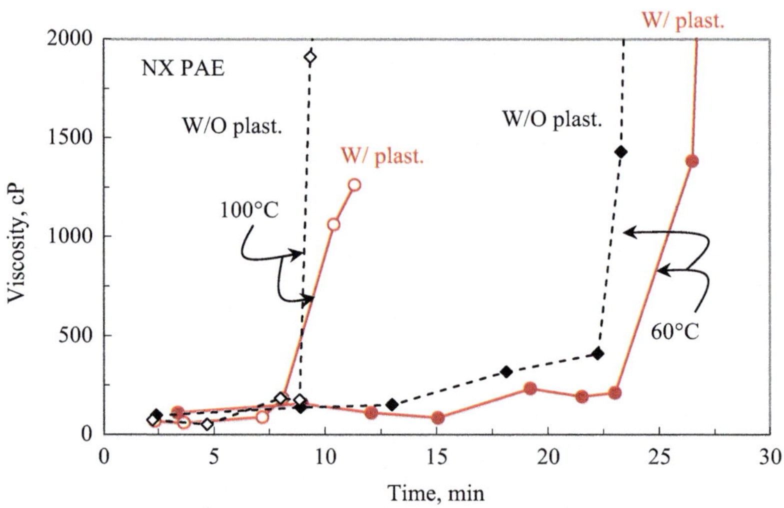

Fig. 5.12 Viscosity as a function of time for a NX PAE film without a plasticizer (*black diamonds*) and with a plasticizer (*red circles*) at 60 and 100 °C. Notice that all of the films' starting total concentration is 15%. The addition of the plasticizer is such that it replaces the adhesive, NX PAE, while maintaining the total concentration constant (1 cP = 10^{-3} Pa·s)

Consequently, the similarly shaped curves (9% NX PAE and 15/85 plasticizer/NX PAE) for films containing different amounts of water lead to the question of whether the mechanism by which polyol plasticizes the NX PAE film is through more water being held within the polymer structure.

Effect of Crosslinking Potential

Figure 5.13 compares the viscosity increase with time for the low crosslinking potential solution based on NX PAE (A in Table 5.1) with the solution containing a lower molecular weight but higher crosslinking potential polymer, X PAE (pH = 6.4, E in Table 5.1). The higher crosslinking potential polymer, X PAE, exhibits a shorter setting time compared with NX PAE. However, this faster setting does not occur until after the X PAE viscosity has crossed the NX PAE viscosity. In this case, the lower starting viscosity of the X PAE is likely dictated by the lower

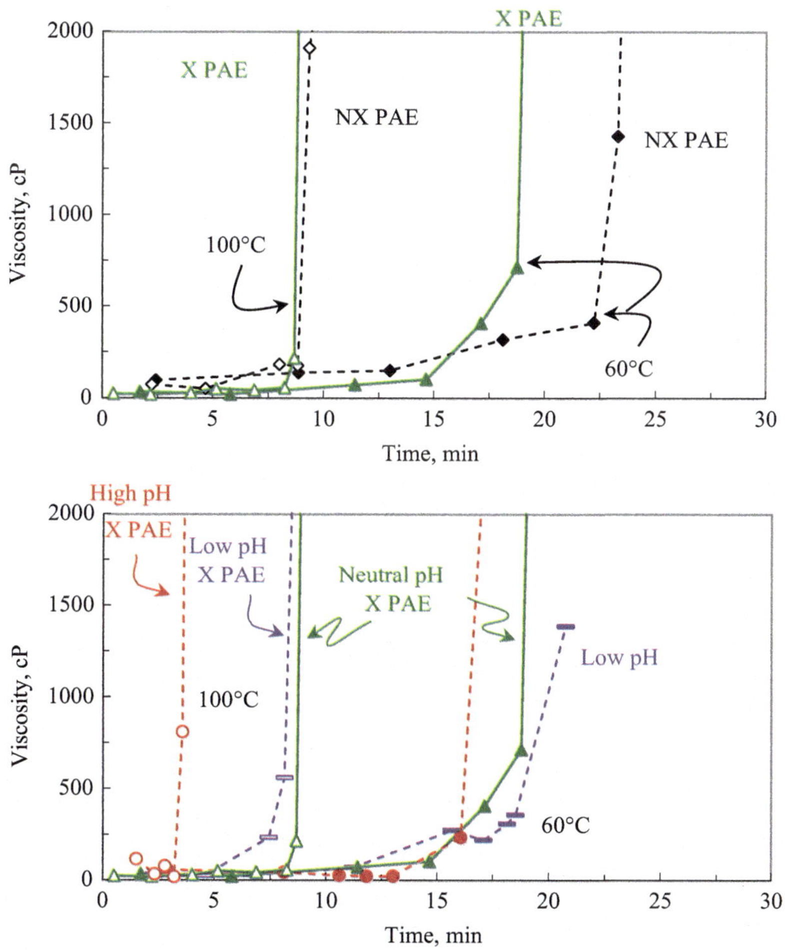

Fig. 5.13 Viscosity as a function of time for (*top*) a NX PAE film (*black diamonds*) and a X PAE film (*green triangles*) at 60 and 100 °C at similar pH of about 6.2–6.4 and (*bottom*) X PAE films at high pH = 8.7 (*red circles*), neutral pH = 6.4 (*green triangles*), and low pH = 4.2 (*purple horizontal lines*) (1 cP = 10^{-3} Pa·s)

molecular weight compared with the higher molecular weight of the NX PAE. The difference in starting viscosities together with the faster viscosity development of the X PAE is significant because it highlights that a starting lower viscosity does not guarantee a later setting time. Moreover, the shorter setting time of the X PAE would then have to be due to the most important difference with the NX PAE: the greater crosslinking potential of the X PAE.

The bottom panel in Fig. 5.13 shows the setting times of the X PAE at low (pH = 4.2, F in Table 5.1), close-to-neutral (pH = 6.4, E in Table 5.1), and high pH (pH = 8.7, D in Table 5.1). The high pH solutions at both temperatures set faster by at least 2 min than either the close-to-neutral or low pH solution. This faster setting time is consistent with the known enhanced crosslinking kinetics of the X PAE at greater pH, which is reflected in the tendency of this product to gel under normal room conditions. The differences between low and close-to-neutral pH are less consistent. At 60 °C, the low pH curve crosses the other curves at least twice and results in a longer setting time as expected; however, at 100 °C, the low pH solution unexpectedly has a somewhat shorter setting time as compared to the neutral. This result suggests that there are limitations in the ability of microrheology to discern differences in crosslinking based on pH in the lower pH range.

Summary

Magnetic microrheology provides a means of characterizing the viscosity of liquid applied coatings. The method makes use of small (~1–10 μm) magnetic probe particles in dilute solution in a coating liquid. An applied magnetic field drives the probe particles through the coating liquid; a digital microscope is used to capture particle positions from which velocity is determined by image analysis. If a particle is far enough from the substrate and the liquid is Newtonian (frequently a good assumption given the low shear rate imposed by a moving particle), the local viscosity can be calculated from the measured velocity and the magnetic parameters of the particle and instrument. The technique can be used to monitor the local viscosity as a function of time during drying or curing and as a function of position through the thickness. When large thickness changes occur during solidification, the technique is difficult because the free surface position as well as the particle position should be known throughout the measurement. In curing coatings that do not shrink appreciably, spatial variations in viscosity can be measured. A case study on the viscosity development of a coating used in tissue making was presented to demonstrate the utility of the method in discerning the effects of temperature and coating solution composition on the coating solidification.

Acknowledgments The authors would like to thank the industrial supporters of the Coating Process Fundamentals Program (CPFP) of the Industrial Partnership for Research in Interfacial & Materials Engineering (IPRIME) at the University of Minnesota. Parts of this work were carried out in the Characterization Facility, University of Minnesota, which receives partial support from NSF through the MRSEC program.

References

1. Bangert, C., Detiveaux, S.: The state of the global coatings industry. http://www.pfon-line.com/articles/the-state-of-the-global-coatings-industry (2014). Accessed 9 June 2016
2. Overdiep, W.S.: The levelling of paints. Prog. Org. Coat. **14**(2), 159–175 (1986)
3. Wu, S.: Rheology of high solid coatings. I. Analysis of sagging and slumping. J. Appl. Polym. Sci. **22**, 2769–2782 (1978)
4. Lade Jr., R.K., Song, J.-O., Musliner, A.D., Williams, B.A., Kumar, S., Macosko, C.W., Francis, L.F.: Sag in drying coatings: prediction and real time measurement with particle tracking. Prog. Org. Coat. **86**, 49–58 (2015)
5. Waigh, T.A.: Microrheology of complex fluids. Rep. Prog. Phys. **68**(3), 685–742 (2005)
6. Gardel, M.L., Valentine, M.T., Weitz, D.A.: Microrheology. In: Breuer (ed.) Microscale Diagnostic Techniques, pp. 1–49. Springer, Berlin (2005)
7. Cicuta, P., Donald, A.M.: Microrheology: a review of the method and applications. Soft Matter. **3**(12), 1449–1455 (2007)
8. Squires, T.M., Mason, T.G.: Fluid mechanics of microrheology. Annu. Rev. Fluid Mech. **42**, 413–438 (2010)
9. Komoda, Y., Leal, L.G., Squires, T.M.: Local, real-time measurement of drying films of aqueous polymer solutions using active microrheology. Langmuir. **30**(18), 5230–5237 (2014)
10. Gu, Y., Kornev, K.G.: Ferromagnetic nanorods in applications to control of the in-plane anisotropy of composite films and for *in situ* characterization of the film rheology. Adv. Funct. Mater. **26**(22), 3796–3808 (2016)
11. Bausch, A.R., Möller, W., Sackmann, E.: Measurement of local viscoelasticity and forces in living cells by magnetic tweezers. Biophys. J. **76**(1), 573–579 (1999)
12. Yagi, K.: The mechanical and colloidal properties of amoeba protoplasm and their relations to the mechanism of amoeboid movement. Comp. Biochem. Physiol. **3**(2), 73–91 (1961)
13. Ortega, F., Ritacco, H., Rubio, R.G.: Interfacial microrheology: particle tracking and related techniques. Curr. Opin. Colloid Interface Sci. **15**(4), 237–245 (2010)
14. Shahin, G.T.: The stress deformation interfacial rheometer. Dissertation, University of Pennsylvania (1986)
15. Brooks, C.F., Fuller, G.G., Frank, C.W., Robertson, C.R.: An interfacial stress rheometer to study rheological transitions in monolayers at the air-water interface. Langmuir. **15**(7), 2450–2459 (1999)
16. Ding, J., Warriner, H.E., Zasadzinski, J.A.: Magnetic needle viscometer for Langmuir monolayers. Langmuir. **18**(7), 2800–2806 (2002)
17. Anguelouch, A., Leheny, R.L., Reich, D.H.: Application of ferromagnetic nanowires to interfacial microrheology. Appl. Phys. Lett. **89**(11), 111914 (2006)
18. Lee, M.H., Lapointe, C.P., Reich, D.H., Stebe, K.J., Leheny, R.L.: Interfacial hydrodynamic drag on nanowires embedded in thin oil films and protein layers. Langmuir. **25**(14), 7976–7982 (2009)
19. Dhar, P., Cao, Y., Fischer, T.M., Zasadzinski, J.A.: Active interfacial shear microrheology of aging protein films. Phys. Rev. Lett. **104**(1), 016001 (2010)
20. Edwards, D.E., Brenner, H., Wasan, D.T.: Interfacial Transport Processes and Rheology, p. 134. Butterworth-Heinemann, Boston (1991)
21. Moschakis, T.: Microrheology and particle tracking in food gels and emulsions. Curr. Opin. Colloid Interface Sci. **18**(4), 311–323 (2013)
22. Song, J.-O., Henry, R.M., Jacobs, R.M., Francis, L.F.: Magnetic microrhoemeter for *in situ* characterization of coating viscosity. Rev. Sci. Instrum. **81**(9), 093903 (2010)
23. Thomas, R.H., Walters, K.: The unsteady motion of a sphere in an elastic-viscous liquid. Rheol. Acta. **5**(1), 23–27 (1966)
24. Song, J.-O.: *In situ* characterization of dynamic structures of coatings. Dissertation, University of Minnesota Twin Cities (2012)

25. Svåsand, E., Skjeltorp, A.T., Akselvoll, J., Helgesen, G.: Local viscosity measurements using oscillating magnetic holes. J. Appl. Phys. **101**(5), 054910 (2007)
26. Panton, R.L.: Incompressible Flow, pp. 609–611. Wiley, New Jersey (2005)
27. Ceylan, K., Herdem, S., Abbasov, T.: A theoretical model for estimating of drag force in the flow of non-Newtonian fluids around spherical solid particles. Powder Technol. **103**(3), 286–291 (1999)
28. Chang, I.D.: On the wall effect correction of the stokes drag formula for axially symmetric bodies moving inside a cylindrical tube. Z. Angew. Math. Phys. **12**(1), 6–14 (1961)
29. Berdan II, C., Leal, C.G.: Motion of a sphere in the presence of a deformable interface: I. Perturbation of the interface from flat: the effects on drag and torque. J. Colloid Interface Sci. **87**(1), 62–80 (1982)
30. Cairncross, R.A., Francis, L.F., Scriven, L.E.: Predicting drying in coatings that react and gel: drying regime maps. AICHE J. **42**(1), 55–67 (1996)
31. Basu, S.K., Scriven, L.E., Francis, L.F., McCormick, A.V.: Mechanism of wrinkle formation in curing coatings. Prog. Org. Coat. **53**(1), 1–16 (2005)
32. Kim, J.C., Hillmyer, M.A., Francis, L.F.: Magnetic microrheology of block copolymer solutions. ACS Appl. Mater. Interfaces. **5**(22), 11877–11883 (2013)
33. Song, J.-O., McCormick, A.V., Francis, L.F.: Depthwise viscosity gradients in UV-cured epoxy coatings. Macromol. Mater. Eng. **298**(2), 145–152 (2013)
34. Biermann, C.J.: Handbook of Pulping and Paper Making, pp. 244–245. Academic Press, San Diego (1996)
35. Smook, G.A.: Handbook for Pulp and Paper Technologists, pp. 319–324. Angus Wild Publications, Inc., Vancouver (2003)
36. Allen, A.J., Lock, G.: Creping release agents. US Patent 5,660,687, 26 Aug 1997
37. Furman, G.S., Li, X.H., Su, W., Grigoriev, V.A.: Modifying agent for yankee coatings. US Patent 8,101,045 B2, 24 Jan 2012

Chapter 6
CryoSEM: Revealing Microstructure Development in Drying Coatings

Kyle Price, Alon V. McCormick, and Lorraine F. Francis

Introduction

The final microstructure of a coating is one of its most important properties as it often dictates the coating performance. Every aspect of the coating process including formulation, deposition, and processing has the potential to influence how the microstructure develops within the coating. Many characteristics of the coating microstructure including porosity, particle distribution, and particle orientation develop as the coating dries. Additionally, during the drying process the solvent concentration, coating thickness, and coating stress change dramatically and drive changes in the coating microstructure. The process of microstructure development during drying can be inferred from measurements made on the final coating, predicted using analytical models, or it can be characterized directly as it develops using the few experimental techniques available to study wet coatings.

Cryogenic scanning electron microscopy (cryoSEM) is a technique that was first developed by biologists as a method to image cells without the need for chemical fixation, which dehydrates the cells and destroys the native cellular structure [1–3]. CryoSEM has also found increasing applications in materials science [4–7] and coating characterization [8–14]. CryoSEM has proven to be a powerful technique in characterizing microstructure development in coatings during drying as it provides high-resolution images of the wet coating microstructure. CryoSEM can

K. Price (✉)
Department of Chemical Engineering and Materials Science, University of Minnesota, Minneapolis, MN, USA

Axalta Coating Systems, Coatings Technology Center, Wilmington, DE, USA
e-mail: kyle.price@axaltacs.com

A.V. McCormick • L.F. Francis
Department of Chemical Engineering and Materials Science, University of Minnesota, Minneapolis, MN, USA

M. Wen, K. Dušek (eds.), *Protective Coatings*, DOI 10.1007/978-3-319-51627-1_6

also be used to characterize initial coating dispersions and phenomenon like freeze–thaw stability. To prepare coating samples for cryoSEM analysis, liquid samples are first vitrified at cryogenic temperatures before the sample is imaged. As a solid, the vapor pressure of the vitrified sample is low and can survive the high vacuum environment of the electron microscope. The vitrification process arrests the dynamic processes occurring in the sample, allowing high-resolution imaging of intermediate or transient states in the coating during microstructure development. Another benefit of vitrification is that the solid sample can be fractured, allowing characterization of the sample's cross section and observation of microstructure development throughout the coating thickness.

The goal of this chapter is to introduce cryoSEM as a characterization tool for studying microstructure development in coatings. A brief description of a typical cryoSEM sample preparation procedure, which highlights the vitrification, cryo-fracture, and sublimation processes, is provided. Lastly, several examples are presented to show how cryoSEM has been used to characterize coatings for practical applications.

Scanning Electron Microscopy

Electron microscopy is a useful characterization tool and one of the most widely used methods for obtaining high-resolution images of solid materials. A scanning electron microscope (SEM) produces images of the sample surface by focusing a beam of electrons and scanning across a specified area of the sample. The primary electrons in the electron beam interact with the sample emitting secondary electrons and backscattered electrons from the specimen surface that are collected by special detectors [15]. Images are produced by mapping the detected signals with the position of the electron beam.

Secondary electrons are low-energy electrons (~50 eV) expelled when the primary electrons from the electron beam scatter inelastically after interaction with valence electrons of atoms in the sample. Backscattered electrons are electrons that have interacted elastically with the sample and can have energy close to that of the primary beam electrons. Contrast in SEM is largely achieved through the topography of the sample and to a lesser extent the electron density (atomic number) of the atoms in the sample. The secondary electron signal is more sensitive to topographic contrast, whereas the backscattered electron signal is more sensitive to atomic number contrast in the sample. Most modern SEM instruments have both secondary electron and backscattered electron detectors, and often the two signals can be mixed to optimize the quality of the captured image.

The SEM is a versatile instrument that can accept a wide variety of sample types. However, due to the high vacuum requirements of the sample chamber (~ 1×10^{-6} Pa), liquid-containing samples cannot be imaged with a conventional SEM. It is possible to image "wet" or liquid-containing samples in an environmental scanning electron microscope (ESEM) or variable pressure SEM. Specialized detectors and differential pumping systems allow the sample chamber in an ESEM to have a higher pressure allowing the control of solvent evaporation during the imaging process [16]. However, the resolution attainable by an ESEM is lower than that of a conventional SEM, and only surface images can be obtained using an ESEM because the sample cannot be fractured to reveal the sample cross section [17, 18]. An alternative to this restriction is cryoSEM, where liquid samples are first vitrified, immobilizing the developing microstructure, and imaged in a high vacuum environment using a conventional SEM equipped with a cold stage.

Cryogenic Scanning Electron Microscopy (cryoSEM)

The principle of cryoSEM relies on the fact that fast freezing or vitrification of the sample immobilizes the kinetic processes within the sample, allowing intermediate or transient states in the coating to be studied. The low vapor pressure of the solidified sample resolves the high vacuum problem, and low temperature imaging can reduce radiation damage from the electron beam for beam-sensitive materials. Researchers at the University of Minnesota have pioneered the use of cryoSEM to study the microstructure development in a wide variety of coating systems. For example, cryoSEM has been used to characterize film formation in latex coatings [8–10], study microstructure development latex composites [14], investigate particle distributions in drying coatings [11, 12] and reveal the origins of stress development in drying particulate coatings [19].

Sample preparation for a typical cryoSEM sample involves fast freezing, cryofracture, sublimation, and imaging. Each step of the procedure is conducted at low temperature to prevent recrystallization of the vitrified water in order to maintain the original microstructure of the sample. The fracture, sublimation, and imaging steps are all performed under vacuum to protect the freshly formed fracture surface from exposure to atmospheric moisture that can condense on the sample surface, creating ice and frost contamination. Once sample preparation is complete, samples are transferred to a conventional SEM equipped with a cold stage capable of maintaining a temperature of approximately −150 °C during imaging.

Vitrification

One of the most important steps in a cryoSEM experiment is the vitrification of the liquid sample so it can be imaged under high vacuum conditions. Vitrification of the liquid, rather than freezing, is desired to ensure that the original microstructure of the sample is accurately represented. Proper vitrification of the solvent produces an amorphous solid, whereas freezing produces a crystalline solid. During freezing, the formation and growth of crystals in the sample rearranges the native microstructure due to the density changes that accompany the phase transformation.

Whether a liquid vitrifies or freezes depends on the cooling rate of the sample and the crystallization kinetics (nucleation and growth rates) [20].

Although preferred, vitrification is not strictly necessary so long as any crystallization that occurs upon cryofixation is small enough in scale that it does not disrupt the microstructure that needs to be visualized. For instance, ice crystals that are much smaller than the particles in the coating may not interfere with the characterization of the relevant microstructure.

Pure water is one of the most difficult liquids to vitrify and requires a very high cooling rate of approximately 10^6 K·s^{-1} [21]. Dissolved species, such as surfactants, depress the freezing point of water, aiding in vitrification [22]. Solutions are also easier to vitrify than pure water as most solutes act as cryo-protectants that delay the nucleation and growth of ice crystals [23, 24].

There are two main methods employed to cool the sample, high-pressure freezing and plunge freezing into a liquid cryogen. The plunge freezing technique can produce fast cooling rates of up to 10^4 K·s^{-1}, while the high-pressure freezing technique can only achieve rates of ~10^2–10^3 K·s^{-1} [25]. Achieving rapid cooling rates complicates cryoSEM sample preparation; however, it is possible to produce cryoSEM images that accurately represent the state of the liquid sample [8–14, 19].

High-pressure freezing, or freezing at 210 MPa, slows the kinetics of crystallization by delaying the onset of ice nucleation and slowing down crystal growth. Both the freezing point and supercooling point of pure water reach a minimum at 210 MPa; the phase diagram for ice is shown in Fig. 6.1 [26]. The supercooling point refers to the temperature at which homogeneous ice crystal nucleation starts. Additionally, the viscosity of supercooled water is roughly 1000 times higher than the viscosity of room temperature water. This increased viscosity slows crystal

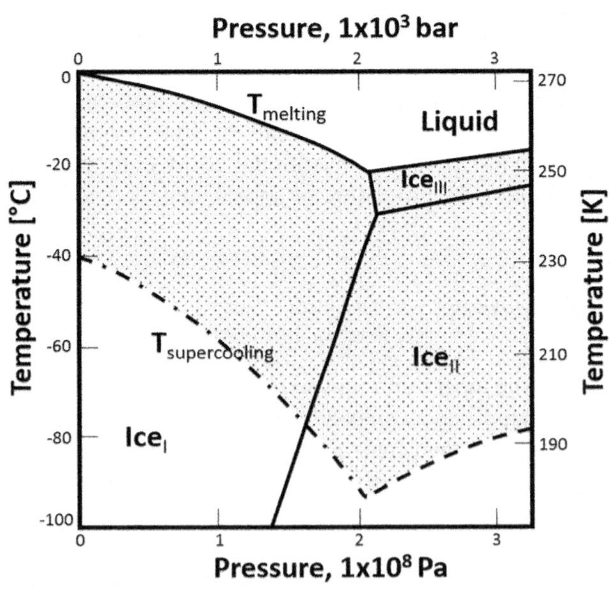

Fig. 6.1 Phase diagram of ice (adapted from [26]). The solid line represents the solid–liquid equilibrium curve, and the dashed line is the supercooling curve. Ice_I, Ice_{II}, and Ice_{III} represent different phases of solidified water

growth by lowering the diffusion rate of water molecules to the nucleated ice crystal. For proper vitrification, the Ice_{I} phase is required as both Ice_{II} and Ice_{III} have a higher density than liquid water. Using the high-pressure freezing technique, it is possible to vitrify thick aqueous samples, while plunge freezing is commonly used for thinner samples. High-pressure freezing is most often used to characterize solutions and dispersions in their initial state before drying.

Plunge freezing a sample into a cryogen cools the sample by removing heat from the sample surface through conduction and convection. Immediately after the sample is plunged into the cryogen, conduction takes place between the sample surface and the adjacent cryogen. Heat is carried away from the sample to the bulk cryogen by convection. Conductive and convective heat transfer controls the overall cooling rate for very thin samples (~20 μm). For thin samples, the cooling rate can be optimized by controlling the plunge velocity, cryogen depth, and insertion angle [27, 28]. However, for thicker samples, heat conduction from the interior of the sample to the sample–cryogen interface is the limiting step. For samples thicker than 20 μm, the cooling rate cannot be improved by optimization of the plunge freezing technique, and the extent of well-frozen zones is independent of the quench method [27, 29]. Another factor that must be considered is whether the surface of the sample will be disturbed by the plunging conditions.

Successful cryoSEM sample preparation using the plunge freezing technique begins with the proper choice of cryogen. To optimize the heat transfer from the sample, a good cryogen should have a low melting point, high boiling point, high thermal conductivity, and low viscosity [30]. If the cryogen has a boiling point close to its melting point, bubbles will immediately nucleate on the sample surface during the cooling process, creating an insulating layer of gas between the sample surface and the cryogen. This effect should be avoided to maximize the sample cooling rate. Liquid ethane represents the most effective cryogen as the cooling rate can reach $1.2 \times 10^4\ \mathrm{K \cdot s^{-1}}$, which was measured by plunging bare thermocouples into the liquid [31]. Using the same method, the efficiency of different cryogens can be ranked in descending order as ethane, propane, propane:pentane mixtures, Freon 22, and liquid nitrogen [31]. Liquid ethane removes heat from the sample surface through a subcooled nucleate boiling mechanism, the most effective method of heat transfer. Bubbles nucleate on the sample surface after being plunged into the cryogen; these bubbles are carried to the bulk cryogen by convection and quickly collapse due to the low temperature of the bulk cryogen [30]. This circulation repeats itself removing heat quickly and efficiently from the sample.

Once a sample is frozen, the dynamic processes occurring within the sample are halted. To characterize the evolution of the coating microstructure, a "time-sectioning" technique is often employed where a series of samples are prepared identically but dried for successively longer times before freezing the sample [7]. Using this method, it is possible to image the changes that develop in the coating microstructure as a function of time. The time-sectioning technique can be very powerful in revealing the complex processes that occur within a coating during drying.

Cryo-fracture

Cryo-fracture or freeze-fracture is a necessary step in cryoSEM sample preparation in order to visualize the internal structure of the sample. Cryo-fracture can expose the internal structures present in a dispersion and reveal the particle distribution throughout a cross section of a coating. There are two basic modes of fracture: ductile fracture and brittle fracture. Brittle fracture involves very little plastic deformation of the sample, whereas ductile fracture involves both deformation and fracture. For cryoSEM, brittle fracture is preferred to prevent any structural changes that would accompany plastic deformation. It is even possible for ductile materials like polymers to undergo brittle fracture below the ductile-to-brittle temperature of the sample. For polymers, the glass transition temperature is often used as an estimate of whether the sample will undergo ductile or brittle fracture.

During freeze-fracture, the sample breaks or fractures along the direction that is least resistant to an external force. When characterizing coatings using cryoSEM, a substrate is necessary to support the coating, as the sample is normally fluid or wet before it is frozen. Therefore, it is necessary for both the sample and the substrate to fracture, which necessitates the need to use brittle substrates like glass or silicon wafers. These substrates can be scored using a sharp tool prior to sample preparation, offering some control over the location and direction of the fracture. Substrates can also be processed with specific cleaning procedures before sample application to aid in substrate-sample adhesion during subsequent freeze-fracture.

Freeze-fracture can also be a useful tool in characterizing film formation in latex coatings. When latex, a waterborne dispersion of high molecular weight polymer particles is freeze-fractured; the latex particles embedded in the frozen matrix often undergo plastic deformation forming an artifact referred to as a "pullout" (Fig. 6.2) [32]. The formation of these artifacts is unusual because fracture is initiated at temperatures well below the nominal glass transition temperature of the bulk polymers, and one would expect brittle fracture to occur. The appearance of these artifacts is often an indicator of the degree of film formation in the coating [32].

For example, if a latex particle was fully dispersed in water when it was frozen, it will have strong adhesion to the surrounding frozen matrix. As a fracture surface encounters the embedded particle, the fracture travels through the center of the particle and the polymers are stretched into "horn"- or "dog-bone"-type shapes as the fracture surface forms [32]. The cryoSEM micrograph in Fig. 6.2 shows the appearance of pullouts formed from a latex that was frozen in its fully dispersed state. Some particles may be removed from the surface during the fracture process, resulting in crater-like structures on the frozen fracture surface.

After further drying, the once dispersed latex particles become consolidated and achieve touching contact. Very little water remains between the particles at this stage of film formation, and interparticle adhesion between is weak. When the sample is fractured during this stage of film formation, the fracture surface will preferentially travel along the weakly bonded particle boundaries, and pullouts will not form [32]. Some pullouts are still possible depending on the amount of particle coalescence and interdiffusion.

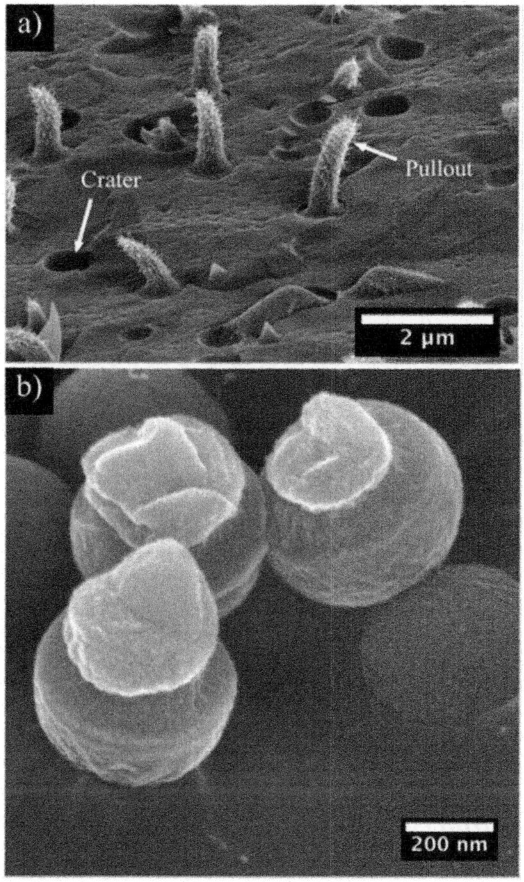

Fig. 6.2 (**a**) CryoSEM micrographs showing the formation of pullouts from a frozen dispersion of latex particles. The fine hairlike structure on the pullouts is believed to be a form of ice contamination. (**b**) CryoSEM image of latex particle "pullouts" that have been revealed through sublimation of the frozen water matrix

Later in the film formation process, the latex particles will have undergone extensive deformation, and the polymer chains have had sufficient time to diffuse across particle boundaries, increasing the cohesive strength of the network. Consequently, pullouts that form during this stage of drying will bridge multiple particles and form weblike structure on the fracture surface [32]. When film formation process is complete, particle boundaries no longer exists, and the resulting facture surface will appear smooth, without pullouts.

Sublimation

It can be beneficial during SEM characterization to sublime away some of the ice matrix surrounding the sample in order to expose more of the topography of the fracture surface and expose more of the coating solids [33]. In other situations, no intentional sublimation is performed so that the original concentration and location of water in the sample can be preserved [19]. If too much sublimation occurs, particles that were once embedded in the ice matrix can become dislodged and fall onto the sample surface. This technique can be beneficial for characterization of the coating solids; however, the equilibrium arrangement and concentration of particles are no longer preserved.

Frozen water, or any other frozen solvent, sublimes when it is subjected to a lower pressure than its equilibrium vapor pressure at the sublimation temperature. The rate of sublimation can be controlled by adjusting the sample temperature or the pressure within the preparation chamber. Increasing either the sample temperature or the vacuum increases the sublimation rate; however, if the temperature is raised too high, recrystallization of the frozen water can occur, and ice crystals can form within the sample. Therefore, the sublimation temperature is usually set just below the recrystallization temperature to allow for reasonable sublimation rates while protecting against recrystallization.

Sublimation of water from the frozen fracture surface of a coating is more complicated than the sublimation of pure ice. Particles, surface morphology, cosolvents, and various solutes affect the sublimation rate. During sublimation, solutes are left behind and concentrate on the sample surface, blocking pathways through which water vapor molecules can escape [30]. Likewise, particles protruding from the sample surface will slow the sublimation rate compared to a smooth frozen solvent surface. After sublimation, samples can be coated with a thin conductive layer (e.g., platinum) to help reduce charging effects.

Artifacts of CryoSEM Sample Preparation

As discussed previously, when the freezing rate of a sample is too slow, ice crystals nucleate and grow within the sample and foreign solids are pushed to the grain boundaries as the crystal grows. Rearrangement of the coating solids destroys the microstructure present prior to freezing. Poor freezing is identifiable by the presence of particle-free zones surrounded by a particle-rich zone where the excluded particles are deposited. Poor freezing typically occurs when the sample is too thick, or the sample is frozen in a poor cryogen (Fig. 6.3). Even if no ice crystals form during the freezing process, crystallization of the frozen water can occur if the temperature of the sample increases beyond its recrystallization temperature.

For proper sample vitrification, it is necessary to use a very small sample size; however, the drying of very small samples is dominated by edge effects. Lateral or edge-in drying can alter the microstructure development of the drying sample and may not be representative of microstructure development in larger samples. Altering the substrate by including borders or walls limits lateral drying effects and promotes drying uniformity for very small samples [11].

If the frozen sample is exposed to ambient atmosphere, moisture in the air is likely to condense onto the cold sample surface, resulting in frost contamination.

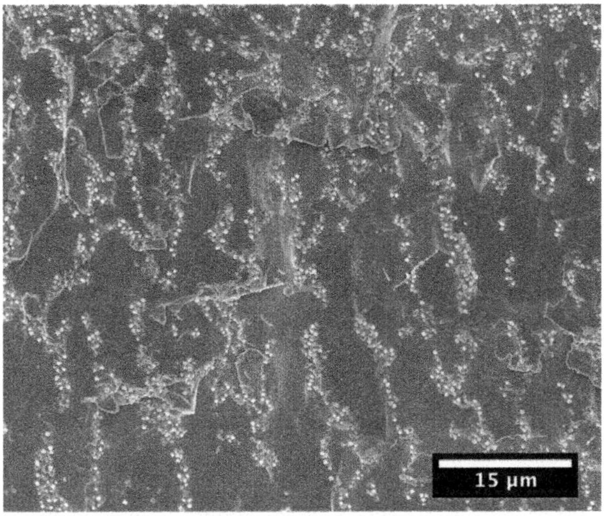

Fig. 6.3 CryoSEM micrograph of a sample in which poor freezing resulted in ice crystal growth and particle rearrangement in the frozen sample. Particle-free and particle-rich zones are clearly evident throughout the sample

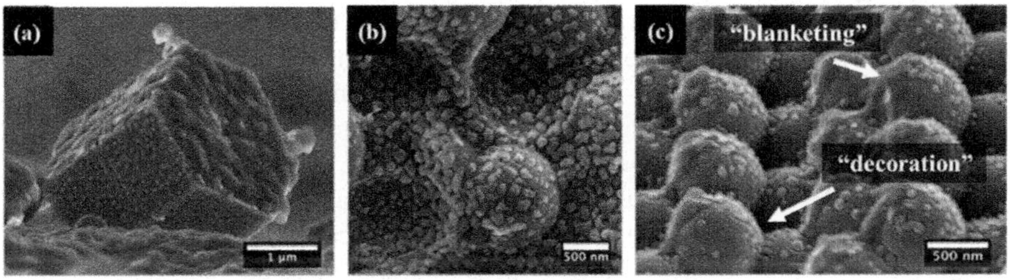

Fig. 6.4 CryoSEM micrographs showing the different appearance of ice and frost contamination on the sample surface: (**a**) large hexagonal ice crystal on the sample surface, (**b**) speckled frost "decoration" covering the fracture surface, (**c**) directional frost "blanketing" and speckled frost "decoration"

Frost contamination can take many forms and has a wide variety of appearances, including large ice crystals on the sample surface, a thin layer of ice covering the sample surface, or small specks of contamination that appear uniformly distributed across the sample surface (Fig. 6.4). Identifying frost contamination and differentiating frost from particles in the sample can be difficult depending on the type of contamination and the size and morphology of particles being investigated. Frost contamination on the sample surface is not necessarily detrimental to the original microstructure of the coating, as this contamination occurs after the sample has been frozen. However, extensive frost contamination may make it difficult to accurately characterize the microstructure of the sample.

Examples of CryoSEM Characterization in Coatings Research

Many important microstructural features of particle-based coatings like porosity, particle order, and particle orientation are established during drying. Understanding how particle distribution varies within the coating during drying is useful in designing coating formulations and defining processing conditions.

As an example, Cardinal et al. [12] developed a model predicting the particle distribution through the depth of a coating at various drying times considering factors including particle sedimentation, particle diffusion, and evaporation. The results from the model have been condensed into a drying regime map that can be used to predict whether evaporation, sedimentation, or diffusion dominates the drying behavior for a specific set of drying conditions.

To verify the model predictions, cryoSEM was used to image the intermediate stages of drying. For a given set of conditions, the model predicts that the particles will sediment at a speed that is slightly slower than the velocity of the descending free surface. In this example, both a particle sediment layer near the substrate and a particle skin along the free surface were predicted to form simultaneously during drying. This behavior was observed directly by freezing the coating at an intermediate drying time and imaging the frozen cross section using cryoSEM (Fig. 6.5). A well-ordered particle packing is present along the free surface of the coating, as well as a disordered particle packing near the substrate from sedimentation effects.

CryoSEM has also been used to image variations in particle concentration laterally within a coating. Near the edges of a discrete coating or around the perimeter of a drying drop, particle accumulation near the edges leads to a

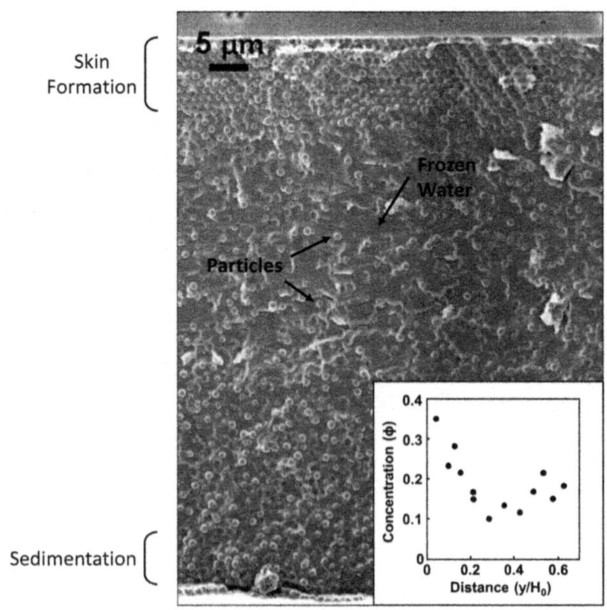

Fig. 6.5 CryoSEM cross section image of a coating showing the simultaneous formation of a particle skin along the free surface and a sedimentation layer along the substrate. The inset shows the volume fraction of particles as a function of depth ($y/H_0 = 0$ represents the free surface). Reprinted with permission from [12]. Copyright (2010) American Institute of Chemical Engineers

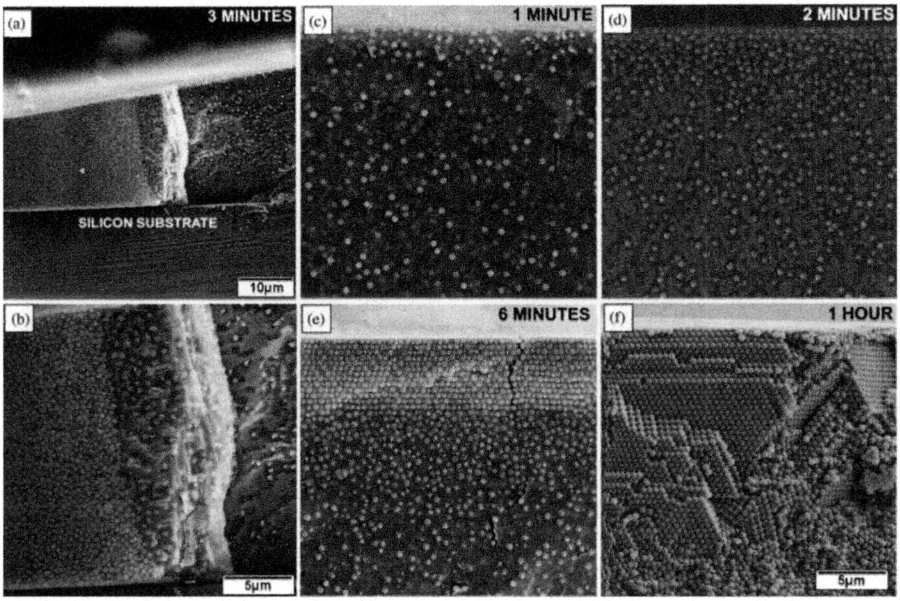

Fig. 6.6 CryoSEM micrographs of the formation and progression of lateral (**a** and **b**) and top-down consolidation fronts (**c–f**). The image shown in (**b**) is a higher magnification view of the fracture step shown in (**a**) that occurs between the suspension and the consolidated material. The fracture step appears as a vertical bright stripe in both images. Reprinted with permission from [8]. Copyright (2005) Progress in Organic Coatings

phenomenon commonly known as lateral drying or the coffee ring effect [34–36]. A coating or drop on a planar surface is thinnest near the edges where the liquid meets the substrate. At the edges, particle consolidation occurs quickly, and further drying from the consolidated edges results in the formation of curved menisci between adjacent particles and the development of a pressure gradient in the coating. Subsequently, liquid and particles flow from the bulk of the coating to the edges, depositing particles near the edges and growing the particle packing from the edges inward. The boundary between the consolidated particle packing and the bulk coating is commonly referred to as the "consolidation front." The presence of a consolidation front can be inferred by observing changes in the visual appearance of the coating due to changes in light scattering efficiency stemming from variations in particle concentration. This phenomenon has been observed directly using cryoSEM to study changes in particle distribution within the coating [8, 9, 11]. Figure 6.6 shows the formation of both a lateral consolidation front and a top-down consolidation front in the coating. As time progresses the lateral consolidation front moves from the edges toward the center of the coating, while the thickness of the top-down accumulation front grows with drying time.

CryoSEM has also been used to investigate how microstructures of coatings develop from the starting dispersions. Electrostatic interactions between ceramic particles and latex particles in a coating can be predicted using DLVO theory-based

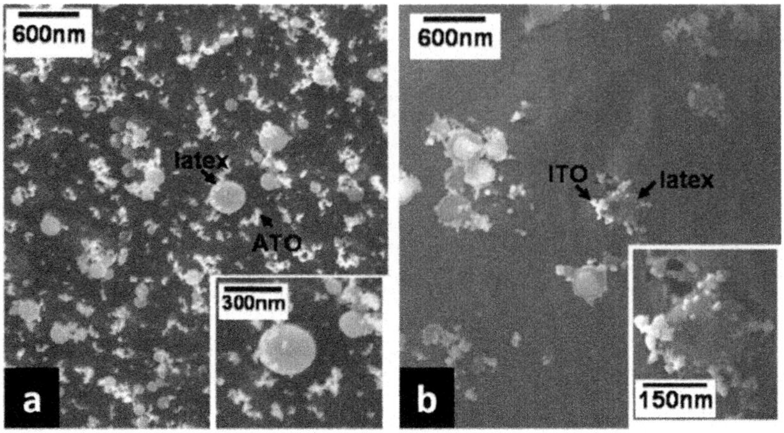

Fig. 6.7 CryoSEM images of (**a**) a frozen ATO/latex dispersion (pH = 3), showing individual latex particles and ATO clusters distributed homogeneously in the frozen water, and (**b**) a frozen ITO/latex dispersion (pH = 3), showing ITO particles adsorbed onto surfaces of latex particles in frozen water. Reprinted with permission from [14]. Copyright (2007) Journal of Colloid and Interface Science

calculations [37]. These predictions were confirmed using cryoSEM by imaging particle distributions in mixtures of ceramic and latex particles [14]. CryoSEM images show that small clusters of antimony-doped tin oxide (ATO) particles and latex particles distribute themselves homogeneously in the dispersion. At a pH value of 3, both the ATO particles and the latex particles are negatively charged and electrostatically repel each other in the dispersion. Alternatively, the isoelectric point of indium tin oxide (ITO) particles is predicted to be around a pH value of 6; at a pH value of 3 the ITO particles are positively charged and the latex particles are negatively charged. Electrostatic attraction causes the ITO particles to adsorb onto the surface of the latex particles. Figure 6.7 shows the dispersion microstructures for ATO/latex and ITO/latex dispersions imaged using cryoSEM [14]. The observed microstructures confirm the theoretical predictions made by the calculation of total interaction potentials from DLVO theory [37].

In the final example, cryoSEM has been used to show the microstructural origin of stress development that occurs during drying for coatings made from aqueous dispersions of silica particles [19]. Past studies indicated that stress development in the coating was related to the degree of saturation; however, these predictions were made without high-resolution imaging of the coating microstructure [35, 38, 39]. Stress evolution was monitored using a cantilever beam bending technique with modified walled cantilevers to increase drying uniformity and limit lateral drying [19]. At various drying times, the coatings were frozen, and the coating microstructure was characterized using cryoSEM as shown in Fig. 6.8.

The cryoSEM micrographs reveal that stress development in the coating is dependent on the development of capillary pressure and the degree of saturation of the particle network. In the early stages of drying, there is no stress development

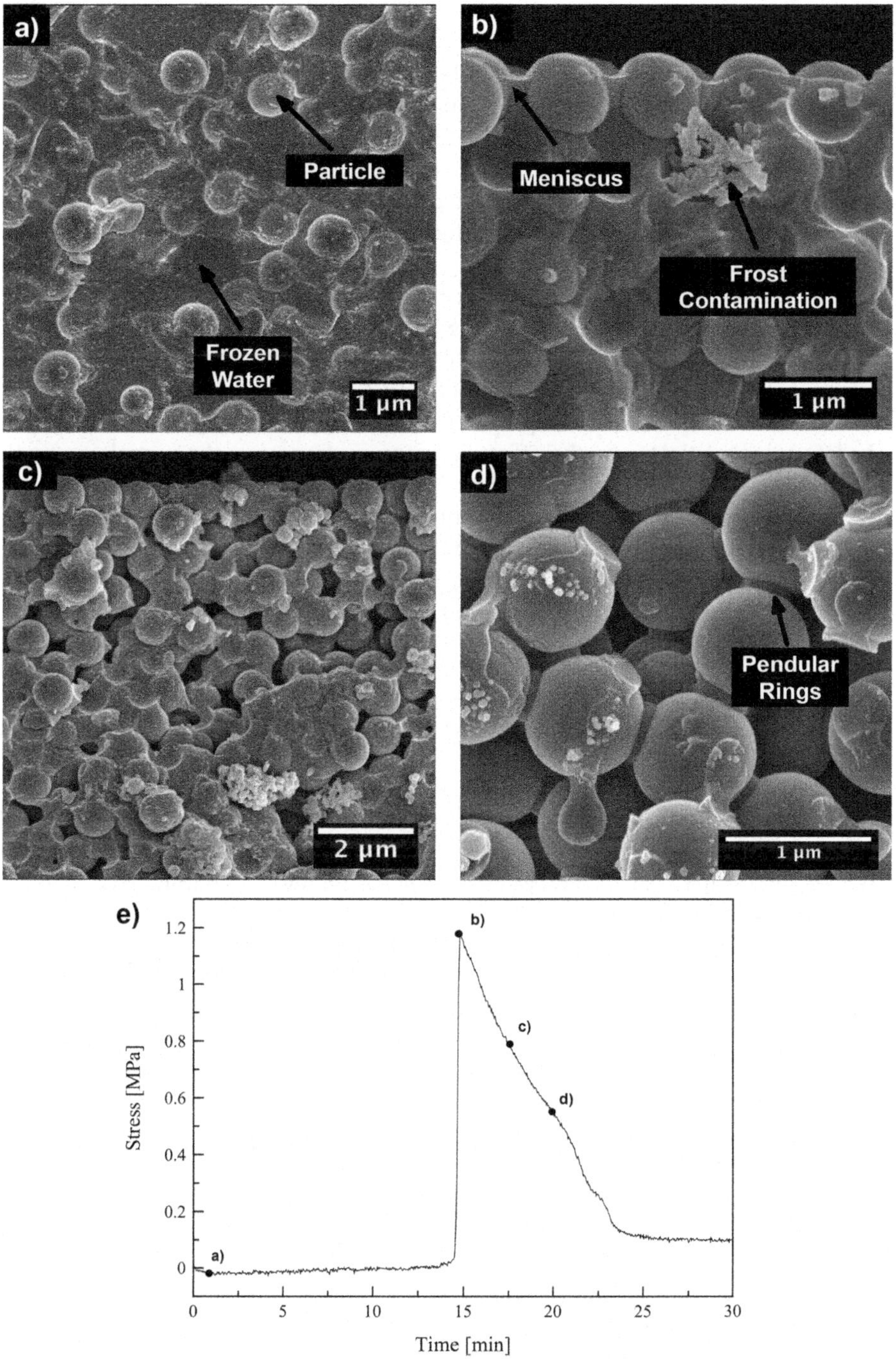

Fig. 6.8 CryoSEM cross-sectional micrographs of aqueous silica particle coatings dried for (**a**) 1 min, (**b**) 14 min, (**c**) 17 min, and (**d**) 20 min. Stress development curve for the silica particle coating measured using a cantilever beam bending technique (**e**), the corresponding microstructures are marked on the plot. Reprinted with permission from [19]. Copyright (2015) Journal of the American Ceramic Society

in the coating as the coating is still fluid and no consolidated particle network has formed (Fig. 6.8a). Once enough drying has occurred, a consolidated particle network forms, and curved menisci develop between adjacent particles along the free surface of the coating (Fig. 6.8b). Once these menisci form, the capillary pressure acts to further consolidate the particle network into a smaller volume. Constrained shrinkage of the coating by the substrate leads to the development of an in-plane tensile stress.

With further drying, the particle packing begins to empty, and the liquid–air interface moves into the coating, creating air-filled pores (Figs. 6.8c and 6.8d). The effective pressure acting on the particle network lowers as the volume fraction of water-filled pores decreases [40]. The effective capillary pressure acting on the particle network is the sole driving force for shrinkage for coatings made from hard, nondeforming particles. As the effective pressure decreases, the particle network relaxes to its equilibrium position and the measured stress in the coating decays. The combination of cryoSEM visualization of the wet-coating microstructure and in situ stress measurement has revealed that stress development is directly related to the state of saturation or extent of drying in the coating. Direct visualization of the coating during the intermediate stages of drying is instrumental in understanding the fundamental mechanism of stress development in these hard particle coatings.

Conclusions

The goal of this chapter was to provide a brief introduction to cryoSEM and to highlight its utility in characterizing the microstructure of coatings. To achieve desirable coating properties, it is critical to understand the initial dispersion microstructure and also the process in which microstructure develops as the coating dries or solidifies. CryoSEM has proven to be a powerful tool to explore the microstructure of wet coating samples and can reveal the dynamic processes of microstructure development with time. However, cryoSEM sample preparation can be quite challenging, and care is required when interpreting cryoSEM images. Poor freezing, ice contamination, and numerous other artifacts of cryoSEM sample preparation can all influence the ability to properly interpret the microstructure present in the coating. There are several commercial cryoSEM systems available that aim to make sample preparation easier and more reproducible, which should advance the use of this powerful characterization technique. As one of the few experimental techniques that can study the microstructure of wet coatings, cryoSEM promises to be a useful tool in coatings research.

Acknowledgments The authors would like to gratefully acknowledge support from the National Science Foundation under Award No. CBET 0967348, and the industrial supporters of the Coating Process Fundamentals Program (CPFP), part of the industrial partnership IPrime at the University of Minnesota. Parts of this work were carried out in the Characterization Facility, University of Minnesota, which receives partial support from NSF through the MRSEC program.

Acknowledgments also go to Chris Frethem for all of his contributions to the work presented here. The authors are also grateful to the late Professors L. E. Scriven and H. T. Davis and the past graduate student researchers in CPFP for their pioneering work in developing cryoSEM techniques for coatings characterization.

References

1. Apkarian, R.P.: High resolution Cryo-SEM in the biological sciences. Microsc. Microanal. **9**, 272 (2003)
2. Bastacky, J., Wodley, C., LaBrie, R., Backhus, C.: A bibliography of low-temperature scanning electron microscopy (LTSEM, Cryo-SEM) and scanning electron microscopy of frozen hydrated biological systems. Scanning. **9**, 219–225 (1987)
3. Erlandsen, S.L., Ottenwaelter, C., Frethem, C., Chen, Y.: Cryo field emission scanning electron microscopy. BioTechniques **31**, 300–302, 304–305 (2001)
4. Wyss, H.M., Hutter, M., Muller, M., Meier, L.P., Gauckler, L.J.: Quantification of microstructure in stable and related suspensions from Cryo-SEM. J. Colloid Interface Sci. **248**, 340–346 (2002)
5. Apkarian, R.P., et al.: In-Lens cryo-high resolution scanning electron microscopy: methodologies for molecular imaging of self-assembled organic hydrogels. Microsc. Microanal. **9**, 286–295 (2003)
6. Vermant, J., Cioccolo, G., Golapan Nair, K., Moldenaers, P.: Coalescence suppression in model immiscible polymer blends by nano-sized colloidal particles. Rheol. Acta. **43**, 529–538 (2004)
7. Prakash, S.S., Francis, L.F., Scriven, L.E.: Microstructure evolution in dry cast cellulose acetate membranes by cryo-SEM. J. Membr. Sci. **283**, 328–338 (2006)
8. Ma, Y., Davis, H.T., Scriven, L.E.: Microstructure development in drying latex coatings. Prog. Org. Coat. **52**, 46–62 (2005)
9. Sutanto, E., Ma, Y., Davis, H.T., Scriven, L.E.: In: Provder, T., Urban, M. (eds.) Film Formation in Coatings: Mechanisms, Properties, and Morphology, p. 174. Oxford University Press, Washington, DC (2001)
10. Gong, X., Davis, H.T., Scriven, L.E.: Role of van Der Waals force in latex film formation. J. Coat. Technol. Res. **5**, 271–283 (2008)
11. Price, K.K., McCormick, A.V., Francis, L.F.: CryoSEM investiration of latex coatings dried in walled substrates. Langmuir. **28**, 10329–10333 (2010)
12. Cardinal, C.M., Jung, Y.D., Ahn, K.H., Francis, L.F.: Drying regime maps for particulate coatings. AIChE J. **56**, 2769–2780 (2010)
13. Roberts, C.C., Francis, L.F.: Drying and cracking of soft latex coatings. J. Coat. Technol. Res. **10**, 441–451 (2013)
14. Luo, H., Scriven, L.E., Francis, L.F.: Cryo-SEM studies of latex/ceramic nanoparticle coating microstructure development. J. Colloid Interface Sci. **316**, 500–509 (2007)
15. Goldstein, J.: Scanning Electron Microscopy and X-Ray Microanalysis: A Text for Biologists, Materials Scientists, and Geologists, p. 820. Plenum Press, New York (1992)
16. Danilatos, G.D.: Foundations of environmental scanning electron microscopy. Adv. Electronics Electron Phys. **71**, 109–250 (1988)
17. Donald, A.M.: The use of environmental scanning electron microscopy for imaging wet and insulating materials. Nat. Mater. **2**, 511–516 (2003)
18. Meredith, P., Donald, A.M.: Study of 'wet' polymer latex systems in environmental scanning electron microscopy: some imaging considerations. J. Microsc. **181**, 23–25 (1996)
19. Price, K., Wu, Y., McCormick, A.V., Francis, L.F.: Stress development in hard particle coatings in the absence of lateral drying. J. Am. Ceram. Soc. **98**, 2214–2222 (2015)

20. Turnbull, D.: Under what conditions can a glass be formed. Contemp. Phys. **10**, 473–488 (1969)
21. Hobbs, P.V.: Ice Physics, p. 837. Clarendon Press, Oxford (1974)
22. Devireddy, R.V., Leo, P.H., Lowengrub, J.S., Bischof, J.C.: Measurement and numerical analysis of freezing in solutions enclosed in a small container. Int. J. Heat Mass Transf. **45** (9), 1915–1931 (2002)
23. Steinbrecht, R.A., Zierold, D.: Cryotechniques in Biological Electron Microscopy, p. 297. Springer, Berlin (1987)
24. Angell, C.A., Choi, Y.: Crystallization and vitrification in aqueous systems. J. Microsc. **141**, 251–261 (1986)
25. Severs, N.J., Shotton, D.: Rapid Freezing, Freeze Fracture and Deep Etching, p. 372. Wiley-Liss, New York (1995)
26. Robards, A.W., Sleytr, U.B.: Low Temperature Methods in Biological Electron Microscopy, p. 551. Elsevier, New York (1985)
27. Elder, H.Y., Gray, C.C., Jardine, A.D., Chapman, J.N., Biddlecombe, W.H.: Optimum conditions for cryoquenching of small tissue blocks in liquid coolants. J. Microsc. **126**, 45–62 (1982)
28. Ge, H., Suszynski, W.J., Davis, H.T., Scriven, L.E.: New controlled environment vitrification system for preparing wet samples for cryo-SEM. J. Microsc. **229**, 115–126 (2008)
29. Bald, W.B.: Optimizing the cooling block for the quick freeze method. J. Microsc. **131**, 11–23 (1983)
30. Echlin, P.: Low-Temperature Microscopy and Analysis, p. 539. Plenum Press, New York (1992)
31. Ryan, K.P., Purse, D.H., Robinson, S.G., Wood, J.W.: The relative efficiency of cryogens used for plunge-cooling biological specimens. J. Microsc. **145**, 89–96 (1987)
32. Ge, H.: Microstructure development in latex coatings: high-resolution cryo-scanning electron microscopy. Ph.D. thesis, University of Minnesota, Minneapolis, MN (2005)
33. Fukuta, N., Gramada, C.M.: Vapor pressure measurement of supercooled water. J. Atmos. Sci. **60**, 1871–1875 (2003)
34. Deegan, R., Bakajin, O., Dupont, T., Huber, G.: Capillary flow as the cause of ring stains from dried liquid drops. Nature. **389**, 827–829 (1997)
35. Chiu, R.C., Cima, M.J.: Drying of granular ceramic films: II, drying stress and saturation uniformity. J. Am. Ceram. Soc. **76**, 2769–2777 (1993)
36. Routh, A.F., Russel, W.B.: Horizontal drying fronts during solvent evaporation from latex films. AIChE J. **44**, 2088–2098 (1998)
37. Sun, J., Velamakanni, B.V., Gerberich, W.W., Francis, L.F.: Aqueous latex/ceramic nanoparticle dispersions: colloidal stability and coating properties. J. Colloid Interface Sci. **280**, 387–399 (2004)
38. Guo, J.J., Lewis, J.A.: Aggregation effects on the compressive flow properties and drying behavior of colloidal silica suspensions. J. Am. Ceram. Soc. **82**, 2345–2358 (1999)
39. Wedin, P., Martinez, C.J., Lewis, J.A., Daicic, J., Bergström, L.: Stress development during drying of calcium carbonate suspensions containing carboxymethylcellulose and latex particles. J. Colloid Interface Sci. **272**, 1–8 (2004)
40. Coussy, O., Dangla, P., Lassabatère, T., Barohel-Bouny, V.: The equivalent pore pressure and the swelling and shrinkage of cement-based materials. Mater. Struct. **37**, 15–20 (2004)

Chapter 7
Film Formation Through Designed Diffusion Technology

Zhenwen Fu, Andy Hejl, Andy Swartz, Kebede Beshah, and Gary Dombrowski

Introduction

Polymeric-based coatings can be classified into two major classes, solution resin or polymer based and emulsion latex polymer based. For solution polymers, the polymer is typically dissolved in organic and often volatile solvents. For emulsion latex polymers, many polymer chains are entangled and packed together on a nanometer scale and suspended typically in water as colloidal particles.

Figure 7.1 shows the differences between the film formation process for both solvent-based solution polymers and water-based thermoplastic latex polymers. For solvent-based solution resins or polymers, solvents are often needed to manage viscosity. For water-based thermoplastic, high molecular weight latex polymers, coalescing solvents are often required to aid film formation. The details of latex film formation processes can be found in reference [1].

The properties for latex polymer-based coatings are affected not only by polymer composition, molecular weight, and chain sequence as in solution polymer-based coatings but also by the film formation process unique to the latex particle nature. The latex film formation process includes particle packing, particle deformation, and polymer interdiffusion across the latex particle boundary to form a uniform polymer film at the molecular level. This process is controlled by the above basic polymer properties and also by the latex colloidal particle properties such as particle size, morphology, and stabilization methods. It is often aided by adding a certain percentage of organic solvents as coalescent. Due to the relentless effort in reducing the volatile organic compounds (VOC) in the industry and scientific interest in the academia, the topic of latex film formation has received extensive attention since the late 1980s and the early 1990s. Over the past 20 years or so, the

Z. Fu (✉) • A. Hejl • A. Swartz • K. Beshah • G. Dombrowski
The Dow Chemical Company, Collegeville, PA, USA
e-mail: ZFu@dow.com

M. Wen, K. Dušek (eds.), *Protective Coatings*, DOI 10.1007/978-3-319-51627-1_7

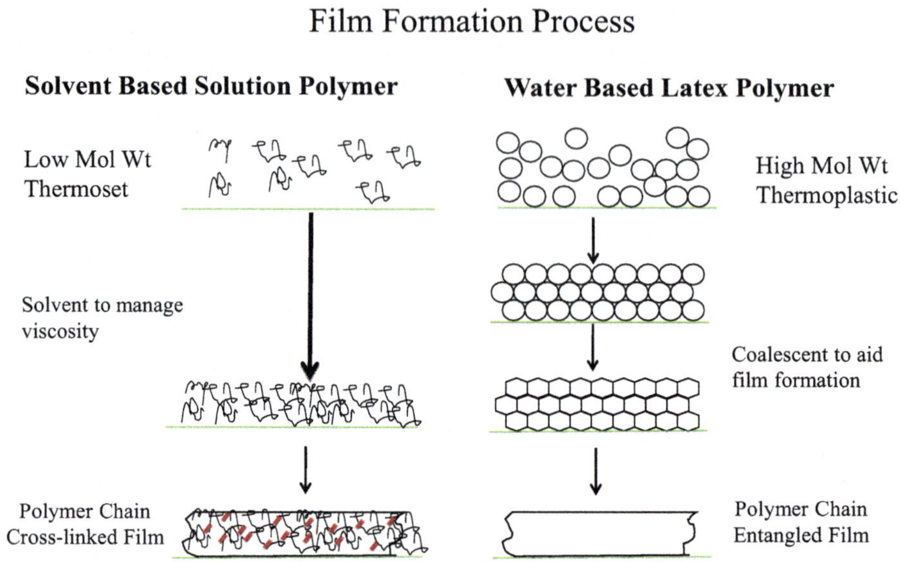

Fig. 7.1 Film formation processes, the red segments are indications for cross-linking functionalities

research has advanced the many years' very rich empirical knowledge across the industry in understanding latex-based organic coatings to a new and more comprehensive theoretical level. It provides a framework for us to organize our knowledge to identify opportunity in research and to analyze and identify the root cause of real-world problems. It has unified and created a set of terminology that we speak today in latex-based coatings.

Traditional methods to lower VOC and maintain film hardness include lowering the dominant phase polymer T_g and blending in small amounts of hard polymer to reinforce a soft film or manipulating latex morphology. These approaches may not be enough to fully satisfy the needs of industrial coatings formulators. In this chapter, a new technology, Designed Diffusion™ technology, is presented that significantly improves the early property development of waterborne latex coatings while addressing the challenges of the low VOC formulation landscape, which was first published in European Coatings Journal, June Issue of 2009 [2].

Designed Diffusion™ technology uses a two-polymer system, in which a small amount of a soft, specially designed Polymer DD is added to a high T_g dominant phase Polymer A to accelerate early property development. The polymers (Polymer A and Polymer DD) have been chosen so that the coalescent partitions selectively into Polymer A in the wet state. Film formation triggers a change in the solubility of the coalescent in the two polymers. Transport of the coalescent into Polymer DD facilitates the removal of the coalescent from the dominant phase (Polymer A), and thus the coating develops properties at a faster rate. This chapter presents detailed evidence of this Designed Diffusion™ technology with analytical, physical, and application data.

Model and Mechanism of Designed Diffusion™ Technology

The Designed Diffusion™ technology relies on the interplay of the dominant phase Polymer A, the Designed Diffusion™ minor phase Polymer DD, and the selected coalescent, as illustrated in Fig. 7.2.

The dominant phase (65–85% by weight) Polymer A is a high T_g (>20 °C) latex polymer, which ultimately controls the film hardness and related properties when the coalescent is removed. The film formation of this high T_g polymer is facilitated by addition of a small amount (15–35% by weight) of a specially designed Polymer DD and properly selected coalescent. This enables a reduction of up to 30% or more of the VOC relative to the 100% latex Polymer A formulations without compromising film formation and final film properties. Polymer DD is designed to be a good film former without coalescing aid. The composition of Polymer DD also dictates that in the wet state the selected coalescent preferentially partitions into the latex Polymer A.

As the film formation continues, water in the bulk of the film is lost, triggering a change in the relative solubility of the coalescent in the two polymers. The coalescent now prefers the Polymer DD phase and migrates out of the dominant phase Polymer A into the Polymer DD domains. The Polymer DD releases the coalescent at a much faster rate than the dominant phase Polymer A. This significantly accelerates the development of properties such as hardness, block resistance, print resistance, dirt pickup resistance, and early water resistance. The unique feature of the Designed Diffusion™ technology is its ability to reduce VOC, to ensure film formation, and to accelerate hardness-related property development simultaneously. In the next section, we will focus on the design and synthesis of the Designed Diffusion™ polymers.

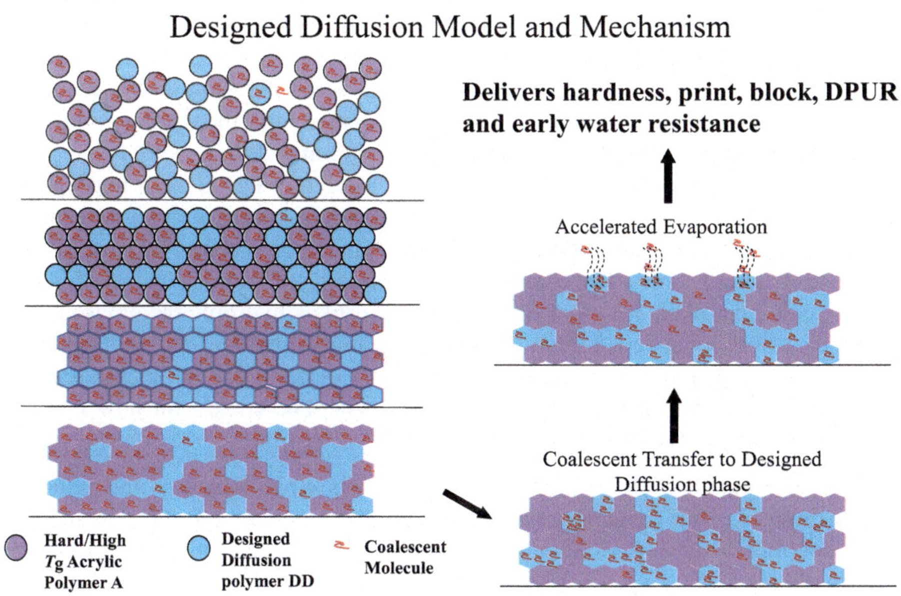

Fig. 7.2 Designed Diffusion™ model and mechanism

Design and Synthesis of the Designed Diffusion™ Polymers

The dominant latex Polymer A can be any high T_g (>20 °C) and relatively hydrophobic latex polymer used in industrial coating formulations. The requirements for the design of the Designed Diffusion™ Polymer DD are the following:

1. The T_g of the Designed Diffusion™ Polymer DD needs to be significantly lower than the T_g of the dominant phase Polymer A and the drying temperature. For ambient application conditions, the T_g of the Polymer DD needs to be less than 5 °C, preferably less than −5 °C. Low T_g is critical for ensuring good film formation at reduced VOC and for providing an easier and faster pathway for the coalescent to escape from the dominant Polymer A phase once a film is formed to accelerate property development. Coalescent can diffuse or permeate through polymers with high free volume faster by several orders of magnitude than through polymers with low free volume [3]. The free volume of a polymer increases significantly as the temperature goes above its glass transition temperature [4].
2. The Designed Diffusion™ Polymer DD should form a percolating network within the Polymer A dominant phase at low levels (<35%). The formation of a percolating network is important in order to have a continuous pathway through the film to facilitate faster coalescent evaporation. The low percentage ensures that the low T_g Designed Diffusion™ Polymer DD does not compromise the bulk film properties of the hard Polymer A. The percolation threshold largely depends on the packing lattice, distribution, and the relative particle sizes of Polymer A and the Designed Diffusion™ Polymer DD (see reference [5] for details). For particles of equal size and random distribution, the percolation threshold is 19.8% for face-centered cubic (FCC) and hexagonal close packed (HCP) packing, 24.6% for body-centered cubic (BCC), and 31.2% for simple cubic packing [5]. For identical packing modes, a minor phase with smaller particles may reach the percolation threshold faster. Packing and distribution depend on many properties of the colloidal system, and the drying conditions and may depend only indirectly on the polymer composition.
3. It is preferable that the Designed Diffusion™ Polymer DD is more hydrophilic than the Polymer A, when it is hydro-plasticized, meaning a good percentage (>1%) of water is dissolved in Polymer DD in the wet state. This makes the coalescent partitioning more favorably into the Polymer A in the wet state, since the coalescent used are more hydrophobic and likes hydrophobic environment better, where it is needed the most to aid the film formation. As film formation continues, the bulk of the water is lost. This triggers a change in the relative solubility of the coalescent in the two polymers. The coalescent now is more soluble in the Polymer DD.

Since we cannot hold requirement #2 constant while we vary the polymer composition, T_g, and particle size, it is very difficult to map out all of the details in the polymer design space. Nonetheless, we have discovered and synthesized both acrylic emulsion polymers and polyurethane dispersions that deliver the synergy in reducing VOC, maintaining film formation and accelerating property development.

Model emulsion polymers were prepared using standard emulsion polymerization techniques found in the literature [6, 7]. Acrylic or styrene-acrylic polymer compositions were found to function as Designed Diffusion™ polymers effectively through one-stage, two-stage, and resin-supported emulsion processes. Low T_g (<-5 °C) compositions made through a resin-supported emulsion process were found to be most effective; see reference [8] for the detailed composition and process to make the polymer.

Model polyurethane dispersions (PUD) were prepared using standard prepolymer mixing processes found in the literature [9, 10]. A unique set of PUD compositions with a defined type and molecular weight polyol [11] at a certain level (>40% by weight) were found to function as Designed Diffusion™ polymers effectively.

Physical and Analytical Results and Discussion

Acceleration of Hardness Development

Clear coatings (30% solids by weight) were prepared according to the recipes given in Table 7.1. Coating #1 was prepared with 100% Polymer A (STY/EHA/MMA, STY is styrene, EHA is 2-ethylhexyl acrylate, and MMA is methyl methacrylate with 45% solid, about 150 nm and T_g, 35 °C). Coatings #2 and #3 were prepared with 70% (based on the solids of the binders) Polymer A and 30% Designed Diffusion™ Polymer DD PUD type and acrylic type, respectively. Texanol™ ester alcohol is the coalescent, Zonyl® FS-610 is a wetting agent, and Acrysol™ RM-825 is a rheology modifier. The film hardness was tested via König pendulum damping (according to the ASTM method ANS/ISO 1522, 1998(E)) as a function of drying time. The dried films had thickness 40 ± 3 μm.

Table 7.1 Clear coating formulation (parts by weight, wet)

Ingredient	Coating #1	Coating #2	Coating #3
Polymer A (45% solids)	46.67	46.67	46.67
Polymer DD (PUD type, 32% solids)		28.13	
Polymer DD (acrylic type, 42% solids)			21.43
Water	23.33	21.21	27.90
Texanol™ ester alcohol	3.6	3.6	3.6
Zonyl® FS-610	0.2	0.2	0.2
Acrysol™ RM-825	0.2	0.2	0.2

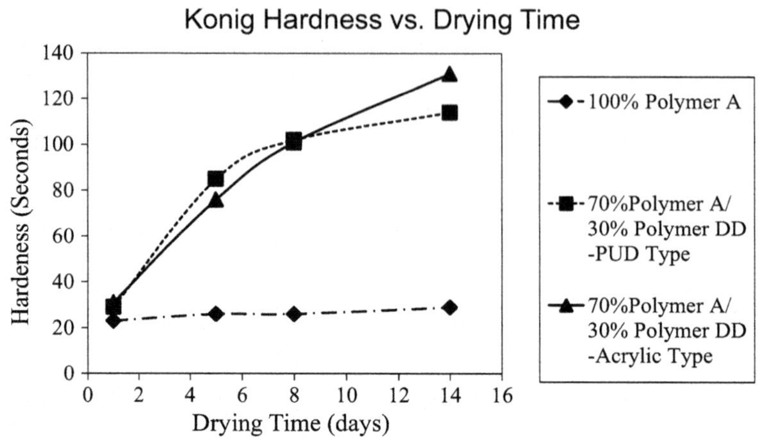

Fig. 7.3 Hardness development as a function of drying time

Table 7.2 Pigmented gloss paints film formation property

Paint ID	Texanol (%)	DPM (%)	NMP (%)	Total (%)	VOC reduction	MFFT(°C)	
						Visual	Tape Pull
Paint #1	5.40	1.80		7.20		−3.2	6.7
Paint #2	3.78		2.25	6.03	16%	−5.0	4.1
Paint #3	3.78	1.26		5.04	30%	−2.2	9.0

The hardness development as a function of drying time is shown in Fig. 7.3. It is clear that the Designed Diffusion™ Polymer DD of both acrylic and PUD type accelerated the hardness development of the coatings significantly.

VOC Reduction Without Compromising Film Formation

Use of the Designed Diffusion™ approach to reduce VOC levels was tested using an 18% PVC TiO_2-based gloss formulation at 35% volume solids. Three pigmented paints were made using the same binder combinations as in Table 7.1 for the clear coating formulation. The paints were prepared from dry powder TiO_2 (Ti-Pure® R-706) using Tamol™ 2001 (Dow Chemical Company, ~1.5% w/w on TiO_2) dispersant in the mill base. The coalescent package was a combination of Texanol™, Dowanol™ DPM, and N-methylpyrrolidone (NMP). The NMP in Paint #2 was from the PUD polymer.

Both visual and mechanical (Tape Pull) Minimum Film Formation Temperatures (MFFT) were measured and listed in Table 7.2. VOC levels (calculated based on the formulation) were reduced by 16% for Paint #2 and 30% for Paint #3 relative to Paint #1 without compromising film formation.

Coalescent (Texanol) Partitioning between Polymer A and Designed Diffusion™ Polymer DD

An NMR technique was used to determine the partitioning of Texanol coalescent between the two polymer phases in the wet latex. Near the polymer (Polymer A) T_g, the linewidth of the NMR signal at half height of the high T_g Polymer A is inversely proportional to the Texanol level in the polymer (Fig. 7.4). As the Texanol level in the polymer increases, the mobility of the polymer chains is increased, sharpening the NMR signal. A near-perfect linear calibration line was obtained relating line width to Texanol level in Polymer A. This technique selectively measures the high T_g Polymer A since the signal corresponding to Polymer DD is quite sharp at the testing temperature (room temperature) so far above Polymer DD's T_g. The calibration curve was used to determine the Texanol level in Polymer A in a latex blend of Polymer A, Polymer DD, and Texanol. It was found that 81% of the Texanol partitioned into Polymer A with 19% of the Texanol partitioning into the Designed Diffusion™ Polymer DD when the Polymer DD is PUD type (Table 7.1, Coating #2). When the Polymer DD was the acrylic type (Table 7.1, Coating #3), 97% of the Texanol partitioned into Polymer A, and 3% of the Texanol partitioned into the Designed Diffusion™ Polymer DD.

This difference in Texanol partitioning is likely due to hydroplasticization of the Designed Diffusion™ Polymer DD. The rationalization is supported by the following calculation using the group contribution method developed through the

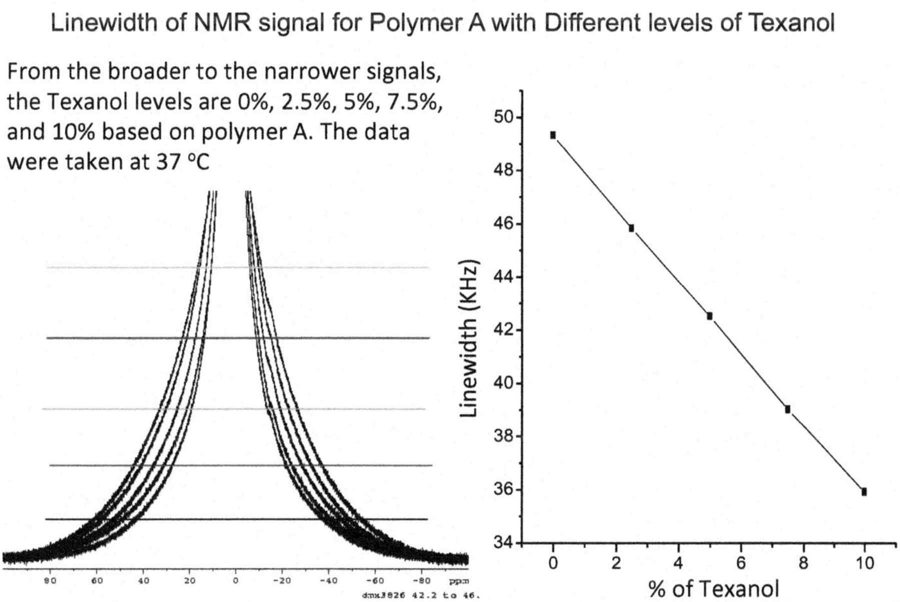

Fig. 7.4 NMR calibration for measuring Texanol partitioning

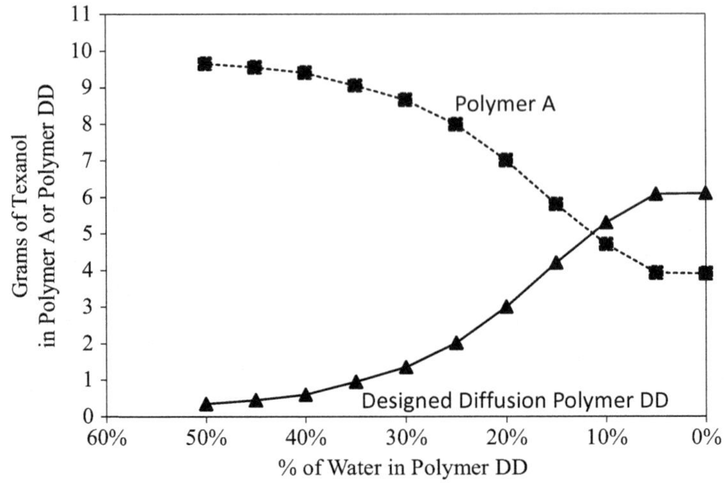

Fig. 7.5 Calculated Texanol partitioning between Polymer A and Polymer DD as a function of Polymer DD's hydroplasticization level

UNIFAC Consortium founded at the Carl von Ossietzky University of Oldenburg (Fig. 7.5).

In the calculation, 10 g of Texanol was forced to partition between 50 g of Polymer A and 50 g of a Designed Diffusion™ Polymer DD, similar to the one used in coating #3 of Table 7.1, but with 10% MAA, methacrylic acid, at different levels of hydroplasticization. The levels of hydroplasticization of the Designed Diffusion™ Polymer DD were varied in the calculation by simply adding different numbers of water "groups" in the group contribution computer program. In the wet state, the Designed Diffusion™ Polymer DD is highly hydroplasticized, the Texanol partitions highly into Polymer A, where it is needed to aid film formation. As the coating dries, the level of hydroplasticization of the Designed Diffusion™ Polymer DD decreases, and by the time the film is nearly completely dry, the Texanol partitioning reverses and slightly favors Polymer DD. This is likely due to formation of hydrogen bonding between the hydroxyl group on the Texanol molecule and the acid groups on the Designed Diffusion™ Polymer DD. The change in partitioning removes some level of the Texanol away from the dominant Polymer A phase and helps the hardness development but may not be enough for the performance difference in demanding applications. We will see below that enhancing coalescent evaporation is a critical role of the Designed Diffusion™ Polymer DD.

Clear coatings #4–#6 with the same Polymer A and Polymer DD combinations as coatings #1–#3 in Table 7.1 were prepared with either 12% dibutyl phthalate or 12% Texanol (on polymer solids). The T_g of the dominant Polymer A phase was measured by differential scanning calorimetry (DSC) in the films after 5 days of drying (Table 7.3). For the 100% Polymer A film (Coating #4), the measured T_g were 8.0 °C and 8.7 °C for dibutyl phthalate and Texanol coalescent, respectively.

Table 7.3 Effect of Designed Diffusion™ Polymer DD on the T_g of Polymer A with nonvolatile and low volatile coalescent

Coatings ID	T_g (°C) 12% Dibutyl phthalate	12% Texanol
Coating #4	8.0	8.7
Coating #5	8.7	38.8
Coating #6	7.3	32.9

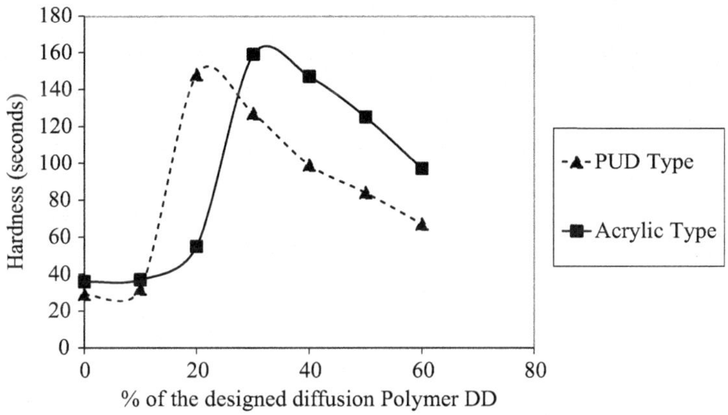

Fig. 7.6 Hardness development of Polymer A as a function of percentage of the Designed Diffusion™ Polymer DD added

Those measured glass transitions were severely depressed from the true T_g of Polymer A (35 °C) and indicates that a significant amount of plasticizer remains in the film. When the Designed Diffusion™ Polymer DD was added, the T_g of the dominant Polymer A phase remained depressed when nonvolatile dibutyl phthalate was used as coalescent. However, when the low volatility Texanol was used as coalescent, blending in the Designed Diffusion™ Polymer DD allowed the T_g of the dominant Polymer A phase to completely recover to the original T_g of the neat Polymer A. The Designed Diffusion™ Polymer DD accelerated the removal of Texanol from Polymer A phase dramatically from months to days.

Percolation Threshold and Accelerated Diffusion or Permeation Pathways

Coatings were formulated with varying percentage of the Designed Diffusion™ Polymer DD (both PUD and acrylic type) with Polymer A. The hardness development was measured on films after 7 days of drying, and the results are displayed in Fig. 7.6. For the PUD-type Designed Diffusion™ Polymer DD, the hardness improvement began between 10 and 15% and peaked around 23%, while for the acrylic-type Designed Diffusion™ Polymer DD, the hardness improvement began around 20–25% and peaked around 32%. The onset of hardness development and the peak of hardness correlate well with the percolation theory as discussed in the Design and Synthesis of the Designed Diffusion™ Polymers section; please refer to reference [5] for more details. The earlier onset for the PUD type than the acrylic type is likely due to their difference in particle size. In this example, the particle diameter for the PUD type is about 50 nm, while the acrylic type DD polymer is about 150 nm, while dominant Polymer A is about 150 nm too, and the smaller size of PUD particles would lead to earlier onset of the percolation threshold (lower percentage of Polymer DD needed). The fact that no significant hardness improvement was observed before the onset point strongly suggests that it is necessary for the Designed Diffusion™ Polymer DD to form accelerated diffusion or permeation pathways in order to enhance the hardness development.

In Fig. 7.6, the hardness decreases at levels of the Designed Diffusion™ Polymer DD beyond the peaks are due to the fact that as the soft Designed Diffusion™ Polymer DD level in the polymer blend increases, the overall hardness of the film starts to be influenced by the large portion of the soft Polymer DD.

Effect of Coalescent Boiling Point

The effect of boiling point of various coalescents (12% by weight based on total polymer solids) on the hardness development of clear films based on coating # 3 is shown in Table 7.4. It can be seen that the early hardness development is critically dependent on the boiling point of the coalescent. Early hardness development is greatly improved for all of the coalescent with boiling points below 300 °C, and, generally, higher hardness is obtained more readily for lower boiling point coalescent. Coalescents with boiling point greater than 300 °C show essentially no improvement in hardness. Again, it suggests that addition of the Designed Diffusion™ Polymer DD provides a pathway to accelerate coalescent evaporation and allows the dominant hard phase Polymer A to approach its maximum hardness (about 160 s) much faster.

Table 7.4 Effect of coalescent on hardness development

Coalescent	Boiling point (°C)	Hardness (s) after 7 days
Diethyl adipate	250–252	152.1
Exxate [12] 1200	240–285	151.2
Texanol [10]	255	124.1
TXIB [10]	280	118.5
Dibutyl adipate	305	14.9
Dibutyl phthalate	340	31.3

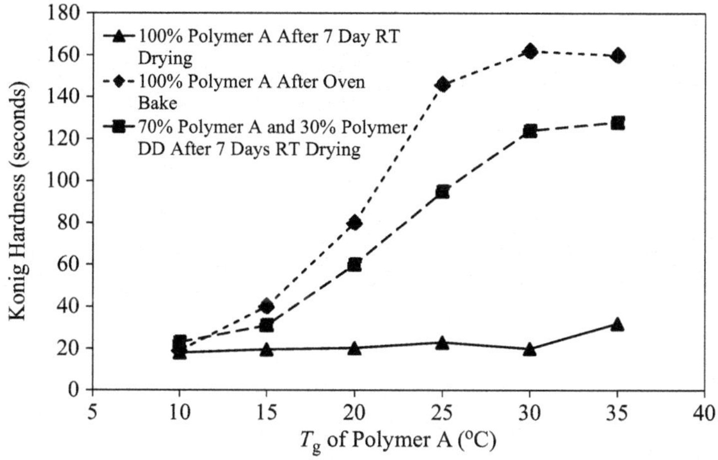

Fig. 7.7 Ultimate hardness as a function of the T_g of the dominant phase Polymer A for 100% Polymer A after oven baking

Ultimate Hardness of the Dominant Phase Polymer A as a Function of its T_g

All of the results we discussed so far are based on formulations with a dominant phase Polymer A of 35 °C T_g. The T_g effect of the dominant phase Polymer A on König pendulum hardness is displayed in Fig. 7.7. The top curve is the hardness of films made with 100% Polymer A of different T_g without any coalescent. The films were made by drawdown of the neat polymer on aluminum panels and dried in 80 °C oven for 2 days. The variation of T_g was achieved by varying the monomer ratios of STY, EHA, or MMA in the STY/EHA/MMA copolymer latex. These are the ultimate hardness that the Polymer A can achieve with different T_g values. König pendulum hardness of films made with 100% Polymer A of different T_g with 4% Texanol on polymer solids for 10 °C T_g polymer, 6% for 15 °C, 8% for 20 °C, 10% for 25 °C, 12% for 30 °C, and 12% for 35 °C was plotted as the bottom curve after 7 days of drying in a CTR at 25 °C and 50% relative humidity. The different levels of coalescent for different T_g were to ensure the film formation. König pendulum hardness of films made with the same set of polymers A with identical percentage of Texanol but with 30% of the soft PUD type of Designed Diffusion™ Polymer DD added was plotted as the middle curve also after 7 days of drying in a CTR at 25 °C and 50% relative humidity. The range bar on the last point of the middle curve was based on the variation of coalescent from TXIB™ to diethyl adipate (see Table 7.4). The dominant phase Polymer A determines the ultimate hardness of the films, and the Designed Diffusion™ Polymer DD enables the development of hardness to occur more rapidly.

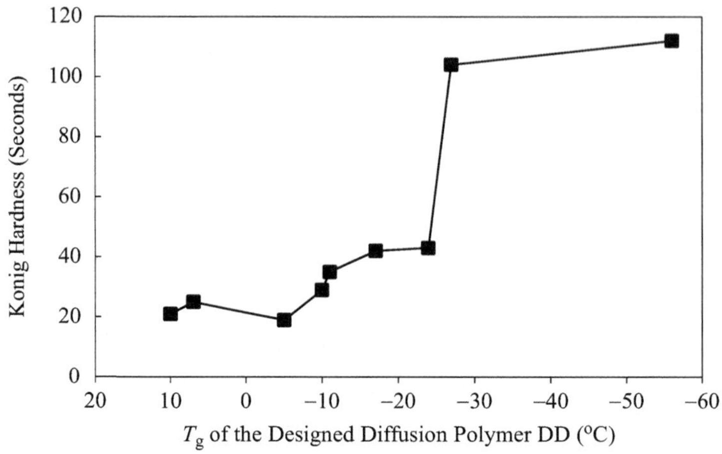

Fig. 7.8 Effect of T_g of the Designed Diffusion™ Polymer DD on hardness development

T_g Effect of Designed Diffusion™ Polymer DD

The T_g effect of a set of selected acrylic-type Designed Diffusion™ Polymer DD on the hardness development of the coatings with a 70% dominant phase 35 °C T_g Polymer A, based on the coating #3 formulation in Table 7.1, is displayed in Fig. 7.8. The hardness was measured after 7 days of drying. The *x*-axis was plotted with the T_g of the Designed Diffusion™ Polymer DD ordered from high to low. In the T_g range between −5 °C and −30 °C, the lower the T_g of the Designed Diffusion™ Polymer DD, the harder the films develop at a given drying time. This is likely due to the fact that the lower T_g polymer has more free volume and therefore a faster diffusion rate for the coalescent.

It must be pointed out that not all low T_g polymers work as Designed Diffusion™ polymers. This could be due to a couple of reasons. If the Designed Diffusion™ Polymer DD does not pack and distribute well with the dominant phase Polymer A and cannot reach the percolation threshold at a reasonable low level, it will not work. If the coalescent does not have a good solubility in the Designed Diffusion™ Polymer DD, even if Polymer DD reaches its percolation threshold and the coalescent has a high diffusion rate, the system will not work. For the Designed Diffusion™ Polymer DD to form an effective pathway for coalescent evaporation, it must have high permeability—a product of diffusion coefficient and solubility [12].

Application Results and Discussion

In the previous sections, we focused on the most direct physical and analytical results to understand the system. Here, we will discuss some of the properties that are mostly affected by the hardness development: block, print, and dirt pickup resistance.

Paints #4–#6 in Table 7.5 were made in similar ways as Paints #1–#3 in Table 7.2. Films of these paints were drawn down over Leneta test charts with a 75 μm Bird Film Applicator. The coated charts were allowed to dry in a constant temperature room (CTR) at 25 °C and 50% relative humidity for 1 day or 7 days, depending on the test requirement. Four 4 cm × 4 cm sections (to run duplicates) of each type of paint film were prepared. The cut sections were placed with the paint surfaces face to face and then placed on a flat metal plate. Each individual specimen was topped with a number eight rubber stopper with the narrow side down, and a 1000 g weight was placed on each stopper. The block resistance of each paint was evaluated under two sets of conditions. These are for face-to-face contact for: (1) 24 h at room temperature and (2) 30 min at 50 °C. After the test period, the stoppers and weights were removed, and the sections were separated with a slow and steady force at an angle of approximately 180°. The samples were then rated for block resistance on a scale of 0 to 10 as described below:

10 = no tack, perfect
9 = trace tack, excellent
8 = slight tack, very good
7 = slight tack, good
6 = moderate tack, good
5 = moderate tack, fair
4 = severe tack, no seal, fair
3 = 5–25% seal, poor
2 = 25–50% seal, poor
1 = 50–75% seal, poor
0 = complete seal, very poor

Table 7.5 Block resistance of Designed Diffusion™ coatings

Paint ID	Binder system	1-day block		7-day block		7-day König pendulum hardness (s)
		24-h room temp.	30-min 50 °C oven	24-h room temp.	30-min 50 °C oven	
Paint #4	100% Polymer A	2	0	7	0	22
Paint #5	70% Polymer A and 30% Polymer DD-PUD type	7	4	10	5	112
Paint #6	70% Polymer A and 30% Polymer DD-acrylic type	9	8	10	7	105

Table 7.6 Dirt pickup resistance of Designed Diffusion™ coatings

Paint ID	Binder system	Dirt pickup resistance		
		Y-ref. before	Y-ref. after	Ratio after/before
Paint #4	100% Polymer A	88.54	84.43	0.954
Paint #5	70% Polymer A and 30% Polymer DD-PUD type	90.37	87.59	0.969
Paint #6	70% Polymer A and 30% Polymer DD-acrylic type	89.25	87.95	0.985

Table 7.7 Pencil hardness and gloss of Designed Diffusion™ coatings

Paint ID	Binder system	König hardness, s (7 days)	Pencil hardness (7 days)	Gloss	
				20°	60°
Paint #4	100% Polymer A	22	2B	62	85
Paint #5	70% Polymer A and 30% Polymer DD-PUD type	112	HB	56	82
Paint #6	70% Polymer A and 30% Polymer DD-acrylic type	105	2B	65	85

The results are listed in Table 7.5 along with the Designed Diffusion™ binder system identification and 7-day König pendulum hardness data.

The block resistance is consistent with the König hardness development. A harder coating leads to better block resistance. It is remarkable to see that the addition of 30% low T_g (<-5 °C) Polymer DD to a 35 °C T_g Polymer A paint dramatically changed the block resistance from a complete seal contact to a slight tack or no tack in most of the cases.

Dirt pickup resistance (DPUR) is a measurement of a coating's surface resistance to the accumulation of dirt, which may not be easy to remove. In the lab, aluminum "Q" panels were coated with a 75 μm Bird Film Applicator and dried in a constant temperature room (CTR) at 25 °C and 50% relative humidity for 7 days. Before contamination of the coating, its Y reflectance was measured (recorded as Y-ref. before). Then, a uniform coating of homemade Mapico 422 iron oxide slurry was brushed over the panel. The panel was dried for 3 h under the ambient condition and then dried for 1 h at 140 °F. The panel was cooled to room temperature and washed under running tepid water while being rubbed lightly and evenly with a cheesecloth pad. Then, after drying under air for 4 hours, the Y reflectance was measured again (recorded as Y-ref. after). Y-ref. before, Y-ref. after, and their ratios are listed in Table 7.6. Higher ratios (closer to one) indicate better dirt pickup resistance. In general, harder and/or more hydrophobic surfaces are more resistant to dirt pickup. The Designed Diffusion™ Polymer DDs are more hydrophilic than the Polymer A, but they make the films harder faster. Overall, the Designed Diffusion™ Polymer DD improved the DPUR in these formulations.

The pencil hardness tests were performed for the same set of paints on aluminum panels coated with a 125 μm Block Applicator and dried in a CTR at 25 °C and 50% relative humidity for 7 days (Table 7.7). With the PUD type of Designed

Diffusion™ Polymer DD, pencil hardness was improved by about 2 units, while with the acrylic type of Designed Diffusion™ Polymer DD, no improvement was observed.

It is clear that the pencil hardness measures a different property of the film than the König hardness. The pencil hardness improvement for the film modified by the PUD type of Designed Diffusion™ Polymer DD is likely due to the toughness of the PUD rather than the acceleration of removal of coalescent from Polymer A. As pencil hardness is also an empirical test, a less empirical test such as indentation hardness should be pursued in the future to understand the difference.

The gloss at 20° and 60° specular angles for the same set of paints was measured using the same panels as those used for the pencil tests and listed in Table 7.7. It is known that PUD binders tend to yield low gloss in pigmented systems.

The effect of the PUD-type Designed Diffusion™ Polymer DD on gloss was minimized when the right type of acrylic Polymer A was used. The acrylic-type Designed Diffusion™ Polymer DD can sometimes enhance the gloss.

The Designed Diffusion™ Polymer DD enhances film formation and accelerates hardness development, and it is typically more hydrophilic than the dominant phase Polymer A. Due to these two competing factors, variations in chemical resistance, early water resistance, and corrosion resistance have been observed. Nevertheless, these properties could be balanced through formulation work.

Conclusions

When a low T_g (< -5 °C) Designed Diffusion™ Polymer DD is added to a higher T_g (>20 °C) Polymer A dominant phase coating formulation at or slightly above the percolation threshold (typically between 20 and 35% by weight), a continuous phase of Polymer DD is formed. This continuous phase of Polymer DD serves as a pathway for selected coalescent to rapidly escape from the Polymer A dominant phase. This polymer phase structure or morphology enhances film formation, reduces VOC, and accelerates the development of hardness and hardness-related properties such as block, print, and dirt pickup resistance at a much faster rate without significantly compromising other properties. It was found that both PUD and acrylic polymers can be made to function as the Designed Diffusion™ Polymer DDs. The Designed Diffusion™ Polymer DD is preferably slightly more hydrophilic than the dominant phase Polymer A. In this way, the coalescent partitions selectively into the high T_g Polymer A phase in the wet state. Water evaporation during film formation triggers the acceleration of removal of the coalescent from the Polymer A phase. This work offers a totally new strategy toward simultaneously enhancing film formation, reducing VOC, and accelerating hardness-related property development.

Acknowledgment The authors would like to express their appreciation to Dr. S. Fitzwater for her help in using UNIFAC program, to Ms. H. Ray, Ms. J. McKenna, and Ms. J. Lemon for preparing the polymers, coatings, and performing the applications tests, and to Dr. W. Rosano, Dr. G. Larson, Ms. K. Korzybski, and Dr. A. Sheppard for their early work on PUDs. We also would like to thank Dr. M. Johnson, Dr. L. Procopio, and Dr. T. Tepee for their input and simulating discussions and to Mr. T. Ferrari, Mr. M. Yu, and Mr. D. Chen for their interests and evaluation of the technology.

References

1. Boczar, E.M., Dionne, C.B., Fu, Z., Kirk, A.B., Lesko, P.M., Koller, A.D.: Spectroscopic studies of polymer interdiffusion during film formation. Macromolecules. **26**(21), 5772–5781 (1993)
2. Fu, Z., Hejl, A., Swartz, A.J.: Designed diffusion technology: a paradigm shift in film formation. Eur. Coat. J. **06**, 1–12 (2009)
3. Vrentas, J.S., Duda, J.L., Ling, H.C., Hou, A.C.: Free-volume theories for self-diffusion in polymer–solvent system. II. Predictive capabilities. J. Polym. Sci. Polym. Phys. Ed. **23**, 289–304 (1985)
4. Sperling, L.H.: Introduction to Physical Polymer Science, 2nd edn. Wiley, New York (1992)
5. Stauffer, D., Aharony, A.: Introduction to Percolation Theory. Taylor & Francis, London (1992)
6. Piirma, I.: Emulsion Polymerization. Academic, New York (1982)
7. Gilbert, R.G.: Emulsion Polymerization—A Mechanistic Approach. Academic, New York (1995)
8. Fu, Z., Hejl, A., Sheppard, A.C., Swartz, A.J.: Aqueous compositions comprising a blend of emulsion polymers, Patent No. US 9,012,561 (2015)
9. Szycher, M.: Szycher's Handbook of Polyurethanes. CRC Press, Boca Raton, FL (1999)
10. Müller, B., Poth, U.: Coatings Formulation: An International Textbook. Vincentz Network, Hannover (2006)
11. Fu, Z., Hejl, A., Sheppard, A.C., Swartz, A.J.: Aqueous compositions comprising a blend of emulsion vinyl copolymer and polyurethane dispersion, Patent No. US 8,653,180, (2014)
12. Brandrup, J., Immergut, E.H.: In: Polymer Handbook, vol. 6, 3rd edn, pp. 435–449. Wiley, New York (1989)

Chapter 8
In Situ FTIR Study of Cure Kinetics of Coatings with Controlled Humidity

Mei Wen and Karlis Adamsons

Background

In situ Fourier transform infrared spectroscopy (FTIR) is a technique widely used for the kinetic study of organic coatings. Recently, a small humidity generator was created as an add-on device for microscopic FTIR and other analytical tools for humidity control [1]. The humidity generation system generates a small flow of nitrogen or air with the desired humidity. In the FTIR spectrophotometer employed, a special sample chamber with an IR-transparent window is used to allow an IR beam to pass through and to enclose the controlled humidity nitrogen or air. Nevertheless, controlling humidity in transmission mode of FTIR has not been reported. It is the objective of this work to control the humidity in an FTIR transmission mode instrument for the kinetic study of organic coatings.

It is important to be able to determine the kinetics under controlled humidity since many reaction systems in coating applications are affected by moisture. Moisture can serve as a reactant or a catalyst for a coating system. In the category of moisture as a reactant, examples are coatings that contain isocyanates [2–4], silanes [5, 6], and ketimines [2, 7, 8]. In the category of moisture as a catalyst, examples are aspartate coatings [2, 9]. Moisture can further affect the drying process of coatings, such as waterborne coatings. In this case, high levels of moisture in the air suppress the evaporation of water in the coatings. Therefore, analyzing kinetics with controlled humidity is of great value in order to better understand the coating film formation process for many different applications.

The system selected in this work is an aliphatic isocyanate reacting with moisture during the curing of polyurethane coatings. In urethane coatings, the main reaction (shown in Scheme 1) involves the OH groups of a polyol reacting

M. Wen (✉) • K. Adamsons
Axalta Coating Systems, Coatings Technology Center, Wilmington, DE, USA
e-mail: mei.n.wen@axaltacs.com

M. Wen, K. Dušek (eds.), *Protective Coatings*, DOI 10.1007/978-3-319-51627-1_8

Scheme 1

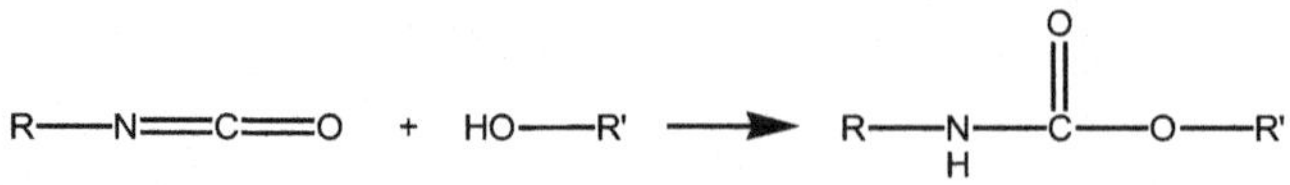

Scheme 2

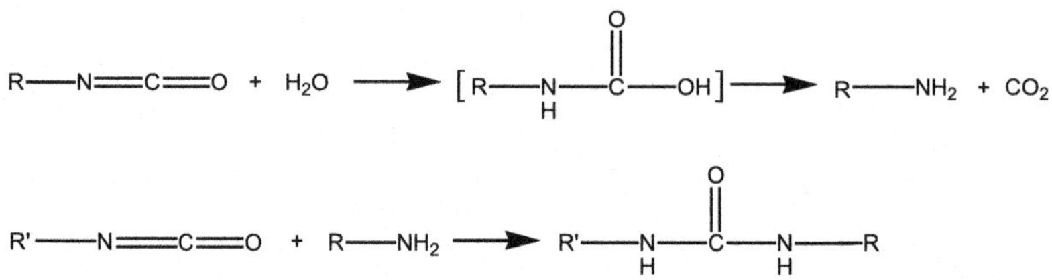

with isocyanate groups to form urethane linkages. When water is present, it reacts with isocyanate to form a carbamic acid intermediate, which then dissociates to form an amine and carbon dioxide. The amine further reacts with another isocyanate to form urea. This is shown in Scheme 2. Due to the fact that both reaction schemes occur at the same time during curing of a polyurethane coating, it is often hard to determine the reaction kinetics of isocyanate reacting with moisture alone. Therefore, in this work, we decided to determine the kinetics of isocyanate reacting with moisture first and then the kinetics of an acrylic polyol reacting with an isocyanate.

Most of the kinetic studies of isocyanate reacting with moisture reported in the literature are based on bulk systems, where isocyanate reacts with added water in a solution [10–12]. The consumption of isocyanate functional groups is considered to follow a second-order kinetic equation, where the consumption rate of isocyanate is proportional to the product of isocyanate concentration and the water concentration [10, 11]. In addition, some reported work suggests that the second-order kinetics are only valid at low conversion, after which an autocatalytic effect of the newly formed urea can speed up the reaction [12]. Limited work has been done on studying the kinetics of curing of isocyanate through moisture in the air. One nice report in this area shows the application of FTIR to determine the curing kinetics of isocyanate-ended urethane prepolymers under fixed humidity conditions [13]. The isocyanate used was an aromatic isocyanate based on polymeric diphenylmethane diisocyanate. There is a need to understand the curing kinetics of an aliphatic isocyanate with moisture in air under different levels of humidity.

On the other hand, a lot of work has been reported on the kinetic study of an aliphatic isocyanate reacting with a polyol to form urethane. The simplest way to describe the reaction is that it follows a second-order reaction [14–16]. A more complicated model includes the autocatalytic effect of the urethane formed, i.e., the formed urethane further catalyzing the reaction, as well as an organotin catalyst effect [17, 18]. The effect of humidity on the curing of polyurethane coatings has also been determined through an ex situ ATR analysis of coating films cured at different humidity levels. A higher humidity leads to a faster rate of isocyanate group consumption due to its faster reaction with water [19, 20]. Also, there is more urea relative to urethane formed when humidity increases [20]. However, the analysis focuses on the film–air and film–substrate interfaces instead of the whole film. It is also limited to slower reaction systems due to the ex situ approach. Therefore, it would be interesting to study the effect of humidity on curing of a polyurethane coating under different humidity levels through a transmission FTIR system in situ.

Different forms of urethane and urea, such as free and hydrogen-bonded (H-bonded) forms, can exist during formation of a polyurethane or polyurea. Most of the work done in this area focuses on aromatic isocyanate reacting with polyol and water, such as in polyurethane–urea foams, where carbonyl absorption can be used to interpret different forms of urethane or urea [21–22, 13]. Free urethane, H-bonded urethane, free urea, and H-bonded urea coexist in the material (see examples of structures below). Among the H-bonded urethane or urea, different interactions for the H-bonding can separate the urethane or urea into finer structures. For instance, the H-bonded urea exists as H-bonded, less ordered urea and H-bonded, ordered urea [13, 22]. However, for aliphatic isocyanate, particularly the commonly used 1,6-hexamethylene diisocyanate (HDI) trimer, no work has been reported to investigate the different forms of urea when it reacts with water. In this work, exploration of the different forms of urea formed is carried out through FTIR peak subtraction analysis.

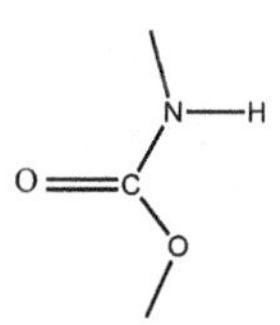

Free urethane

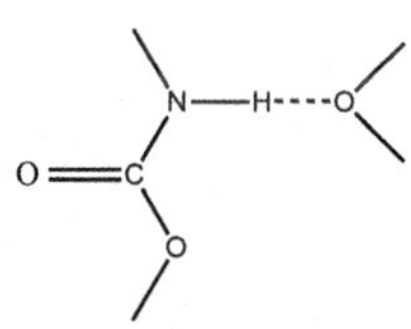

H-bonded, less ordered urethane

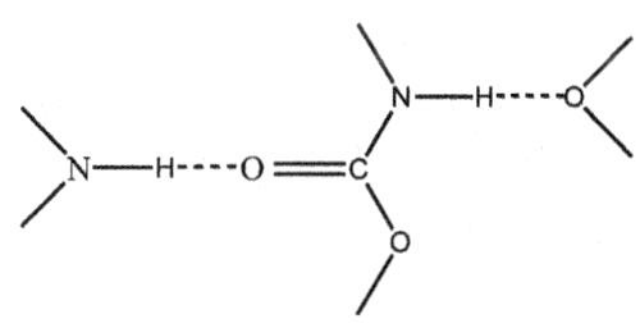

H-bonded, ordered urethane

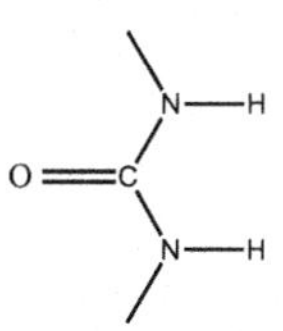

Free urea

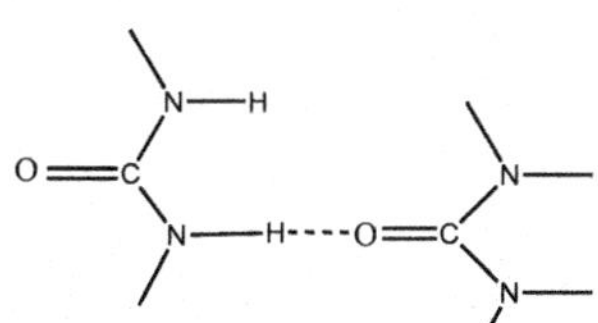

H-bonded, less ordered urea

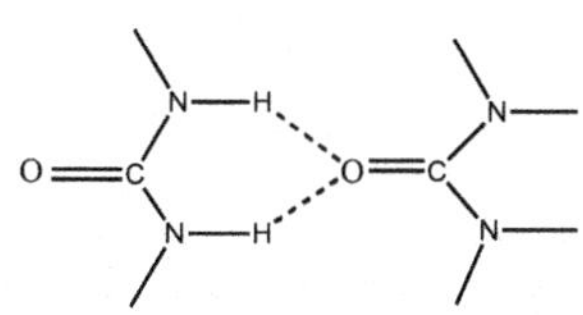

H-bonded, ordered urea

In order to control humidity in a transmission FTIR chamber during an in situ kinetic study, several challenges need to be solved. One is the large sample compartment, which can significantly slow down the purging process when the chamber is opened and closed to insert a sample. A small sample chamber that can allow the IR laser to pass through would be needed to overcome this issue. Other issues also need to be addressed, such as designing and setting up a moisture generation system that can maintain a controlled humidity for over 24 h, moisture absorption, and potential condensation of water onto KBr crystal plates at high humidity levels.

In this work, a novel humidity chamber was designed and installed in a transmission mode-configured FTIR chamber to conduct in situ kinetic studies. Considerations have been taken to achieve good humidity control over 24 h and to achieve consistent kinetic data. The system was applied to follow the kinetics of important reactions in polyurethane and polyurea coatings as discussed above: (1) an isocyanate reacting with moisture and (2) an acrylic polyol reacting with the isocyanate. Three relative humidity levels were used: 15, 45, and 75% RH at room temperature. The effect of humidity on the kinetics of these two reaction systems is discussed.

Experimental

Materials

An aliphatic polyisocyanate, 1,6-hexamethylene diisocyanate (HDI) trimer (Desmodur® N 3300), was used. It was mixed with a tetrahydrofuran (THF) solution of polystyrene. Polystyrene was used as a reference material for FTIR analysis. In the mixture, the weight ratio of polystyrene to the HDI trimer is 0.05 and the solids level is 50% by weight. THF was chosen due to its good solubility of both polystyrene and HDI trimer. Dibutyltin dilaurate (DBTDL) was added to give a concentration of 2400 ppm based on the total solids in the solution.

In the urethane system, an acrylic polyol with primary OH groups was used to react with the HDI trimer. The acrylic polyol was mixed with methyl n-amyl ketone, butyl acetate, and dibutyltin dilaurate (DBTDL) to form part A of each coating solution. Then part A was mixed with an HDI trimer solution (part B) for drawdown. The molar ratios of NCO to OH are 1.05 for sample 1 and 2.0 for sample 2, respectively. The DBTDL level is 2400 ppm based on the amount of total solids in a solution. Table 8.1 shows the composition of the solutions for sample 1 and sample 2.

Sample Preparation

Either the HDI trimer solution by itself or the mixture of HDI trimer and the acrylic polyol solution (sample 1 or sample 2) was drawn down on a 41 mm × 23 mm × 6 mm KBr plate (International Crystal Labs) for each experiment. See Fig. 8.1a for

Table 8.1 Compositions of coating solutions used (wt.%)

		Sample 1	Sample 2
	NCO/OH ratio	1.05	2.00
Part A	Acrylic polyol (100% solids)	31.21	23.30
	Methyl n-amyl ketone	12.69	9.47
	Butyl acetate	18.81	14.31
	DBTDL	0.12	0.12
Part B	HDI trimer	17.83	25.33
	Polystyrene	0.89	1.27
	THF	18.44	26.20
Sum		100.00	100.00

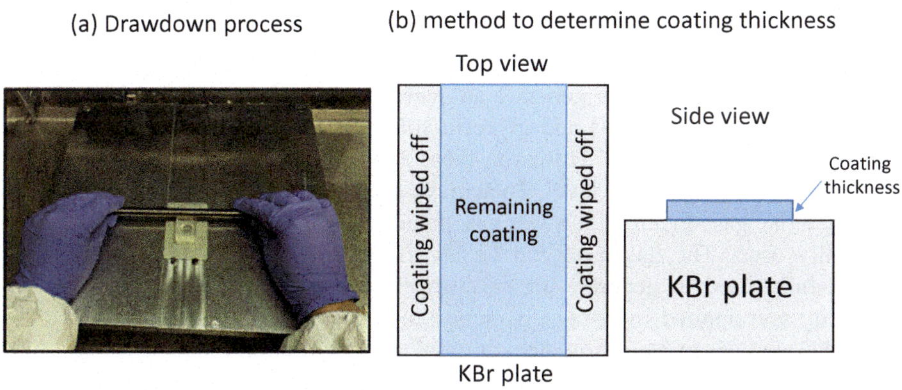

Fig. 8.1 (**a**) Coating drawdown process with a drawdown table designed to allow for convenient application of a coating solution onto a KBr salt plate. The table was placed inside a hood. (**b**) Schematic to illustrate how to measure coating thickness by creating steps of uncoated areas

illustration. A #2.5 and a #5 wire coil drawdown bar were used to apply a relatively thin layer of sample. To achieve consistent sample thickness, the maximum absorption peak of a given coating system was targeted at a specified level. The actual film thickness of the coating was determined through a calibration curve built for each coating system based on measured film thickness and absorption of the polystyrene peak at 700 cm^{-1}.

To build the calibration curve for a coating system, a series of coatings with different thickness were drawn down. Immediately following the drawdown, the liquid coating on both sides of each KBr plate was wiped off for every sample to form steps of coated and non-coated areas for thickness measurement. The uncoated areas were used as the zero point for film thickness. After 1 day of drying and curing of the coating, the center of the coating was scanned with FTIR for the absorption of polystyrene peak at 700 cm^{-1}. Then a profilometer was used to measure the dry film thickness. A calibration curve was then built based on the absorption of the polystyrene peak vs. film thickness. Once a calibration curve was established for a coating system, the dry film thickness of a coating was estimated by its polystyrene peak intensity at 700 cm^{-1}.

Due to the different amounts of HDI trimer used in the HDI trimer solution and samples 1 and 2, it is hard to maintain the same thickness for all the samples. The carbonyl groups in the HDI trimer have very strong absorption at ~2272 cm^{-1} due to the isocyanurate ring structure. When the HDI trimer concentration is high, a thinner film is desired to avoid saturation of this peak absorption. In the coatings applied here, the coatings containing an isocyanate reacting with moisture were about 3-micron thick, and the coatings containing the acrylic polyol reacting with the isocyanate (both samples 1 and 2) were about 9 microns.

Humidity Control System

There are different ways to generate air with a specific humidity level. Different saturated salt solutions [23] and glycerin solutions [24] can be used to generate different levels of humidity. However, these approaches generally take a long time to reach a desired humidity level. To achieve a faster response of humidity control, mixing a dry gas stream with a wet gas stream that contains saturated moisture is generally used. The gas used can be air or nitrogen. This approach has been successfully used to generate air or nitrogen with precisely controlled relative humidity and applied to X-ray diffraction analysis of crystal structures [25, 26] and DSC analysis of hydration, dehydration, and rehydration processes of pharmaceutical compounds [27] under various relative humidity conditions [25, 26]. In some cases, temperature control of the gas stream is also available [26, 27]. There are also commercially available stand-alone humidity generators such as an SRG series humidity generator (Shinyei) based on the same principle.

The humidity generation system used in this work is similar to the one used in reference [25]; however, no controller was used and the change of humidity is done manually here. Dry air and wet air streams are combined to create a desired % RH. The dry air stream is from the house source, which contains ~5% RH. The wet air stream is created by bubbling the dry air stream through a jar containing deionized (DI) water. An air stone is used to efficiently incorporate a high level of humidity (near 100% RH). The flow rate for each stream can be adjusted independently prior to blending. The final % RH is determined by the flow rates of the dry and wet air streams. The bubbler is placed inside a controlled temperature bath to maintain the temperature of the DI water inside the bubbler. If this is not done, it is hard to maintain the absolute humidity of the wet air. As a result, the mixed air humidity will change over time as discussed below. The water bath temperature should be set at ~1 °C above room temperature to maintain the bubbler water temperature.

A humidity control chamber (Fig. 8.2) located in the FTIR sample chamber is specially designed to limit the volume for controlled air. A similar sample chamber has been reported for X-ray diffraction analysis (see Japanese patent

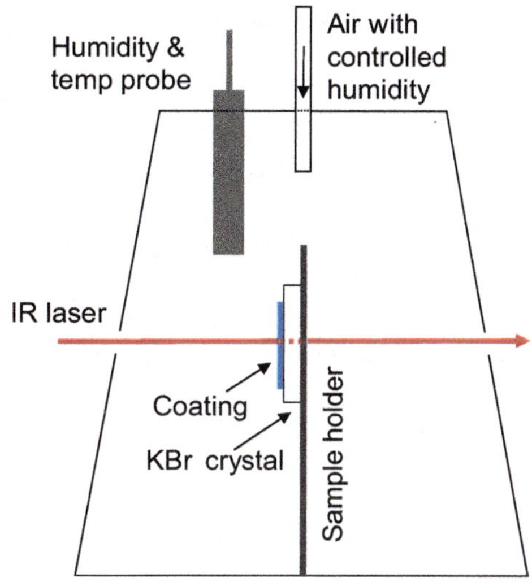

Fig. 8.2 Schematic of the humidity control chamber used for FTIR kinetic study

JPH07280748). The volume is reduced to about one fifth of the transmission FTIR chamber. There are two holes, one on each side, of the humidity control chamber to allow the IR laser beam to pass through. The size of the holes (about 1 cm in diameter) is just big enough for the IR source beam to pass through without touching the chamber wall. The chamber is locked in position on the bottom plate of the FTIR chamber. The incoming air with controlled humidity enters the humidity control chamber through the top center and comes out through the two holes created for the IR laser beam.

A Vaisala HUMICAP® temperature and humidity meter HM70 with a probe HMP77 was used to monitor temperature and relative humidity inside the humidity control chamber where the system cure was taking place. The probe is small enough to remain inside the sample chamber. It also allows in situ monitoring of humidity and temperature change on a computer. The meter can measure humidity in the range of 0–100% RH. It has an accuracy of measuring humidity of ±1% RH from 0–90% RH and ±1.7% RH from 90–100% RH. The response time is 50 s for 90% RH at 20 °C in still air. Another advantage of the Vaisala system probe design is the rapid response to changing conditions, which makes fine-tuning the desired % RH settings rapid and efficient. The meter also measures temperature well, with an accuracy of ± 0.2 °C at 20 °C.

Duct tape was used inside the FTIR sample compartment to prevent a humid atmosphere from entering the area under the instrument hood, which contains sensitive optics and electronics. This approach effectively isolates the sample compartment, avoiding water damage to the optics and electronics.

FTIR Kinetic Study

A Nicolet/Thermo model 4700 FTIR spectrometer was used. Each spectrum was collected by using 32 co-added scans in the range of 650–4000 cm^{-1} at a resolution of 4 cm^{-1}. IR transmission mode experiments were run using an OMNIC Macro program, where spectra were obtained on a periodic basis during the curing of a coating under controlled humidity. Spectra were typically obtained at 5, 10, 15, 30, 60, 90, 120, 150, 180, 210, 240, 270, 300, 330, 360 min and then once every hour up to 21–24 h in automated (i.e., Macro controlled) fashion. From the point when each sample was drawn down to the desired film thickness to the point the first spectrum was collected, it took about 3 min. The reaction in this period of time was neglected.

Results and Discussion

FTIR with Humidity Control

In the following section, factors that affect humidity or the FTIR kinetic study are discussed. These factors include the bubbler DI water temperature effect, temperature control and condensation, and background collection. Then the humidity data controlled at 15, 45, and 75% are discussed.

Bubbler DI Water Temperature Effect

If the bubbler used to generate the wet air is placed in an ambient environment, the DI water in the bubbler becomes cooler as the air carries moisture from it. Figure 8.3 shows the relative humidity and temperature of the humidity-controlled chamber when the bubbler just sits on a bench under an ambient environment. The wet air flow rate was set at 0.28 standard cubic meters per hour (SCMH), whereas the dry air flow rate was zero. The relative humidity rose quickly once the wet air was turned on. It reaches a maximum of 79.5% RH. However, the humidity dropped to 62.4% within 4.3 h. Even though the temperature of the chamber did not change, as shown in Fig. 8.3, the temperature of the water inside the bubbler dropped from the initial 20.5 to 12.5 °C. Continuous bubbling through the evaporation of water into the gas stream cools down the water. Cooling of water further reduces its evaporation rate, leading to lowered humidity. Therefore, to maintain humidity control, the water bubbler should be put inside a water bath. This is particularly needed when the flow of air going through the bubbler is high. A water bath set to about 21 °C was used to successfully maintain the bubbler DI water temperature at room temperature.

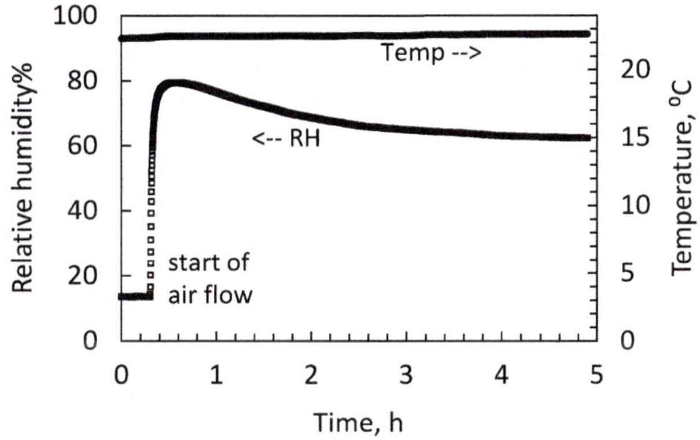

Fig. 8.3 Relative humidity and temperature of the humidity chamber when the bubbler was not kept in a water bath and only the humid air flew into the humidity chamber at a rate of 10 SCFH

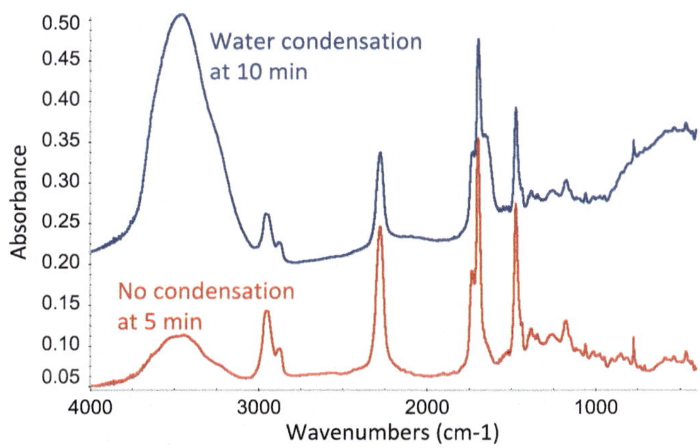

Fig. 8.4 Spectrum change caused by water condensation due to high humidity for a polyurethane coating. The two spectra were collected at 5 and 10 min from the beginning of the reaction at 85% RH and 21 °C

Temperature Control and Condensation

It is critical to monitor the temperature of the humid air in the FTIR humidity chamber. Under high humidity conditions, such as above 80% RH, water vapor can condense onto the KBr plate if its temperature is higher than that of the plate. When water condensation occurs, a large, broad water peak at 3000–3600 cm^{-1} can be seen clearly on the collected FTIR spectrum (Fig. 8.4). This leads to nonrepresentative kinetic data. The condensed water also makes the crystal fuzzy, particularly on the other side of the crystal where coating is not applied.

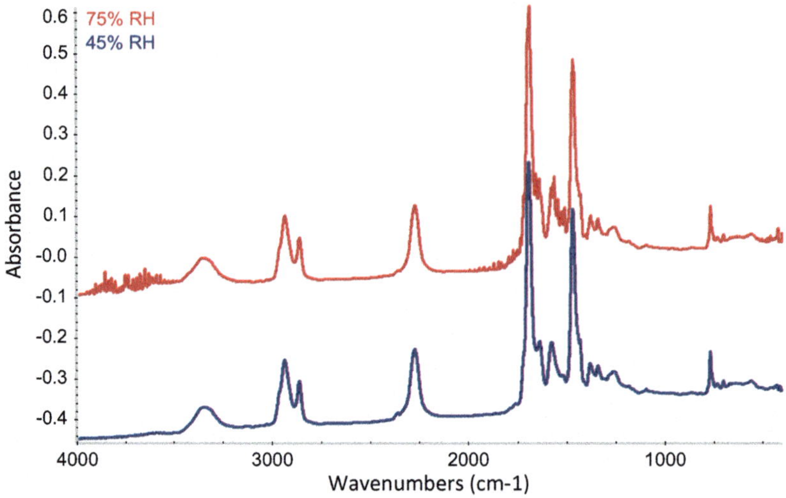

Fig. 8.5 Effect of humidity on spectra. The background was collected at 45% RH. The sample spectrum at 45% is smooth; however, the sample spectrum at 75% has a lot of small and sharp peaks due to presence of moisture

Background Collection

The background spectrum for each FTIR sample was collected through scanning the KBr plate before application of the coating. Due to the fact that spectra are collected with a certain background spectrum, it is important to collect the background spectrum at the same humidity as that maintained during the coating curing process. If a background is collected at a humidity level quite different from that of the coated sample, the difference in water absorption will appear. Figure 8.5 shows the spectra of a cured sample (HDI trimer reacting with moisture at 24 h) when the humidity was changed from 45% RH to 75% RH. The same background spectrum was used for both spectra and was collected at 45% RH. When the sample spectrum is collected at the same humidity as the background spectrum, i.e., 45% RH, no water peaks appear. However, when the sample spectrum is collected at 75% RH, water absorption peaks in the 3400–4000 and 1300–2000 cm^{-1} ranges are evident. It should be noted that what is shown in Fig. 8.5 is an example of an extreme case. In a real application process, a variation in humidity of a few percent can introduce some tiny water absorption peaks in the 1300–2000 cm^{-1} region if the coating is thin, making the absorption envelope not smooth. This may not be a problem if the peaks of interest are outside this region. However, when analyzing the urethane and urea peaks in the carbonyl region of 1600–1760 cm^{-1} is important, maintaining a good humidity level is a must. Furthermore, a peak subtraction technique can be used to get rid of the moisture absorption if the background is collected at a slightly different humidity than the sample. It should be noted that in the study of isocyanate conversion, the NCO peak at ~2272 cm^{-1} is not affected by the moisture in the environment.

Table 8.2 Flow rates of the dry and wet air streams used to achieve the target relative humidity at room temperature in the humidity chamber. The pressure of dry air before split into the two streams was set at 1.22 atm

Relative humidity (%)	Dry air flow rate, SCMH	Wet air flow rate, SCMH
15	0.280	0.042
45	0.238	0.084
75	0.199	0.123

Humidity Control

With the adjustment of flow rates of the wet and dry streams, the mixed air can have relative humidity generated reliably from 5% to 80%. The dry air coming from the supply line has a relative humidity less than 5% at room temperature. The wet air stream has a relative humidity close to 100%. With the leakage from the holes on the humidity control chamber wall, this range of relative humidity can be easily achieved. In this work, three relative humidity levels were used: 15% to represent low humidity, 45% for medium humidity, and 75% for high humidity. Table 8.2 shows the flow rates of the dry and wet air streams needed to generate these humidity levels. The total volume flow was set to be the same each time. Note that the pressure of dry air before split into the two streams was set at 1.22 atm. Inside the humidity control chamber, the pressure is slightly higher than 1 atm causing a steady outward flow. The flow is low enough that the coating only experiences a laminar flow of air. This was estimated by determining the Reynolds number of the air flow on a flat plate. Once a desired humidity is achieved, further adjustment of the flow rates is minimal.

Figure 8.6 shows the humidity control for 15, 45, and 75% RH during in situ kinetic study. The points at which the humidity chamber is opened to insert a sample are also marked in Fig. 8.6a. At the beginning of each kinetic study, the sample chamber is maintained at the desired relative humidity. Opening the chamber to insert the sample causes the change of humidity when the environment humidity is different. Nevertheless, the humidity level can go back to the target humidity within 2–3 min. This fast-response feature allows accurate kinetic study for a rapid reaction system. For a longer period of time, such as 21–24 h, the humidity control is within ± 3% RH at each target relative humidity. Overall, with the setup, good humidity control is possible for at least 24 h. A period of time longer than 24 h is still feasible by adding DI water to the bubbler to maintain its level.

Kinetics of HDI Trimer Reacting with Water

A coated layer of HDI trimer reacts with moisture in the air over time. Moisture in the air diffuses into the film and reacts with isocyanate to form primary amine and carbon dioxide first, and then the primary amine rapidly reacts with isocyanate to form urea.

Figure 8.7 shows the spectra of this system at initial time, 2 h, and 23 h of curing at 75% RH. There is a large change of chemical composition over the course of

Fig. 8.6 Humidity control inside the sample chamber during in situ FTIR analysis: (**a**) initial 5 min of data, (**b**) over 21–24 h of data

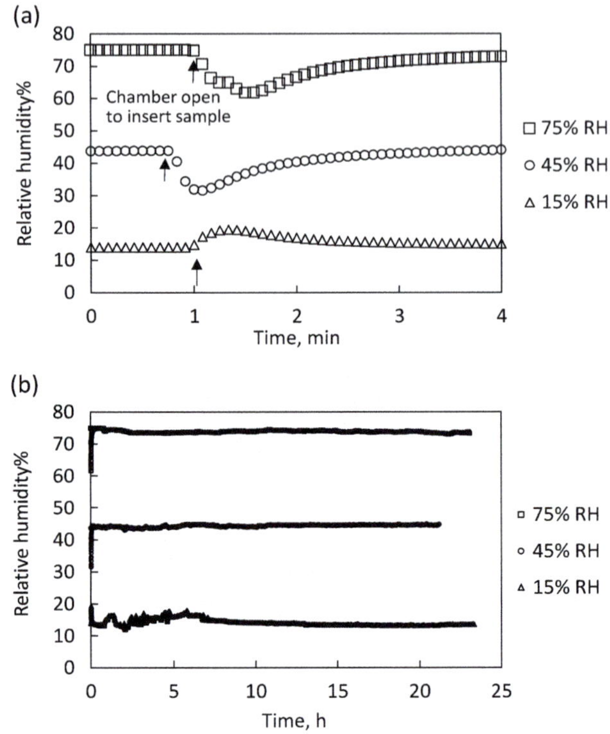

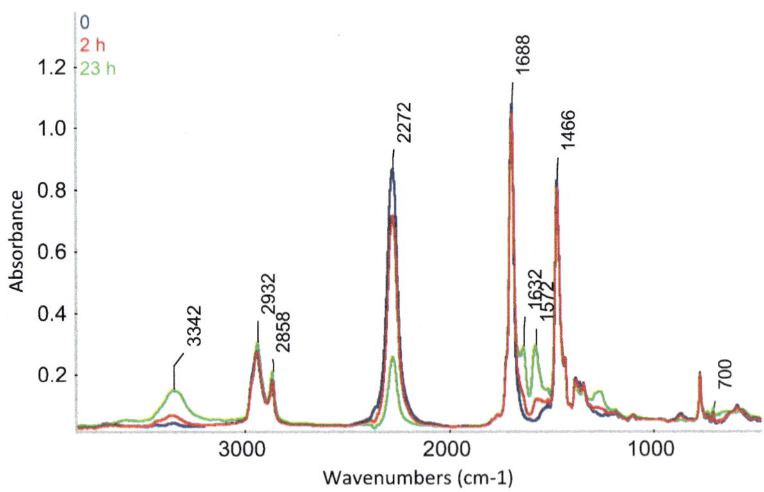

Fig. 8.7 FTIR spectra of HDI trimer reacting with moisture in air at 75% RH at $t = 0$, 2, and 23 h

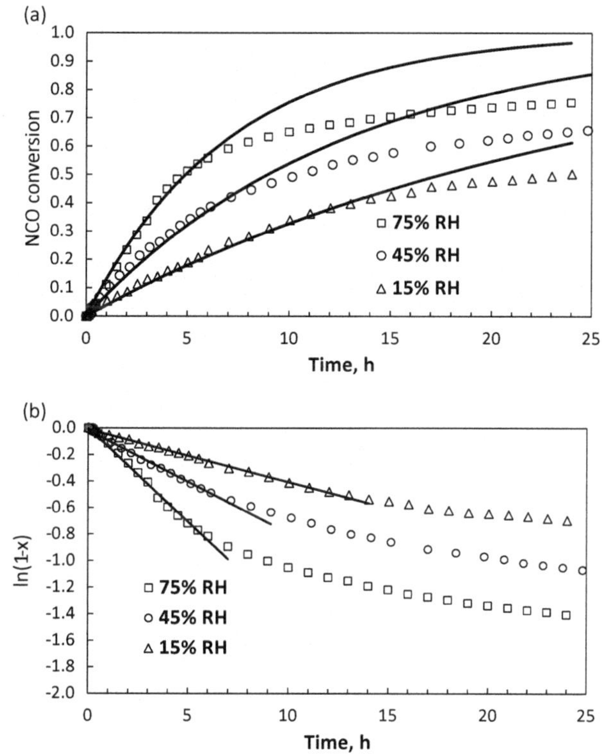

Fig. 8.8 (**a**) NCO conversion as a function of time during its reaction with moisture in the air at three humidity levels: 15, 45, and 75% RH. Dotted points are experimental data and the curved lines are model predictions. (**b**) first-order kinetic curve fitting for the initial data points on ln $(1-x)$ vs. time based on Eq. (8.3)

23 h. The isocyanate peak at ~2272 cm^{-1} decreased in intensity, and new urea peaks formed at ~3342 cm^{-1} (NH stretching), ~1632 cm^{-1} (C=O stretching), and ~1572 cm^{-1} (C–N stretching and N–H bending). Carbon dioxide (CO_2) generation can not be clearly identified in this open system. Generated CO_2 simply diffuses out of the coating. However, in a closed system formed by two KBr plates of HDI trimer with added water, CO_2 generation is present (data not shown). The conversion of isocyanate functional groups is determined based on the change of the NCO peak height at ~2272 cm^{-1}. The styrene peak at ~700 cm^{-1} was used as the reference.

Figure 8.8a shows the conversion of NCO groups as a function of time at different humidity levels. The reaction of isocyanate with moisture is significant at this DBTDL catalyst level, 2400 ppm by solids. Over the course of 20 h, conversion of NCO groups reached 0.65 at 45% RH. Even at a low humidity level of 15% RH, conversion reached 0.48 at 20 h.

As shown in Fig. 8.8a, at a higher humidity level, the reaction is faster than that at a lower humidity. It took 4.3 h to reach 0.50 conversion of NCO groups at 75% RH, whereas it took 10.7 h to reach 0.50 conversion of NCO groups at 45% RH and 23 h at 15% RH. Higher humidity in the air allows more moisture to diffuse into the coating, leading to higher concentrations of water to react with the isocyanate.

The reaction of isocyanate and moisture can be considered as a second-order reaction [10–12], where the rate of consumption of NCO groups can be written as

$$-\frac{d[\mathrm{NCO}]}{\mathrm{d}t} = k[\mathrm{NCO}][\mathrm{H_2O}] \qquad (8.1)$$

where [NCO] is the molar concentration of the NCO groups, $[H_2O]$ is the molar concentration of water in the coating, and k is the rate "constant" depending on the state of the coating film. As the film goes from a liquid to rubbery and then to glassy state, mobility of the NCO groups and water drops, particularly in the glassy state. Therefore, the rate constant depends on the state of the coating. However, in this simple treatment, the diffusion limitation is not included. Any deviation of the model from the experimental data will indicate the diffusion limitation. Moisture in the air diffuses into the film and reacts with isocyanate. To simplify the kinetic modeling of this reaction process, it is assumed there is no concentration gradient of NCO groups or water along the film thickness direction. This is likely valid due to the thin film nature of the coatings used here and when the diffusion rate of the water molecules is much higher than the water reaction rate. It is also assumed that the concentration of moisture in the coating is equilibrated with the moisture in air and maintained at a constant level during curing. Therefore, Eq. (8.1) can be simplified to the following pseudo-first-order kinetic relation:

$$-\frac{d[\mathrm{NCO}]}{\mathrm{d}t} = k_{\mathrm{app}}[\mathrm{NCO}] \qquad (8.2)$$

where k_{app} is the apparent rate constant ($k_{\mathrm{app}} = k[\mathrm{H_2O}]$). If we consider k_{app} as a constant, which means that k and $[H_2O]$ are constant, throughout the reaction, Eq. (8.2) can be integrated to obtain the conversion of NCO groups as

$$\ln(1 - x) = -k_{\mathrm{app}}t \qquad (8.3)$$

$$\text{or} \quad x = 1 - \mathrm{e}^{-k_{\mathrm{app}}t} \qquad (8.4)$$

where t is the time of curing, x is the conversion of NCO groups, $[\mathrm{NCO}] = [\mathrm{NCO}]_0(1 - x)$, and $[\mathrm{NCO}]_0$ is the initial NCO group concentration. Note that the coating solution contains THF as a solvent, which evaporates rapidly on a thin film after drawdown and before the collection of kinetic data. Therefore, concentration change due to solvent evaporation can be neglected.

Figure 8.8b shows the curve fit of the kinetic data with the linear function described in Eq. (8.3). The initial part of the kinetic data for each humidity system can be fitted well with the first-order kinetic equation. The slope of each line determines the apparent rate constant k_{app} as listed in Table 8.3. The apparent rate constant k_{app} increases as the relative humidity goes up. Due to the definition that $k_{\mathrm{app}} = [\mathrm{H_2O}]k$, at low NCO conversions, it is expected that the apparent rate constant grows linearly with the relative humidity in the air. For comparison, the ratios of the actual average relative humidity during the curve fitting region to that at 15% RH and the ratios of the obtained k_{app} to that at 15% RH are compared in Table 8.3 for 45 and 75% RH, respectively. To the first approximation, the ratios of k_{app} and relative humidity are close at each humidity level. Further detailed

Table 8.3 Apparent first-order kinetic rate constants obtained through curve fitting for the HDI trimer reacting with moisture in air. At 45% and 75% RH, the ratios of the actual average relative humidity during the curve fitting region to that at 15% RH and the ratios of the obtained k_{app} to that at 15% RH are also listed

RH (%)	Average RH for the linear curve fitting part (%)	k_{app}, 1/min	Ratio of average RH	Ratio of k_{app}
15	15	6.60E-04	n/a	n/a
45	43	1.29E-03	2.9	2.0
75	72	2.35E-03	4.8	3.6

comparison shows that the increase of k_{app} from 15 to 45% RH is less than the increase of relative humidity. The same trend holds true for 75% RH. The difference indicates that a more detailed kinetic model is needed to describe the moisture level in the coating and diffusion of moisture into the coating. For instance, the solubility of water inside the coating can be incorporated into the model.

The rate constants obtained were used to predict the kinetic curves at each humidity as shown in Fig. 8.8a. An overall good fit of experimental data is achieved at the early stages of the reaction at all three humidity levels. After that, the model overpredicts the conversion of NCO groups. The deviation could be due to the onset of vitrification of the coating. Vitrification of the coating not only slows down water diffusion but also limits the diffusion of isocyanate functional groups available to react with amine functional groups after water reacts with an isocyanate. Therefore, at this stage, moisture can react with isocyanate to form primary amine, but formed primary amine groups have difficulty to find and react with isocyanate functional groups, leading to a slower reaction. More thorough kinetic modeling, taking into account the decrease in the rate constant at the onset of vitrification and diffusion of moisture into the coating, should be pursued to better predict the NCO conversion.

The reaction of isocyanate with moisture leads to formation of urea. Due to the large peak at ~1687–1690 cm^{-1} from the carbonyl on the isocyanurate ring structure, it is difficult to directly determine information about different forms of urea (i.e., free urea, H-bonded but less ordered urea, and H-bonded and more ordered urea). However, through peak subtraction, i.e., subtracting the initial spectrum from the spectra at later times, formation of different urea forms can be extracted. In the peak subtraction, instead of subtracting the initial spectrum at $t = 0$, subtracting of the spectrum at $t = 5$ min was used. This spectrum has similar relative humidity to the ones in later times. Therefore, the subtracted peaks obtained will be smoother as tiny moisture peaks, if any, can be subtracted out.

As an example, the spectra collected at 75% RH were processed with the peak subtraction method and shown in Fig. 8.9. As reaction proceeds forward, three distinct peaks in the C=O stretching region grow at ~1698, ~1672, and ~1630 cm^{-1}. The three C=O stretching peaks probably represent free urea, H-bonded but less ordered urea, and H-bonded and ordered urea [13, 22].

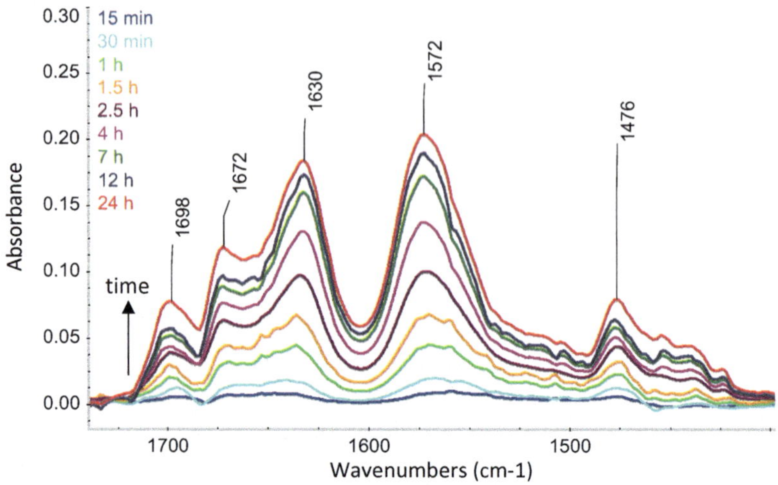

Fig. 8.9 Spectra of isocyanate reacting with moisture in the air obtained by subtracting the spectrum at a certain time to the spectrum at 5 min. Data collected at 75% RH

All the C=O urea peaks grow with time as the conversion of NCO groups increases. The H-bonded and more ordered urea at ~1632 cm^{-1} is the dominant one. The free urea form is present at the lowest level, and it remains as a smaller fraction throughout the reaction. The H-bonded and less ordered urea is in the middle. At ~1572 cm^{-1}, one amide II (C–N stretch and N–H bend) peak due to urea formation also grows.

Kinetics of HDI Trimer Reacting with Acrylic Polyol

Figure 8.10 shows the NCO group conversion when isocyanate reacts with acrylic polyol under different levels of humidity for sample 1, NCO/OH = 1.05. This urethane system has the same level of DBTDL based on solids as the above urea system, where isocyanate reacts with moisture in air. The urethane reaction looks much faster initially than the system where isocyanate reacts with moisture. For example, at 45% RH, in 1 hour, the NCO group conversion is at 0.28 in this urethane reaction system, whereas the conversion is only 0.1 in the urea reaction system. Nevertheless, the reaction quickly becomes diffusion-controlled and as a result the reaction rate slows down. The conversion of NCO groups at 20 h is 0.59, which is less than the conversion of 0.65 in the system with isocyanate reacting with moisture. It indicates that the urethane system will enter into a vitrified state earlier in terms of NCO conversion than the case of isocyanate reacting with moisture.

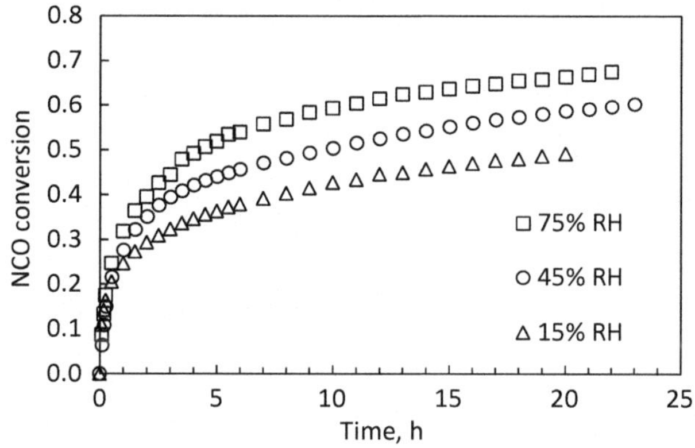

Fig. 8.10 Conversion of isocyanate group as a function of time in the isocyanate and acrylic polyol reaction at NCO/OH = 1.05 and three humidity levels: 15, 45, and 75% RH

This is likely caused by the relatively high rigidity and glass transition temperature of the polyol resin in the urethane system.

When the relative humidity goes up, the reaction rate increases and a higher conversion of NCO groups is reached. The total consumption of NCO groups can be estimated by

$$-\frac{d[\mathrm{NCO}]}{\mathrm{d}t} = k_1[\mathrm{NCO}][\mathrm{OH}] + k_{\mathrm{app}}[\mathrm{NCO}] \tag{8.5}$$

where k_1 is the rate constant for the acrylic polyol and isocyanate and [OH] is the concentration of the hydroxyl groups on the acrylic binder. The reaction is considered as a second-order reaction for a simple illustration. Here, it is assumed that the NCO groups only react to form urethane and urea linkages. Formation of biuret and allophanate is not present at this room temperature reaction. This assumption is likely valid as reported work indicates that allophanate and biuret formation are not observed even at 60 °C when bis(4-isocyanatophenyl)methane (MDI) is reacted with a diol [28]. A more complicated kinetic model can be considered by taking into account the autocatalytic effect on the reaction by formed urethane at later stages of the reaction and the DBTDL catalytic effect [17, 18]. The second term in Eq. (8.5) is the rate of NCO groups reacting with moisture coming from air. Equation (8.5) can further be written as the following equation based on the conversion of NCO groups x:

$$-\frac{\mathrm{d}x}{\mathrm{d}t} = k_1(1-x)[\mathrm{OH}] + k_{\mathrm{app}}(1-x) \tag{8.6}$$

Here again, we assume $[\mathrm{NCO}]=[\mathrm{NCO}]_0\ (1-x)$. This relation assumes there is no volume change or drying of solvents during curing. In analyzing the spectrum data collected at different conditions, the CH_2 peak envelop area at ~2800–3050 cm^{-1} changes no more than 7% from the beginning of reaction to the end of data collection. Therefore, this assumption is fair. When humidity is increased, k_{app}

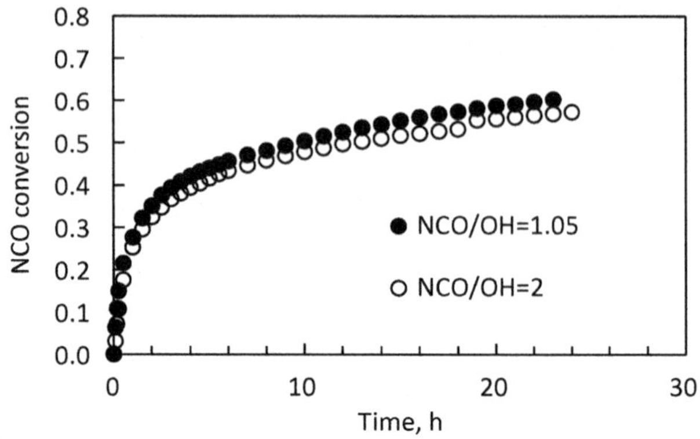

Fig. 8.11 Conversion of isocyanate group as a function of time in the isocyanate and acrylic polyol reaction at 45% RH for coatings with different NCO/OH ratios: 1.05 and 2

increases as discussed above. Consequently, the overall reaction rate based on conversion of NCO groups increases.

Figure 8.11 compares the kinetic curves of the urethane reaction of samples 1 and 2, i.e., at two NCO/OH ratios, 1.05 and 2, at 45% RH. Both coatings have similar film thickness. When the NCO/OH ratio is increased, the reaction rate drops and reaches a lower conversion of NCO groups at later stages of the reaction. This drop is due to the lower concentration of OH groups, which reduces the value in the first term in Eq. (8.6).

In an effort to estimate how much urethane vs. urea groups are formed under different levels of humidity, peak subtraction was used to eliminate the big carbonyl peaks from the acrylic ester groups and from the isocyanurate ring structure. With this approach, the carbonyl peaks from the formed urethane and urea become clear (accessible). In the peak subtraction, the spectrum at 5 min of curing was used to subtract from a spectrum at a later time. Again, the point of 5 min was selected to obtain a stabilized humidity level, so that the subtracted peaks are not affected by the humidity absorbance. Furthermore, at 5 min of data collection, the drying of the coating has gone through a total of about 8 min. The majority of solvents have evaporated by this time. In the subtraction, a factor was applied to the spectrum at 5 min to achieve zero absorption at the CH_2 stretching and bending peaks at ~2952 and ~2875 cm^{-1}.

Figure 8.12 shows the subtracted spectra in the region of ~1475–1780 cm^{-1} at the three humidity levels at NCO/OH = 1.05. In the region of ~1610–1760 cm^{-1}, the peaks are the carbonyl peaks coming from urethane and urea groups. Recall that different forms of the carbonyl groups could exist, including free urethane, H-bonded urethane, free urea, and H-bonded urea [13, 22]. In the figure, different forms of urethane C=O peaks can be seen at 75% RH, i.e., ~1736, ~1727, and ~1711 cm^{-1}, which could be free urethane, H-bonded but less ordered urethane (e.g., one H-bond on each amide group), and ordered, H-bonded urethane (e.g., two H-bonds on each amide group) [13]. Under other humidity conditions, the ~1727 cm^{-1} peak and ~1711 cm^{-1} shoulder can be seen clearly. The urea C=O peaks have quite different shapes compared to that in Fig. 8.9. They have broad

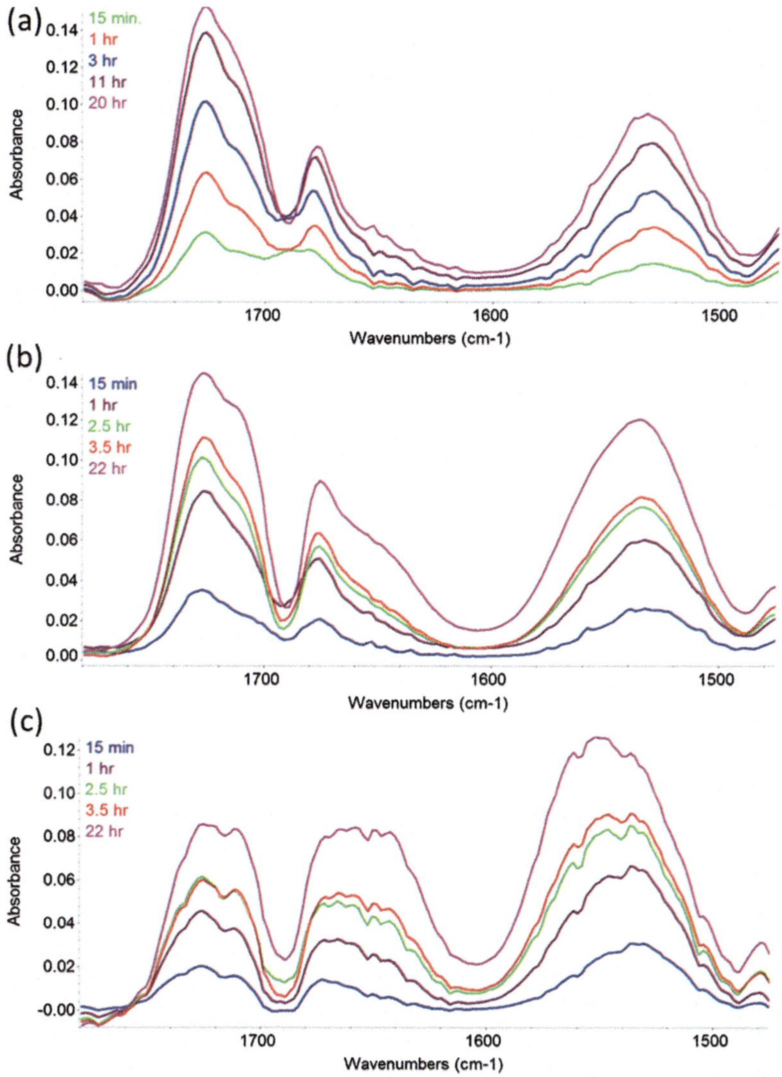

Fig. 8.12 Spectra obtained by peak subtraction of the spectra at 5 min at each humidity level: (**a**) 15% RH, (**b**) 45% RH, and (**c**) 75% RH

peaks ranging from ~1632 to ~1679 cm^{-1}, and the free urea peak is not visible at ~1698 cm^{-1}, which is likely due to its low concentration. Due to the complexity of peak deconvolution with multiple urethane and urea carbonyl peaks, further efforts at deconvolution were not pursued. Rather, it is approximated that the peak area in the region from ~1693 to 1760 cm^{-1} is due to the urethane carbonyl groups, and the peak area in the region from ~1610 to 1691 cm^{-1} is due to the urea carbonyl groups.

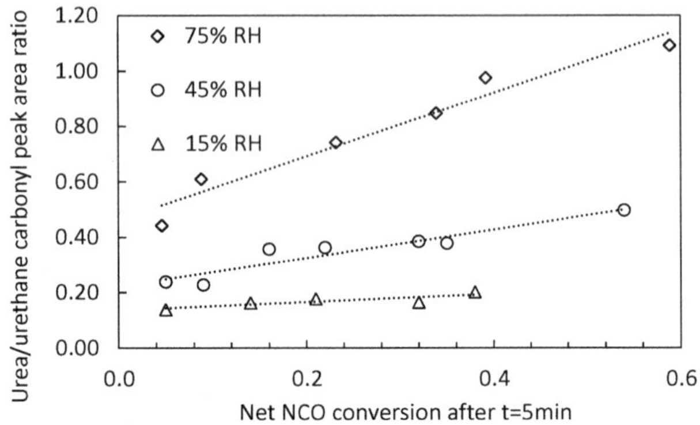

Fig. 8.13 Peak area ratio of the urea/urethane carbonyls as a function of net NCO conversion after 5 min for the coating with NCO/OH = 1.05

The spectra in Fig. 8.12 indicate that when humidity goes up, more urea is formed and less urethane is formed. Also, more ordered, H-bonded urea is formed as reflected from the shift of the urea C=O peak absorption to the lower frequency. Also, the amide II peak, which represents both formed urea and urethane, shifts to the higher wavenumber direction at a higher humidity, indicating more urea (or less urethane) is formed [20].

Here, we focus on the total amount of urea vs. urethane formation, particularly their ratios. The ratio of the two carbonyl peak areas, urea vs. urethane, is plotted against the net conversion of NCO groups after 5 min in Fig. 8.13. The net conversion is the conversion of NCO groups at a certain time minus the conversion at 5 min. In other words, the ratio indicates the relative amount of urea vs. urethane formation after 5 min.

The peak ratio increases with the net NCO conversion. As the reaction proceeds, there is a higher amount of NCO reacting with moisture to form primary amine and consequently urea than early on. Furthermore, when the humidity increases, there is more urea formation. At a low humidity condition (15% RH), NCO mainly reacts with acrylic polyol to form urethane, and the ratio of the urea carbonyl to the urethane carbonyl is small. This ratio also grows slowly as conversion increases. As humidity rises, there is more moisture available, and thus there is more urea formation compared to that at 15% RH throughout the reaction. Also, the slope becomes higher as humidity increases, indicating that considerably more urea is formed as the reaction proceeds at higher humidity. If we assume the urea carbonyl peak and the urethane carbonyl have similar absorptivity [29], the peak ratio here is the ratio of urea vs. the urethane formed. The results indicate a significant amount of urea formation for the coating at NCO/OH = 1.05. As the urea/urethane peak ratio is proportional to the urea/urethane molar ratio, the fraction of urea linkages over the total urea and urethane linkages can be calculated. At NCO conversion of 0.49, the fraction of urea linkages in the total urea and urethane linkages formed is about 17% at 15% RH, 30% at 45% RH, and 49% at 75% RH. These apparently high levels of urea formation could be due to early onset of diffusion control, i.e., vitrification, in the system of the acrylic polyol reacting with HDI trimer. In the vitrified state, moisture diffusion is much faster than diffusion of the acrylic polyol to get close to isocyanate and react.

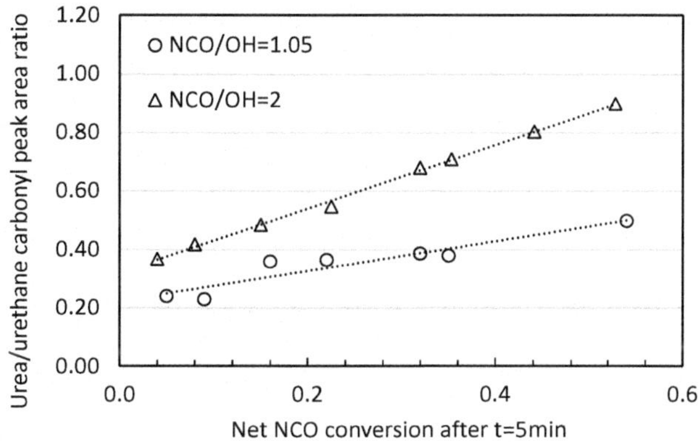

Fig. 8.14 Peak area ratio of the urea/urethane carbonyls as a function of net NCO conversion after 5 min for the coatings with NCO/OH = 1.05 and 2 at 45% RH

When the NCO/OH index is changed from 1.05 to 2, more urea is formed as shown in Fig. 8.14. Clearly, when the index is raised, there is a higher fraction of urea formed at the beginning of the reaction, and the increase of urea content is faster as a function of conversion, leading to more urea formation at the end. For instance, at 0.54 NCO conversion and NCO/OH = 1.05, the urea linkages make up about 33% of the total of urea and urethane bonds and about 47% at NCO/OH = 2.

The amount of urea and urethane formed at the beginning of the reaction can be analyzed further among the different conditions. From Eq. (8.5), the formation of urethane and urea linkages can be described by the following equations. Note that the second term in Eq. (8.5) is the consumption of NCO groups to form urea and every two moles of NCO groups convert into one mole of urea:

$$\frac{d[\text{urethane}]}{\mathrm{d}t} = k_1[\text{NCO}][\text{OH}] \tag{8.7}$$

$$\frac{d[\text{urea}]}{\mathrm{d}t} = \frac{1}{2}k_{\text{app}}[\text{NCO}] \tag{8.8}$$

where [urethane] and [urea] are the molar concentrations of urethane and urea linkages, respectively, at a certain time t. From Eqs. (8.7) and (8.8), we further get

$$\frac{d[\text{urea}]}{d[\text{urethane}]} = \frac{k_{\text{app}}}{2k_1[\text{OH}]} \tag{8.9}$$

By integrating Eq. (8.9) from the beginning of the reaction to a certain point of reaction and assuming no volume change, i.e., $[\text{OH}] = [\text{OH}]_0 - [\text{urethane}]$, we get the following relationship:

$$[\text{urea}] = -\frac{k_{\text{app}}}{2k_1}\ln\left(1 - \frac{[\text{urethane}]}{[\text{OH}]_0}\right) \tag{8.10}$$

where $[OH]_0$ is the initial OH group molar concentration. At the beginning of the reaction, when $[urethane]/[OH]_0$ is small, Eq. (8.10) can be simplified further to

$$\frac{[\text{urea}]}{[\text{urethane}]} = \frac{k_{app}}{2k_1[\text{OH}]_0} = \frac{k[\text{H}_2\text{O}]}{2k_1[\text{OH}]_0} \tag{8.11}$$

Equation (8.11) simply illustrates the ratio of urea and urethane formation at the beginning of a reaction. When the moisture level goes up or k_{app} increases, or when $[OH]_0$ drops, the urea/urethane ratio increases. Here if we extrapolate the linear curve fitting functions shown in Figs. 8.13 and 8.14, we get the estimates of [urea]/[urethane] for net NCO conversion = 0, i.e., $t = 5$ min, as shown in Tables 8.4 and 8.5.

Table 8.4 Estimates of the [urea]/[urethane] ratio at $t = 5$ min for NCO/OH = 1.05

Relative humidity (%)	[Urea]/[urethane]	Ratio of [urea]/[urethane] to that at 15% RH	Ratio of k_{app} to that at 15% RH
15	0.14	n/a	n/a
45	0.22	1.6	2
75	0.46	3.3	3.6

Table 8.5 Estimates of the [urea]/[urethane] ratio at $t = 5$ min for NCO/OH = 1.05 and 2 at 45% RH

NCO/OH ratio	[Urea]/[urethane]	Ratio of [urea]/[urethane] at NCO/OH = 2 to that at NCO/OH = 1.05	Ratio of $[OH]_0$ at NCO/OH = 1.05 to that at NCO/OH = 2
1.05	0.22	–	n/a
2	0.32	1.5	1.3

Table 8.4 also compares the estimated [urea]/[urethane] ratio at different levels of relative humidity. To the first approximation, the [urea]/[urethane] ratio is proportional to the measured k_{app}. For example, when the [urea]/[urethane] ratio is increased by a factor of 3.3 as the humidity is changed from 15% to 75% RH, k_{app} increases by a factor of 3.6. Both factors are very similar. Table 8.5 further compares the estimated [urea]/[urethane] ratio between NCO/OH = 2 vs. 1.05 at 45% RH. The ratio is close to the inverse of the ratio of the $[OH]_0$ for the two samples.

Conclusions

A broad range humidity control chamber was successfully designed to permit in situ FTIR transmission mode measurements throughout the cure studies, some as long as 24 h. The setup is capable of adjusting the humidity from 5 to 80% RH by varying and combining the dry and wet air flow rates. The wet air flow is generated by a water bubbler where dry air is passed through an air stone. To maintain the bubbler water temperature and the wet air humidity level, the bubbler is placed in a water bath. With this setup, it only takes 2–3 min to reach the desired humidity. For proper humidity control, furthermore, care should be taken during FTIR data collection to avoid water condensation at high humidity, which affects the transparency of the KBr crystal, and to avoid moisture absorption due to potential variation between the background and sample collection.

The studies described herein detail the effects of humidity at various humidity levels, including 15, 45, and 75% RH for two reaction systems: (a) isocyanate reacting with moisture in air and (b) acrylic polyol reacting with isocyanate. In the case of isocyanate reacting with the moisture in air, the rate of NCO conversion increases as moisture increases. This reaction is modeled with an apparent first-order kinetic equation, and it fits the kinetic curve well at the beginning of the reaction. The apparent rate constant goes up as moisture increases. Furthermore, through peak subtraction, it is found that the urea generated has different forms: free urea, H-bonded but less ordered urea, and H-bonded and more ordered urea. The H-bonded and more ordered urea is the most abundant form.

In the case of isocyanate reacting with acrylic polyol, isocyanate goes through parallel reaction pathways to form urethane when NCO groups react with OH groups or urea when NCO groups react with moisture in air. When the air humidity increases, the overall reaction rate goes up. Furthermore, estimation of the urea and urethane formation indicates that the ratio of urea to urethane linkages increases as the moisture rises. At higher humidity levels, more urea is formed initially, and further increase of urea is more pronounced as the reaction proceeds.

When the NCO/OH molar ratio changes from 1.05 to 2, there is also more urea formed at the beginning of the reaction, and further increase of urea is more pronounced as the reaction proceeds.

The humidity control within the transmission FTIR chamber has turned out to be a useful tool to monitor the effect of humidity on the kinetics and mechanisms of organic coatings during cure.

Acknowledgments We would like to thank John Perry for running numerous cure studies with controlled humidity, Kyle Price for determining thicknesses of drawdown coatings on KBr salt plates with a profilometer, Adrienne Blevins for conducting the preliminary cure studies with controlled humidity, Joanne Paci for suggestions on the humidity control chamber design, and Justin Ditchfield for creation of the OMNIC Macros used to automatically follow the cure reactions over selected time periods.

References

1. Surface Measurements Systems, GenRH-A Ambient. http://surfacemeasurementsystems.com/genrh.-family/genrh-ambient/
2. Wicks, D.A., Yeske, P.: Amine chemistries for isocyanate-based coatings. Prog. Org. Coat. **30** (4), 265–270 (1997)

3. Frisch, K.C., Rumao, L.P., Mickiewicz, A.J.: Effect of isocyanate variation on physical and environmental properties of moisture-cure urethane coatings. J. Paint Technol. **42**(547), 461–465 (1970)
4. Lowe, A.: Utilization of isocyanates in the surface coatings industry. In: Buist, J.M. (ed.) Adv. Polyurethane Technol., pp. 251–298 (1968)
5. Baghdachi, J.; Li, D.: Soybean oil-based non-isocyanate containing moisture cure resins and coatings. In: Proceedings of 28th International Waterborne, High-Solids, and Powder Coatings Symposium, pp. 231–233. Department of Polymer Science, The University of Southern Mississippi and the Southern Society for Coatings Technology, New Orleans (2001)
6. Ni, H., Skaja, A.D., Soucek, M.D.: Acid-catalyzed moisture-curing polyurea/polysiloxane creamer coatings. Prog. Org. Coat. **40**(184), 175 (2000)
7. Sathiyanarayanan, S., Maruthan, K., Muthukrishnan, S., Venkatachari, G.: High performance polyaniline containing coating system for wet surfaces. Prog. Org. Coat. **66**(2), 113–117 (2009)
8. Squiller, E.P., Wicks, D.A.: The direct reaction between ketimines and aliphatic isocyanates. In: Proceedings of 20th Waterborne, High-Solids, and Powder Coatings Symposium, pp. 15–25. Department of Polymer Science, The University of Southern Mississippi and the Southern Society for Coatings Technology, New Orleans (1993)
9. Wicks, D.A., Yeske, P.E.: Control of the reaction between polyaspartic esters and aliphatic polyisocyanates. In: Waterborne, High Solids, and Powder Coatings Symposium, pp. 1–8. Department of Polymer Science, The University of Southern Mississippi and the Southern Society for Coatings Technology, New Orleans, LA (1993)
10. Aleksandrova, Y.V., Tarakanov, O.G.: Some kinetic relationships of the water-isocyanate reaction. Polym. Sci. U.S.S.R. **14**(1), 25–32 (1972)
11. Ni, H., Nash, H.A., Worden, J.G., Soucek, M.D.: Effect of catalysts on the reaction of an aliphatic isocyanate and water. J. Polym. Sci. A Polym. Chem. **40**, 1677–1688
12. Han, J.L., Yu, C.H., Lin, Y.H., Hsieh, K.H.: Kinetic study of the urethane and urea reactions of isophorone diisocyanate. J. Appl. Polym. Sci. **107**, 3891–3902 (2008)
13. Daniel-da-Silva, A.L., Bordado, J.C.M., Martin-Martinez, J.M.: Moisture curing kinetics of isocyanate ended urethane quasi-prepolymers monitored by IR spectroscopy and DSC. J. Appl. Polym. Sci. **107**, 700–709 (2008)
14. Wissman, H.G., Rand, L., Frisch, K.C.: Kinetics of polyether polyols-diisocyanate reactions. J. Appl. Polym. Sci. **8**(6), 2971–2978 (1964)
15. Rand, L., Thir, S.L., Reegen, S.L., Frisch, K.C.: Kinetics of alcohol-isocyanate reactions with metal catalysts. J. Appl. Polym. Sci. **9**(5), 1787–1795 (1965)
16. Vessot, S., Andrieu, J., Laurent, P., Galy, J., Gerard, J.F.: Air convective drying and curing of polyurethane-based paints on sheet modeling compound surfaces. J. Coat. Technol. **70**(882), 67–76 (1998)
17. Wicks, Z.W., Jones, F.N., Pappas, S.P., Wicks, D.A.: Organic Coatings—Science and Technology, 3rd edn. Wiley, Hoboken, NJ (2007)
18. Kiil, S.: Quantification of simultaneous solvent evaporation and chemical curing in thermoset coatings. J. Coat. Technol. Res. **7**(5), 569–586 (2010)
19. Kaminski, A.M., Urban, M.W.: Interfacial studies of crosslinked polyurethanes; Part I. Quantitative and structural aspects of crosslinking near film-air and film-substrate interfaces in solvent-borne polyurethanes. J. Coat. Technol. **69**(872), 55–66 (1997)
20. Urban, M.W., Allison, C.L., Finch, C.C., Tatro, B.A.: Interfacial studies of crosslinked urethanes: Part III. Structure–property relationships in polyester waterborne polyurethanes. J. Coat. Technol. **71**(888), 75–85 (1999)
21. de Haseth, J.A., Andrews, J.E., McClusky, J.V., Priester Jr., R.D., Harthcock, M.A., Davis, B. L.: Characterization of polyurethane foams by mid-infrared fiber/FT-IR spectrometry. Appl. Spectrosc. **47**(2), 173–179 (1993)
22. Ning, L., De-Ning, W., Sheng-Kang, Y.: Hydrogen-bonding properties of segmented polyether poly(urethane urea) copolymer. Macromolecules. **30**, 4405–4409 (1997)

23. Wexler A., Hasegawa S.: Relative humidity-temperature relationships of some saturated salt solutions in the temperature range 0° to 50 °C. Journal of Research of the National Bureau of Standards 53(1), Research Paper 2512, July 1954
24. Forney, C.F., Brandl, D.G.: Control of humidity in small controlled environment chambers using glycerol-water solutions. Hort. Technol. **2**(l), 52–54 (1992)
25. Watanabe, T., Sato, T.: Expansion characteristics of montmorillonite and saponite under various relative humidity conditions. Clay Sci. **7**(3), 129–138 (1988)
26. Hashizume, H., Shimomura, S., Yamada, H., Fujita, T., Nakazawa, H., Akutsua, O.: An X-ray diffraction system with controlled relative humidity and temperature. Powder Diffract. **11**(4), 288–289 (1996)
27. Kishi, A., Toraya, H.: Simultaneous measurements of X-ray diffraction (XRD) and differential scanning calorimetry (DSC) data under controlled humidity condition: Instrumentation and application to studies on hydration, dehydration, and rehydration processes of pharmaceutical compounds. JCPDS—International Centre for Diffraction Data. Adv. X-ray Anal. **47**, 240–248 (2004)
28. Dusek, K., Spirkova, M., Ilavsky, M.: Network formation in polyurethanes due to allophanate and biuret formation: gel fraction and equilibrium modulus. Makromol. Chem. Macromol. Symp. **45**, 87–95 (1991)
29. He, Z.A., Blank, W.J., Picci, M.E.: A selective catalyst for two-component waterborne polyurethane coatings. J. Coat. Technol. **74**(930), 31–36 (2002)

Chapter 9
Shrinkage in UV-Curable Coatings

Yong He, Miao Yao, and Jun Nie

Introduction

Among the presently available technologies, UV curing is one of the most promising technologies being used in a wide range of applications, such as coatings, adhesives, inks, and other high-tech domains including: optoelectronics [1], laser imaging [2], stereolithography [3], and nanotechnology [4]. UV-curable coatings have some advantages including easy and a wide range of applications, space and time control, room-temperature operation, high curing rate, no emission of volatile organic compounds (VOCs), excellent mechanical properties, and chemical resistance. In contrast, this coating process possesses some drawbacks such as toxicity, difficult operability for the surface of complex structure, and shrinkages. The fundamentals of UV-curable coatings and the character, measurements, and controlling methods of its shrinkage with developing prospects are included in this chapter.

The following list is very subjective and is of course by no means complete. We hope that this text will be a good and practical guide for newcomers and help researchers improve their own standard procedures.

Fundamentals and Applications of UV-Curable Coatings

UV-curing technologies are expanding rapidly on industrial scale, and a number of textbooks and review articles have documented the progress in this area [5–8]. This advanced technology uses light irradiation to initiate photochemical and chemical

Y. He (✉) • M. Yao • J. Nie
College of Materials Science and Engineering, Beijing University of Chemical Technology, Beijing, China
e-mail: heyong@mail.buct.edu.cn

M. Wen, K. Dušek (eds.), *Protective Coatings*, DOI 10.1007/978-3-319-51627-1_9

reactions of organic compounds, leading to the formation of new polymeric materials that have direct impact on commercial end-use applications. Some of the earliest and most significant breakthroughs in UV-curable coatings were made in the wood-finishing industry [9–12]. Thereafter, the experience gained in this area was extended to the application of surface coatings on a large variety of substrates (plastics, metals, wire, pipe, and vinyl flooring), optical and electrooptical materials [13, 14], adhesives, dental materials, printing inks, and composites [5–8, 15]. UV technologies have been developing rapidly, driven by a number of economic and environment advantages over traditional thermal-curing processes, such as rapid through cure, low energy requirements, room-temperature treatment, nonpolluting and solvent-free formulations, and low cost. Simply stated, UV-curing technology can be considered the most rapid and effective way of transforming a solvent-free liquid monomer into a cross-linked polymer at room temperature [16]. On the other hand, a large number of new applications are currently emerging in the fields of graphic arts and coating industries that demand ever-higher performance characteristics.

In general, a photopolymerizable formulation contains a photoinitiator or a more complex photoinitiating system (PIS), a mixture of monomers, some additives, and fillers. Photoinduced polymerizations proceed by a chain reaction mechanism involving the propagation of an active center by the interaction with a monomer. The active center can be a radical, a cation [17], or more rarely an anion species. The photogenerated active species is the "true" initiator that is able to initiate the polymerization of a reactive monomer, usually by addition to a multiple bond or by a ring-opening process. Therefore, the only factor that distinguishes an active center photopolymerization from classical radical, cationic, or anionic polymerization is the use of light to generate the active species. Compared with the thermal polymerization, the initiation, chain growth, and chain termination processes of photopolymerization seem to be the same, but the mechanism of generating active species is different. Photoinitiator absorbs radiation of a certain range of wavelengths and produces primary active species that are able to polymerize multifunctional monomers into a cross-linked network.

Character and Influence of Shrinkage on UV-Curable Coatings

For different applications, the final properties of the cross-linked network have to match precise specifications adapted for the final end use and precision achieved by a selection of highly specified resin systems [18]. Especially, acrylate and methacrylate monomers are favored in nearly all radically polymerizable resins due to their synthetic ease, good processability, and high chemical variability [15]. However, their observed shrinkage is pronounced and inevitable. The shrinkage caused by the change of distance in the polymerization is shown in Fig. 9.1. The formation of C–C

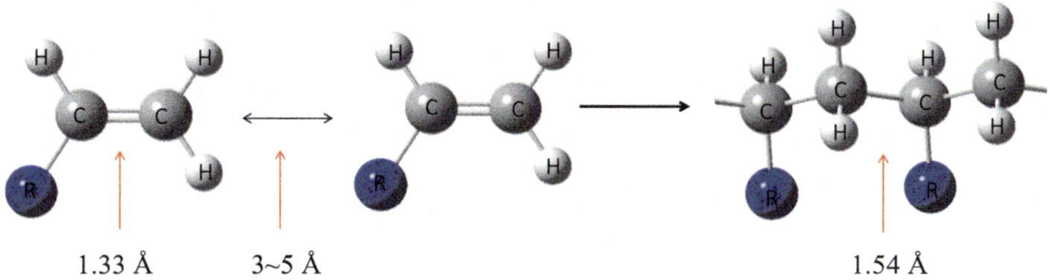

Fig. 9.1 Change of distance in the polymerization

covalent bonds (~1.54 Å in length) from C=C double bonds (~1.33 Å in length) brings a level of expansion. But at the same time, the transformation from intermolecular distances (~3–5 Å in length) into the typical C–C covalent distance leads to shrinkage, whose value is found to be bigger than the expansion and results in the shrinkage observed [19]. The observed value of shrinkage also depends on the rate and extent of physical relaxation. In the melt or rubbery state, the relaxation can be fast enough, so that "maximum" shrinkage at 100% conversion can be reached experimentally. When the system passes into the glassy state during polymerization, the observed shrinkage is always smaller due to the delay of conversion and the freezing in of extra free volume. Usually, two terms are used to characterize the shrinkage: measured shrinkage and normalized shrinkage. The measured shrinkage is the real value determined in experiment, and the normalized shrinkage is the measured shrinkage divided by the conversion of the same reaction system. Since time is involved in shrinkage determination methods in different forms (reaction kinetics and volume relaxation by physical aging) and the methods differ in duration, the test results must be interpreted with extreme care. For common acrylate monomers, the measured shrinkage in photopolymerization is usually 8–17%, and the normalized shrinkage at complete conversion (maximum shrinkage) is much higher, which can achieve 10–27% [20].

As a result, photopolymerization shrinkage causes a lot of adverse impacts on photolithography [21], UV-curable coatings [22], and dental materials [23], such as the low precision of fine structure, low adhesion, and poor mechanical properties. Shrinkage of patterns becomes a problem in the fabrication of large-scale integration (LSI) circuits, where severe critical dimension (CD) control is required. For example, photo nano-imprinting [21] is a patterning method using a mold by which its reversal structure is replicated in a moldable material. However, in reality, the size of the replica turns out to be smaller than the ideal one of the mold because of the shrinkage of UV-curable resin during photopolymerization. The polymerization shrinkage also causes a large buildup of internal stress in the networks formed [24], and then the stress is transmitted to the bonding interface and leads to a reduction in adhesion between the coating layer and the substrate. For example, photopolymerization of composites, which have proved to be very useful in dental restoration and other biomedical applications, can develop stress at the interface between the tooth and the composite adhesive in a bonded cavity preparation.

Clinically, the stress buildup can damage the integrity of this interface, which is crucial to the success of the composite restoration. As a result, a number of potential clinical problems, such as postoperative hypersensitivity, secondary caries, and pulpal inflammation [23], can rise.

Measurement and Evaluations of Shrinkage in UV-Curable Coatings

The shrinkage problem in UV-cured systems has gained a lot of attention in the research field. To evaluate it precisely and easily becomes a need. Different methods have been developed and used in various technical fields. Early investigations of the polymerization shrinkage of UV-cured materials were carried out by dilatometry [25, 26], pycnometry [27], and buoyancy [28] methods, which are also applied for the shrinkage measurement of other curing technologies.

Dilatometry Method

Measurements of shrinkage are usually conducted indirectly in a dilatometer, which was reported by Bekkedahl in 1949 [29]. It monitors the height changes of a column of fluid (commonly mercury or water) connected to a reservoir surrounding the test substance (Fig. 9.2). Normally, the initial volume of a sample is directly calculated by the difference between the volume of surrounding liquid, obtained from its weight and density, and the total volume of sample and the liquid. In order to achieve high precision, the weight of the specimen should be measured nearest to 1 mg. The mercury, and thus the dilatometer too, should be weighed to the nearest 10 mg. Utilizing an analytical balance having a sensitivity of 0.1 mg could achieve reproducible data with accuracy to about 0.0002 g/cm^3 or better [29]. The readings of the capillary of the dilatometer should be estimated to the closest 0.5 mm, which is equivalent to about 0.002 mL. The accuracy of the temperature should be within 0.1 °C, as an error of 0.1 °C is equivalent to about 0.002 mL volume change for the contents of the dilatometer. However, because of the opacity of mercury, it is difficult for light to go through and to arrive at the sample. Thus it is more difficult for such methods to be applied to light-curable polymers. Water-based dilatometry also depends on the bath temperature stabilization, which is difficult to maintain, because the thermostability is constantly affected by the exothermic reaction of the resin polymerization and heat released by the curing light tip. Temperature fluctuations lead to variations of fluid volume, which affect accuracy of the contraction measurements.

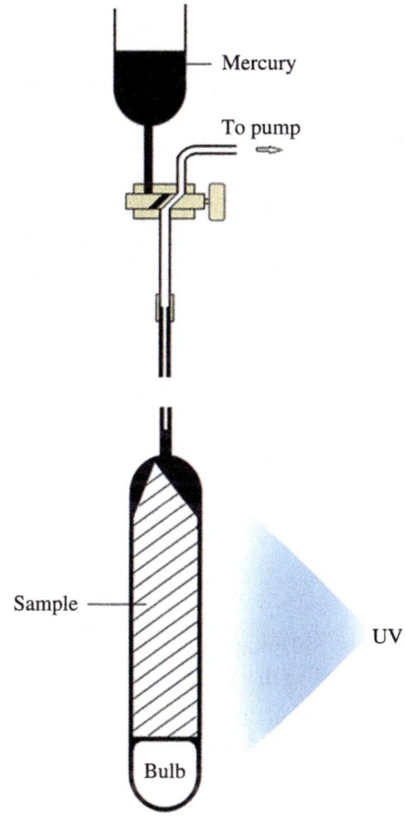

Fig. 9.2 Schematic diagram of a dilatometer [29]

Pycnometer Method

In early research, a pycnometer was often used to characterize the volume shrinkage by measuring the density of the investigated system before and after polymerization [19]. For a liquid sample, the density can be calculated from the weight of the liquid and the volume of the pycnometer (V_0) by adding the liquid to fill the pycnometer full. V_0 could be measured by the weight of a liquid with a known density with the same method. For a solid sample, the density is measured by the hydrostatic weighing method. It is applied by suspending grounded sample pieces in a degassed nonswelling liquid, whose density could be measured and is equal to the solid sample. The liquid must have good wetting properties with the sample, and the sample cannot swell in it. In general, if the density of a sample is between 0.8 and 1 g/cm^3, a water/ethanol mixture can be chosen; if the density is higher than 1 g/cm^3, an aqueous solution of salt can be used.

The volume shrinkage ε_v is calculated by Eq. (9.1) ρ_i and ρ_f are the initial and final density values of the sample, respectively. However, this method is complex

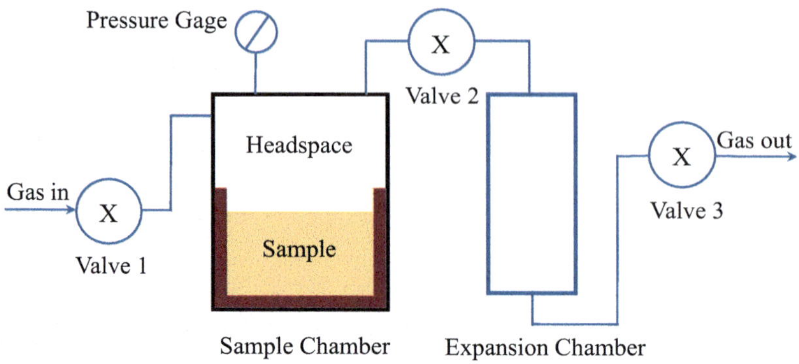

Fig. 9.3 Schematic diagram of a pycnometer with a gas displacement technique [27]

and sensitive to air bubbles attached to the sample surface, which leads to poor accuracy.

$$\varepsilon_V = 1 - \frac{\rho_{\mathrm{i}}}{\rho_{\mathrm{f}}} \tag{9.1}$$

The shrinkage can also be measured by using a gas pycnometer [27] through monitoring the volume of the resin sample as it cures at ambient temperature. This instrument works by measuring the amount of displaced gas. As shown in Fig. 9.3, valves 2 and 3 are first closed, and valve 1 is open to charge the sample chamber with a gas till its pressure reaches a certain pressure, p_1. Then valve 1 is closed, and valve 2 is open to let the gas diffuse from the sample chamber to the expansion chamber. The equilibrium pressure of sample and expansion chamber is p_2. The volume of sample (V) is calculated by equation:

$$V = V_{\mathrm{s}} - V_{\mathrm{e}} \frac{p_2}{p_1 - p_2} \tag{9.2}$$

where V_{s} and V_{e} are the volume of the sample chamber and that of the expansion chamber, respectively. The sample is cured outside of the sample chamber under ambient conditions, and the sample volumes before and after curing are measured, and then the shrinkage ε_v is calculated by equation:

$$\varepsilon_V = 1 - \frac{V_{\mathrm{f}}}{V_{\mathrm{i}}} \tag{9.3}$$

where V_{i} and V_{f} are the initial and final volumes of the sample, respectively. Calibration of the pycnometer and initial error determination tests suggest a calibration error of only 0.003% [27]; thus the technique ensures high-precision volume measurements. This method is suitable for unsaturated polyester (UP), whose shrinkage is 9.35% concurring with the literature value of 7–10%

[30]. The gas pycnometer offers similar accuracy with dilatometer and is less labor intensive.

The dilatometry and pycnometer methods are simple and do not require a special apparatus but can only give the final shrinkage, while a continuous volume change is difficult to observe. A real-time technique of measuring the shrinkage becomes important and should be developed because the final shrinkage does not illustrate the whole shrinkage process and the rate of volume changes during the UV cure. Moreover, the shrinkage profile as a function of time shows the levels of shrinkage during the whole curing process and the time when major shrinkage occurs, which can help to analyze the reaction mechanism of complex polymerization systems and to provide theoretical basis of controlling the shrinkage. Nevertheless, the real-time shrinkage characterization for UV curing is challenging because for a rapidly shrinking system, the detected rate and accuracy need to be comparable with the curing technologies. And for some special systems, curing can be so fast that shrinkage cannot occur (initial volume is frozen), and the product can be porous or microporous. In this case, measuring shrinkage seems to be impossible.

Buoyancy Method

The buoyancy [31] method is one of real-time measurements in the early days although it is not a good candidate for the moisture sensitive or fluid systems. The sample is placed in an aluminum pan loaded on the arm of a balance and set to the null position. The end of the light guide is positioned near the sample pan (Fig. 9.4). While the light-curing unit is irradiating, the sample gradually decreases in volume as a result of polymerization shrinkage. Accordingly, the change in the volume of the sample results in buoyancy change of the distilled water due to the Archimedes' principle. The change of balance is detected by the position-sensitive photo detector (PSPD), which generates a continuous voltage signal in real time, The signal is recorded in constant time intervals to measure the change in buoyancy and then

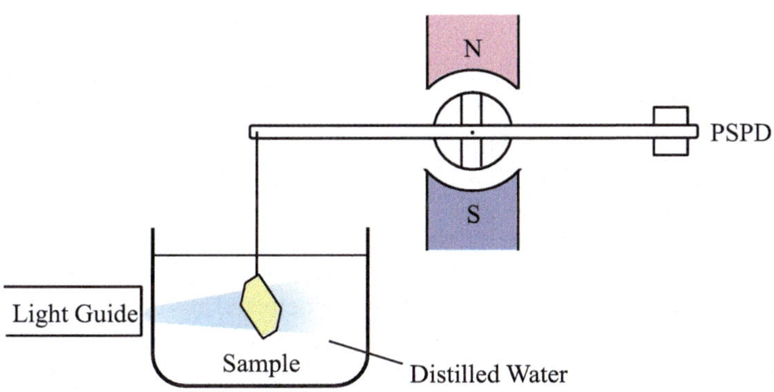

Fig. 9.4 Schematic diagram of the buoyancy method [31]

converted to volume change (ΔV_t). The initial volume is calculated from the sample weight and the predetermined density of the uncured sample by a pycnometer. Polymerization shrinkage in vol.% is calculated using the following equation: $100\times\Delta V_t/V_{\text{initial}}$ (%), where ΔV_t is the volumetric shrinkage at t after the beginning of light exposure and V_{initial} is the initial volume of the uncured sample. Unlike the dilatometry, which depends on the bath temperature, the buoyancy method is reasonably insensitive to temperature changes, but it is very sensitive to air bubbles attached to the sample surface or to the suspending wire.

It seems that methods for measuring the shrinkage of the abovementioned techniques are complex because the density or volume of sample must be obtained in advance with a series of difficult processes. So a simplification was developed. Feilzer et al. [32] and Watts et al. [33–35] studied the volume shrinkage of a film sample with cylindrical shape sandwiched between two glass slides. They concluded that if a C-factor (a configuration factor, the area ratio of the bonded surface to the free surface) is high and the adhesion between the substrate and sample is sufficient, the axial shrinkage (along the height direction of cylindrical) value is approximate to the total shrinkage because no radial contraction is allowed and the level of axial contraction is intensified. Extending to the general cases, if the sample is sandwiched by substrates as a film, no matter what shape of the film is, and the adhesion between the substrates and sample is sufficient, the normal shrinkage (shrinkage along the thickness direction) is approximately equal to the total shrinkage. After simplifying the principle of measurement, the operational process becomes much easier because only two variables need to be considered during photopolymerization: the dimensional change of a coating in the thickness direction and time. Then a lot of simplified methods, which measure linear shrinkage real time, have been developed such as bonded-disk method [31, 35, 36], direct laser displacement measurement [20, 37], interferometer [38], and optical scanning method [39].

Bonded-Disk Method

Bonded-disk method is a technique that is widely used for dental composites, which was built as a real-time method to measure the shrinkage by Watts and Cash [33]. The prepared specimen is placed on the horizontal metal plate; the linear variable differential transformer (LVDT) probe tip is positioned at the center of the cover glass, which is in contact with the cover glass under its weight (Fig. 9.5). The probe is set at the null position by using a micrometer on the vertical stage. After obtaining a baseline without light exposure, the sample is irradiated. As polymerization proceeds, the axial linear shrinkage is detected by the LVDT, which has sensitivity greater than 0.1 μm in a moving range of ±1 mm. After the measurement, the diameter and thickness of each specimen are measured, and the axial shrinkage and C-factor are obtained using the equation, $\Delta h/(h+\Delta h)\times 100(\%)$ and

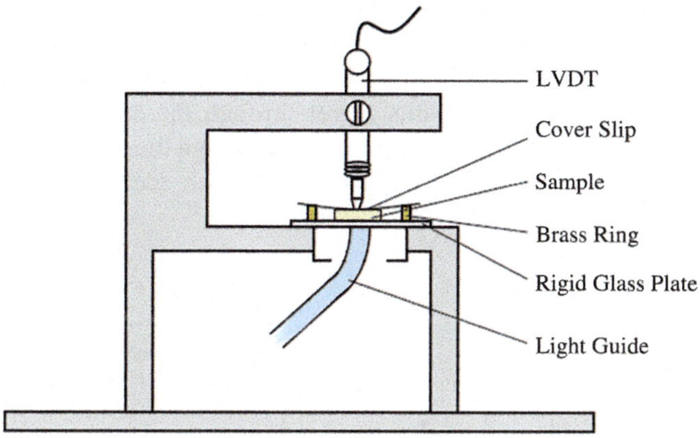

Fig. 9.5 Schematic diagram of the bonded-disk method

$D/[2\times(h+\Delta h)]$, respectively, where h is the thickness of the cured specimen, Δh is the measured shrinkage by the LVDT, and D is the diameter of the specimen.

Optical measurements using lasers allow the determination of the shrinkage with higher precision than does the mechanical approach. The two most common methods are interferometry and direct laser displacement method; their resolutions are in the range of 10 nm–0.5 μm [20, 38, 40–42]. The value of the volume shrinkage obtained by this method is somewhat different than the value obtained for free specimen.

Interferometer Method

The interferometer method employs a Michelson interferometer to measure the linear shrinkage of dental materials [38]. An optical flat is mounted firmly on top of the sample in place so that the distance between this optic and beam splitter remains constant. The optical flat on the bottom side of the sample is a first-surface mirror, where the mirrored side of the flat is opposite the sample. A small amount of liquid coating is placed on the bottom optical flat and subsequently translated upward using a manual vertical micrometer stage until the material is sandwiched against the upper optical flat with the desired initial film thickness. During photopolymerization, the interference intensity measured at the stabilization detector is constant, and the intensity measured at the path length detector varies as a result of the photopolymerization shrinkage (Fig. 9.6). The following expression is used to calculate the percentage of linear shrinkage (ε_v) at any point during the curing reaction:

$$\varepsilon_V = \frac{N\lambda}{2l} \tag{9.4}$$

where N represents the total number of oscillations in the path length interferogram up to the particular time of interest, λ is the wavelength of laser, and l is the final thickness of the cured restorative. A factor of 2 appears in the denominator of the above expression because the beams travel through the sample cell twice in a Michelson interferometer optical arrangement. Although the interferometer method has high resolution, the process to create a light path seems difficult and time-consuming.

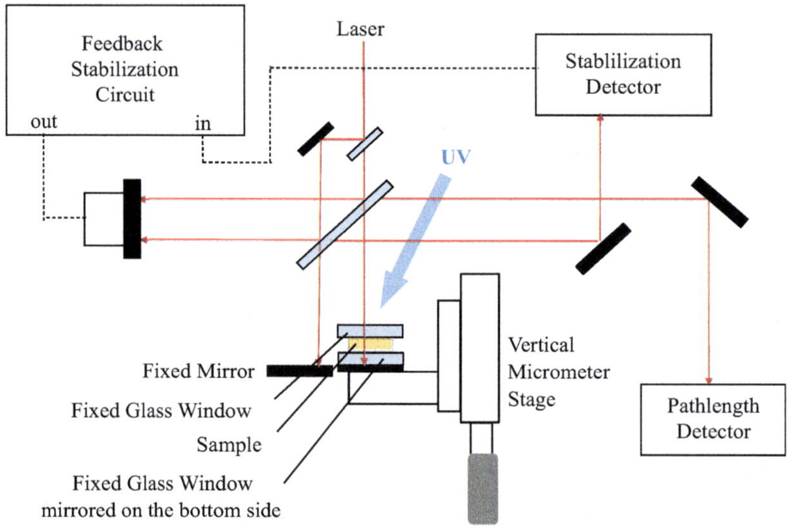

Fig. 9.6 Schematic diagram of the interferometer method [40]

Laser Displacement Method

The laser displacement method [20, 37] is more convenient and also has high resolution. The direct photo-cure shrinkage measurement setup is shown in Fig. 9.7. A Keyence LK-G10 was chosen as a laser displacement sensor (LDS) due to its fast sampling speed of 50 KHz and good resolution of 0.02 μm. The laser beam 30 μm wide in diameter was initially zeroed on the glass slice and then reflected off from the surface of the sample to give the thickness of the sample layer. In all measurements, the thickness of the sample layer was around 100 μm. The LDS detects the displacement of the sample (dimensional change) due to the shrinkage of photopolymerization by recording the motion of mirror reflecting laser on the charge-coupled device (CCD).

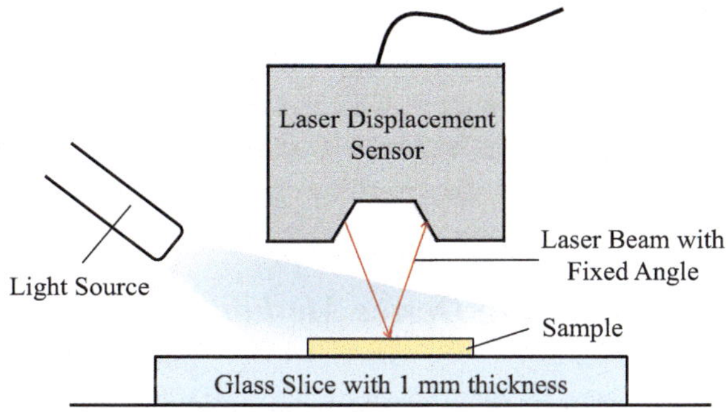

Fig. 9.7 Schematic diagram of the laser displacement method

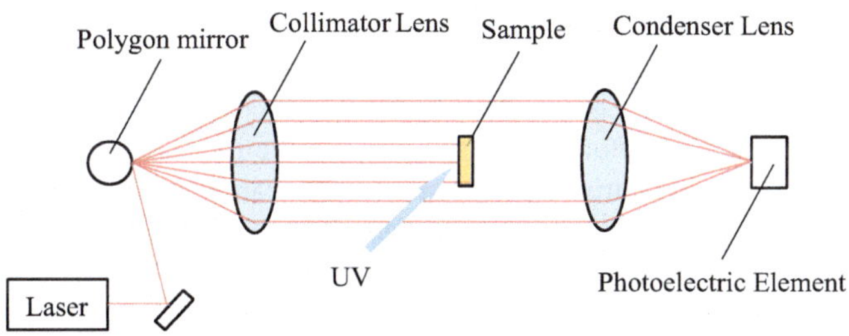

Fig. 9.8 Schematic diagram of laser scanning method [39]

Some other methods based on optical measurements for determination of UV-curing shrinkage are directly used to measure the shape change of materials, such as laser scanning method [39] and video-imaging device method [42].

Laser Scanning Method

A new noncontact method using a laser scanning micrometer was built by Fano et al. [39]. It performs a simple and accurate measurement of soft materials and prevents distortion due to the chemical or mechanical dimensional variations caused by contact between the materials and the measuring apparatus. Unlike the light emitted by a conventional source, the laser provides extremely fine parallel beams which do not diffuse. Therefore, a scanning laser beam can be cast over the sample and the time the specimen obstructs the beam is measured.

The He-Ne laser beam reflected by a rotating polygonal mirror is then collimated by a lens that creates a parallel light flux which scans the specimen (Fig. 9.8). The scanning beam passes through a condenser lens onto a photoelectric element and generates a voltage which is proportional to the intensity of the light. The voltage allows the number of pulses generated, while the beam is obstructed by the specimen to be counted. The number of pulses is provided by a clock pulse and computed by a counter, which is proportional to the length of the specimen. This method makes it possible to analyze small sample with a low error (± 1 μm).

Video-Imaging Device Method

In the video-imaging method, a sample is shaped into a semisphere (with volume averaging 12 μL) and placed on the Teflon pedestal in the light path. While using the multi-view volumetric reconstruction mode, the sample was rotated, and 20 different camera angle shots were taken to be averaged for a single point measurement. In the single-view volumetric reconstruction mode, only one angle picture was used for each measurement. The specimen was allowed to rest for several minutes to eliminate the influence of slumping on the measurement, and the visual image of sample was captured by the unit's CCD camera. The specimen volume was measured, and the specimen was then cured in the light-curing unit. The post-curing volume was measured and recorded several minutes after light activation in order to get specimen temperature equal to room temperature. The difference between the pre-curing volume and the post-curing volume was used to calculate percentage of shrinkage. The precision of 0.1% for determining shrinkage by video-imaging is comparable to other mentioned techniques, such as dilatometry [43], but the video-imaging method is only suitable for almost nonfluid system.

The polymerization shrinkage of UV-cured materials is not only closely related to the chemical composition of the resin formulation but also strongly depends on the conversion of active groups achieved during and possibly after UV irradiation. Consequently, the indication of shrinkage values can be understood further when we know the extent of reaction such as conversion of functional groups as a function of time or the maximum conversion achieved in specified systems at the end of curing. However, none of the abovementioned methods are able to relate the extent of shrinkage to the conversion of functional groups at the same time. Obtaining both types of data requires a combination of a volume shrinkage technique with another analytical technique to determine the conversion. The optimum method for direct determination of the conversion in real time is infrared spectroscopy.

Very few attempts have been reported so far to develop such kind of combined real-time techniques. Neo and Chan-Park [40] mounted a laser displacement sensor used for triangulation above the crystal of a diamond ATR unit placed in an FTIR spectrometer, which enabled simultaneous measurement of shrinkage and conversion. Unfortunately, no data of the shrinkage and dependence on the conversion have been reported in their article.

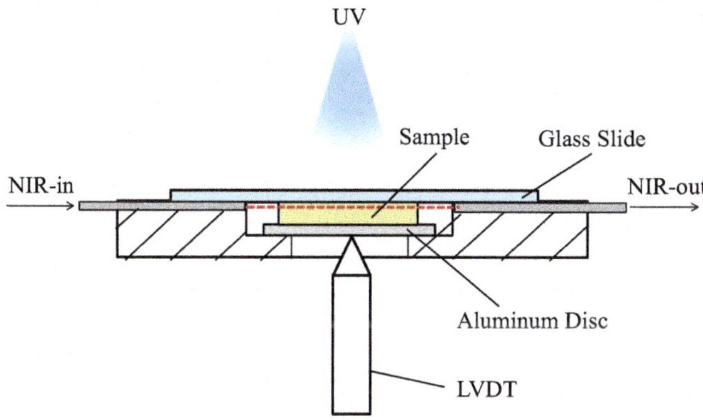

Fig. 9.9 Schematic diagram of the linometer-NIR method

Linometer and NIR Spectroscopy Combination Method

Szczepanski et al. [44] reported an accurate method to measure shrinkage and conversion simultaneously by combining a linometer and an NIR spectrometer. The sample was sandwiched between a glass slide and an aluminum disc that was placed on top of a noncontact probe (Fig. 9.9). A light guide was positioned so that the sample was irradiated from upward side through the glass slide. As the material polymerized and shrunk, the aluminum disc was lifted, and differences in potential sensed by the probe were recorded by the instrument software. Two near-IR fiber optic cables were positioned oppositely and let the signal pass through the sample film, and the conversion was monitored simultaneously.

Photorheometry and NIR Spectroscopy Combination Method

In recent years, another technique was designed by combining a photorheometer with an NIR spectrometer [45]. A rheometric method can also be used to measure the linear shrinkage of a photopolymerization system using an oscillation mode with a UV cell. In this cell, the lower plate of the conventional parallel plate system was replaced by a quartz plate, which allows UV irradiation of the sample, and a disposable aluminum plate was used as the upper plate (Fig. 9.10). Moreover, the temperature of the sample can be controlled by both Peltier heating (or cooling) of the quartz plate and convection heating of the environmental chamber.

The experiments were subdivided into several phases with different experimental conditions. In the first step, the state of the sample before UV curing was analyzed by acquiring rheometric and NIR data simultaneously. A sinusoidal strain of 10% with a frequency of 20 Hz was applied to the sample, and the rheometric data were recorded at a rate of one value per second. In the second step, UV illumination of the sample was started, and the applied strain was decreased from

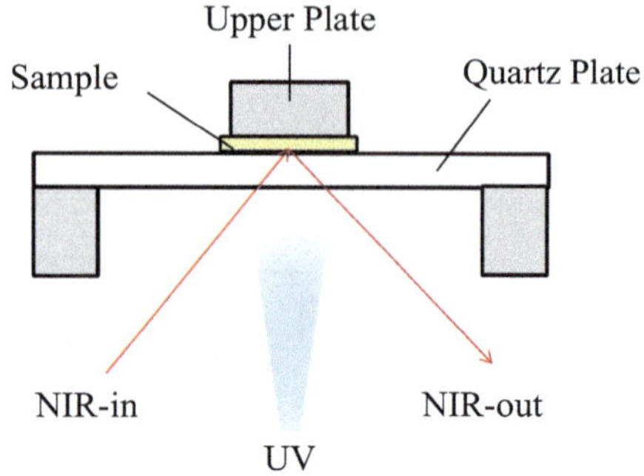

Fig. 9.10 Schematic diagram of photorheometry [45]

10 to 0.05% using a logarithmical ramp and a sampling rate of 10 s^{-1}. This ramping of the applied strain is necessary because of the strong change of the modulus of the sample during network formation. If the strain would not be reduced, the maximum allowable torque of the instrument could be exceeded. Moreover, the decrease of the strain ensures that the deformation of the sample remains in the linear visco-elastic range. Later, the sampling rate was reduced to 1 s^{-1} and then to 0.1 s^{-1}. Finally, the UV illumination was switched off; however, rheometric and NIR data were still recorded for several minutes.

During the experiment, the automatic gap control of the rheometer was activated, which compensated for changes of the thickness due to the shrinkage of the sample during polymerization by continuous readjusting the normal force F_N applied to the upper plate of the rheometer to $F_N=0$. The extent of shrinkage was determined from the ratio of the gap at a certain time to the initial gap. Because of the combined control of rheometry and spectroscopy via one control panel, each NIR spectrum (or conversion value) could be directly related to the corresponding rheometry data (i.e., shrinkage value). In this way, the temporal development of shrinkage could be monitored in dependence on the conversion.

Methods of Controlling the Shrinkage

As mentioned above, the polymerization shrinkage causes a large buildup of internal stress in polymer networks, which in turn leads to a reduction in adhesion between the coating layer and its substrate in coating applications. It also causes the shape change of materials and resolution reduction, which has a negative impact on high-tech patterning or shape formation fields due to the difficulty to make ideal structures. The methods of controlling and adjusting the shrinkage in UV-curing coating have been studied from different perspectives, such as changing the condition of photopolymerization, addition of inert components, variation of monomer structures, and application of different polymerization methods. In this section, these aspects are introduced in detail.

The Condition of Photopolymerization

In the early days, adjusting the condition of photopolymerization was the easiest way to reduce the shrinkage. Obviously, temperature is the most important factor, which is controlled by irradiation and cooling instruments and can be increased by the exotherm of polymerization. The coatings could transfer from a rubbery state to a glassy state during the polymerization, and then volume relaxation would slow down remarkably. So higher temperature can increase the photopolymerization rate and accelerate the volume relaxation, especially when temperature goes over the glass transition point of systems. But photopolymerization usually occurs under room temperature with satisfied cooling, and then irradiation intensity becomes a pretty effective method to control the shrinkage. Some studies demonstrated that higher irradiation intensity could lead to higher conversion [46, 47]. Lecamp et al. [47] explained the mechanism that the initial polymerization rate (R_{pi}) is described as Eq. (9.5), where $[M]_0$ and $[A]_0$ are initial monomer concentration and initial photoinitiator concentration, respectively; k_p and k_t are chain propagation rate constant and chain termination rate constant, respectively; φ is the quantum yield for chain initiation; ε is the photoinitiator molar extinction coefficient; and I_0 is the light intensity:

$$R_{pi} = -\frac{d[M]_0}{dt} = k_p[M]_0\left(\frac{\phi\varepsilon I_0[A]_0}{k_t}\right)^{1/2} \tag{9.5}$$

As described in Eq. (9.5), higher light intensity would cause higher initial polymerization rate. Due to the fact that volume shrinkage rate is much slower than the chemical reaction rate, the cross-linking systems cannot be in volume equilibrium. This difference generates a temporary excess of free volume which increases the mobility of the residual double bonds and help to reach higher conversion. The shrinkage is closely related to conversion, and the intensity also plays a key role in the dynamics of shrinkage. Polymerization shrinkage at the end of curing was found to increase with irradiation intensity [48, 49].

It was reported that after irradiation at two different intensity levels (750 and 200 mW/cm^2), the shrinkage of an investigated system decreased with reduction in light intensity and so did conversion (Table 9.1) [49]. However, on the one hand,

Table 9.1 The degree of conversion and maximum volume shrinkage of composite (Tetric Ceram, TRC) [49]

	Mode of irradiation	Conversion (%)	Shrinkage (%)
I	40 s standard (750 mW/cm^2)	60.3	2.66
II	40 s two-step (10 s at 200 mW/cm^2 + 30 s at 750 mW/cm^2)	61.0	2.55
III	40 s low (4 × 10 s at 200 mW/cm^2)	47.8	2.09
IV	10 s low (200 mW/cm^2)	36.7	1.44

higher conversion may lead to superior physical and mechanical properties, and on the other hand, low intensity may be favorable for marginal adaptation [50, 51] and the integrity of restoration-cavity interface [52]. The "soft-start" polymerization was brought as a new technology, which employs an initially low intensity followed by a high-intensity irradiation, resulting in relatively low shrinkage and similar conversion comparing with the high-intensity irradiation case [53]. A lower intensity of light can be effective in terms of delaying the formation of a too rigid network, thus leaving the coating more time to flow and stress relieve [54].

Addition of Inert Component

The shrinkage stress developed during cure can lead to interfacial defects, which could have a detrimental effect on the structure of the materials. For composite materials, the addition of fillers, such as glass fiber, quartz fiber, ceramics, and silica, which do not contribute to shrinkage, can reduce shrinkage strain [55–57]. However, addition of these ceramic fillers results in an increase in stiffness [58] and the internal stress [59]. Therefore, the end effect is not straightforward. For the composites containing nano-sized particles SiO_2 in an acrylate polymer matrix, shrinkage reduction of ~33% was achieved by adding 20 vol.% of filler [60].

In addition, a series of thermoplastic materials, such as poly(methyl methacrylate) (PMMA), poly(vinyl acetate) (PVAc), and polystyrene (PS), can be used as low profile additives (LPA) to reduce the shrinkage [30]. The addition of LPA reduces the concentration of double bonds (this will be discussed in detail in the next part) and induces partial phase separation between LPA- and UV-cured polymer network. Micro-voids are formed at the interface. These micro-voids cause a little expansion in local areas and finally result in lower shrinkage. Liu et al. [61] examined the shrinkage behavior of unsaturated polyester (UP) resin with varying amounts of LPA (PVAc). A shrinkage of approximately 10% was found for UP without any LPA. Increasing the content of LPA from 0 to 35% resulted in a decrease in the shrinkage from 10 to 2%, respectively. Szczepanski et al. [44] controlled the shrinkage of triethylene glycol dimethacrylate (TEGDMA) by addition of prepolymers, poly(ethyl methacrylate). Addition of 20 wt% of prepolymers led to 7.3% of shrinkage, which is much lower than the value of TEGDMA by itself (13.0%).

Adjusting the Structure of Monomers

The Density of Functional Groups

Due to the mechanism that shrinkage comes from reduction of the distance between active groups from the van der Waals distance to a single bond, reducing the functional group concentrations of a coating system is an impactful way to decrease

Table 9.2 The molecular weight of resins and shrinkage of composites made by these resins [63]

Monomer	Molecular weight (g/mol)	Curing shrinkage (%)
THMPE	733	2.13
THMPM	719	2.53
BisGMA	513	2.74

Table 9.3 Photopolymerization shrinkage and kinetics: the oligomer (Desmolux XP 2513)/monomer ratio was 50/50 by weight [65]

Monomer	Final shrinkage (%)	Final conversion (%)	C_{db} (mol/L)
TMPTA	7.42	67.75	11.24
TMPT(EO)3A	6.00	87.16	7.73
TMPT(EO)9A	4.83	96.76	4.81
TMPT(EO)15A	4.05	98.58	3.49

the shrinkage. Prepolymers and oligomers with lower reactive functional group density than monomers can be used to reduce shrinkage [62]. Chung et al. [63] reported that adding high molecular weight oligomers to methacrylate monomers could reduce the shrinkage. The composite with 1,1,1-tris[4-(2′-hydroxy-3′-methacryloyloxypropoxy)phenyl]ethane (THMPE) as monomer revealed lower shrinkage than those with tris[4-(2′-hydroxy-3′-methacryloyloxypropoxy)phenyl] methane (THMPM) or 2,2-bis[4-(2′-hydroxy-3′-methacryloyloxypropoxy)phenyl] propane (BisGMA) due to its highest molecular weight (Table 9.2). The following equation [20] can explain the phenomenon:

$$S_{\mathrm{m}} = -2.18 + 3100 \times \frac{F}{M} \tag{9.6}$$

where S_{m} is the shrinkage at 100% double bond conversion, F is the functionality, and M is the molecular weight of monomer.

Jian et al. [64] studied the relationship between the maximum shrinkage and the concentration of double bonds (C_{db}) of a series of acrylates and showed the linear result:

$$S_{\mathrm{m}} = 23.98 C_{\mathrm{db}} \tag{9.7}$$

The ratio of S_{m} and C_{db} (23.98 mL/mol) is defined as the shrinkage factor, which can be used to evaluate the shrinkage capability of double bonds for photopolymerization of acrylates. Jian et al. [64, 65] also reported that introduction of an ethoxylated (inserting ethoxyl groups into molecules) structure is a promising way to lower the shrinkage by increasing the chain length and molecular weight. The shrinkage of a series of ethoxylated trimethylolpropane triacrylate was studied and compared (Table 9.3). It demonstrates that the ethoxylated monomers of lower C_{db} lead to lower shrinkage and much higher conversion than TMPTA (Fig. 9.11).

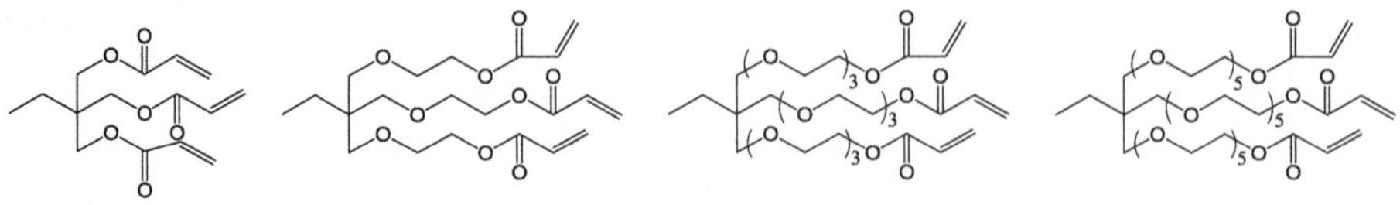

TMPTA TMPT(EO)3A TMPT(EO)9A TMPT(EO)15A

Fig. 9.11 The structure of monomers in Table 9.3

Structure Rigidity and Steric Hindrance

In the light of shrinkage formation mechanism, it is reasonably inferred that the presence of rigid groups, which could obstruct the shrinkage, would lead to lower shrinkage. Liu et al. [66] synthesized a polyurethane acrylate prepolymer with rigid alicyclic structure (PUA) and then mixed it with a commercial polyurethane acrylate oligomer CN9001. When the content of PUA changes from 0 to 80%, the shrinkage decreases from 4.20 to 1.75%. Unfortunately, these values of shrinkage are measured shrinkage, and the normalized shrinkage at complete conversion or the maximum shrinkage was not given, but they can be easily calculated by using the conversion of each system under the same conditions. While the content of PUA changes from 0 to 80%, the maximum shrinkage decreases from 4.20 to 2.13%. Because of the higher double bond concentration of PUA than CN9001, it was concluded that its higher rigidity would be the reason of this tendency.

Besides the rigidity, the steric hindrance could result in a similar effect. Qin et al. [67] synthesized a series of polyurethane acrylate (PUAs) with long carbon chain substitutions which have higher double bond density (1.323–1.171 mol/L) than CN9001 (0.711 mol/L). The results showed that their shrinkage decreased from 4.36 to 3.59%, while the carbon chain was varied from butyl to octadecyl groups. The shrinkage of these modified PUAs were all lower than CN9001 (5.24%), which explained that the flexibility of long carbon chain is the main cause. But it is reasonable to believe that long-chain group can conform and occupy the free volume produced by polymerization and hinder polymer chains getting close each other, leading to less shrinkage.

Ring-Opening Polymerization

An elegant route to avoid or to reduce shrinkage during photopolymerization is the ring-opening polymerization. During the ring-opening polymerization, it happens that for every bond that changes from a van der Waals distance to a covalent distance, at least one bond change from a covalent distance to a near van der Waals distance (Fig. 9.12). This is the reason why the ring-opening photopolymerization occurs essentially with low or negative shrinkage. Most epoxy monomers polymerized by a cationic process shrink about 2–6% during photopolymerization, which is lower than the acrylate system [68].

Another kind of valuable monomers is spiroorthocarbonates (SOC) showed in Fig. 9.13. These double cyclic acetals are stable under basic conditions but easily undergo ring-opening polymerization catalyzed by cationic catalysts [69].

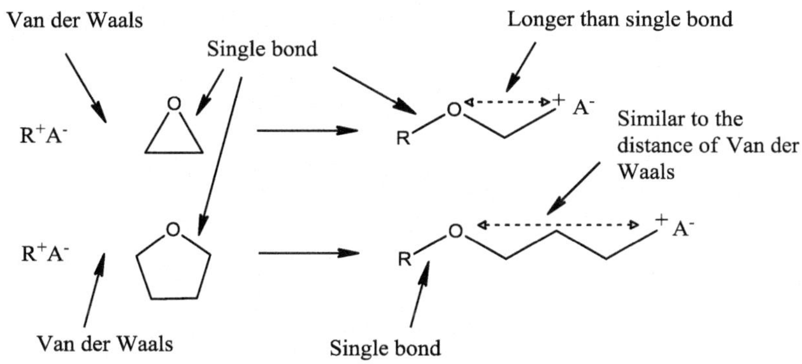

Fig. 9.12 Change of bond distances of epoxy during ring-opening polymerization

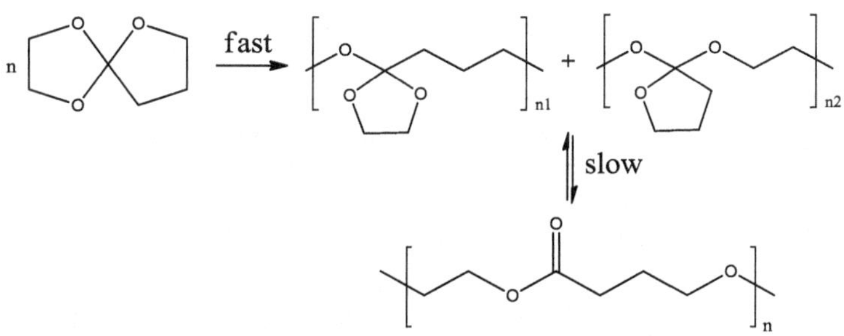

Fig. 9.13 The mechanism of double ring-opening of spiroorthocarbonates [69]

Fig. 9.14 Change of bond distances in SOC during ring-opening polymerization

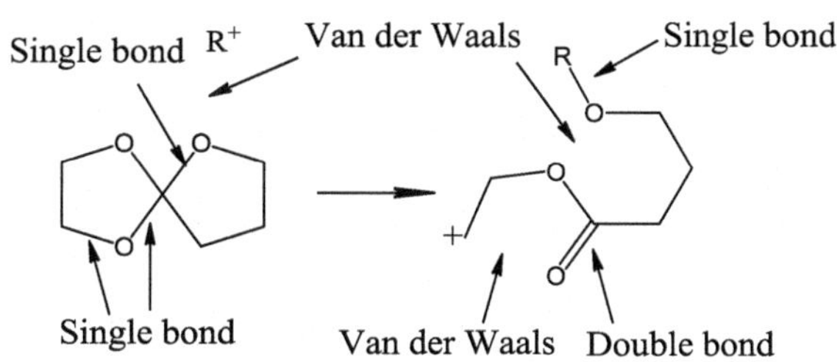

Compared with the epoxy compounds, which are three-membered ring monomers, spiroorthocarbonates with two five-membered rings may make both ends of breakpoint farther after ring-opening polymerization. Moreover, the spiro compound can open the other ring slowly after the opening of the first one, which means that one van der Waals distance and two covalent distance are replaced by two van der Waals distances and one covalent distance after photopolymerization (Fig. 9.14). This double ring-opening process makes the molecules looser than the very compact structure of the starting monomers and results in the volume expansion.

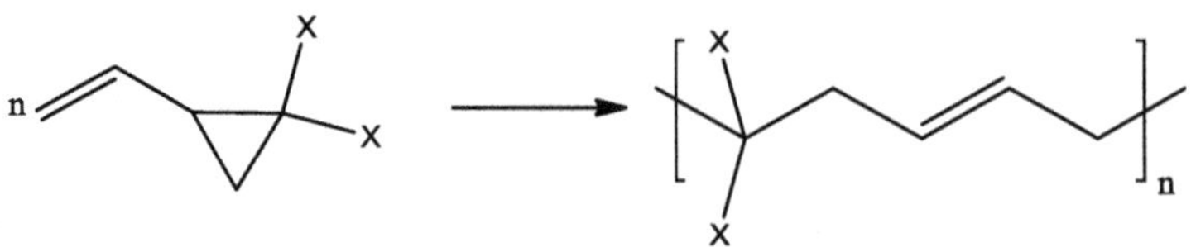

Fig. 9.15 The mechanism of ring opening of VCPs

Due to the volume expansion, spiroorthocarbonate is used as a low-shrinkage additive in cationic UV curing of epoxy resin system [68, 70–73]. A study demonstrated that the shrinkage of the homopolymerization of an epoxy monomer (3,4-epoxycyclohexylmethyl-3′,4′-epoxycyclohexyanecarboxylate, CE) was 3.93%. In order to reduce the shrinkage, an epoxy functionalized bicyclo-spiroorthocarbonate (BSOC) was added, while the content of BSOC was 7%, and the shrinkage was only 1.10%. Furthermore, when the content of BSOC increased to 10%, it even showed a little bit expansion (−1.12% of shrinkage) [68]. Similarly, epoxy system with bicyclo-orthoester (BOE) as a low-shrinkage additive [70] had a significant effect on reducing shrinkage. With the content of BOE increased from 0 to 30%, the shrinkage went down from 3.3% to 0.5%, and finally, it resulted in expansion (−1.6% of shrinkage) when the content of BOE was 50%. Zero shrinkage is the most ideal result that researchers are eager to achieve, which means that the precise structure and great mechanical properties can be obtained. When the amount of expansion needed is smaller than the shrinkage, the system still shrinks; when the amount of expansion needed is bigger than the shrinkage, the system may expand. However, when the values of expansion and shrinkage are the same, the system encounters neither expansion nor shrinkage. It is deduced that zero shrinkage can be achieved by adjusting the content of spiroorthocarbonate in a formulation. Nagai et al. [73] reported that adding 32% spiroorthocarbonate to 1,4-bis(((3-ethyl-3-oxetanyl)methoxy)methyl)benzene can obtain a zero shrinkage coating after curing.

2-Vinylcyclopropane-1,1-dicarboxylates (VCPs) are another kind of attractive monomers which display less polymerization shrinkage than acrylate monomers [74, 75], and the polymerization mechanism is shown as Fig. 9.15. A study demonstrated that the shrinkage of VCPs after polymerization is 5–5.5%, which is much lower than the shrinkage of methacrylate (21.4%) and styrene (15.7%) [75].

Thiol-Containing System

High activity of thiol with carbon-carbon double bonds, or simply "ene", was well known in the early 1900s [76]. During the last century, two thiol reactions emerged: thiol-ene free-radical addition to electron-rich/electron-poor carbon-carbon double bonds and catalyzed thiol-Michael addition to electron-deficient carbon-carbon double bonds [77]. The exceptional versatility and propensity for proceeding to quantitative conversions under even the mildest conditions make thiol-ene chemistry applicable to applications ranging from high-performance protective polymer networks to processes that are important in the optical [78], biomedical [79], sensing [80], and stereolithography [81] fields.

Light-mediated thiol-ene radical reactions effectively combine the classical benefits of thiol-ene system with the advantages of a photoinitiated process, which can be activated at specific times and locations and could result in a powerful method for chemical synthesis and tailorable material fabrication [82]. Thiol-ene photopolymerizations have several significant and unique advantages compared with (meth)acrylate polymerizations. The cross-linking of thiol-ene proceeds very rapidly but will not reach the gel-point until a relatively high functional group conversion is reached [83].

Lu et al. [83] studied the relationship between the shrinkage stress and double bond conversion of the pentaerythritol tetra(3-mercaptopropionate) (PETMP) and triallyl-1,3,5-triazine-2,4,6-trione (TATATO) system and compared it with that of the 2,2-bis[4-(2-hydroxy-3-methacryloyloxypropoxy)-phenyl]propane (BisGMA)/triethylene glycol dimethacrylate (TEGDMA) control system under the same conditions. The thiol-ene PETMP/TATATO system had significantly reduced stress comparing to that of the chain-growth polymerized BisGMA/TEGDMA system. This is attributed to lower shrinkage and higher gel point conversion with the thiol-ene system [83]. For the PETMP/TATATO system, shrinkage stress did not start to build up until the conversion exceeded the gel-point conversion (theoretical value is 40.8% for this system [84]). This result agrees well with their prediction that any shrinkage that occurs prior to the gel-point would not result in shrinkage stress development, as it can readily be accommodated by viscous flow. Only the shrinkage that occurs after the gel point will contribute to the ultimate shrinkage stress. It is not clear where vitrification occurs in the thiol-ene system, where significant stress development could occur due to the difficulty of stress relaxation and high modulus developed [24].

In theory, thiol-ene reaction provides stoichiometric consumption of both thiol and carbon-carbon double bond, but due to the free-radical homopolymerization of ene, both the thiol-ene click reaction and the chain polymerization of ene coexist in the real system. It was found that for thiol-acrylate polymerizations, the acrylate propagation kinetic constant was 1.5 times greater than the hydrogen abstraction from the thiol. This difference causes the conversion of the acrylate functional groups roughly twice of the conversion of thiol functional groups for stoichiometric thiol-acrylate [7] and increase of the shrinkage and monomer residual, which may lead to detrimental effect on the properties of cured materials. Depending on monomer, catalysts, and additives, the mechanism can vary from pure Michael addition to almost pure C=C polymerization (with chain transfer). One way to inhibit the free-radical homopolymerization of ene is to use photobase instead of free-radical photoinitiator to catalyze the Michael addition reaction, but due to the low absorption wavelength of photobase, the addition of photosensitizer may produce only a small amount of free radicals. Another way is to add free-radical trapping agent, such as 2,2,6,6-tetramethylpiperidinooxy (TEMPO), which can suppress the radical-mediated side reactions and provide stoichiometric and complete conversion of both thiol and acrylate functional groups efficiently [85].

Hybrid System

Hybrid photopolymerization is used to overcome the limitations of single type of polymerization systems. Acrylate free-radical photopolymerization suffers from oxygen inhibition and relatively high polymerization shrinkage, while ring-opening cationic polymerization is characterized by water inhibition and low reaction rates with less shrinkage. Combining the mechanisms of cationic and free-radical polymerizations into the same system seems to be an effective approach to reduce the shrinkage. Such free-radical/cationic hybrid systems were studied [86–90]. There are two approaches applied: combining two different kinds of functional groups to form a single monomer and mixing two different kinds of monomers together. Jiang et al. [91] synthesized a series of monomers containing both acrylate group and vinyl ether group. The lowest shrinkage achieved was 6.7%, which was much lower than that of the acrylate monomer with similar structure (8.5%). Hexanediol diacrylate (HDDA) and 3,4-epoxycyclohexylmethyl-3′,4′-epoxycyclohexylcarboxylate (CE) were mixed in different ratios with both free-radical and cationic photoinitiators [86], and some shrinkage decrease upon UV curing was observed with the increase of the epoxy resin content. When the content of CE was 50%, the shrinkage of the hybrid system was 9.89%, which was lower than the neat HDDA system (13.25%).

Mixing thiol-ene and methacrylate photopolymerization has also been studied to achieve relative low shrinkage. A thiol-ene/methacrylate hybrid system was studied by Schreck et al. [92], and they found that both thiol-ene click reaction and free-radical chain reaction of methacrylate occurred during photopolymerization and the shrinkage stress of the hybrid system was much lower than the methacrylate system alone. However, the thiol-ene systems have obvious drawbacks of compromised mechanical properties, such as lower storage modulus, flexural modulus, and surface hardness, which are due to the formation of a flexible thioether structure, low cross-link density, and less rigid monomer structure. Considering these issues, Schreck et al. [92] prepared hybrid organic/inorganic networks by a two-step process, and the hybrid materials showed low-shrinkage, high glass transition temperature, and high rubbery storage modulus. The first step was to synthesize thiol-contained (or ene-contained) polysiloxane through an acid-catalyzed sol-gel reaction, which then reacted with ene (or thiol) in the second step. However, this sequential method seems much more complex and may cost more time than a simultaneous method. Belon et al. [93] prepared a hybrid organic/inorganic material by using a kind of monomer containing both siloxane and acrylate functional groups through simultaneous free-radical chain polymerization and sol-gel reactions. Both acrylate and siloxane cause high shrinkage. However, by simultaneous thiol-ene click and sol-gel reactions, low-shrinkage organic/inorganic hybrid materials can be obtained.

Solid-State Photopolymerization

For common acrylate, the shrinkage at the end of polymerization is usually higher than 8% [20], but by narrowing the intermolecular distance of monomers before photopolymerization, a lower volume shrinkage can be achieved. This is a promising way to reduce shrinkage because a huge amount of shrinkage occurs before the photopolymerization. Solid-state photopolymerization, especially in the crystalline state, became a novel method to reduce the shrinkage in recent years. During the crystallization process, monomers pack into crystal cells quite closely and the pre-shrunk system forms. Photopolymerization in the crystalline state can reduce the shrinkage efficiently. However, in general, photopolymerization in the solid state leads to slower polymerization rates and lower functional group conversions due to mobility restrictions [94]. Thus many solid monomers, which cannot be polymerized by UV light, must be polymerized by high-energy irradiation such as γ-rays to achieve reasonable or even rapid photopolymerization rates [95–101].

Nevertheless, some monomers with long alkyl chains were found to polymerize well in the solid state irradiated by UV light. Octadecyl acrylate [102] and octadecyl methacrylate [103] showed very quick UV-initiated free-radical solid-state polymerization at temperatures under their melting points. The double bond conversion could reach more than 50% in 1 min at 5 mW/cm^2 light intensity and then continue to grow with time. However, the polymerization reactivity was influenced by the molecular arrangement of monomers. Triclinic packing (β-form) had little tendency to polymerize, while hexagonal packing (α-form) of monomer molecules with a particular orientation and more rotational freedom in the layered structure gave much better result. The maximum shrinkage of octadecyl acrylate for solid-state polymerization was 1.5%, which was much lower than the shrinkage value in the liquid-state polymerization (8.0%).

Not only the free-radical type, the cationic photopolymerization can also occur in the solid-state condition. Octadecyl vinyl ether showed good photopolymerization reactivity below its melting point. Although it had a little lower reaction rate than the free-radical type monomers, its double bond conversion in the solid-state polymerization could reach over 40% in just 10 min at 5 mW/cm^2 light intensity and could be further improved to more than 70% in 10 min by simply increasing the light intensity to 20 mW/cm^2 [104]. The maximum shrinkage of octadecyl vinyl ether for liquid-state photopolymerization was 5.0%; in contrast, the maximum shrinkage for solid-state photopolymerization decreased to 3.1%.

It was clear that the solid-state photopolymerization shrinkage is much lower than the liquid-state photopolymerization because of the more ordered structure of monomers in the solid state and consequently smaller intermolecular distance before the photopolymerization. Compared with a crystalline state, a semicrystalline state and a glassy state have less ordered arrangement of monomers. The lower the degree of crystallinity, the lower the degree of order. The less ordering leads to longer initial distance between monomers. A liquid crystalline state has relative higher degree of orientation, but the relative looser packing than a solid crystal could result in much longer distance between monomers too. Although the initial distance of monomers influences the shrinkage during photopolymerization, the shrinkage is also affected by other factors such as the structure of monomers, density of functional groups, and mechanism of polymerization. It is inaccurate to analyze the shrinkage by only one of these factors. Solid-state chain reaction photopolymerization is a kind of new method to control the shrinkage, and the development of stress during and after the solid-state photopolymerization has not been studied yet and deserves further investigation.

Summary and Developing Prospects of Shrinkage Research in UV Curing

In this chapter, shrinkage, one of the most important issues in UV-curing technology, was firstly described from its formation mechanism and the associated disadvantages. After that a series of methods to determine shrinkage were introduced including dilatometry, pycnometer, buoyancy, bonded-disk, interferometer, laser displacement, laser scanning, video-imaging device, and some combination methods. Finally, several methods to control shrinkage to overcome the mentioned disadvantages were summarized. Changing the UV-curing condition is the simplest one, but the effect is limited. Adding inert components can also reduce shrinkage, and organic/inorganic composites can be applied by the addition of inorganic components. Adjusting the structure of monomers is the most flexible and effective method, which can even achieve zero shrinkage or expansion. Lastly, the solid-state photopolymerization is a new method that needs to be continuously studied.

As UV-curable coatings become more and more widely applied, the improvement of their comprehensive properties is becoming a significant research interest. What is needed is a deeper understanding of the relationship between shrinkage and bulk properties and monitoring, controlling, and utilizing the shrinkage. Firstly, more combined real-time testing methods need to be developed, which allow collection of more synchronous information related to shrinkage and other properties such as modulus, optical transmittance, microstructure, surface morphology, glass transition temperature, stress, adhesion, and volume relaxation. The information will enrich our knowledge of the shrinkage process, shrinkage mechanism, and its influence on coating properties. Secondly, more new chemical structures and new approaches should be developed to control the shrinkage and balance the shrinkage with other bulk properties, such as stiffness, flexibility, adhesion, hardness, elasticity, and resistance (such as chemical resistance, moisture resistance, and weathering resistance). Finally, attentions need to be paid to more applications of UV coatings with different levels of shrinkage. Although low or zero shrinkage is one of the most favorable aims, controlled shrinkage from normal to high levels could also find some potential applications. For example, higher resolution could be obtained in the photolithography process with suitable shrinkage as long as it is repeatable and does not affect the adhesion and other mechanical properties. Overall, there is still a long way to go in the research of shrinkage in UV coatings, and it needs more intelligent thinking and diligent work.

References

1. Kodate, K., Takenaka, H., Kamiya, T.: Fabrication of high numerical aperture zone plates using deep ultraviolet lithography. Appl. Opt. **23**(3), 504–507 (1984)
2. Fouassier, J.-P., Morlet-Savary, F.: Photopolymers for laser imaging and holographic recording: design and reactivity of photosensitizers. Opt. Eng. **35**(1), 304–312 (1996)
3. He, Y., Wu, W.B., Fu, J.Z.: Rapid fabrication of paper-based microfluidic analytical devices with desktop stereolithography 3D printer. RSC Adv. **5**(4), 2694–2701 (2015)
4. Lee, J.W., Cho, D.W.: Development of a resin curing model for UV nanoimprint. J. Nanosci. Nanotechnol. **5**(10), 1637–1642 (2005)
5. Decker, C.: The use of UV irradiation in polymerization. Polym. Int. **45**, 133–141 (1998)
6. Crivello, J.V.: The discovery and development of onium salt cationic photoinitiators. J. Polym. Sci. A Polym. Chem. **37**, 4241–4254 (1999)
7. Cramer, N.B., Bowman, C.N.: Kinetic of thiol-ene and thiol-acrylate photopolymerizations with real-time Fourier transform infrared. J. Polym. Sci. A Polym. Chem. **39**, 3311–3319 (2001)
8. Berchtold, K.A., Hacioglu, B., Lovell, L., Nie, J., Bowman, C.N.: Using changes in initiation and chain transfer rates to probe the kinetics of cross-linking photopolymerizations: effects of chain length dependent termination. Macromolecules. **34**, 5103–5111 (2001)
9. Chen-Yang, Y.W., Chuang, J.R., Yang, Y.C., Li, C.Y., Chiu, Y.S.: New UV-curable cyclotriphosphazenes as fire-retardant coating materials for wood. J. Appl. Polym. Sci. **69**, 115–122 (1998)
10. Jung, S.-J., Lee, S.-J., Cho, W.-J., Ha, C.-S.: Synthesis and properties of UV-curable waterborne unsaturated polyester for wood coating. J. Appl. Polym. Sci. **69**, 695–708 (1998)
11. Chang, C.-W., Lu, K.-T.: Epoxy acrylate UV/PU dual-cured wood coatings. J Appl. Polym. Sci. **115**, 2197–2202 (2010)
12. Herrera, R., Muszyńska, M., Krystofiak, T., Labidi, J.: Comparative evaluation of different thermally modified wood samples finishing with UV-curable and waterborne coatings. Appl. Surf. Sci. **357**, 1444–1453 (2015)
13. Sponsler, M.B.: Refractive index and orientation profiles of holographic graftings written in liquid crystalline monomers. J. Phys, Chem. **99**, 9430–9436 (1995)
14. Löchel, B., Maciossek, A., Quenzer, H.J., Wagner, B.: Ultraviolet depth lithography and galvanoforming for micromachining. J. Electrochem. Soc. **143**, 237–244 (1995)
15. Decker, C.: Photoinitiated crosslinking polymerisation. Prog. Polym. Sci. **21**, 593–650 (1996)
16. Decker, C.: Kinetic study and new applications of UV radiation curing. Marcomol. Rapid Commun. **23**, 1067–1093 (2002)
17. Sangermano, M., Razza, N., Crivello, J.V.: Cationic UV-Curing: Technology and Applications. Macromol. Mater. Eng. **299**, 775–793 (2014)
18. Anseth, K.S., Bowman, C.N., Peppas, N.A.: Polymerization kinetics and volume relaxation behavior of photopolymerized multifunctional monomers producing highly crosslinked networks. J. Polym. Sci. A Polym. Chem. **32**, 139 (1994)
19. Lu, B., Xiao, P., Sun, M., Nie, J.: Reducing shrinkage by low-temperature photopolymerization. J. Appl. Polym. Sci. **104**, 1126–1130 (2007)
20. Jian, Y., He, Y., Jiang, T.Z., Li, C.G., Yang, W.T., Nie, J.: Polymerization shrinkage of (meth)acrylate determined by reflective laser beam scanning. J. Polym. Sci. Part B: Polym. Phys. **50**, 923–928 (2012)

21. Hiroshima, H., Suzuki, K.: Study on change in UV nanoimprint pattern by altering shrinkage of UV curable resin. Jpn. J. Appl. Phys. **50**, 60GK0901–660GK095 (2011)
22. Wang, Q.K., Huang, B.Q., Wei, X.F., Shen, H.C.: Study on shrinkage of cured volume for UV-curing coatings. Appl. Mech. Mater. **731**, 588–592 (2015)
23. Kim, R.J.-Y., Kim, Y.-J., Choi, N.-S., Lee, I.-B.: Polymerization shrinkage, modulus, and shrinkage stress related to tooth restoration interfacial debonding in bulk-fill composites. J. Dent. **43**, 430–439 (2015)
24. Wen, M., Scriven, L.E., McCormick, A.V.: Differential scanning calorimetry and cantilever deflection studies of polymerization kinetics and stress in ultraviolet curing of multifunctional (meth)acrylate coatings. Macromolecules. **35**, 112–120 (2002)
25. de Gee, A.J., Davidson, C.L., Smith, A.: A modified dilatometer for continuous recording of volumetric polymerization shrinkage of composite restorative materials. J. Dent. **9**, 36–42 (1981)
26. Lai, J.H., Johnson, A.E.: Measuring polymerization shrinkage of photo-activated restorative materials by a water-filled dilatometer. Dent. Mater. **9**, 139–143 (1993)
27. Shah, D.U., Schubel, P.J.: Evaluation of cure shrinkage measurement techniques for thermosetting resins. Polym. Test. **29**, 629–639 (2010)
28. Rueggeberg, F., Tamareselvy, K.: Resin cure determination by polymerization shrinkage. Dent. Mater. **11**, 265–268 (1995)
29. Bekkedahl, N.: Volume Dilatometry. J. Res. Natl. Bur. Stds. **42**, 145–156 (1949)
30. Haider, M., Hubert, P., Lessard, L.: Cure shrinkage characterization and modeling of a polyester resin containing low profile additives. Composites: Part A. **38**, 994–1009 (2007)
31. Lee, I.-B., Cho, B.-H., Son, H.-H., Um, C.-M., Lim, B.-S.: The effect of consistency, specimen geometry and adhesion on the axial polymerization shrinkage measurement of light cured composites. Dent. Mater. **22**, 1071–1079 (2006)
32. Feilzer, A.J., de Gee, A.J., Davidson, C.L.: Increased wall-to-wall curing contraction in thin bonded resin layers. J. Dent. Res. **68**, 48–50 (1989)
33. Watts, D.C., Cash, A.J.: Determination of polymerization shrinkage kinetics in visible-light-cured materials: methods development. Dent. Mater. **7**, 281–287 (1991)
34. Watts, D.C., Hindi, A.A.: Intrinsic 'soft-start' polymerization shrinkage kinetics in a acrylate-based resin composite. Dent. Mater. **15**, 39–45 (1999)
35. Watts, D.C., Marouf, A.S.: Optimal specimen geometry in bonded-disc shrinkage-strain measurements on light-cured biomaterials. Dent. Mater. **16**, 447–451 (2000)
36. Atai, M., Watts, D.C., Atai, Z.: Shrinkage strain rates of dental resin monomer and composite systems. Biomaterials. **26**, 5015–5020 (2005)
37. Koseki, K., Sakamaki, H., Jeong, K.-M.: In situ measurement of shrinkage behavior of photopolymers. J. Photopolym. Sci. Technol. **26**, 567–572 (2013)
38. Fogleman, E.A., Kelly, M.T., Grubbs, W.T.: Laser interferometric method for measuring linear polymerization shrinkage in light cured dental restoratives. Dent. Mater. **18**, 324–330 (2002)
39. Fano, V., Ortalli, I., Pizzi, S., Bonanini, M.: Polymerization shrinkage of microfilled composites determined by laser beam scanning. Biomaterials. **18**(6), 467–470 (1997)
40. Neo, W.K., Chan-Park, M.B.: A new model and measurement technique for dyamic shrinkage during photopolymerization of multi-acrylate. Macromol. Rapid Commun. **26**, 1008–1013 (2005)
41. Zhou, W.X., Chan-Park, M.B.: Effect of oligomer length on the buckling of long and high aspect ratio microwalls UV embossed from oligomer/monomer mixtures. Sens. Actuat. B. **128**, 12–22 (2007)
42. Dudi, O., Grubbs, W.T.: Laser interferometric technique for measuring polymer cure kinetics. J. Appl. Polym. Sci. **74**, 2133–2142 (1999)
43. Sharp, L.J., Choi, I.B., Lee, T.E., Sy, A., Suh, B.I.: Volumetric shrinkage of composited using video-imaging. J. Dent. **31**, 97–103 (2003)

44. Szczepanski, C.R., Stansbury, J.W.: Stress reduction in phase-separated, cross-linked networks: Influence of phase structure and kinetics of reaction. J. Appl. Polym. Sci. **131**, 40879/1–40879/10 (2014)
45. Schmidt, C., Scherzer, T.: Monitoring of the shrinkage during the photopolymerization of acrylates using hyphenated photorheometry/near-infrared spectroscopy. J. Polym. Sci. Part B: Polym. Phys. **53**, 729–739 (2015)
46. Schmidt, L.E., Leterrier, Y., Schmäh, D., Månson, J.-A.E., James, D., Gustavsson, E., Svensson, L.S.: Conversion analysis of acrylated hyperbranched polymers UV cured below their ultimate glass transition temperature. J. Appl. Polym. Sci. **104**, 2366–2376 (2007)
47. Lecamp, L., Youssef, B., Bunel, C., Lebaudy, P.: Photoinitiated polymerization of a dimethacrylate oligomer 1. Influence of photoinitiator concentration, temperature and light intensity. Polymer. **38**, 6089–6096 (1997)
48. Asmussen, E., Peutzfeldt, A.: Polymerization contraction of resin composite vs. energy and power density of light-cure. Eur. J. Oral. Sci. **113**, 417–421 (2005)
49. Silikas, N., Eliades, G., Watts, D.C.: Light intensity effects on resin-composite degree of conversion and shrinkage strain. Dent. Mater. **16**, 292–296 (2000)
50. Uno, S., Asmussen, E.: Marginal adaptation of a restorative resin polymerized at reduced rate. Scand. J. Dent. Res. **99**, 440–444 (1991)
51. Unterbrink, G.L., Muessner, R.: Influence of light intensity on two restorative systems. J. Dent. **23**, 183–189 (1995)
52. Feilzer, A.J., de Gee, A.J., Dooren, L.H., Davidson, C.L.: Influence of light intensity on polymerization shrinkage and integrity of restoration-cavity interface. Eur. J. Oral. Sci. **103**, 322–326 (1995)
53. Neves, A.D., Discacciati, J.A.C., Oréfice, R.L., Yoshida, M.I.: Influence of the power density on the kinetics of photopolymerization and properties of dental composites. J. Biomed. Mater. Res. B Appl. Biomater. **72**, 393–400 (2005)
54. Vaessen, D.M., Ngantung, F.A., Palacio, M.L.B., Francis, L.F., Mccormick, A.V.: Effect of lamp cycling on conversion and stress development in ultraviolet-cured acrylate coatings. J. Appl. Polym. Sci. **84**, 2784–2793 (2002)
55. Ferracane, J.L., Mitchem, J.C., Condon, J.R., Todd, R.: Wear and marginal breakdown of composites with various degrees of cure. J. Dent. Res. **76**, 1508–1516 (1997)
56. Turssi, C.P., Ferracane, J.L., Vogel, K.: Filler features and their effects on wear and degree of conversion of particulate dental resin composites. Biomaterials. **26**, 4932–4937 (2005)
57. Spinell, T., Schedle, A., Watts, D.C.: Polymerization shrinkage kinetics of dimethacrylate resin-cements. Dent. Mater. **25**, 1058–1066 (2009)
58. Kahler, B., Kotousov, A., Borkowski, K.: Effect of material properties on stresses at the restoration-dentin interface of composite restorations during polymerization. Dent. Mater. **22**, 942–947 (2006)
59. Condon, J.R., Ferracane, J.L.: Assessing the effect of composite formulation on polymerization stress. J. Am. Dent. Assoc. **131**, 497–503 (2000)
60. Geiser, V., Leterrier, Y., Månson, J.-A.E.: Conversion and shrinkage analysis of acrylated hyperbranched polymer nanocomposites. J. Appl. Polym. Sci. **114**, 1954–1963 (2009)
61. Liu, C.J., Kiasat, M.S., Nijhof, A.H., Blokland, H., Marissen, R.: The effect of the addition of a low profile additive on the curing shrinkage of an unsaturated polyester resin. Polym. Eng. Sci. **39**(1), 18–25 (1999)
62. Carioscia, J.A., Lu, H., Stanbury, J.W., Bowman, C.N.: Thiol ene oligomers as dental restorative materials. Dent. Mater. **21**, 1137–1143 (2005)
63. Chung, C.-M., Kim, M.-S., Kim, J.-G., Jang, D.-O.: Synthesis and photopolymerization of trifunctional methacrylates and their application as dental monomers. J. Biomed. Mater. Res. **62**, 622–627 (2002)
64. Jian, Y., He, Y., Zhao, L.Y.: Effect of monomer structure on real-time UV-curing shrinkage studied by a laser scanning approach. Adv. Polym. Technol. **32**(1), 21331/1–21331/9 (2013)

65. Jian, Y., He, Y., Jiang, T.Z., Li, C.G., Yang, W.T., Nie, J.: Volume shrinkage of UV-curable coating formulation investigated by real-time laser reflection method. J. Coat. Technol. Res. **10**(2), 231–237 (2013)
66. Liu, B.H., Nie, J., He, Y.: From rosin to high adhesive polyurethane acrylate: synthesis and properties. Int. J. Adhes. Adhes. **66**, 99–103 (2016)
67. Qin, L.L., He, Y., Liu, B.H., Jian, Y., Li, C.G., Nie, J.: Preparation and properties of polyurethane acrylates modified by saturated alcohols. Prog. Org. Coat. **76**, 1594–1599 (2013)
68. Sangermano, M., Ortiz, R.A., Urbina, B.A.P., Duarle, L.B., Valdez, A.E.G., Santos, R.G.: Synthesis of an epoxy functionalized spiroorthocarbonate used as low shrinkage additive in cationic UV curing of an epoxy resin. Eur. Polym. J. **44**, 1046–1052 (2008)
69. Nuyken, O., Böhner, R., Erdmann, C.: Oxetane Photopolymerization-A system with low shrinkage. Macromol. Symp. **107**, 125–138 (1996)
70. Sangermano, M., Malucelli, G., Delleani, G., Priola, A.: Bicyclo-orthoester as a low-shrinkage additive in cationic UV curing. Polym. Int. **56**, 1224–1229 (2007)
71. Sangermano, M., Giannelli, S., Ortiz, R.A., Berlanga Duarte, M.L., Rueda Gonzalez, A.K., Garcia Valdez, A.E.: Synthesis of an Oxetane-functionalized hemispiroorthocarbonate used as a low-shrinkage additive in the cationic ultraviolet curing of oxetane monomers. J. Appl. Polym. Sci. **112**, 1780–1787 (2009)
72. Moon, E.J., Lee, J.Y., Kim, C.K., Cho, B.H.: Dental restorative composites containing 2,2-Bis-[4-(2-hydroxy-3-methacryloyloxy propoxy) phenyl] propane derivatives anf spiro orthocarbonates. J. Biomed. Mater. Res. B: Appl. Biomater. **73B**, 338–346 (2005)
73. Nagai, D., Nishida, M., Nagasawa, T., Ochiai, B., Miyazaki, K., Endo, T.: Non-shrinkage networked materials from the cross-linking copolymerization of spiroorthocarbonate with bifunctional oxetane. Macromol. Rapid Commun. **27**, 921–925 (2006)
74. Contreras, P.P, Tyagi, P., Agarwal, S.: Low shrinkage of polymers by photopolymerization of 1,1-bis(ethoxycarbonyl)-2-vinylcyclopropanes. Polym. Chem. 6, 2297-2304 (2015)
75. Sanda, F., Takata, T., Endo, T.: Radical ring-opening polymerization of novel vinylcyclopropanes designed as low shrinkage monomers. Structure of the polymer, mechanism of the polymerization, and volume change on the polymerization. Macromolecules. **28**, 1346–1355 (1995)
76. Posner, T.: Contributions for the knowledge of the unsaturated compounds. Ber. Dtsch. Chem. Ges. **38**, 646–657 (1905)
77. Hoyle, C.E., Bowman, C.N.: Thiol-ene click chemistry. Angew. Chem. Int. Ed. **49**, 1540–1573 (2010)
78. Senyurt, A.F., Warren, G., Whitehead Jr., J.B., Hoyle, C.E.: Matrix physical structure effect on the electro-optic characteristics of thiol-ene based H-PDLC films. Polymer. **47**, 2741–2749 (2006)
79. Rydholm, A.E., Bowman, C.N., Anseth, K.S.: Degradable thiol-acrylate photopolymers: polymerization and degradation behavior of an in situ forming biomaterial. Biomaterials. **26**, 4495–4506 (2005)
80. Cramer, N.B., Reddy, S.K., Lu, H., Cross, T., Raj, R., Bowman, C.N.: Thiol-ene photopolymerization of polymer-derived ceramic precursor. J. Polym. Sci. A Polym. Chem. **42**, 1752–1757 (2004)
81. Khire, V.S., Yi, Y., Clark, N.A., Bowman, C.N.: Formation and surface modification of nanopatterned thiol-ene substrates using step and flash imprint lithography. Adv. Mater. **20**, 3308–3313 (2008)
82. Hoyle, C.E., Lee, T.Y., Roper, T.: Thiol-enes: chemistry of the past with promise for the future. J. Polym. Sci. A Polym. Chem. **42**, 5301–5338 (2004)
83. Lu, H., Carioscia, J.A., Stansbury, J.W., Bowman, C.N.: Investigations of step-growth thiol-ene polymerizations for novel dental restoratives. Dent. Mater. **21**, 1129–1136 (2005)
84. Odian, G.: Principles of Polymerization, 4th edn. Wiley, Hoboken, NJ (1991)
85. Chatani, S., Gong, T., Earle, B.A., Podgórski, M., Bowman, C.N.: Visible-light initiated thiol-Michael addition photopolymerization reactions. ACS Macro. Lett. **3**, 315–318 (2014)

86. Sangermano, M., Carbonaro, W., Malucelli, G., Priola, A.: UV-cured interpenetrating acrylic-epoxy polymer networks: preparation and characterization. Macromol. Mater. Eng. **293**, 515–520 (2008)
87. Decker, C., Viet, T.N.T., Decker, D., Weber-Koehl, E.: UV radiation curing of acrylate epoxide systems. Polymer. **42**, 5531–5541 (2001)
88. Mineart, K., Dillman, B., Jessop, J.L.P.: Search for dual initiator synergy in UV initiated acrylate epoxy hybrid polymerization systems. Paper presented at the 241st ACS National Meeting & Exposition, Anaheim, CA, United States, 27-31 March 2011
89. Fouassier, J.P., Lalevée, J.: Photochemical production of interpenetrating polymer networks simultaneous initiation of radical and cationic polymerization reactions. Polymers. **6**, 2588–2610 (2014)
90. Cho, J.-D., Hong, J.-W.: UV initiated free radical and cationic photopolymerizations of acrylate epoxide and acrylate vinyl ether hybrid systems with and without photosensitizer. J. Appl. Polym. Sci. **93**, 1473–1483 (2004)
91. Jiang, T.Z., He, Y., Jian, Y., Nie, J.: Exploration for decreasing the volume shrinkage for photopolymerization. Prog. Org. Coat. **75**, 398–403 (2012)
92. Schreck, K.M., Leung, D., Bowman, C.N.: Hybrid organic/inorganic thiol-ene-based photopolymerized networks. Macromolecules. **44**, 7520–7529 (2011)
93. Belon, C., Chemtob, A., Croutxé-Barghorn, C., Rigolet, S., Houérou, V.L., Gauthier, C.: Combination of radical and cationic photoprocesses for the single-step synthesis of organic/inorganic hybrid films. J. Polym. Sci. A Polym. Chem. **48**, 4150–4158 (2010)
94. Berchtold, K.A., Hacioglu, B., Nie, J., Cramer, N.B., Stansbury, J.W., Bowman, C.N.: Rapid Solid-State Photopolymerization of Cyclic Acetal-Containing Acrylates. Macromolecules. **42**, 2433–2437 (2009)
95. Shibasaki, Y., Nakahara, H., Fukuda, K.: Solid-state polymerization of long-chain vinyl compounds. I. Effect of molecular arrangement on polymerizability of octadecyl methacrylate. J. Polym. Sci.: Polym. Chem. Edit. **17**, 2387–2400 (1979)
96. Shibasaki, Y., Fukuda, K.: Solid-state polymerization of long-chain vinyl compounds. II. Effect of molecular arrangement on polymerizability of octadecyl acrylate. J. Polym. Sci.: Polym. Chem. Edit. **17**, 2947–2959 (1979)
97. Shibasaki, Y.: Solid-state polymerization of long-chain vinyl compounds. III. Mechanism of γ-ray-initiated postpolymerization in layered structures of octadecyl methacrylate and acrylate. Polym. J. Sci.: Polym. Chem. Edit. **18**, 1693–1709 (1980)
98. Shibasaki, Y., Fukuda, K.: Solid-state polymerization of long-chain vinyl compounds. IV. Effects of chain length on the polymorphic behavior and postpolymerization of n-alkyl methacrylate. J. Polym. Sci.: Polym. Chem. Edit. **18**, 2437–2449 (1980)
99. Restaino, A.J., Mesrobian, R.B., Morawetz, H., Ballantine, D.S., Dienes, G.J., Metz, D.J.: γ-ray initiated polymerization of crystalline monomers. J. Am. Chem. Soc. **78**, 2939–2943 (1956)
100. Morosoff, N., Morawetz, H., Post, B.: Polymerization in the crystalline state. VII. A crystallographic study of the radiation -initiated polymerization in single crystals of vinyl stearate. J. Am. Chem. Soc. **87**(14), 3035–3040 (1965)
101. Fujimori, A., Saitoh, H., Shibasaki, Y.: Influence of molecular arrangement on the γ-ray-irradiation solid-state polymerization of 1-octadecyl vinyl ether with a characteristic polymorphism. J. Polym. Sci. A Polym. Chem. **37**, 3845–3853 (1999)
102. Jian, Y., He, Y., Wang, J., Yang, W.T., Nie, J.: Rapid solid-state photopolymerization of octadecyl acrylate: low shrinkage and insensitivity to oxygen. Polym. Int. **62**, 1692–1697 (2013)
103. Jian, Y., He, Y., Wang, J., Xu, B.B., Yang, W.T., Nie, J.: Rapid photopolymerization of octadecyl methacrylate in the solid state. New J. Chem. **37**, 444–450 (2013)
104. Wang, J., Jian, Y., Nie, J., He, Y.: Solid photopolymerization and polymer properties of octadecyl vinyl ether. J. Photoch. Photobio. A Chem. **27**, 105–110 (2013)

Chapter 10
Measurements of Stress Development in Latex Coatings

Kyle Price, Wenjun Wu, Alon V. McCormick, and Lorraine F. Francis

Introduction

Waterborne latex dispersions underlie a wide range of industrial coating applications including architectural paints, floor coatings, automotive paints, paper coatings, carpet backing, and adhesives [1]. Latex dispersions are also frequently added to other types of coating formulations to increase mechanical integrity and to reduce the susceptibility to cracking. The use of latex suspensions in various coatings formulations continues to grow due to their low cost, ease of use, and their environmental friendliness. However, many challenges still exist in developing latex coatings that balance the need for superior film properties with complete film formation.

The complex mechanisms by which discrete latex particles transform into a continuous polymer film are often referred to collectively as latex film formation, as shown schematically in Fig. 10.1. Latex film formation is commonly divided into three stages: consolidation, compaction, and coalescence [1–3]. During the consolidation stage, water evaporates and the particle concentration increases until a consolidated close-packed particle network is formed. After consolidation, particle deformation begins and the void space in the particle network is eliminated. Once

K. Price (✉)
Department of Chemical Engineering and Materials Science, University of Minnesota, Minneapolis, MN, USA

Axalta Coating Systems, Coatings Technology Center, Wilmington, DE, USA
e-mail: kyle.price@axaltacs.com

W. Wu
Arkema Coating Resins, Cary, NC, USA

A.V. McCormick • L.F. Francis
Department of Chemical Engineering and Materials Science, University of Minnesota, Minneapolis, MN, USA

M. Wen, K. Dušek (eds.), *Protective Coatings*, DOI 10.1007/978-3-319-51627-1_10

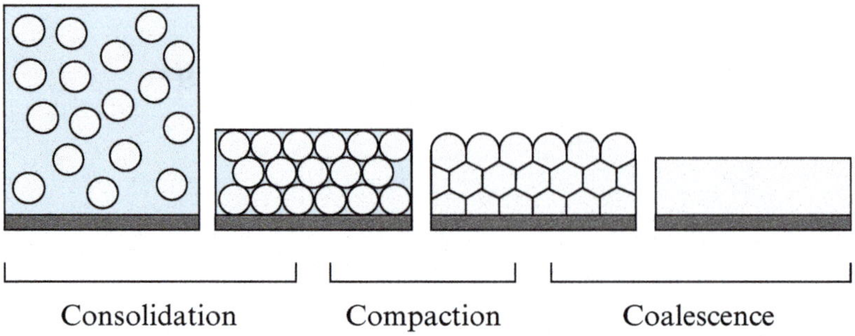

Fig. 10.1 Schematic diagram of the three stages of the latex film formation process

intimate contact is achieved between adjacent particles, polymer diffusion begins along the particle–particle contacts. If complete film formation occurs, the particle boundaries become indistinguishable and a continuous polymer film forms. However, the extent of film formation depends on many variables including polymer glass transition temperature (T_g), drying temperature, latex particle size, and particle morphology [4].

Particles comprised of low T_g polymers easily undergo film formation, but the resulting films are often tacky and lack film hardness. Conversely, films made from high T_g polymer particles can exhibit far superior mechanical properties, but complete film formation is often difficult without elevated temperature drying or the use of plasticizers or coalescing aids. The conflicting requirement between film formation and performance is often called the "film formation paradox" [1]. Traditionally, volatile coalescing aids have been used to temporarily plasticize higher T_g latex particles, allowing them to undergo film formation at lower temperatures [4]. Once the film formation is complete, the coalescing aid leaves the film by evaporation, and the polymer regains its original T_g and hardness. However, the use of volatile organic compounds (VOCs) in coating formulations is being phased out due to increasing government regulation and consumer awareness. Other strategies for balancing the requirements of film formation and mechanical performance include the use of nonvolatile or low-VOC coalescing aids, latex blends, or composite latex particles [1, 2]. Understanding how these variables affect the film formation behavior is fundamental in the ability to engineer coatings capable of achieving the desired performance balance.

During the film formation process, the volume of the coating is reduced due to particle coalescence and loss of volatiles. Stress develops in the coating due to constrained shrinkage of the film imposed by the substrate. This stress has the potential to initiate coating defects, such as cracks, which are detrimental to the coating's decorative and protective functions. As shown in Fig. 10.2, assuming good adhesion of the coating to the substrate, a non-zero biaxial tensile stress develops in the coating due to the constrained in-plane shrinkage of the coating during film formation [5, 6]. For the bulk of the coating, there is no stress development normal to the substrate as the coating may shrink freely in this

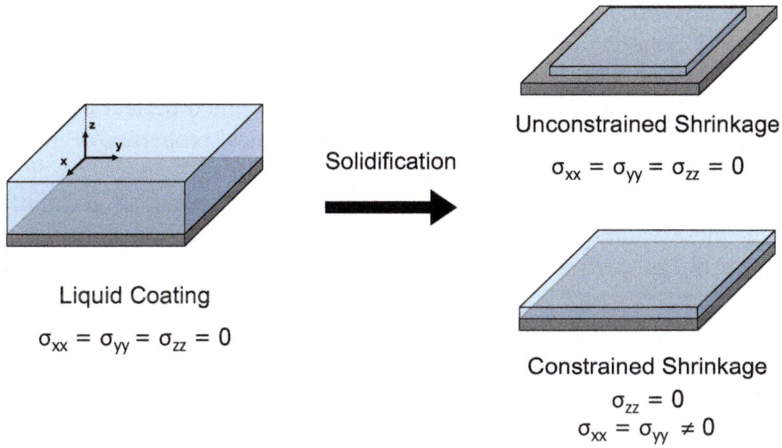

Fig. 10.2 Stress formation during solidification of a liquid coating. Unconstrained coating shrinkage during solidification results in an ideal stress-free state, whereas constrained shrinkage results in a non-zero tensile stress state in the plane of the substrate

direction in response to volumetric changes. Near the edges of the coating, shear and peeling stresses are present; however, these stresses are concentrated at the edge and can be assumed to be negligible when the lateral dimensions of the coating are much greater than the coating thickness [7, 8]. Thermal stresses may also arise but are negligible when experiments are conducted at room temperature.

Internal stress development in coatings has been investigated by many researchers using cantilever deflection or beam bending techniques. These techniques determine the internal stress in a coating from a measurement of the curvature or deflection of a coated beam or cantilever. This curvature or deflection is typically measured by optical interference methods [9, 10], capacitive deflection measurements [11], with a laser–photodiode combination [12–16], or by other means [17]. Stoney [18], using classical beam theory, derived the following expression relating coating stress, σ, to the deflection of a coated cantilever beam:

$$\sigma = \frac{E_s t^2 d}{3cL^2} \tag{10.1}$$

where E_s is the Young's modulus of the substrate, t is the substrate thickness, d is the deflection of the cantilever, c is the coating thickness, and L is the free length of the cantilever. Corcoran [19] improved upon Stoney's model, by developing an equation for the biaxial stress in the coating using plate theory:

$$\sigma = \frac{E_s t^3 d}{3cL^2(1-\nu_s)(t+c)} + \frac{E_c d(t+c)}{L^2(1-\nu_c)} \tag{10.2}$$

where ν_s is the Poisson's ratio of the substrate and E_c and ν_c are the Young's modulus and Poisson's ratio of the coating, respectively. The second term of Eq. (10.2) accounts for the stress relaxation in coating that results from the bending action of the cantilever. In this analysis, Corcoran assumed perfect coating adhesion, spherical cantilever deflection, isotropic mechanical properties of the coating and substrate, and uniform stress in the coating [19]. If the Young's modulus and Poisson's ratio of the coating are known, the full form of the equation can be used. However, these values are often difficult to characterize for thin films and may differ from the properties of the bulk material. Corcoran showed that when the coating thickness is small relative to the substrate thickness ($c << t$) and when rigid substrates are used ($E_s >> E_c$), the second term in the equation can be neglected without introducing significant error into the calculation [19].

Perera and Van den Eynde [20] were the first to investigate stress development in latex-based paints using the cantilever deflection method. The authors described stress development in latex coatings as a competition between two opposing processes: a stress development due to constrained volume shrinkage of the coating as a result of particle coalescence and loss of volatiles and stress relaxation due to the viscoelastic behavior of the polymers. In other studies, the effects of pigment particles [21] and coalescing aids [22] were also investigated in relation to stress evolution in the coating. When no coalescing aid was present, stress increased with drying time until a maximum value was achieved, followed by a slight decay in stress from the maximum value. For paint formulations containing a coalescing aid, though, a variety of behaviors were observed after the maximum stress was achieved [22].

In more recent studies, Martinez and Lewis [16] investigated stress development in latex-only coatings as a function of polymer T_g. They found that the maximum stress increased only slightly as the T_g of the particles was increased; however, they only investigated latex particles with T_gs below room temperature. The authors did recognize that the evolution of stress in the film varied greatly with particle T_g. For lower T_g particles, stress developed gradually over time, with no apparent maximum in the stress developed. For higher T_g particles, stress rose to the maximum value, followed by a slight decay in stress to the final residual value. In a similar study, Routh and coworkers [23] observed comparable stress evolution behavior in latex-only coatings, where stress development was found to correlate with the minimum film formation temperature (MFFT) of the latex.

One of the most common techniques to study stress evolution in coatings is the cantilever beam bending technique [12–16]. This technique yields an average value for the internal stress throughout the entire coating. However, this technique cannot account for variations in stress within the coating. Particle-based coatings cast on small discrete substrates typically exhibit lateral or nonuniform drying from the edges inward [9, 17, 24–26]. For coatings that dry nonuniformly, the degree of film formation varies across the coating along with the stress [17]. Nonuniform drying complicates the interpretation of stress measurements obtained using the cantilever beam bending technique, and efforts should be taken to suppress lateral drying front when this technique is used. One strategy for suppressing lateral drying is the use of modified walled cantilevers which have been shown to promote drying uniformity and deliver measurements of stress that more accurately represent the true stress in the coating [27].

In this study, stress development in latex coatings was characterized by the cantilever beam bending technique using walled cantilevers to suppress lateral drying. In a related study, this technique was used to investigate the influence of particle size and particle morphology on film formation and stress development of multiphase composite latex coatings [28]. In order to more fully understand the fundamental relationships between the extent of film formation and internal stress development for films containing homogeneous, single-phase latex particles, a series of acrylic latexes with varying T_g were synthesized. Furthermore, coalescing aids (both convention and low VOC) were added in various concentrations to the latex dispersion to further explore their role in film formation and stress development [29]. This work provides insights into the general process of stress development for industrially relevant latex coatings plasticized with a variety of widely used coalescing aids.

Experimental

Stress Measurement

The cantilever beam bending technique was used to characterize the evolution of stress in the coating throughout the drying process. The stress measurement apparatus is shown schematically in Fig. 10.3. To measure the cantilever deflection, a

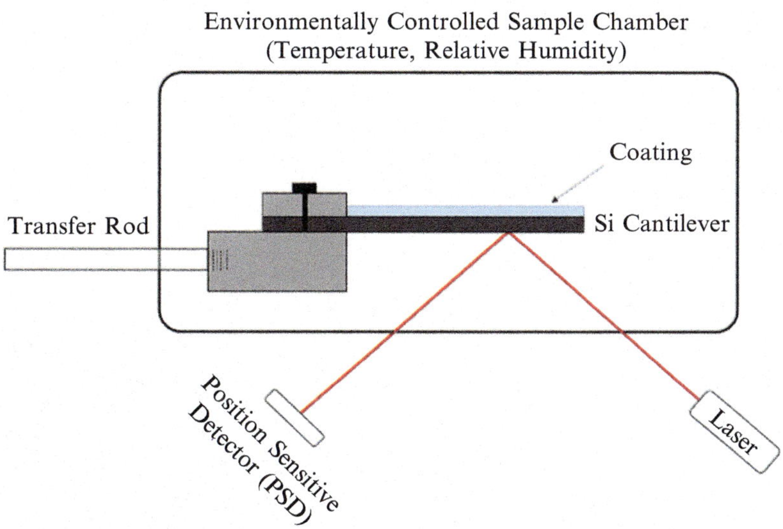

Fig. 10.3 Cantilever beam bending stress measurement apparatus used to measure in situ stress development in drying coatings

laser beam is reflected off of the bottom of the cantilever and onto a position sensitive detector (PSD). Before the measurement, a calibration curve is generated for each cantilever by applying a series of known displacements to the end of the cantilever (using a micrometer) and monitoring the movement of the laser spot on the PSD. Using the slope of the calibration curve, the position data recorded during the experiment (on the PSD) can be transformed into a cantilever deflection. After calibration, the coating liquid is applied to the free end of the cantilever by depositing a controlled volume of the suspension using a micropipette. Data acquisition begins momentarily after the coating is applied. During the experiments, the temperature and relative humidity were maintained at 21 ± 1 °C and 40% RH ± 2% RH, respectively. A video camera positioned above the cantilever recorded the appearance changes of the coating during drying to monitor drying uniformity and cracking.

Modified cantilevers with polydimethylsiloxane (PDMS) walls were used in this work to suppress the lateral drying front and to promote a uniform drying condition by eliminating thin edges and ensuring a uniform coating thickness. Walled cantilevers were fabricated by molding PDMS walls onto silicon cantilevers (6 mm × 60 mm) cut from (100) silicon wafers. The process used to fabricate the PDMS-walled cantilevers has been described in detail previously [27]. The PDMS walls used in this study were 0.5 mm wide and semicircular in shape. The walls form a perimeter around the free end of the cantilever (45 mm × 6 mm) (see Fig. 10.4). The Young's modulus and Poisson's ratio of the silicon cantilevers were taken to be $E_s = 169$ GPa, $\nu_s = 0.064$, respectively [30].

Corcoran [19] developed a useful expression (Eq. (10.2)) relating the curvature or deflection of a coated substrate to the average stress in the film. The deflection, d, in Eq. (10.2) assumes that the cantilever is fully coated. However, finite element simulations have shown that the deflection of a coated walled cantilever is smaller than the deflection of a fully coated cantilever due to the reduction in coated area because of the wall [27]. Therefore, using Eq. (10.2) to calculate the coating stress from deflection values measured for walled cantilevers results in an underestimation of the true coating stress. To account for this, a modified form of Eq. (10.2) was derived with a term that compensates for the reduction in coated area caused by the wall [27]:

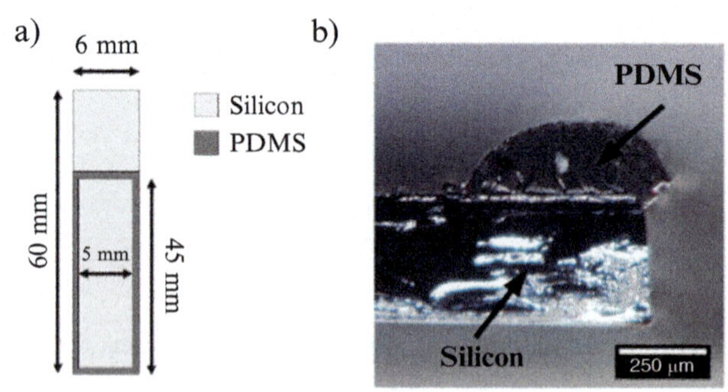

Fig. 10.4 (**a**) Schematic diagram of the modified PDMS-walled cantilevers used for stress measurements. (**b**) Photograph of a cross section of a walled cantilever showing the semicircular shape of the PDMS wall

$$\sigma_{\text{walled}} = \frac{Et^3d}{3cL^2(1-\nu)(t+c)}\left(\frac{w_s}{w_c}\right) \tag{10.3}$$

where σ_{walled} is the average stress in the coating for walled cantilevers, w_s is the width of the cantilever, and w_c is the width of the coating. This relationship is essentially the first term of Eq. (10.2) with the additional correction factor (w_s/w_c). Considering the experimental conditions used for these experiments (i.e., $E_s >> E_c$, $t >> c$), the second term of Eq. (10.2) was omitted for this analysis. The correction factor introduced in Eq. (10.3), w_s/w_c, is similar to those used by others conducting stress measurements on standard cantilevers that are not fully coated [23, 31]. The correction factor accounts for the reduced area of the coating along the long edges of the cantilever, but does not account for the reduced area along the short edges. Additionally, the coating thickness, c, used to calculate the coating stress was measured with a micrometer after completion of the experiment. The use of the dry film thickness is reasonable because stress development begins only after a consolidated particle network forms, and only small changes in thickness are expected after consolidation. For these experiments, the measurements of the final coating thickness ranged from 70 to 80 μm.

Minimum Film Formation Temperature

The minimum film formation temperature (MFFT) measurement is a common technique used to study the film formation behavior of latex coatings. The MFFT defines the lowest temperature at which a latex coating forms a homogeneous film upon drying. The cloudy-clear transition in the dry coating serves as an indicator of the extent of particle deformation in the film. To determine the MFFT, a latex suspension is drawn over an aluminum temperature gradient bar; an MFFT90 (Rhopoint) was used for this study (Fig. 10.5). The film was allowed to dry

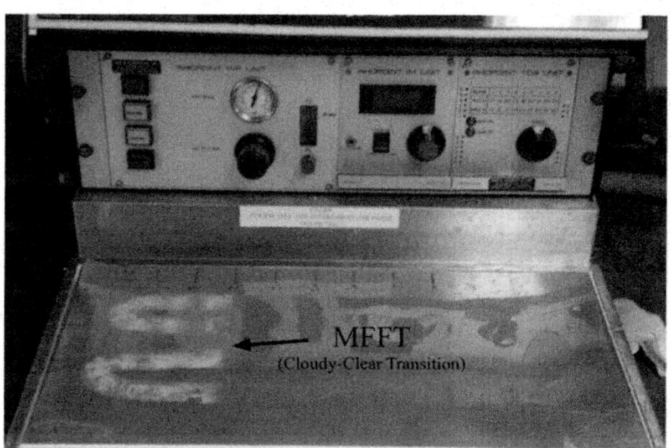

Fig. 10.5 Photograph of the minimum film formation temperature bar, MFFT90 (Rhopoint). A PET film covers the metal temperature gradient bar, which is cooler on the left side and warmer on the right. A latex coating has been fully dried, and the cloudy-clear transition is clearly visible

completely as moisture was removed from the sample by a constant flow of nitrogen over the wet film. After drying was complete, the MFFT was identified as the lowest temperature where the coating formed a clear, crack-free film.

Latex

The monodisperse acrylic latexes used in these studies were provided by Arkema Coating Resins (Cary, NC). The latex samples were synthesized from three common acrylic monomers: butyl acrylate (BA), methyl methacrylate (MMA), and methacrylic acid (MAA). The characteristics of the latex samples used in this study are compiled in Table 10.1. The volume-averaged (D_v) and number-averaged (D_n) particle sizes were measured using NANOTRAC® UPA 150 (Microtrac) particle size analyzer. The T_gs of the samples were determined by differential scanning calorimetry (Q2000 DSC, TA Instruments). Four coalescing aids having different water solubility and vapor pressure were investigated in order to explore the effect of the additives on the MFFT and the stress development of a mid-T_g latex. The details of the coalescing aids are listed in Table 10.2.

Table 10.1 Sample characteristics for acrylic latexes

	Composition [wt%]					Particle size	
Sample Name	BA	MMA	MAA	T_g [°C]	MFFT [°C]	D_v [nm]	D_n [nm]
MBA-High	23.5	75	1.5	68	>60	182	156
MBA-Mid	46.1	52.4	1.5	22	18.4	177	154
MBA-Low	53.3	43.2	1.5	11	5.2	169	150

Table 10.2 Characteristics of coalescing aids

Coalescing aid	Classification	Water solubility [%]	Boiling point [°C]	Vapor pressure at 20 °C [mmHg]	Rate of evaporation (butyl acetate = 1)
Texanol®[a]	Conventional	0.1	254	0.01	0.002
Butyl Carbitol®[b]	Conventional	∞	230	0.03	0.004
Optifilm® Enhancer400[a]	Low VOC	0	344	<0.0001	0.000017
Benzoflex® LC-531[a]	Low VOC	Slightly soluble	>350	<0.0001	No data

[a]Texanol®, Optifilm®, and Benzoflex® are all registered trademarks of Eastman Chemical and the data are from their product technical data sheets (TDS)
[b]Butyl Carbitol® is a registered trademark of the Dow Chemical Company and the data are from its TDS

Results and Discussion

Effect of MFFT on Stress Development

The first objective of this study was to understand the fundamental relationships between the extent of latex film formation and the evolution of stress in the coating. In this work, the MFFT was used to characterize the film formation behavior, while the cantilever beam deflection technique was used to monitor in situ stress development during drying. The MFFT of a latex depends on many variables including the polymer composition, particle size, particle T_g, and additives like plasticizers or coalescing aids [4]. For the latex materials used in this study, the MFFT was found to be slightly below the polymer T_g in all cases, which is expected as the polymer is slightly plasticized by the water phase of the suspension [32].

When a latex coating is dried below the MFFT, the resulting film is opaque or cloudy and is often cracked, indicating incomplete film formation. At temperatures below the MFFT, the polymer is well below its T_g, and the stiff latex particles do not undergo extensive deformation upon drying. Under these conditions, the extent of particle deformation is not sufficient to close the void spaces in the particle network. When the coating is dry, any remaining air voids exhibit strong light scattering, giving the coating an opaque appearance. Likewise, cracks often form below the MFFT because the cohesive strength of the coating is low due to limited polymer diffusion across particle–particle boundaries.

Latex coatings dried above the MFFT are optically clear and crack-free. At temperatures above the MFFT, the polymer is typically above its T_g, and the softer particles have a greater capacity to deform during film formation. Particle deformation eliminates the void space in the particle packing, reducing the amount of light scattering from voids in the film. A clear or transparent appearance of the coating does not necessarily guarantee a void-free coating, only that the sizes of the voids are smaller than the wavelength of visible light [33]. Additionally, above the MFFT there is more extensive polymer diffusion across particle boundaries leaving the resulting films with greater mechanical toughness to resist crack formation and propagation.

Constrained shrinkage of the coating during film formation gives rise to the development of a biaxial tensile stress in the film. Figure 10.6 shows the stress measurement results for latex coatings with various MFFTs dried at room temperature in the absence of lateral drying fronts. With an MFFT greater than 60 °C, MBA-High did not undergo film formation at room temperature, and the film cracked extensively upon drying. The stress development profile for MBA-High has an initial stress-free period followed by a sharp increase in coating stress. During the stress-free period, the coating is still wet, and the latex particles have not formed a consolidated particle network. In this stage, the coating is fluid and flows in response to volume changes; therefore, no stress develops in the coating at this stage.

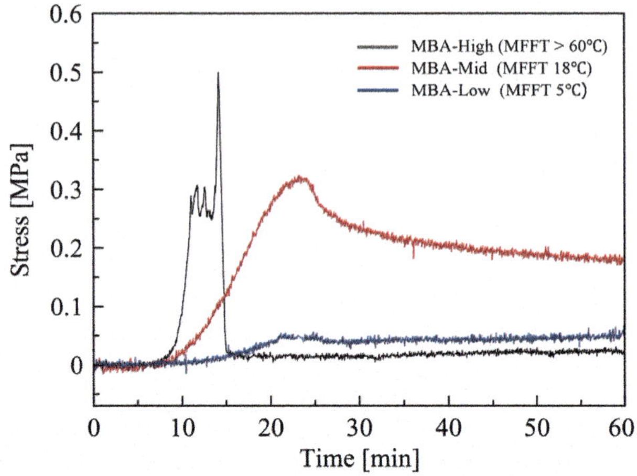

Fig. 10.6 Stress development curves comparing latex coatings with varying MFFT without coalescing aids. MBA-Mid and MBA-Low form continuous films upon drying at room temperature, whereas MBA-High cracks extensively upon drying

After consolidation, the particle network is capable of supporting a stress and stress development begins. For MBA-High, there is a sharp rise in stress followed by plateau period where there are several small increases and decreases before the measured stress rises to the maximum value. A video recording of the cantilever during the experiment revealed that throughout the plateau period, small cracks developed and the coating began delaminating along the PDMS wall. Cracking and delamination relieve stress locally; however, stress continued to increase where the coating had good adhesion to the cantilever. When the measured stress reached the maximum value, large cracks formed throughout the coating, and the measured stress decayed rapidly to a negligible value. For coatings that form cracks upon drying, the measured stress is less accurate as the stress distribution in the coating is no longer uniform. Cracking, like lateral drying, complicates the interpretation of stress measurements made using the cantilever beam bending technique.

Stress development curves for samples MBA-Mid and MBA-Low are also shown in Fig. 10.6. Both samples have MFFTs below room temperature and form continuous, crack-free films upon drying. The stress development curve for MBA-Mid has an initial stress-free period followed by a gradual rise in stress to a maximum value. After the maximum stress is achieved, there is a modest stress relaxation in the coating. For the MBA-Low sample, stress develops in the film gradually with time and plateaus at the final residual stress.

As shown in Fig. 10.6, during the period of stress rise, the rate of stress development increases with the latex MFFT. As the MFFT increases, the latex particles are more rigid and resist deformation during drying. Evaporation causes curved menisci to form between adjacent particles along the free surface of the coating. This generates a capillary pressure in the liquid that acts to further consolidate the particle network. For nondeforming particles, the menisci quickly reach the maximum curvature and thus the highest capillary pressure allowed by the pores. The sharp rise in stress is a direct result of the increased capillary pressure in the liquid acting to compress and shrink the network of consolidated particles. More details of the relationship between stress evolution and menisci formation in hard particle coatings can be found in reference [27], where the relationship was investigated by combining stress measurements and cryoSEM imaging. When the latex particles are less rigid (i.e., lower MFFT), stress development occurs more gradually with time. During the drying process, capillary stresses develop in the liquid after the particles achieve a close-packed structure. The stresses in the coating are sufficient enough to cause deformation of the softer particles, minimizing the remaining pore volume and limiting the maximum achievable capillary pressure.

In the work presented here, latex coatings with lower MFFTs developed less maximum stress compared to those with higher MFFTs. One explanation for this behavior is that particle deformation limits the maximum capillary pressure in the coating, therefore limiting the maximum achievable stress in the film. Also contributing to this behavior are the elastic modulus of the polymer (smaller with lower MFFTs) and the viscoelastic relaxation of stress in the coating (faster for lower MFFTs). Similar trends in stress evolution were reported in previous studies; however, when films were dried at or above the MFFT, the maximum stress was found to be essentially independent of the MFFT [16, 23]. It should also be noted that the stress measurements in those studies were conducted in the presence of lateral drying, which may have complicated the interpretation of the measured stress.

Effect of Coalescing Aids on Film Stress Development

One of the common strategies employed to enhance latex film formation and lower the MFFT is the addition of a coalescing aid or plasticizer to the formulation. The addition of a coalescing aid not only lowers the T_g of the polymer but also increases the diffusivity of the polymer chains by increasing the free volume. Juhue et al. [34] demonstrated this effect by showing the diffusivity of the polymer chains of poly (butyl methacrylate) increased by one order of magnitude when three weight percent of Texanol® coalescing aid was added to the latex suspension; the diffusivity increased by four orders of magnitude when 12 weight percent was added. More recently, Schroeder et al. [35] used fluorescence resonance energy transfer (FRET) to show that the addition of Texanol promotes an earlier onset of diffusion during the film formation process in common acrylic latex coatings. Polymer diffusion possibly begins in the consolidation stage before all voids disappear.

The effectiveness of a coalescing aid depends on several factors. First, the volatility or evaporation rate of the coalescent dictates how long the plasticizing molecules are present in the film. Second, the compatibility of the coalescing aid with the polymer dictates how efficiently the coalescent lowers the T_g. Third, the solubility of the coalescing aid determines how the coalescent molecules partition themselves in the suspension. For example, Texanol has low solubility in water;

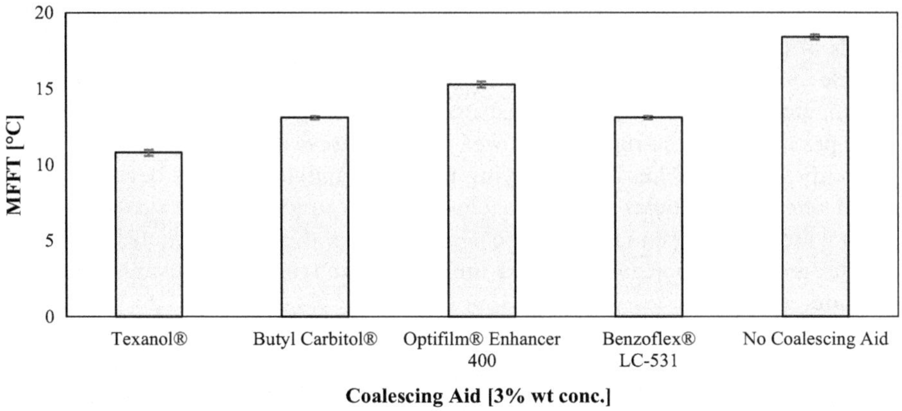

Fig. 10.7 Effectiveness of coalescing aids at lowering the MFFT for MBA-Mid at a concentration of 3% on the solids in the coating

therefore, the coalescent should have a higher concentration in the polymer particles. This allows the coalescing aid to promote polymer diffusion during the early stages of film formation when the particles make initial contact. A hydrophilic coalescing aid should partition itself preferentially in the water phase; hence, the coalescing aid becomes more effective during the later stages of film formation when the water content is low.

To investigate the effectiveness of the coalescing aid, MBA-Mid was used as it has an intermediate MFFT of 18 °C and formed a continuous film upon drying at room temperature. The effects of various coalescing aids on the MFFT of the MBA-Mid latex are shown in Fig. 10.7. Each of the coalescing aids investigated was effective at lowering the MFFT at a concentration of three weight percent. From these measurements, Texanol was found to be the most efficient in lowering the MFFT, while Optifilm® Enhancer 400 coalescing aid was the least effective. The two hydrophilic coalescing aids investigated, Butyl Carbitol® and Benzoflex® LC-351 coalescing aids, showed comparable efficiency in lowering the MFFT. However, it should be noted that the MFFT is not the most precise experimental measurement. Measurements of the MFFT can be combined with measurements of the plasticized T_g to estimate the efficiency of the coalescing aid; however, these measurements were not performed in this study.

Stress development curves for the MBA-Mid latex with three weight percent coalescing aid are shown in Fig. 10.8. From the plot, it is clear that the addition of the coalescing aids decreases the maximum stress measured in the film. At a concentration of three weight percent, the maximum stress values are comparable for the latex samples with Texanol, Optifilm, and Benzoflex added; however, the measured stress is significantly lower for the film containing the Butyl Carbitol. The apparent effectiveness of the Butyl Carbitol at reducing stress and MFFT may seem unusual given its high water solubility; the results reported in this chapter are, however, consistent with the findings of other researchers and the fact that historically Butyl Carbitol has been one of most widely used coalescing aids for many years (before the tightened VOC regulation). Besides a low evaporation rate, both

Fig. 10.8 Stress development curves for the MBA-Mid sample (MFFT = 18.4 °C) with the addition of various coalescing aids at 3% by weight on the total solids level

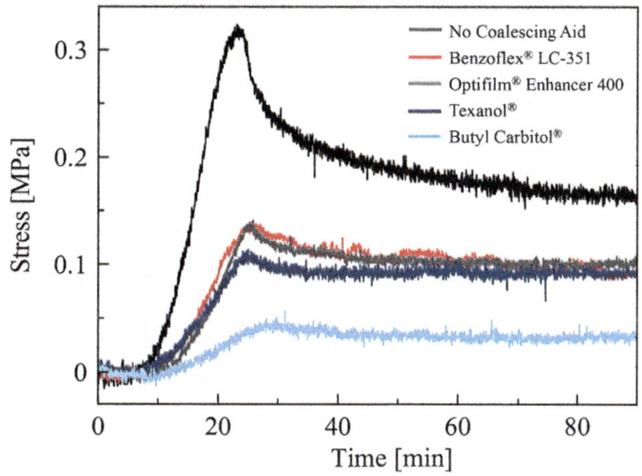

theoretical calculations and experimental data demonstrate that Butyl Carbitol partitions predominantly into the polymer particle despite being 100% soluble in water [36–39]. It was suggested that Butyl Carbitol is preferentially located in the hydrophilic outermost thin layer of particle and enhances the miscibility of the particle shells during coalescence. Due to the high water solubility of Butyl Carbitol, more water may be retained in the system that will plasticize the film and result in lower stress development; however, more work is needed to better understand the effectiveness of Butyl Carbitol at reducing stress.

The concentration of the coalescing aid was also found to influence stress development in the coating. Texanol and Optifilm were used to investigate the influence of additive concentration on stress development. As shown in Fig. 10.9, the maximum stress is lowered as the concentration of the coalescing aid increases. This may be explained by the fact that the particles become softer and can deform to a greater extent as the T_g is lowered with higher concentrations of the coalescing aid. The maximum stress in the film decreases dramatically as concentration is increased to three weight percent; above three weight percent, the differences in maximum stress are less substantial. Additionally, as a general trend, the residual stress in the film was reduced as the amount of coalescing aid was increased. Both of the coalescing aids investigated were found to have comparable performance in lowering stress. Similarly, lowering the polymer T_g was found to be equally effective in reducing stress development as adding coalescing aid to these materials.

Conclusions

The goal of this study was to investigate the film formation behavior and stress development of latex coatings as a function of MFFT. Using the cantilever beam bending technique with modified walled cantilevers, stress measurements were

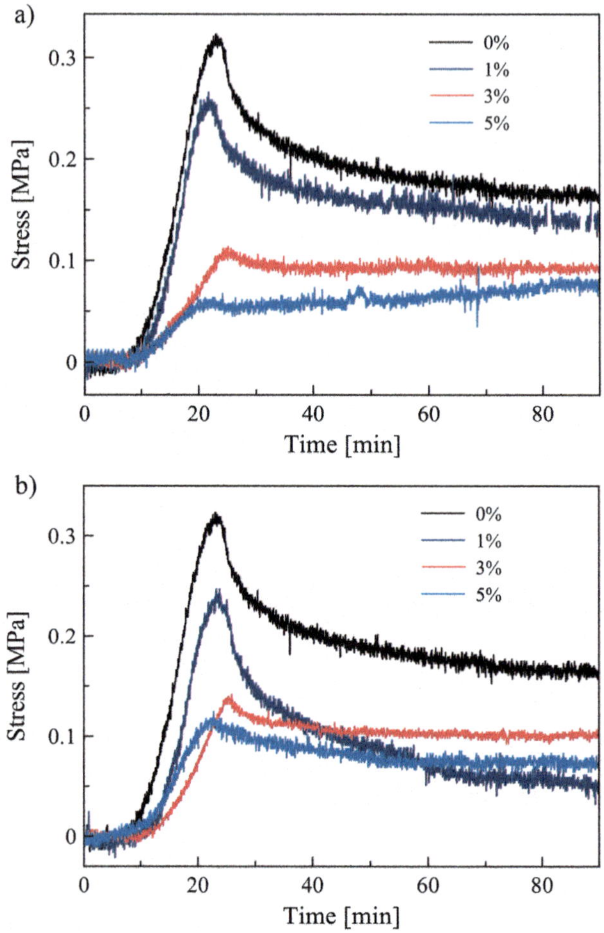

Fig. 10.9 The influence of the concentration of (**a**) Texanol® coalescing aid and (**b**) Optifilm® Enhancer 400 coalescing aid on the stress development in the coating of MBA-Mid

made without the complications of lateral drying. Stress measurements show that latexes with lower MFFTs achieve lower maximum stresses and develop stress more gradually during film formation. Additionally, the effects of various types of coalescing aids were investigated. All of the additives were successful in lowering the MFFT. For these materials, Texanol was found to be the most efficient at lowering the MFFT, while Optifilm was the least efficient of the additives studied. At a concentration of three weight percent, Benzoflex, Optifilm, and Texanol had a similar effect on stress development of the latex with an MFFT of 18 °C, while Butyl Carbitol showed an enhanced ability to decrease stress. In general, the maximum stress in the film decreased as the concentration of coalescing aid was increased. However, above a concentration of three weight percent, the differences in maximum stress were minimal. Combined together, the results from these experiments show how latex film formation and the stress development in the film are interdependent. The results presented here advance the fundamental understanding of latex film formation and how the film formation process relates to stress development in the film.

Acknowledgments The authors would like to gratefully acknowledge support from the National Science Foundation under Award No. CBET 0967348 and the industrial supporters of the Coating Process Fundamentals Program of the IPrime industrial consortium at the University of Minnesota. Parts of this work were carried out in the Characterization Facility, University of Minnesota, which receives partial support from NSF through the MRSEC program.

References

1. Keddie, J.L., Routh, A.F.: Fundamentals of Latex Film Formation, pp. 95–120. Springer, Dordrecht (2010)
2. Steward, P.A., Hearn, J., Wilkinson, M.C.: An overview of polymer latex film formation and properties. Adv. Colloid Interf. Sci. **86**, 195–267 (2000)
3. Holl, Y., Keddie, J.L., Mcdonald, P.J., Winnik, W.A.: Drying modes of polymer colloids. ACS Symp. Ser. **790**, 2–26 (2001)
4. Keddie, J.: Film formation of latex. Mater. Sci. Eng. **21**, 101–170 (1997)
5. Lei, H., Payne, J.A., Mccormick, A.V., Francis, L.F., Gerberich, W.W., Scriven, L.E.: Stress development in drying coatings. J. Appl. Polym. Sci. **81**, 1000–1013 (2001)
6. Hoffman, R.W.: Stress distributions and thin film mechanical properties. Surf. Interface Anal. **3**, 62–66 (1981)
7. Vaessen, D.: Stress and structure development in polymeric coatings. Doctoral dissertation, University of Minnesota (2001)
8. Jindal, K.: Stress development in particulate, nano-composite, and polymeric coatings, University of Minnesota (2009)
9. Chiu, R.C., Cima, M.J.: Drying of granular ceramic film: II, drying stress and saturation uniformity. J. Am. Ceram. Soc. **76**, 2769–2777 (1993)
10. Guo, J.J., Lewis, J.A.: Aggregation effects on the compressive flow properties and drying behavior of colloidal silica suspensions. J. Am. Ceram. Soc. **82**, 2345–2358 (1999)
11. Lampenscherf, S., Pompe, W., Wilkinson, D.: Stress development due to capillary condensation in powder compacts: a two-dimensional model study. J. Am. Ceram. Soc. **83**, 1333–1340 (2000)
12. Payne, J.A., McCormick, A.V., Francis, L.F.: In-situ stress measurement apparatus for liquid applied coatings. Rev. Sci. Instrum. **68**, 4564 (1997)
13. Francis, L.F., McCormick, A.V., Vaessen, D.M., Payne, J.A.: Measurement and development of stress in polymer coatings. J. Mater. Sci. **37**, 4717–4731 (2002)
14. Petersen, C., Heldmann, C., Johannsmann, D.: Internal stresses during film formation of polymer latices. Langmuir. **15**, 7745–7751 (1999)
15. Tirumkudulu, M., Russel, W.: Role of capillary stresses in film formation. Langmuir. **20**, 2947–2961 (2004)
16. Martinez, C.J., Lewis, J.: Shape evolution and stress development during latex−silica film formation. Langmuir. **18**, 4689–4698 (2002)
17. König, A.M., Johannsmann, D.: Stress fluctuations in drying polymer dispersions. Langmuir. **26**, 9437–9441 (2010)
18. Stoney, G.: The tension of metallic films deposited by electrolysis. Proc. R. Soc. Lond. Ser. A Contain. Pap. Math. Phys. Character. **82**, 172–175 (1909)
19. Corcoran, E.M.: Determining stresses in organic coating using plate beam deflection. J. Paint Technol. **41**, 635–640 (1969)

20. Perera, D., Van den Eynde, D.: Considerations on a cantilever (beam) method for measuring the internal stress in organic coatings. J. Coat. Technol. **53**, 39–44 (1981)
21. Perera, D., Van den Eynde, D.: Effect of pigmentation on internal stress in latex coatings. J. Coat. Technol. **56**, 47–53 (1984)
22. Perera, D., Van den Eynde, D.: Effect of organic solvents on internal stress in latex coatings. J. Coat. Technol. **56**, 69–75 (1984)
23. Yow, H.N., Beristain, I., Goikoetxea, M., Barandiaran, M.J., Routh, A.F.: Evolving stresses in latex films as a function of temperature. Langmuir. **26**, 6335–6342 (2010)
24. Deegan, R., Bakajin, O., Dupont, T., Huber, G.: Capillary flow as the cause of ring stains from dried liquid drops. Nature. **389**, 827–829 (1997)
25. Price, K.K., McCormick, A.V., Francis, L.F.: CryoSEM investigation of latex coatings dried in walled substrates. Langmuir. **28**, 10329–10333 (2010)
26. Routh, A.F., Russel, W.B.: Horizontal drying fronts during solvent evaporation from latex films. AIChE J. **44**, 2088–2098 (1998)
27. Price, K.K., Wu, Y., McCormick, A.V., Francis, L.F.: Stress development in hard particle coatings in the absence of lateral drying. J. Am. Ceram. Soc. **98**, 2214–2222 (2015)
28. Price, K.K., Wu, W., Wood, K., Kong, S., McCormick, A.V., Francis, L.F.: Stress development and film formation in multiphase composite latexes. J. Coat. Technol. Res. **11**, 827–839 (2014)
29. Price, K.K: Stress and microstructure development in particle-based coatings. Ph.D. thesis, University of Minnesota. Minneapolis, MN (2014)
30. Hopcroft, M.A., Nix, W.D., Kenny, T.W.: What is the young's modulus of silicon? J. Microelectromech. Syst. **19**, 229–238 (2010)
31. Yow, H.N., Goikoetxea, M., Goehring, L., Routh, A.F.: Effect of film thickness and particle size on cracking stresses in drying latex films. J. Colloid Interface Sci. **352**, 542–548 (2010)
32. Sperry, P.R., Snyder, B.S., O'Dowd, M.L., Lesko, P.M.: Role of water in particle deformation and compaction in latex film formation. Langmuir. **10**, 2619–2628 (1994)
33. Keddie, J.L., Meredith, P., Jones, R.A.L., Donald, A.M.: Kinetics of film formation in latex studied with ellipsometry and environmental-SEM. Macromolecules. **28**, 2673–2682 (1995)
34. Juhue, D., Wang, Y., Winnik, M.A., Haley, F.: Influence of a coalescing aid on polymer diffusion in poly(butyl methacrylate) latex films. Macromol. Chem. Rapid Commun. **14**, 345–349 (1993)
35. Schroeder, W.F., Liu, Y., Tomba, J.P., Soleimani, M., Lau, W., Winnik, M.A.: Effect of a coalescing aid on the earliest stages of polymer diffusion in poly(butyl acrylate-co-methyl methacrylate) latex films. Polymer (Guildf). **52**, 3984–3993 (2011)
36. Hoy, K.L.: Estimating the effectiveness of latex coalescing aids. J. Paint Technol. **45**, 51–56 (1973)
37. Alahapperuma, K., Glass, J.E.: Influence of coalescing aids in associative thickener dispersions. J. Coat. Technol. **63**, 69–78 (1991)
38. Toussaint, A., De Wilde, M.: A method to predict the distribution coefficient of coalescing agents between latex particles and the water phase. Prog. Org. Coat. **30**, 173–177 (1997)
39. Juhue, D., Lang, J.: Latex film formation in the presence of organic solvents. Macromolecules. **27**, 695–701 (1994)

Chapter 11
Stress Development in Reactive Coatings

Jiří Zelenka, Karel Dušek, and Mei Wen

Introduction

Existence of coating stress can have catastrophic consequences for the polymeric coating systems. Coating properties and defect formation can be affected by the development of stress. Stress can lead to weakened coating adhesion to the substrate [1], coating corrosion [2–4], appearance defects such as wrinkling [5], development of micro-cracks or macro-cracks [6, 7], and delamination [8].

A coating develops stress due to its limitation to shrink freely in the direction parallel to the substrate when it adheres to the substrate. The stress caused by relative shrinkage of the coating to the substrate is tensile stress. If the shrinkage is negative, i.e., expansion, the stress is compressive. Coating stress can be generated due to shrinkage caused by curing [9], solvent evaporation [10], cooling (thermal process) [11], and moisture desorption when humidity lowers [11]. Perera [1] used mathematical relations to describe individual stresses produced by these processes, and the total stress is the sum of the individual stresses:

$$\sigma = \sigma^{\mathrm{F}} + \sigma^{\mathrm{T}} + \sigma^{\mathrm{H}} \tag{11.1}$$

J. Zelenka (✉)
SYNPO, akciová společnost, Pardubice, Czech Republic

TOSEDA s.r.o., Pardubice VI-Staré Čívice, Czech Republic
e-mail: jiri.zelenka@toseda.cz

K. Dušek
Institute of Macromolecular Chemistry, Academy of Sciences of the Czech Republic, Prague, Czech Republic

M. Wen
Axalta Coating Systems, Coatings Technology Center, Wilmington, DE, USA

M. Wen, K. Dušek (eds.), *Protective Coatings*, DOI 10.1007/978-3-319-51627-1_11

where σ^{F} is the internal stress produced due to the film-forming process caused by drying and curing, σ^{T} is the thermal stress produced due to temperature change, and σ^{T} is the hygroscopic stress produced due to humidity change. According to Perera [1], these stresses can be calculated based on the following equations:

$$\sigma^{\mathrm{F}} = \frac{E}{1-\upsilon}\frac{\Delta V}{3V} \tag{11.2}$$

$$\sigma^{\mathrm{T}} = \int_{T_1}^{T_2} \frac{E}{1-\upsilon}\left(\alpha_{\mathrm{C}}^{\mathrm{T}} - \alpha_{\mathrm{S}}^{\mathrm{T}}\right)\mathrm{d}T \cong \frac{E}{1-\upsilon} \cdot \left(\alpha_{\mathrm{C}}^{\mathrm{T}} - \alpha_{\mathrm{S}}^{\mathrm{T}}\right)\left(T_{\mathrm{g}} - T\right) \tag{11.3}$$

$$\sigma^{\mathrm{H}} = -\int_{\mathrm{RH}_1}^{\mathrm{RH}_2} \frac{E}{1-\upsilon} \cdot \left(\alpha_{\mathrm{C}}^{\mathrm{H}} - \alpha_{\mathrm{S}}^{\mathrm{H}}\right)\mathrm{dRH} \tag{11.4}$$

In these relations, E is the Young's modulus of the coating and ν is its Poisson's ratio; V and ΔV are coating's volume at solidification and the change in volume after solidification respectively; $\alpha_{\mathrm{C}}^{\mathrm{T}}$ and $\alpha_{\mathrm{S}}^{\mathrm{T}}$ are the thermal expansion coefficients of the coating and substrate, respectively; T_1 and T_2 are the starting and ending temperatures that the coating goes through; T_{g} is the glass transition temperature of the coating and T is the ambient temperature; $\alpha_{\mathrm{C}}^{\mathrm{H}}$ and $\alpha_{\mathrm{S}}^{\mathrm{T}}$ are the hygroscopic expansion coefficients of the coating and substrate, respectively; and RH_1 and RH_2 are the starting and ending humidity conditions that the coating goes through, respectively. These relations are of limited practical use: the coating is not a mechanically linear material, and the stress-determining parameters (E, ν, α) are drying history or time dependent (dependent on the rates of crosslinking, liquid evaporation or absorption, and rubber-to-glass transition).

For a coating cured above its T_{g}, σ^{F} is small because E of a rubbery material is by several orders of magnitude smaller than E of a glassy material. When no further reaction or drying occurs at the point when the coating starts to be cooled down, the total stress based on Eq. (11.1) becomes

$$\sigma = \sigma^{\mathrm{T}} + \sigma^{\mathrm{H}} \tag{11.5}$$

Furthermore, if volume change due to moisture absorption is small (σ^{H} negligible), the resulting stress is dominated by the difference between thermal expansion coefficients of the coating and the substrate.

For a coating cured at a temperature below its T_{g}, the total stress will be a function of shrinkage developed during the baking process (constant temperature), the shrinkage developed during the cooling process, and potentially hygroscopic expansion (or shrinkage) due to humidity changes. If further reaction or drying occurs after cooling down, the stress will continue to develop due to further film formation process at room temperature.

For a coating cured at room temperature without temperature and humidity change, the stress of the coating is

$$\sigma = \sigma^{\mathrm{F}} \tag{11.6}$$

In this case, volume shrinkage comes from solvent evaporation and curing. Once the coating solidifies, further shrinkage brought by either solvent evaporation or curing contributes to the stress development.

The curing process in organic coatings generally involves crosslinking reaction to form a network. Shrinkage due to curing can come from different mechanisms of formation of chemical bonds from functional groups and possible elimination of a low-molecular-weight compound. The magnitude of the shrinkage is inversely proportional to the weight per functional group; preformed precursors with higher weight per functional group shrink less. For instance, in UV-curable acrylic coatings, shrinkage occurs mainly due to shortened distance between reactive monomers during the free-radical polymerization process [12]. In acrylic-melamine formaldehyde coatings, shrinkage occurs due to the formation of a reaction by-product, alcohol, in the condensation polymerization [13]. Furthermore, if a ring-opening reaction mechanism is involved, shrinkage can be reduced or even become negative [12], which can be used to reduce the tensile stress of a coating or even to create compressive stress.

The solidification point occurs when the coating has dried or cured enough, so that it has passed from a liquid to a solid state, has developed elasticity, and can support a stress [9]. After passing through the gel point, the liquid state of the coating transforms into a rubbery state, and on further curing and solvent evaporation, the glass transition temperature of the coating reaches the curing temperature [14–17]. Once the coating becomes glassy, further curing and evaporation of solvent take much longer time than in the rubbery state. The gradual curing and release of solvent can potentially grow stress into a detrimental level. In special cases, liquid-to-glass transition sets in before the gel point is reached. In this case, coatings often do not reach full cure.

Due to the slow drying and curing process happening in the glassy state, the development of stress can take much longer time for the reaction to complete and all the solvent to come out. The reported stress development studies have been limited to several days or weeks [7, 17–19]. To our knowledge, the longest time reported is 550 h [19]. Therefore, it is interesting to determine over much longer time how stress evolves during the drying and curing of a coating in the glassy state.

Work reported to understand the stress development during drying and/or curing of reactive, polymeric coatings includes epoxy coatings [20, 21], UV-curable acrylic coatings [15, 22, 23], and different types of commercial coatings [7, 24]. However, no systematic study has been reported in the stress development of polyurethane coatings, particularly with well-defined polyol structure and composition to determine how they affect the stress development and how stress relates to the extent of reaction and other physical properties.

One goal of this work is to find how acrylic polyol composition and structure affect stress development in polyurethane coatings at room temperature. Acrylic polyols are copolymers with defined structures based on either isobornyl methacrylate, ethylhexyl methacrylate, and hydroxyethyl methacrylate (IBoMA/EHMA/HEMA) or methyl methacrylate, ethylhexyl methacrylate, and hydroxyethyl methacrylate (MMA/EHMA/HEMA). They were prepared by group-transfer polymerization (GTP). These acrylic polyols were mixed with polyisocyanates as polyurethane coating solutions. Stress development during the curing and drying of the polyurethane coatings was monitored. The effects of copolymer structure, such as rigid (IBoMA) vs. less rigid (MMA) segments in the main chain, copolymer molecular weight, and crosslink density of the formed polyurethane network on the internal stress are discussed. Addition of a low-T_g polyester polyol to a polyurethane coating was further studied. Not only stress development in these systems was studied but also pendulum hardness, T_g, and NCO group conversion in most cases were measured in parallel to determine how stress development correlates to them during the film formation process.

The second goal of this work is to investigate other formulation and process parameters on stress development. The effect of solvent addition to a solvent-free polyurea coating system on stress development at room temperature was studied. Lastly, the effect of baking on the stress development of a polyurea/polyurethane coating was also investigated.

Stress measurement of all the systems in this study was taken for a long time, i.e., over 1000 h or even a year, to ensure that the plateau of stress level is achieved at the end.

Stress Measurement

The methods used for measuring the internal stress of a coating are mainly based on the cantilever deflection or radius of curvature of a coated elastic substrate originated by Stoney's work [25]. A coated, elastic substrate curves in the direction that relieves the stress in the coating. The internal stress is proportional to the deflection or curvature of the coated substrate [9]. The cantilever deflection method [20, 26–33] has several variations to measure the deflection of the cantilever: optical microscope method as described in ASTM D 6991–05, optical levers [34], linear variable displacement transducers (LVDT) [35], and capacitive transducers [24, 36]. The curvature method uses strain gauge [18], scanning laser device [37], X-ray diffraction [38, 39], or interferometry [40].

The cantilever deflection method is used in this work. Relations between stress and deflection of a coated substrate can be mathematically described by applying the beam and/or the plate theory [26, 41, 42]. This dependence is described by Corcoran's equation [30]:

$$\sigma = \frac{dE_S t^3}{3L^2 c(t+c)(1-\nu_S)} + \frac{dE_C(t+c)}{L^2(1-\nu_C)} \tag{11.7}$$

where d is the deflection of the substrate, E_S is the elastic modulus of the substrate, E_C is the elastic modulus of the coating, ν_S is the Poisson's ratio of the substrate, ν_C is the Poisson's ratio of the coating, t is the thickness of the substrate, c is the

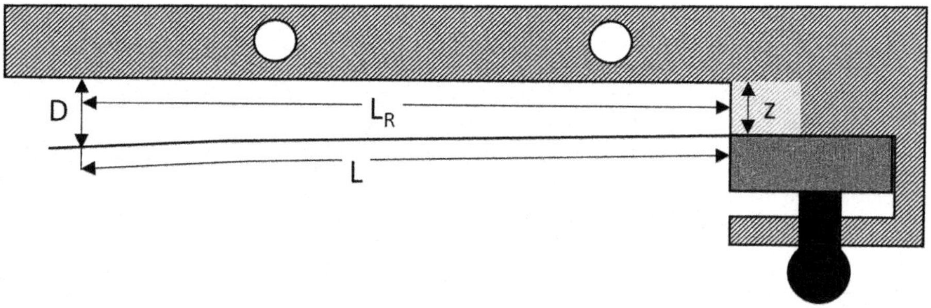

Fig. 11.1 Top view of a coated substrate and clamping setup in the cantilever deflection method to measure coating stress. The substrate is positioned vertically to avoid deflection due to gravity

thickness of the coating, and L is the active length of the substrate (distance between the clamped part and the deflection measuring point).

This equation assumes perfect adhesion of the coating to the substrate, isotropic elastic properties of the substrate and the coating, uniform coating thickness, and uniform stress distribution throughout the coating. The second term in Eq. (11.7) accounts for relaxations of stress that occurs as the cantilever bends. The second term of equation can be neglected without introducing a significant error (1% or less) when $E_S >> E_C$ and $t >> c$ [30]. The new equation after reduction is

$$\sigma = \frac{dE_S t^3}{3L^2 c(t + c)(1 - \nu_S)} \tag{11.8}$$

The cantilever measurement can be affected by other inaccuracies. For example, the active length of the substrate L is approximated as the horizontal distance L_R, where the deflection is measured (Fig. 11.1). The distance L_R is smaller than the active length L and the relation between L_R and L is approximately

$$L^2 = L_R^2 + d^2 \tag{11.9}$$

where the deflection of the cantilever (not shown in Fig. 11.1) $d = D - z$. Based on this approximation, for cases when $L_R >> d$, $L \approx L_R$ can be used. The effect of clamping on the cantilever deflection [41] may be another source of inaccuracy. Clamping generates extra stress in the substrate that can affect the measurement. This effect increases when L is shortened. This effect was observed experimentally and simulated by a finite element method [43]. The influence of clamping on results can be reduced by increasing L, and it is mostly negligible at $L \geq 80$ mm.

The coating thickness must be uniform over the entire free length of the substrate, and curing and drying should be uniform as well. Also, as shown in Eq. (11.1), temperature and humidity should be kept constant or programmed to avoid unintended thermal and hygroscopic stresses.

In this contribution, a microscope technique was used to determine the cantilever deflection in the stress development of polyurethane coatings. This technique is

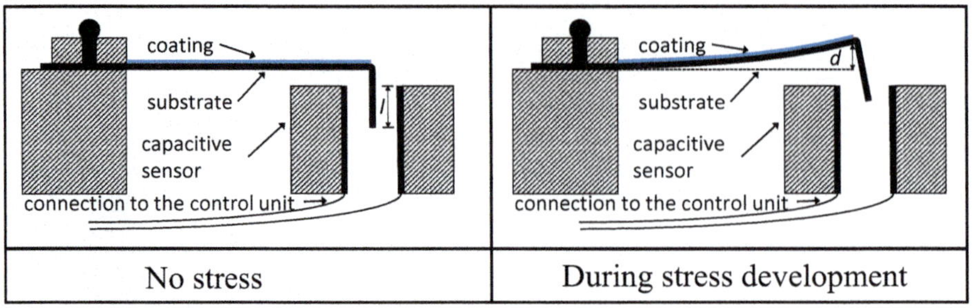

Fig. 11.2 Schematic stress measurement by a capacitive sensor: no stress (left) vs. during stress development

generally used for determination of the deformation or displacement of a specimen [44]. The distance between the cantilever beam and a reference point was measured. Multiple samples are measured to determine reproducibility. This method is limited due to its low sensitivity and manual data recording. To overcome these limitations, another approach to measure the cantilever deflection, a capacitive sensor technique, was applied, and a new setup to measure the deflections of multiple coatings was designed and installed.

The capacitive sensor technique is based on the following principle to measure the deflection of a beam. When an object entering the electrical field of a capacitor formed by two parallel electrodes, it causes a change in the capacitance. The change in capacitance is proportional to the embedding of an object into the capacitor. This phenomenon has been used to monitor changes in position (deflection) of a steel strip with a coating on it. Schematic of stress measurement by means of a capacitive sensor is shown in Fig. 11.2. The metal strip used has a bended tail inserting into the capacitive sensor. As the coating develops stress, the main part of the steel strip (coating substrate) curls up and pulls its tail part inside the capacitive sensor outward, leading to a change of the capacitance of the sensor. Relationship between the deflection of a substrate with a coating and the capacitance of a capacitive sensor can be expressed in the first approximation by the following equation [45]:

$$d = l\left(1 - K\frac{C - C_0}{C_0}\right) \tag{11.10}$$

where l is the starting length of the strip tail in the capacitive sensor, C is the capacitance with the strip tail inside the capacitive sensor (C is affected by d), C_0 is the capacitance when the strip curls up to a point that its tail is just fully out of the capacitor ($d = l$), and K is a constant influenced by construction of the capacitive sensor. The sensitivity of the capacitive sensor decreases with decreasing the relative permittivity of the substrate. With a calibration curve of capacitance vs. deflection of the cantilever established, deflection of the cantilever can be determined based on the capacitance data collected. Consequently, coating stress can be calculated by Eq. (11.8).

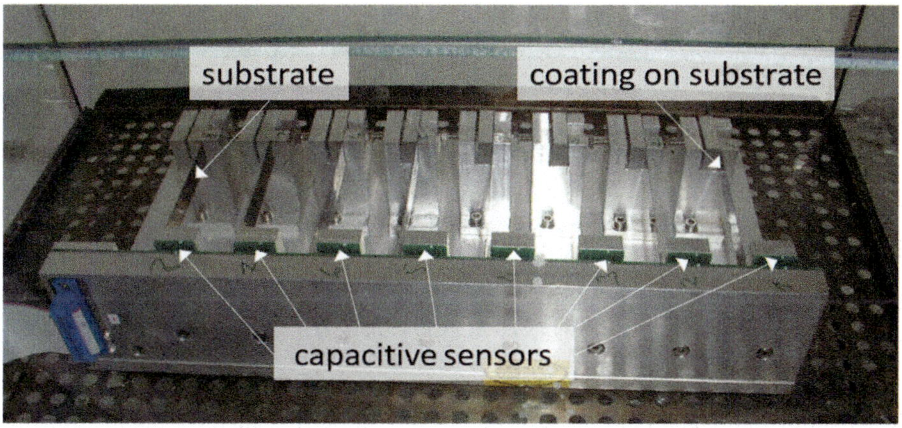

Fig. 11.3 Main part of stress meter based on capacitive sensors. Eight samples can be measured simultaneously by eight sensors

Table 11.1 Comparison of the microscopy and capacitive sensor techniques to measure deflection of a cantilever in coating stress measurement

Feature	Microscopy	Capacitive sensor
Advantage	Simple design	High sensitivity
	Low cost	Automatic data recording
	Not effected by temperature and humidity	Ease of testing multiple samples
	Longtime stability	Works for different substrates
Disadvantage	Low sensitivity	More expensive
	Manual data recording	Calibration affected by temperature and humidity
		Interference by high-frequency magnetic fields

Figure 11.3 shows the picture of the stress meter using the capacitive sensors to measure multiple samples. There are eight capacitive sensors, which allows measuring the stress of eight samples simultaneously. The stress meter is enclosed in a glass box, where temperature and humidity are maintained.

There are some advantages and disadvantages of the microscope method, and the capacitive sensor method to measure coating stress is described in Table 11.1. The microscope method offers simplicity and low cost. It has limited sensitivity; thus a thin substrate and consequently a thin coating can be used. Furthermore, the deflection at each point of measurement is not automatically recorded. Manual reading of the deflection needs to be made. In contrast, the capacitive sensor approach offers higher sensitivity for the deflection measurement and ease to test multiple samples. It can measure a deflection ranging from −1 to 11 mm with a resolution of 0.001 mm (Table 11.2). The deflection is automatically recorded. Different substrates can be used for the coating, such as metals, plastics, glass, and

Table 11.2 Technical data of stress meter based on capacitance sensors

Parameter	Value
Measured deflection	–1 mm to 11 mm
Nonlinearity	<1%
Deflection resolution	0.001 mm
Stability per year	0.1 mm
Measured samples	8 specimens
Power	20 W

other ceramics. The major drawback with the capacitive sensor technique is that the calibration curve needs to be established for the temperature and humidity conditions used for the experiment as the capacitance is affected by them. Therefore, when temperature and humidity change, recalibration is needed for the instrument. Special calibration equipment and procedure were used for it. The stress meter with the microscope method is not affected by the environment. Furthermore, the capacitive sensors are more expensive than an optical microscope. Interference by high-frequency magnetic fields could affect the measurement of capacitance. Therefore, the stress meter should be placed away from other equipment that has high-frequency magnetic fields. Table 11.2 also lists some other parameters of the stress meter based on capacitive sensors. The stress meter with capacitive sensors has excellent longtime stability.

Note that the other optical approaches to measure the deflection of a cantilever beam, such as optical levers or LVDT, are also well reproducible. However, they are more expensive than the capacitive sensor approach. In addition, they are not that flexible when measurement of multiple samples is needed. Simultaneous measurement of multiple samples by the LVDT or optical sensors has not been reported.

In this work, the stress development of all the polyurethane coatings was measured with the microscope method. The stress in the polyurea coating and the polyurea/polyurethane coating was measured with the capacitive sensor method, and the data are presented as continuous curves. In both methods, multiple coatings for each solution were tested to determine the reproducibility of the measurements.

Experimental

Synthesis of Acrylic Polyols

The acrylic polyols for this study were synthesized by GTP, which allows the preparation of polymers with defined structures (statistical or block copolymers) and with desired molecular weights. Acrylate monomers, methyl methacrylate (MMA), hydroxyethyl methacrylate (HEMA), ethylhexyl methacrylate (EHMA), isobornyl methacrylate (IBoMA), and tetrahydrofuran (THF, solvent) used during polymerization were first purified with alumina. Polymerizations were performed in glass reaction flasks under argon atmosphere using syringe techniques to feed the liquid reactants and THF. The mixtures of monomers and THF were stirred by a magnetic stirrer at 60 °C. The initiator used for the polymerization was 1-methoxy-1-trimethylsiloxy-2-methylpropene. Prepared acrylic polyols all had a concentration of 50% by weight in THF.

Before the synthesis, the composition of each acrylic polyol copolymer was designed to achieve a certain level of OH group concentration and a certain glass transition temperature of the copolymer. Whereas the OH group concentration was controlled by the level of HEMA in the copolymer, the glass transition temperature was calculated based on Flory-Fox and Fox equations [46, 47] as shown below. Flory-Fox equation:

$$T_{\mathrm{g}} = T_{\mathrm{g},\infty} + \frac{K}{M_{\mathrm{n}}} \tag{11.11}$$

where $T_{\mathrm{g},\infty}$ is calculated by Fox equation:

$$\frac{1}{T_{\mathrm{g},\infty}} = \frac{w_1}{T_{\mathrm{g},1}} + \frac{w_2}{T_{\mathrm{g},2}} + \frac{w_3}{T_{\mathrm{g},3}} \tag{11.12}$$

Here, T_{g} is the designed glass transition temperature of the copolymer; $T_{\mathrm{g},1}$, $T_{\mathrm{g},2}$, and $T_{\mathrm{g},3}$ are the glass transition temperatures of homopolymers; w_1, w_2, and w_3 are mass fractions of monomers in the copolymer; $T_{\mathrm{g},\infty}$ is the glass transition temperature of polymer with high molecular weight; and K is a constant. $T_{\mathrm{g},1}$, $T_{\mathrm{g},2}$, $T_{\mathrm{g},3}$, and K are predetermined based on earlier experimental data for homopolymers of high molecular weight. With a given composition of the monomers used for the synthesis and the molecular weight of the copolymer, the glass transition temperature can be determined.

Molecular Weight

The molecular weights of the prepared acrylic polyols were characterized by size exclusion chromatography (SEC). The chromatography setup consisted of a Waters® Alliance 2695 separation module, two PLgel Mixed-C columns, and a Waters® refractive index (RI) detector 2414 from Wyatt Technology Corporation. Data were collected and processed with ASTRA software (Wyatt Technology Corporation). Tetrahydrofuran (THF) was used as a mobile phase at a flow rate of 1 mL/min. Samples were injected as solutions in THF in the amount of 100 μL and concentrations of 0.8% w/v. Polystyrene calibration was used for molecular weight determination.

The copolymer compositions, calculated T_{g}, weight average molecular weights, and polydispersity are given in Table 11.3.

Table 11.3 Composition, calculated T_g, M_w, and polydispersity M_w/M_n of acrylic polyols synthesized by GTP

Type of copolymer	Designation	Composition [%]	$T_{g\ calculated}$ [°C]	M_w	M_w/M_n
MMA/HEMA/EHMA	C1	60.5/19.2/20.3	60	24,200	1.34
IBoMA/HEMA/EHMA	C2	48.6/18.8/32.6	60	23,000	1.37
IBoMA/HEMA/EHMA ($R_{IBoMA/EHMA}$= 0.85, by weight)	C3	33.1/28.1/38.8	21	8500	1.3
	C4	33/28/39	33	17,600	1.6
	C5	33.2/28/38.8	42	28,600	1.6
IBoMA/HEMA/EHMA ($R_{IBoMA/EHMA}$ variable)	C6	51.2/15/33.8	40	8200	1.32
	C7	42.9/15.1/42	39	18,400	1.70
	C8	38.4/15.2/46.4	37	26,700	1.90
	C9	44.6/28.3/27.1	42	7800	1.21
	C10	36.6/28.2/35.2	39	16,700	1.58
	C11	33.2/28/38.8	41	28,500	1.48

The composition of the monomers for each copolymer is based on weight

Coating Solutions

Coating solutions were formulated by mixing the acrylic polyols with a polyisocyanate, HDI trimer, Desmodur® 3300 (Covestro). The solids of the polymeric solutions in butyl acetate was 50% by weight. The molar ratio of NCO to OH groups was 1.0. The reaction was catalyzed by dibutyltin diacetate (0.06 wt% on solids). Coatings were dried or cured at room temperature except noted otherwise. A branched, low-T_g polyester (LTP) was used to replace an acrylic polyol IBoMA/HEMA/EHMA at 5%, 10%, 35%, and 100% by weight to determine how the stress of the polyurethane coating could be affected. Furthermore, a 100% solids system was compared with a solvent-borne system in terms of stress development. The 100% solids system was a polyurea coating with an amine binder reacting with HDI trimer. Addition of 25% of n-butyl acetate was applied to study the effect of solvent. Lastly, the effect of baking on the stress development of a reactive polyurea/polyurethane coating cured with a binder and a polyisocyanate at 60 °C and then cooled to 23 °C was investigated.

NCO Conversion

The conversion of isocyanate groups during curing was determined by infrared spectroscopy (FT-IR Nicolet Impact 400D) on KBr sheet. Ultrathin films for conversion study were prepared on KBr crystal surface by a lacquer applicator. The characteristic absorption peak for the NCO groups at 2280 cm^{-1} and an internal standard absorption peaks of C-H bonds at 3000–2800 cm^{-1} were used.

Glass Transition Temperature

Glass transition temperatures of coating films at different stages of drying and curing were determined by means of thermomechanical analyzer TMA 7 (Perkin Elmer). The penetration method with 500 mN load was used; the heating rate was 10 °C/min.

Pendulum Hardness

Pendulum hardness was determined by Persoz pendulum hardness tester according to ISO 1522. The coating films were prepared on glass sheets and cured at 23 °C and a relative humidity of 60%.

Coating Stress

Two stress meters, both based on the cantilever deflection method, were used to characterize coating stress development during drying and/or curing of coatings. In one stress meter, the deflections of coated steel strips were measured by a microscope. The same microscope was used for the measurement of deflections of eight specimens. In the other stress meter, the deflections were measured by capacitive sensors. The stress meter with capacitive sensors contains eight individual sensors (one sensor for one specimen). Coating stress is proportional to the deflection of coated substrate and determined based on Eq. (11.8). The deflection of a non-coated substrate was subtracted from the actual deflection when a coating is applied.

The steel substrate had Young's modulus of 202 GPa and Poisson's ratio of 0.28. The thickness ratio of the steel strip and polymeric coatings t/c was maintained at $t/c \geq 5$ to guarantee that Eq. (11.8) is valid (error $< 1.5\%$ at $E_C = 3$ GPa and $\nu_C = 0.33$). The coatings had thickness about 15–20 μm in the optical sensor method, where the substrate thickness was 100 μm. In the capacitive sensor method, due to the fact that it had higher sensitivity than the optical method, the coating thickness was increased to about 40 μm to mimic real application thickness, and the substrate thickness was changed to 200 μm. Before the capacitive sensor method was initially used, it was verified by using the same coating thickness and substrate thickness with the microscope method. Similar coating stress profiles were achieved with both methods. Stress measurements were conducted at a constant temperature of 23 °C and relative humidity (RH) of 60% unless stated otherwise. The humidity was kept constant using a saturated salt solution. Stress measurements were monitored during the drying and curing process of coatings until stress started to plateau off. The microscope method was used in the first part for polyurethane coatings when different acrylic polyols were used and when a low-T_g polyester was added to a polyurethane coating. In the last part, the capacitive sensor method was used to examine the effect of solvent addition and the effect of baking on coating stress. In both approaches, usually four measurements of stress development were done at the same time. The measurements were repeated when the difference between the measurements was greater than 30%.

Results and Discussion

In the following, first, the effects of acrylic polyol composition and structure on the stress development in polyurethane coatings are discussed. Acrylic polyols made with isobornyl methacrylate, ethylhexyl methacrylate, and hydroxyethyl methacrylate (IBoMA/EHMA/HEMA) or with methyl methacrylate, ethylhexyl methacrylate, and hydroxyethyl methacrylate (MMA/EHMA/HEMA) are compared. The effects of molecular weights and crosslink density (varied by OH group functionality) are examined. Then, the addition of a low-T_g polyester polyol to a polyurethane coating was further studied. Stress development was monitored with the microscope method by determining the cantilever deflection.

Lastly, the effect of other formulation and process parameters on stress development measured by the capacitive sensor method is discussed. The effect of addition of solvent to a polyurea coating that has 100% solids on stress development is studied. Then, the effect of baking on the stress development of a polyurea/polyurethane coating is examined.

Effect of Monomer Type in Acrylic Polyols

The structure of monomers used to synthesize the acrylic polyol copolymers can affect the stress development in polyurethane coatings when the polyols react with the HDI trimer. In this study, two acrylate copolymers with OH-functional groups are compared, one is MMA/HEMA/EHMA (C1 in Table 11.3) and the other is IBoMA/HEMA/EHMA (C2 in Table 11.3). The copolymers were prepared by GTP method, and they had very similar concentrations of HEMA, weight-average molecular weights ($M_w \approx 23{,}000$ g/mol), and calculated glass transition temperatures ($T_g \approx 60$ °C). The copolymer C1 has the less rigid monomer units, MMA, whereas the copolymer C2 has the more rigid monomer units, IBoMA. The time dependences of internal stresses are compared in Fig. 11.4. A very long time (>1000 h) of drying and curing process was followed to ensure that the plateaus of stress were achieved at the end. The measurements for both systems were repeated with excellent reproducibility (Fig. 11.4). It is obvious that the limiting values of internal stress for both copolymers C1 and C2 are practically the same. However, the system with IBoMA developed stress in a very short period of time,

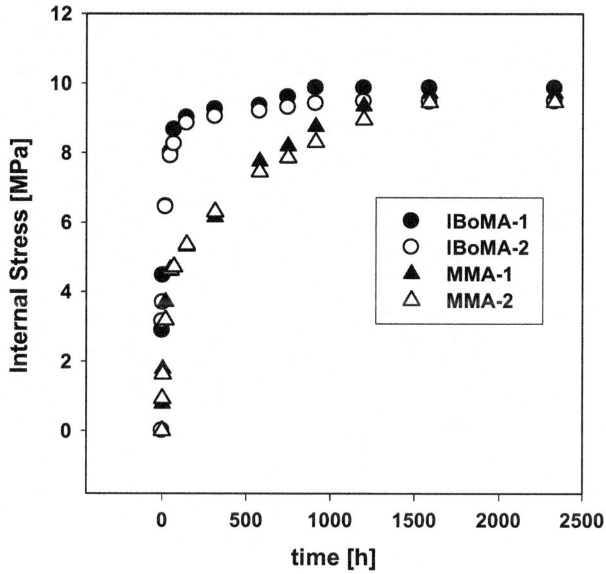

Fig. 11.4 Time dependence of internal stress during the drying and curing of polyurethane coatings made with the two acrylic polyols C1 and C2 at 23 °C and 60% RH. C1 has MMA monomer units and C2 has IBoMA. Two measurements were done for each coating

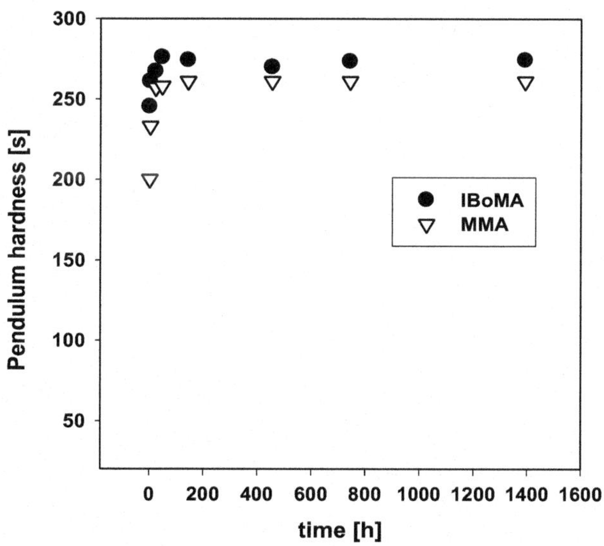

Fig. 11.5 Time dependence of pendulum hardness during the drying and curing of polyurethane coatings made with the two acrylic polyols C1 and C2 at 23 °C and 60% RH. C1 has MMA monomer units, and C2 has IBoMA monomer units

e.g., within 200 h. In contrast, the coating system based on MMA developed the same level of stress after a much longer time, >1000 h.

Figures 11.5 and 11.6 show the pendulum hardness development and NCO group conversion of both systems. The IBoMA-containing coating has a slightly higher pendulum hardness and lower conversion of NCO groups at the plateau regions.

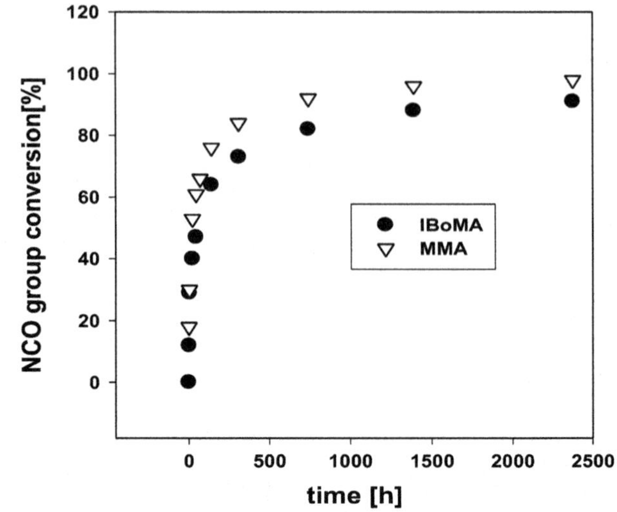

Fig. 11.6 Time dependence of NCO group conversion during the drying and curing of polyurethane coatings made with the two acrylic polyols C1 and C2 at 23 °C and 60% RH. C1 has MMA monomer units, and C2 has IBoMA monomer units

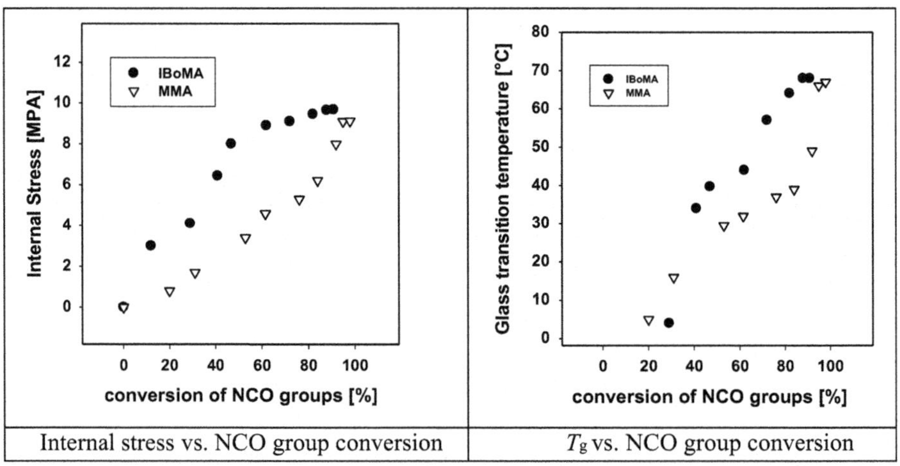

Fig. 11.7 Relationship between internal stress, T_g, and NCO group conversion during the drying and curing of polyurethane coatings made with the acrylic polyols C1 and C2 at 23 °C and 60% RH. C1 has MMA monomer units, and C2 has IBoMA monomer units

Significant differences were observed in the case of the dependence of internal stress on NCO group conversion (Fig. 11.7). At the same conversion of NCO groups, internal stress is much higher with the IBoMA-containing acrylic polyol. The development of the glass transition temperature as a function of the conversion indicates that the IBoMA-containing acrylic polyol leads to higher T_g values at the same NCO group conversion in the majority of the film formation process. This indicates that the higher internal stress in the IBoMA system is associated with the higher glass transition temperature of the coating developed.

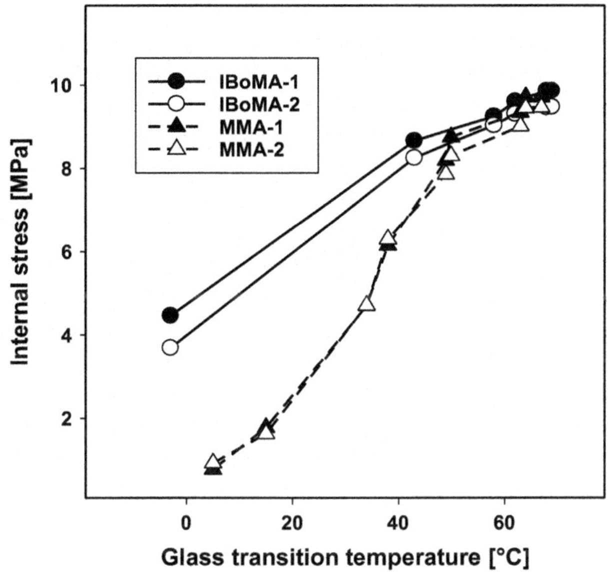

Fig. 11.8 Dependence of internal stress on glass transition temperature during the drying and curing of polyurethane coatings made with two acrylic polyols C1 and C2 at 23 °C and 60% RH. C1 has MMA monomer units, and C2 has IBoMA monomer units. Two measurements were done for each coating

To further determine the correlation between stress and glass transition temperature, these quantities are plotted in the same figure (Fig. 11.8). For both systems, stress grows along with the glass transition temperature of the coating system. The two sets of data did not overlap at low glass transition temperature regions but towards higher glass transition regions; the two sets of data overlapped.

Data shown in Fig. 11.8 indicate that the internal stress starts developing even before the two coating systems enter into their glassy states, the points where their glass transition temperatures become equal to the curing temperature 23 °C. In general, the internal stress of a coating starts developing at the solidification point, after which the coating can no longer flow and further shrinkage produces internal stress in the plane of the coating. During drying of solvent-cast thermoplastic coatings [14] and ultraviolet curing of multifunctional acrylate coatings [15], internal stress is considered to develop when vitrification occurs. The earlier onset of solidification points in these two coatings indicates that there is sufficient crosslink density of the networks formed before vitrification that would slow down stress relaxation. In the rubbery region and around the vitrification point, where T_g is between ~0 °C and 45 °C, the IBoMA-containing acrylic polyol leads to higher stress than the MMA-containing acrylic polyol at an identical glass transition temperature. It would be interesting to investigate further why the IBoMA-containing acrylic polyol leads to higher level of stress, e.g., due to higher stress generated at the first place or due to less stress relaxation around the glass transition region.

Towards the higher T_g region, stress grows with T_g similarly in both systems, indicating the importance of glass transition temperature in affecting stress development. In this region, stress relaxation becomes difficult. The stress developed depends on the volume shrinkage since the solidification point and on the mechanical stiffness (modulus of elasticity) of the film. The overlapped curves indicate that well deep in the glassy state, there are similar volume shrinkage and similar modulus of the coatings. In general, T_g depends on the conversion of functional groups and evaporation of solvent. In the glassy state, further reaction and evaporation of residual solvent often happen simultaneously at much lower rates. As the coating cures and dries further, T_g increases, and more shrinkage occurs since the solidification point. In the meanwhile, the coating develops higher modulus. Therefore, higher T_g is associated with higher internal stress according to Eq. (11.2).

The changes in T_g give us a guidance for estimating the increase of internal stress. However, the increase of T_g with increasing conversion and decreasing concentration of volatiles is not the same for various copolymers and depends on the structure of the chain (interactions and bulkiness of the side groups). IBoMA is known as "T_g builder"; the isobornyl groups have the ability to align one to another, whereas spatially close MMA units are more disordered. This is why the increase in T_g does not follow the same course as a function of conversion of NCO groups for urethane coatings made with acrylic polyols consisting of copolymers with different monomer structure. The higher T_g developed in the IBoMA system also leads to higher pendulum hardness and lower conversion of NCO groups. Onset of vitrification limits the mobility of NCO and OH functional groups to move close enough to react.

Effect of Molecular Weight of Acrylic Polyol

To study the effect of molecular weight of an acrylic polyol on coating stress development, IBoMA/HEMA/EHMA copolymers of varying molecular weights but with a constant weight ratio of IBoMA/EHMA (copolymers C3–C5, Table 11.3) were used to formulate the coating solutions. The concentration of HEMA was held constant at 28 wt%. The time dependence of the internal stress at different molecular weights is shown in Fig. 11.9. Interestingly, all systems display essentially the same time dependence of internal stress, demonstrating that molecular weight does not affect the internal stress development when the monomer compositions are the same.

The coatings also have the same time dependence of T_g (Fig. 11.10). The same conclusion can be made for internal stress vs. NCO group conversion dependences (Fig. 11.11). Overall, molecular weight of the starting acrylic polyols does not affect the stress, T_g, and conversion of NCO groups.

Effect of Crosslink Density

To determine how the crosslink density of the formed polyurethane coatings affects the stress development, two sets of acrylic polyols, IBoMA/HEMA/EHMA

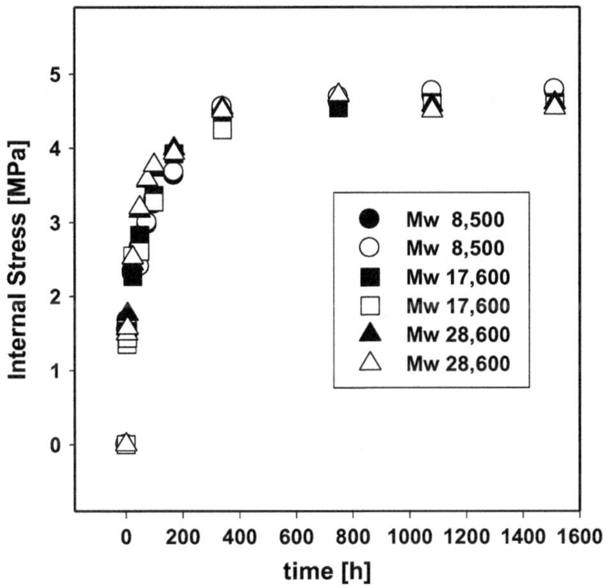

Fig. 11.9 Time dependence of internal stress development of polyurethane coatings using acrylic polyols, IBoMA/HEMA/EHMA copolymers that have different molecular weights, $M_w = 8500$, 17,600, and 28,600 but the same monomer composition (copolymers C3–C5) at 23 °C and 60% RH. For each copolymer, two sets of data are plotted

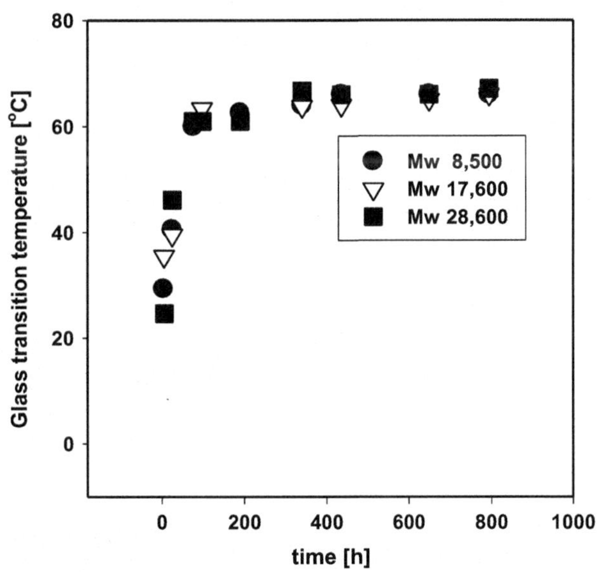

Fig. 11.10 Time dependence of glass transition temperature development of polyurethane coatings using acrylic polyols, IBoMA/HEMA/EHMA copolymers that have different molecular weights, $M_w = 8500$, 17,600, and 28,600 but the same monomer composition (copolymers C3–C5) at 23 °C and 60% RH

copolymers, with different levels of HEMA monomers were synthesized. Higher concentration of HEMA or OH groups would lead to polymer networks with a higher crosslink density. Two concentrations of HEMA, 15 and 28%, were used to make two sets of the copolymers (C6–C11) with similar calculated glass transition temperatures of the copolymers (Table 11.3). In each set of copolymers, the

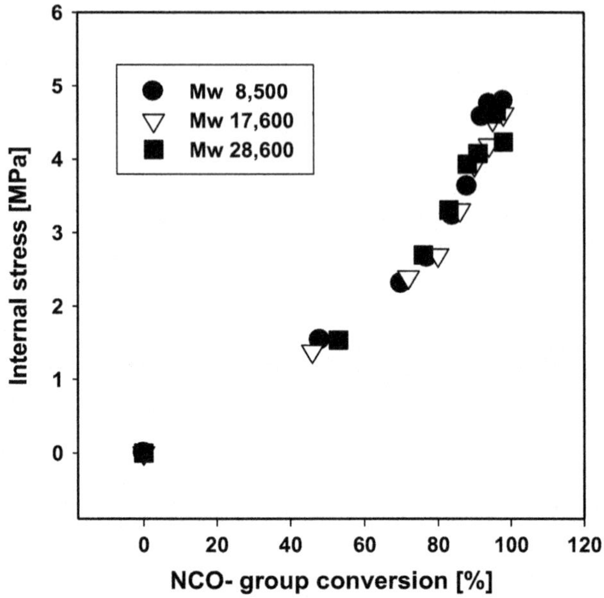

Fig. 11.11 Relationship between internal stress and NCO group conversion of polyurethane coatings using acrylic polyols, IBoMA/HEMA/EHMA copolymers that have different molecular weights, $M_w = 8500$, 17,600, and 28,600 but the same monomer composition (copolymers C3–C5) at 23 °C and 60% RH

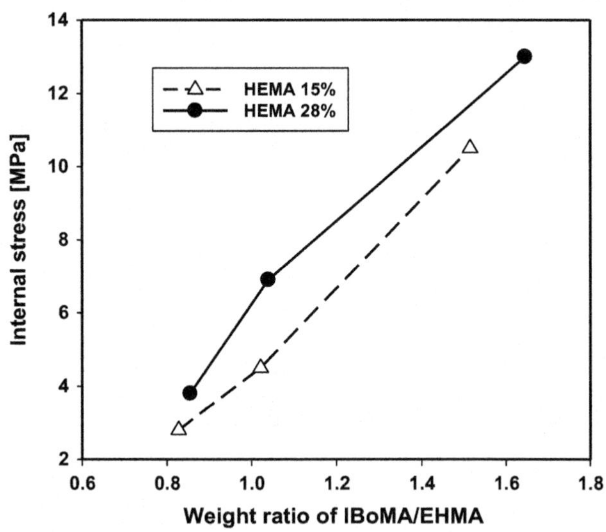

Fig. 11.12 The limiting values of the internal stress of polyurethane coatings vs. the weight ratio of IBoMA to EHMA in the acrylic polyols at two HEMA levels (copolymers C6–C11), at 23 °C and 60% RH

weight ratio of IBoMA to EHMA was varied, and several molecular weights of the copolymers were obtained. Figure 11.12 shows the limiting internal stress (the stress at the plateau region) of the polyurethane coatings as a function of the weight ratio of IBoMA to EHMA at two levels of HEMA. At each HEMA level, the internal stress increases as more rigid monomer IBoMA is used. This is due to the increase of the glass transition temperature of a coating formed with a higher

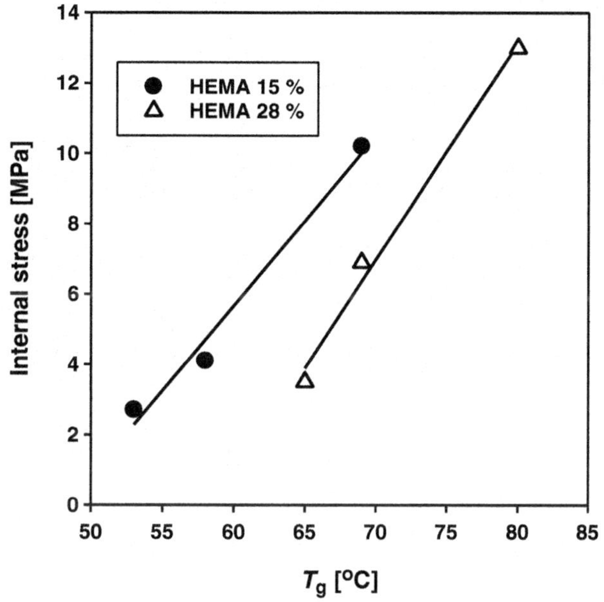

Fig. 11.13 Relationship between the limiting value of internal stress and glass transition temperature of polyurethane coatings using acrylic polyols that have two different HEMA concentrations, 15 and 28%, (copolymers C6–C11), at 23 °C and 60% RH. The lines are curve fitting data through linear regression

IBoMA level. Furthermore, the higher crosslink density of the coating system due to a higher HEMA level increases the internal stress.

The systems with higher crosslink densities also display higher glass transition temperatures. Figure 11.13 shows the limiting value of internal stress developed vs. the glass transition temperature of the polyurethane coatings. Interestingly, the two sets of data do not overlap each other. For each HEMA concentration, there is a linear correlation of internal stress vs. T_g of the coatings; higher T_g of the coating leads to higher stress. At a higher HEMA concentration, the linear curve is shifted to the right. In other words, higher crosslink density has a higher contribution to the glass transition temperatures of the coatings than to the limiting value of internal stress. At present, we are not able to quantify this effect in terms of the concentration of elastically active network chains (EANC); ideally, the content of HEMA 15 and 28 wt% corresponds to 2.2×10^{-3} and 3.1×10^{-3} mol/g, respectively, of EANCs. Higher crosslink density increases the glass transition temperatures of the coatings due to the loss of free volume. In terms of stress development, a higher crosslink density in the glassy state may not translate directly to the volume shrinkage of the coating or the increase of elastic modulus of the coating; therefore, it may have less effect on the stress development vs. T_g change.

From the studies discussed above, we have come to the conclusion that T_g changes are the main factors determining the level of internal stress. The internal stress is also affected by the crosslink density of networks formed.

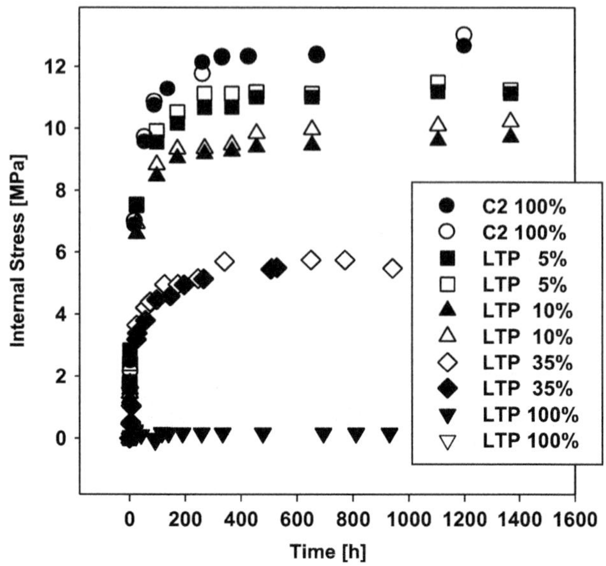

Fig. 11.14 Time dependence of internal stress of polyurethane coatings made with acrylic polyol IBoMA/HEMA/EHMA (C2) modified with low-T_g polyester polyol (LTP) at different weight percentages in the total polyol amount at 23 °C and 60% RH. Two sets of data are plotted for each LTP addition level

Effect of Addition of Low-T_g Polyester

The internal stress can be lowered by the addition of suitable additives, for instance, branched polyester polyols of low T_g (LTP). The decrease of the internal stress with increasing concentration of LTP is shown in Fig. 11.14. In this case, the acrylic polyol used is IBoMA/HEMA/EHMA (C2 in Table 11.3). The polyurethane coating containing this acrylic polyol shows up high stress developed. By adding the low-T_g polyester polyol, even at 5% level, a reduction of stress was observed. Further addition of the low-T_g polyester leads to further reduction of stress in the plateau region. At 100% replacement of the acrylic polyol with the low-T_g polyester polyol, the stress is essentially zero.

However, the addition of low-T_g polyester polyol had a negative effect on the pendulum hardness (Fig. 11.15). The more low-T_g polyester polyol added, the lower the pendulum hardness of the polyurethane coatings formed.

The decrease of internal stress and pendulum hardness with increasing LTP concentration could be attributed to the development of a second, soft phase in the polymer mixtures. The formed polyurethane coatings displayed two glass transition temperatures. The first glass transition temperature was above room temperature, T_{g1}, and the second one below room temperature, T_{g2}. T_{g1} corresponds to the T_g of isocyanate crosslinking with the neat acrylic polyol, and T_{g2} corresponds to the isocyanate crosslinking with the neat LTP (Fig. 11.16). The time dependences of T_{g1} and T_{g2} are shown in Fig. 11.16 for the system with 35% LTP. Due to the

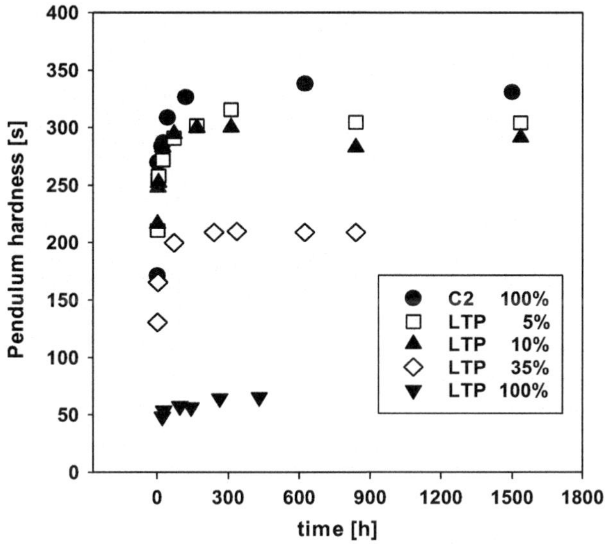

Fig. 11.15 Time dependence of pendulum hardness of polyurethane coatings made with acrylic polyol IBoMA/HEMA/EHMA (C2) modified with LTP at different weight percentages in the total polyol amount at 23 °C and 60% RH

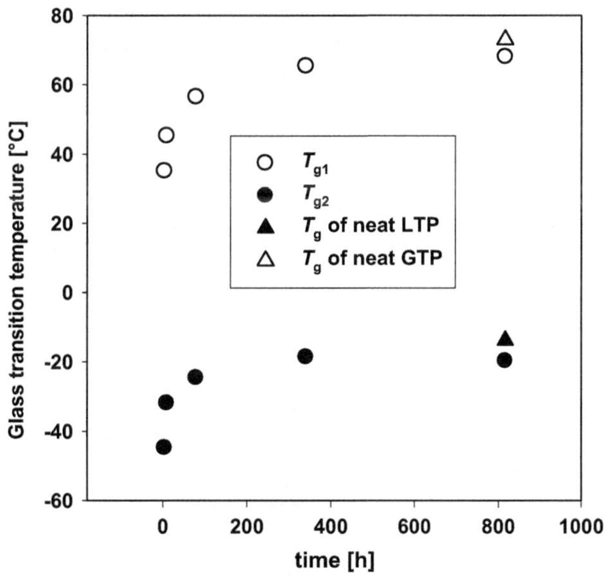

Fig. 11.16 Time dependence of glass transition temperatures of a polyurethane coating made with acrylic polyol IBoMA/HEMA/EHMA (C2) modified with 35% of LTP at 23 °C and 60% RH. The coating had two T_g values, T_{g1} and T_{g2}. The T_g values of polyurethane coatings made with the neat acrylic or LTP polyol are also compared

presence of the extremely low T_{g2} phase present in the coating system, stress can relax more easily, leading to overall lower stress. The soft phase also lowers the pendulum hardness of the coating.

Effect of Solvent Addition

As volume shrinkage caused by solvent evaporation is a significant contributor to the total volume shrinkage of the coating, it is interesting to determine the effect of solvent addition to a solvent-free reactive coating system on stress development. A low-viscosity, solvent-free urea coating was selected for this study, where an amine binder reacts with a HDI trimer during the curing process. The solvent n-butyl acetate was added at 25% by weight to determine its effect on stress development. The comparison of time dependence of internal stress of both systems is shown in Fig. 11.17. The shape of time dependence of internal stress is affected by the solvent addition. In the solvent-containing coating system, tensile stress starts developing very early, e.g., in 3 h, and then it climbs up quickly. In the solvent-free system, interestingly, the system develops some low-level tensile stress first and then some negative (compressive) stress for quite a long time, and finally more tensile stress. The low level of stress developed in the solvent-free system up to 400 h indicates that the volume shrinkage caused by the amine binder and HDI trimer reaction alone is very small. Whereas in the solvent-containing coating system, volume shrinkage caused by the evaporation of solvent leads to significant stress development.

The development of compressive stress observed in the solvent-free system can be caused by the absorption of humidity which is driven by the chemical reaction of NCO groups with water (reaction of one water molecule consumes two isocyanate groups in the formation of urea groups). The solvent-free coating experiences faster moisture diffusion into the coating than the solvent-containing coating. In the solvent-free coating, the gas phase near the surface of the coating has zero

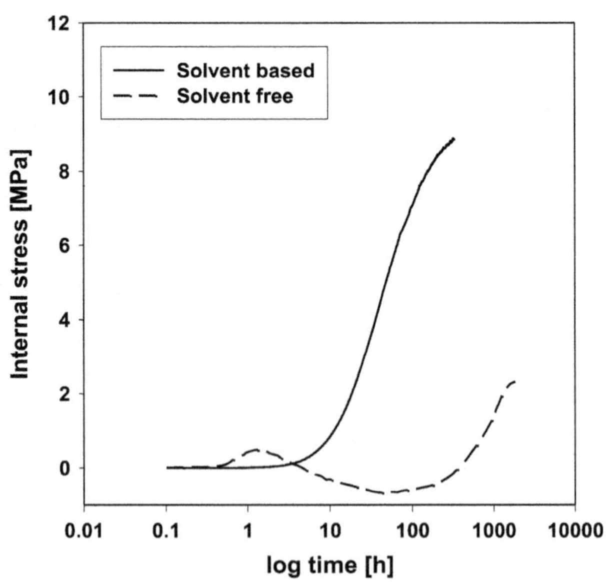

Fig. 11.17 Time dependence of internal stress for solvent-containing (solvent-based) and solvent-free urea coating systems at 23 °C and 60% RH

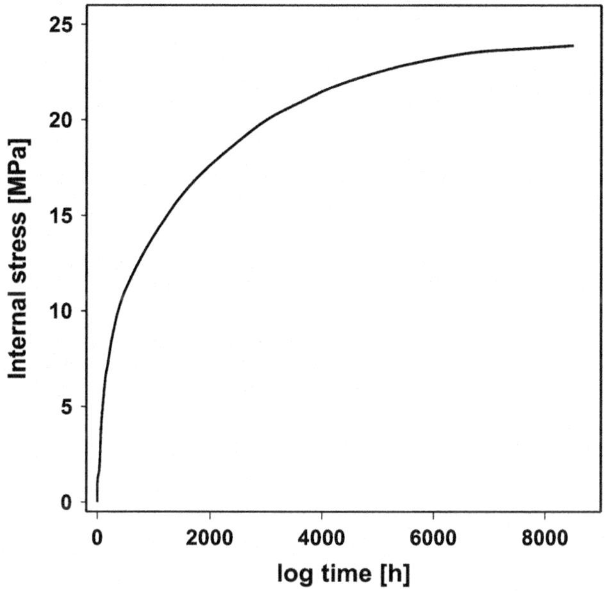

Fig. 11.18 Time dependence of internal stress for the solvent-containing urea coating up to 9500 h at 23 °C and 60% RH

concentration of solvent vapor, increasing the concentration of moisture and thus its partial pressure. This further leads to increased flux of moisture across the coating surface. Therefore, moisture would diffuse into the solvent-free coating more quickly than the solvent-containing coating. Consequently, moisture has more impact in the solvent-free coating than in the solvent-containing coating.

Furthermore, the humidity affects stress not only physically through expansive hygroscopic stress, but also chemically by the formation of new chemical bonds. The hygroscopic stress is negative due to the expansion of volume upon absorption of moisture, whereas the bond formation has a positive effect. In the later stage of the stress development in the solvent-free coating system, the development of tensile stress probably comes from the positive effect of bond formation.

Overall, the solvent-containing system leads to higher stress due to extra shrinkage caused by drying of solvents.

What is interesting in the solvent-based system is that the internal stress climbs up continuously and eventually reaches a steady value. The stress is about 24 MPa after 9500 h (Fig. 11.18).

Effect of Baking During Film Formation

The effect of baking and subsequent cooling process on stress development can be more significant than the internal stress generated by the film forming process at room temperature. A crosslinking polyurea/polyurethane coating is studied here,

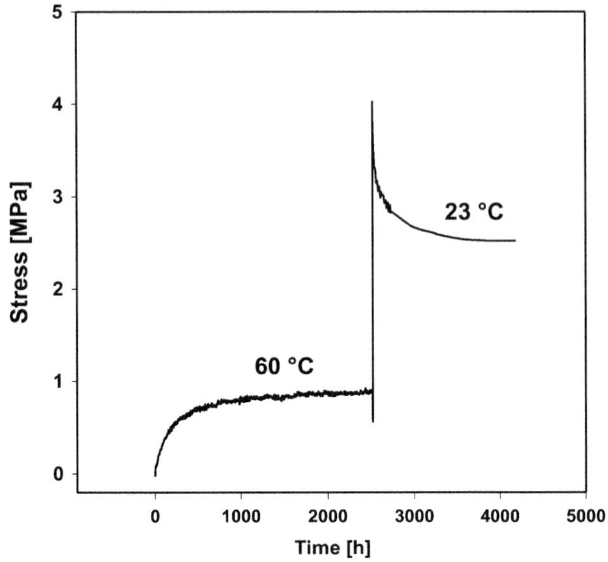

Fig. 11.19 Time dependence of stress of a reactive coating system during the process of baking at 60 °C and cooling to 23 °C

and its stress development during baking (heating) and cooling processes is displayed in Fig. 11.19.

The coating develops some small amount of stress, 0.8 MPa, during the baking process at 60 °C. However, during the cooling process, the film stress suddenly jumps to 4 MPa, which is more than four times higher than the limiting value at 60 °C. This stress gradually relaxes to a lower level, 2.5 MPa, at room temperature.

This sudden increase of stress is caused by the difference in thermal expansion coefficients of the substrate and the coating. This result is very important because all types of coatings cured at a temperature higher than the ambient face stress generated by both the thermal history and the film-forming process. This combined stress can surpass immediate coating strength and may cause undesired phenomena like film cracking, delamination, and other coating defects.

Conclusions

This chapter focuses on the development of internal stress in polyurethane coatings and the influence of acrylic polyol copolymer on the internal stress. The effects of copolymer structure, such as rigid IBoMA vs. less rigid MMA segments in the main chain, copolymer molecular weight, and crosslink density of the formed polyurethane network, on the internal stress are discussed. It is shown that using the more rigid segments in the polyol leads to increased stress mainly in the rubbery region and around the vitrification point of the coating. The limiting stress values of the polyurethane coatings formed by IBoMA-containing polyol and the

MMA-containing polyol are close. Increasing the crosslinking density has a positive effect on the increase of the limiting value of internal stress. On the other hand, molecular weight of the copolymer does not affect the internal stress of the coating when the ratio of the hard to soft segments is fixed and the level of crosslinking is fixed.

Internal stress can be reduced by the addition of a polymer component having a low glass transition temperature. This is demonstrated in the case of adding low-T_g polyester polyol to a polyurethane coating, where even small addition significantly reduces the internal stress of the base coating system. The drop of internal stress is, however, accompanied by a decrease of hardness.

The time dependences of internal stress and the limiting values of internal stress are significantly influenced by the addition of solvent. Comparison of a solvent-free and a solvent-containing coating shows some interesting differences. A negative stress was developed at the beginning of the solvent-free coating system, which likely is due to moisture absorption. Overall, the solvent-free coating system develop lower stress than the solvent-based coating due to less volume shrinkage. Interestingly, the solvent-containing system reached an internal stress about 24 MPa after 9500 h (more than 1 year).

Another important finding is that the temperature changes due to baking, and subsequent cooling can generate higher stress than those developed during drying and curing of a coating at a constant temperature. Coatings cured at higher temperatures than the ambient temperature face combination of stresses generated by the thermal history and film-forming process.

The results of these studies amend the understanding of shrinkage and stress development in organic coatings and provide a guidance to engineering of coatings with superior long-term properties and performance.

Acknowledgments The authors would like to gratefully acknowledge funding support from Axalta Coating Systems of this work.

References

1. Perera, D.Y.: On adhesion and stress in organic coatings. Prog. Org. Coat. **28**, 21–23 (1996)
2. Williams, J.G., Marshall, G.P.: Environmental crack and craze growth phenomena in polymers. Proc. R. Soc. Lond. A. **342**(1628), (1975)
3. Sousa, A.R., Amorim, K.L.E., Medeiros, E.S., Melo, T.J.A., Rabello, M.S.: The combined effect of photodegradation and stress cracking in polystyrene. Polym. Degr. Stab. **91**(7), 1504–1512 (2006)
4. Hansen, C.M.: On predicting environmental stress cracking in polymers. Polym. Degr. Stab. **77**(1), 43–53 (2002)
5. Basu, S.K., Scriven, L.E., Francis, L.F., McCormick, A.V.: Mechanism of wrinkle formation in curing coatings. Prog. Org. Coat. **53**, 1–16 (2005)
6. Oosterbroek, M., Lammers, R.J., van der Ven, L.G.J., Perera, D.Y.: Crack formation and stress development in an organic coating. J. Coat. Technol. **63**(797), 55 (1991)

7. Knudsen, O.Ø., Bjørgum, A., Frydenberg, T., Johnsen, R.: Development of internal stress in organic coatings during curing and exposure. NACE International Corrosion Conference & Expo, #06028 (2006)
8. Croll, S.G.: Adhesion loss due to internal strain. J. Coat. Technol. **52**(665), 35–43 (1980)
9. Francis, L.F., McCormick, A.V., Vaessen, D.M., Payne, J.A.: Development and measurement of stress in polymer coatings. J. Mater. Sci. **37**, 4717–4731 (2002)
10. Lei, H., Payne, J.A., McCormick, A.V., Francis, L.F., Gerberich, W.W., Scriven, L.E.: Stress development in drying coatings. J. Appl. Polym. Sci. **81**, 1000–1013 (2001)
11. Perera, D.Y., Eynde, D.V.: Moisture and temperature induced stresses (hygrothermal stresses) in organic coatings. J. Coat. Technol. **59**(748), 55–63 (1987)
12. He, Y., Yao, M., Nie, J.: Shrinkage in UV curable coatings. In: Wen, M., Dusek, K. (eds.) Protective coatings: film formation and properties. Springer, New York (2017)
13. Lin, J., Shen, J., Zimmer, M.E.: Appearance of automotive coatings. In: Wen, M., Dusek, K. (eds.) Protective coatings: film formation and properties. Springer, New York (2017)
14. Croll, S.G.: The origin of residual internal stress in solvent-cast thermoplastic coatings. J. Appl. Polym. Sci. **23**, 848–858 (1979)
15. Wen, M., Scriven, L.E., McCormick, A.V.: Differential scanning calorimetry and cantilever deflection studies of polymerization kinetics and stress in ultraviolet curing of multifunctional (meth)acrylate coatings. Macromolecules. **35**, 112–120 (2002)
16. Abdelkader, A.F., White, J.R.: Comparison of internal stresses in coatings cured on rigid substrates and on unrestrained thin substrates. Prog. Org. Coat. **44**, 121–129 (2002)
17. Dušková-Smrčková, M., Dušek, K.: Processes and states during polymer film formation by simultaneous crosslinking and solvent evaporation. J. Mater. Sci. **37**, 4733–4741 (2002)
18. Hare, C.H.: Trouble with paint: internal stress: part 1. J. Phys. Chem. Lett. (Jan.) 65–75 (1996)
19. Yan, G., White, J.R.: Residual stresses in marine coatings under simulated service conditions. In: Moalli, J. (ed.) Plastics Failure Analysis and Prevention. Technology & Engineering Series, pp. 79–87. William Andrew, Norwich (2001)
20. Croll, S.G.: Residual stress in a solventless amine-cured epoxy coating. J. Coat. Technol. **51**, 49 (1979)
21. Lange, J., Toll, S., Månson, J.-A.E., Hult, A.: Residual stress build-up in thermoset films cured above their ultimate glass transition temperature. Polymer. **36**(16), 3135–3141 (1995)
22. Vaessen, D.M., Ngantung, F.A., Palacio, M.L.B., Francis, L.F., McCormick, A.V.: Effect of lamp cycling on conversion and stress development in ultraviolet-cured acrylate coatings. J. Appl. Polym. Sci. **84**, 2784–2793 (2002)
23. Stansbury, J.W., Trujillo-Lemona, M., Lu, H., Ding, X., Lin, Y., Ge, J.: Conversion-dependent shrinkage stress and strain in dental resins and composites. Dental Mater. **21**, 56–67 (2005)
24. Islam, M., Repp, J., Ellor, J.A., Shaw, B.: Quantification of curing stresses in paints and coatings. In: The Society for Protective Coatings (SSPC) 2003 Conference Proceedings, pp. 1–15 (2003)
25. Stoney, G.G.: The tension of metallic films deposited by electrolysis. Proc. R. Soc. Lond. A. **82**, 172–175 (1909)
26. Sanzharovskii, A.T., Epifanov, G.I.: Internal stresses in coatings. II. Experimental methods of investigating the internal stresses of polymer, paint, and lacquer coatings. Vysokomolekul. Soedin. **2**, 1703–1708 (1960)
27. Gusman, S.: Studies of the adhesion of organic coatings. Paint Technol. 17–26 (1963)
28. Simpson, W., Boyle, D.A.J.: The observation of stress development in supported [paint] films. J. Oil Colour Chem. Assoc. **46**, 331–338 (1963)
29. Dannenberg, H.: Determination of stresses in cured epoxy resins. Soc. Plastic Eng. J. **21**(77), 669–675 (1965)
30. Corcoran, E.M.: Determining stresses in organic coatings using plate beam deflection. J. Paint Technol. **41**, 635–640 (1969)
31. Nilsson, E.: Internal stresses in paint films. Skandinavisk Tidskrift foer Faerg och Lack **21**(10), 318–20, 323–6, 329, 332 (1975)

32. Saarnak, A., Nilsson, E., Kornum, L.O.: Usefulness of the measurement of internal stresses in paint films. J. Oil Colour Chem. Assoc. **59**, 427–432 (1976)
33. Croll, S.G.: Internal stress in a solvent-cast thermoplastic coating. J. Coat. Technol. **50**, 33–38 (1978)
34. Payne, J.A.: Ph.D. thesis, University of Minnesota (1998)
35. Lu, H., Stansbury, J.W., Dickens, S.H., Eichmiller, F.C., Bowman, C.N.: Probing the origins and control of shrinkage stress in dental resin-composites: I. Shrinkage stress characterization technique. J. Mater. Sci., Mater. Med. **15**(10), 1097–1103 (2004)
36. Lampenscherf, S., Pompe, W., Wilkinson, D.: Stress development due to capillary condensation in powder compacts: a two-dimensional model study. J. Am. Ceram. Soc. **83**(6), 1333–1340 (2000)
37. Flinn, P.A., Gardner, D.S., Nix, W.D.: Measurement and interpretation of stress in aluminum-based metallization as a function of thermal history. IEEE Trans. Electron Devices. **34**, 689–699 (1987)
38. Goldsmith, C.: Measurement of stresses generated in cured polyimide films. J. Vac. Sci. Tech. A. **1**, 407–409 (1983)
39. Nguyen, L.T.: In: Tong, H.-M., Nguyen, L.T. (eds.) New Characterization Techniques for Thin Polymer Films, p. 57. Wiley, New York (1990)
40. Chiu, R.C., Cima, M.J.: Drying of granular ceramic films: II, drying stress and saturation uniformity. J. Am. Ceram. Soc. **76**, 2769 (1993)
41. Chow, T.S., Lin, C.A., Penwell, R.C.: Strains induced upon drying thin films solvent-cast on flexible substrates. J. Polym. Sci., Polym. Phys. Ed. 14, 1311–1316 (1976)
42. Ebbeni, J., Kestens, J.: Experimental analysis of stress in paints. In: Proc. Exper. Stress Anal. Cong., Munich 483 (1978)
43. Kornum, L.-O.: Scandinavian Paint and Printing Ink Research Institute, Report T9-75 M (1975)
44. Zhang, D., Luo, M., Arola, D.D.: Displacement/strain measurements using an optical microscope and digital image correlation. Opt. Eng. **45**(3), 033605 (2006)
45. Tipler, P.A.: Physics Second Edition. Worth Publishers. 653–660 (1982)
46. Fox, T.G., Flory, P.J.: Second-order transition temperatures and related properties of polystyrene. J. Appl. Phys. **21**, 581–591 (1950)
47. Hiemenz, P.C., Lodge, T.P.: Polymer Chemistry. CRC Press, Boca Raton, FL (2007)

Part III
Coating Film Properties and Applications

Chapter 12
Swelling of Coating Films

Karel Dušek, Miroslava Dušková-Smrčková, and C. Brent Douglas

Introduction

Swelling is a process of uptake of a liquid by a polymer film when the film comes into contact with the liquid. Swelling can continue until the film dissolves completely (unlimited swelling), or swelling might stop after a certain amount of liquid has been absorbed (limited swelling). A swelling limit is always reached for a cross-linked polymeric material, but it can also be reached for an uncross-linked polymer when the liquid is a sufficiently poor solvent for it. In this contribution, we will focus on limited, equilibrium swelling of cross-linked polymer films. The equilibrium degree of swelling is a very important quantity which is used to characterize the structure of a film, possibly from its initial formation right after the gel point through its aging up to the final service state. There is one important condition: the swollen film must be in the rubbery state. If it remains in the glassy state after long contact with the liquid, the film will also absorb some solvent, but the swelling degree cannot be used for calculation of various structural parameters.

K. Dušek (✉) • M. Dušková-Smrčková
Institute of Macromolecular Chemistry, Academy of Sciences of the Czech Republic, Prague, Czech Republic
e-mail: dusek@imc.cas.cz; mkdusek@yahoo.com

C. Brent Douglas
Axalta Coating Systems, Coatings Technology Center, Wilmington, DE, USA

M. Wen, K. Dušek (eds.), *Protective Coatings*, DOI 10.1007/978-3-319-51627-1_12

Degree of Swelling and Its Determination

The degree of swelling is determined by measuring changes in weight, dimensions, or volume of the film and, possibly, by determining the solvent concentration in the film. The change in weight (weight degree of swelling, W) is given by the ratio of weights of swollen m_{sw} and dry m_{dry} film:

$$W = \frac{m_{\mathrm{sw}}}{m_{\mathrm{dry}}} \tag{12.1}$$

or expressed as solvent regain, R_{w}

$$R_{\mathrm{W}} = \frac{m_{\mathrm{sw}} - m_{\mathrm{dry}}}{m_{\mathrm{dry}}} = W - 1 \tag{12.2}$$

Alternatively, the volume degree of swelling, B, and volume solvent regain, R_{B}, are determined from changes in volume:

$$B = \frac{V_{\mathrm{sw}}}{V_{\mathrm{dry}}} = \frac{L_x L_y L_z}{L_{x\mathrm{d}} L_{y\mathrm{d}} L_{z\mathrm{d}}}; \quad R_{\mathrm{B}} = \frac{V_{\mathrm{sw}} - V_{\mathrm{dry}}}{V_{\mathrm{dry}}} = B - 1 \tag{12.3}$$

where L_i are dimensions of the swollen film and $L_{i\mathrm{d}}$ are dimensions of the dry film in directions x, y, and z.

The weight and volume degrees of swelling or solvent regains can be converted to one another, if the swollen film *is not (macro) porous,* as shown by Eqs. (12.4):

$$\begin{aligned} B &= \frac{m_{\mathrm{dry}}/\rho_{\mathrm{dry}} + m_{\mathrm{solv}}/\rho_{\mathrm{solv}}}{m_{\mathrm{dry}}/\rho_{\mathrm{dry}}} = 1 + (W-1)\frac{\rho_{\mathrm{dry}}}{\rho_{\mathrm{solv}}} \\ R_{\mathrm{B}} &= B - 1 = \frac{m_{\mathrm{solv}}/\rho_{\mathrm{solv}}}{m_{\mathrm{dry}}/\rho_{\mathrm{dry}}} = (W-1)\frac{\rho_{\mathrm{dry}}}{\rho_{\mathrm{solv}}} \end{aligned} \tag{12.4}$$

where m_{solv} denotes the solvent weight and ρ's denote specific gravities.

Equation (12.3) and interrelations (12.4) are based on the assumption of no volume change on mixing, i.e., additivity of volumes of dry polymer and solvent. This is not fulfilled exactly because there can be a change of volume from mixing in the cases of strong polymer segment–solvent interaction or passage from glassy to rubbery states (devitrification volume). Rigorously, partial quantities should be used. If the volume shrinkage is related to volume of polymer component, the shrinkage will typically be 2–5%. However, it can be more if more free volume was frozen into the glassy state. In practice, the additivity of volumes is assumed (no volume change on mixing). The small error generated by this assumption causes a small change in the interaction parameter χ (see below). Another way of expressing the composition of a swollen film is the volume fraction of polymer, ϕ_2, which is often used in theories of polymer solutions and swollen gels

$$\phi_2 = 1/B \tag{12.5}$$

Experimental determination of W is carried out by simple weighing of swollen and dry samples. The weight of the swollen sample refers to the extracted state, i.e., to the sample equilibrated with excess solvent when solvent has been replaced several times to remove any soluble material that might be present. In order to obtain the swollen weight, the traces of liquid solvent remaining on the surface of the object are removed, usually by gentle blotting with filtration paper or by gentle centrifuging, if the objects are too small. As the weight of the dry sample, one takes the weight of extracted sample after the solvent has been removed by drying to constant weight. The volume degree of swelling is determined by directly measuring the respective volumes or more often by measuring changes of dimensions using, for instance, a traveling microscope. If the sample swells anisometrically, changes of all dimensions (x, y, z) should be measured.

Evolution of Network Structure

In a cross-linking system before the gel point, the molecular weights and polydispersity of the branched polymer are increasing, and at the gel point, the "infinite" structure—the gel—appears in the system for the first time. Beyond the gel point, as the cross-linking reaction advances, the fraction of the gel increases, and the fraction of finite, soluble molecules—the sol—decreases. The connectivity of the gel fraction increases and more elastically active chains are formed. This increase is accompanied by a decrease of the fraction of elastically inactive part of the material (i.e., a decrease in the amount of dangling chains and fraction of sol). Some cross-links are engaged in closing small cycles. They do not contribute to the number of elastically active network chains (EANCs). The concentration of EANCs can approach the ideal value for systems in which all cross-links contribute to EANCs with all of their branches. This limiting value can only be approached, but not reached. For some systems, like the off-stoichiometric alternating $A + B \rightarrow AB$ systems, or moderately or highly diluted systems, the final state of the network can be far away from this ideal situation.

When we talk in practice about cross-link density, we always consider the density (concentration) of *elastically active network chains* (EANCs). We stress here that an elastically active network chain is a covalently connected strand of units between two *elastically active cross-links* (junctions) (EAC). An elastically active cross-link is a junction unit from which at least three bonds issue that have infinite continuation. An active junction issuing i bonds with infinite continuation contributes $(1/2)i$ to the number of EANCs, N_e. Near the gel point, only three bonds with infinite continuation issue from the elastically active cross-links, regardless of their chemical functionality f, and the contribution by each EAC to N_e is equal to 3/2. As the network connectivity increases (conversion increases), the contribution

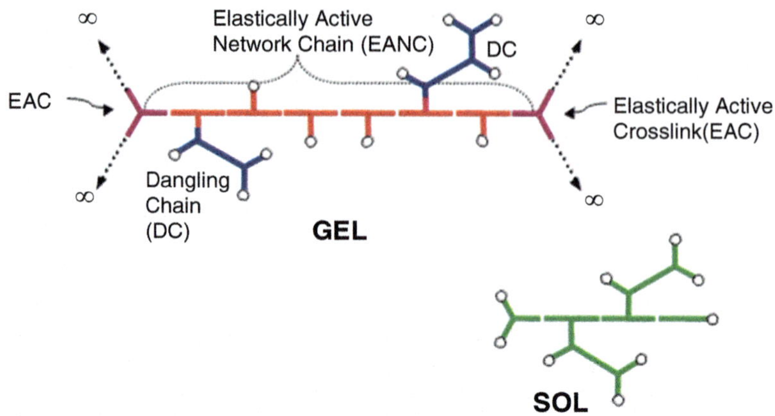

Fig. 12.1 Classification of network elements according to the number of bonds with infinite continuation. For explanation, see text

by each EAC gradually approaches $f/2$ ($f > 3$), provided the other conditions described above are met. Generally,

$$N_{\mathrm{e}} = (1/2) \sum_{i=3}^{f} \sum_{j=3}^{i} N_{i,j} j \tag{12.6}$$

$N_{i,j}$ is the number of cross-links with i bonds (i reacted functional groups) of which j have infinite continuation. The elements of a system undergoing reaction of an A3 monomer (A + A → AA) are shown in Fig. 12.1. The state of units differs in the *number of bonds with infinite continuation*: sol—0, dangling chain—1, backbone of elastically active network chain—2, elastically active cross-link—3.

How can one find the probability that a certain bond has infinite continuation? This is a matter of branching theories—for more details, see, e.g., Refs. [1–5]. The theory also offers a quantity called average functionality of elastically active cross-link, f_{e}, which varies from 3 to f as the system moves from the gel point towards perfect network. It is defined by

$$f_{\mathrm{e}} = \frac{\sum_{i=3}^{f} \sum_{j=3}^{i} N_{i,j} j}{\sum_{i=3}^{f} \sum_{j=3}^{i} N_{i,j}} \tag{12.7}$$

How is the equilibrium swelling degree affected by the network evolution when the gel point is surpassed? First of all, when film is immersed in the solvent, the sol is extracted. The sol can be characterized separately, and its composition in terms of molecular weight and composition distributions can be simulated by branching theories. To get meaningful information on the gel by swelling, the sol must be completely removed. If part of the sol remains in the solvent phase, it decreases solvent activity, and the gel swells less. The swelling degree of the extracted gel decreases as the gel structure develops, but the change is not only due to the increase of the concentration of EANCs. Contributions to this decrease can also come from changes in the polymer–solvent interactions caused by chemical transformations (transformation of functional groups into bonds) and incorporation of sol molecules (of chemical composition different from that of the gel) into the gel.

Theoretical Models of Equilibrium Swelling of Cross-Linked Polymers

For structural evaluation of the degree of swelling data, we need a relation between the degree of swelling and structural parameters of the cross-linked polymer. Swelling occurs because polymer segments have a tendency to mix with solvent molecules. Mixing is associated with a large gain in entropy. In the case of an uncross-linked polymer, mixing of polymer with solvent is affected by the ultimate balance of the binary interactions: segment–segment, solvent–solvent, and segment–solvent. At concentrations at which polymer coils overlap, a temporary, dynamic network having the properties of a gel can be formed; the network chains in such a gel are relaxed. If the interactions are changed to favor segment–solvent interaction, the physical gel can take up more solvent and the chains will rearrange (in time), so that they return to relaxed conformations. Eventually, this temporary gel can be transformed into a solution which can be further diluted.

In a covalently cross-linked network, the mixing tendency of segments and solvent molecules is opposed by stretching of network chains. The equilibrium state was theoretically described first by Flory and Rehner [6], Flory [7, 8] and later by Flory and Erman [9] for the simplest model of polymer chains (random-flight or Gaussian chain). In this model, the change of the Gibbs energy upon mixing dry network with the solvent is considered to be composed of elastic and mixing terms

$$\Delta G_{\mathrm{sw}} = \Delta G_{\mathrm{mix}} + \Delta G_{\mathrm{el,n}} \tag{12.8}$$

$\Delta G_{\mathrm{el,n}}$ is associated with stretching of EANCs and can be described by various models in which the distribution of end-to-end distances of EANCs is Gaussian or non-Gaussian. The mixing Gibbs energy, ΔG_{mix}, is usually expressed by the theory of regular solutions to which a van Laar interaction term is added. For mixing of cross-linked polymer with non-stretched chains

$$\Delta G_{\mathrm{mix}} = kT(N_1 \ln \phi_1 - \Xi_{\mathrm{e}} \ln \phi_2 + g(\phi_2) N_1 \phi_2) \tag{12.9}$$

with $\Xi_{\mathrm{e}} = \frac{f_{\mathrm{e}}-2}{f_{\mathrm{e}}} N_{\mathrm{e}}$, where Ξ_{e} is effective cycle rank and N_{e} is the number of EANCs; k is the Boltzmann constant and $g(\phi_2)$ is the interaction function related to the

Flory–Huggins interaction parameter χ.[1] This formulation of ΔG_{mix} was put forward recently [10, 11].

Further generalization of models of swelling included the consideration of finite extensibility of network chains. The deformation limit of finite extensibility of network chains is important for analysis of swelling of coating films which are as a rule relatively highly cross-linked (see recent review of models in Ref. [10]). Another amendment of the elasticity and swelling theories incorporates the special role of entanglements as sliding cross-links that limit the number of conformations available to entangled chains, thus better explaining the special character of deformation of less cross-linked systems [12]. At higher cross-linking degrees, the entanglement contribution is unimportant.

The equilibrium swelling of coating films can also be affected by external constraints—for instance, when the adhesion to the substrate is strong and the film is not detached from it by swelling. Then, the swelling degree is lower than that for free swelling at the same cross-link density, yet one can calculate the correct value of cross-link density by considering the effect of such a constraint theoretically. In the following sections, we will show and discuss equations that relate equilibrium swelling to the cross-link density and other parameters of the system that determine swelling.

Swelling of Networks of Gaussian Chains

The model of Gaussian chains is the original and the most widely used model of elasticity of polymer networks. For this model, the Helmholtz energy of elastic deformation of EANCs in the affine network $\Delta F_{\mathrm{el,n}} \approx \Delta G_{\mathrm{el,n}}$ reads

$$\begin{aligned} \Delta G_{\mathrm{el}}/\mathrm{k}T &= N_{\mathrm{e}}\left(\lambda_x^2+\lambda_y^2+\lambda_z^2-3\right)/2 \\ \Delta G_{\mathrm{el,n}}/\mathrm{k}T &= N_{\mathrm{e}}\left(\left(\lambda_x^2+\lambda_y^2+\lambda_z^2-3\right)/2-\ln\lambda_x\lambda_y\lambda_z\right) \end{aligned} \tag{12.10}$$

For swelling (Eq. 12.8), it gives

$$\begin{aligned} \Delta G_{\mathrm{sw}}/\mathrm{k}T = {} & N_1\ln\phi_1 + N_1\chi\phi_2 \\ & + N_{\mathrm{e}}\left(\left(\lambda_x^2+\lambda_y^2+\lambda_z^2-3\right)/2-(2/f_{\mathrm{e}})\ln\lambda_x\lambda_y\lambda_z\right) \end{aligned} \tag{12.11}$$

[1] $g=\chi$, if both quantities are concentration independent; otherwise, the concentration dependences can be recalculated; in the examples below, we will be using the concentration-dependent parameter χ, where the concentration dependence will be related to the chemical potentials of the solvent.

where $\lambda_i = L_i/L_{i0}$ are the deformation ratios $(i = x, y, z)$ which are related to isotropic reference state 0; k is the Boltzmann constant; T is the temperature in Kelvin; ϕ_1 and ϕ_2 are volume fractions of solvent and polymer, respectively; N_1 and N_e are the numbers of solvent molecules and elastically active network chains (EANC), respectively; and χ is the Flory–Huggins interaction parameter (χ used instead of g because more data are found in the literature in this form).

The thermodynamic condition for phase equilibrium is defined by equality of chemical potentials of each component in coexisting phases; thus, for component i in a two-phase system (phases I and II)

$$\mu_i^I = \mu_i^{II} \tag{12.12}$$

$i = 1$ is assigned to the solvent; since the activity of pure solvent is by definition equal to unity, $a_1 = 1$, and $\mu_i = RT \ln a_i$, $\mu_1^I = 0$ and

$$RT \ln a_i^{II} = RT \ln a_i^I = 0 \tag{12.13}$$

The chemical potential of the solvent is obtained from ΔG_{sw} (Eq. 12.11) by differentiation with respect to the number of solvent molecules N_1:

$$\mu_1{}' = \frac{\partial G_{sw}}{\partial N_1} \tag{12.14}$$

For free swelling,

$$\lambda_x = \lambda_y = \lambda_z = \lambda = [(V/V_d)(V_d/V_0)]^{1/3} = \phi_2^{-1/3}(\phi_2^0)^{1/3} \tag{12.15}$$

where V, V_d, and V_0 are swollen, dry, and reference volumes, respectively, and the reference state is considered to be the state at network formation. Thus, $1 - \phi_2^0$ is the volume of non-polymerizable diluent at network formation.

Equation (12.11) transforms into

$$\frac{\Delta G_{sw}}{kT} = N_1 \ln\phi_1 + N_1\chi\phi_2 + N_e\left(3(\lambda^2 - 1)/2 - (2/f_e)\ln\lambda^3\right) \tag{12.16}$$

This gives Eq. (12.17) for the chemical potential of the affine network in equilibrium with pure solvent:

$$\Delta\mu_{1,\mathrm{aff}} = RT\left[\ln(1 - \phi_2) + \phi_2 + \chi\phi_2^2 + \nu_e V_{1m}\left(\left(\phi_2^0\right)^{2/3}\phi_2^{1/3} - (2/f_e)\phi_2\right)\right] = 0 \tag{12.17}$$

where V_{1m} is the molar volume of the solvent. By passing to molar quantities, the Boltzmann constant, k, was transformed to gas constant, R, and the number of EANCs, N_e , was transformed to concentration of EANCs in dry network, ν_e.

Equation (12.17) serves as the basis for evaluation of swelling experiments, provided the conditions for its application (free film, not too high cross-link density) are fulfilled.

Considering Finite Extensibility of Network Chains

Imposing a limit on extensibility of network chains puts a serious restriction on stretching of these chains by osmotic forces, especially in the case of short chains and good solvents. To account for the finite extensibility, various models of elasticity have been examined. They are compared in several papers (cf., e.g., refs. [10, 13]). To demonstrate the finite extensibility effect, we use Gent's function. The dependence of elastic energy on deformation closely resembles that of the commonly used three-chain model, but it is simpler in its functional form. The Gibbs energy for the elastic term reads

$$\begin{aligned}&\Delta G_{el}/kT = -\frac{EJ_M}{6}\ln\left[1-\frac{J_1}{J_M}\right];\quad J_1 = I_1 - 3;\quad I_1 = \lambda_x^2+\lambda_y^2+\lambda_z^2\\&J_M = I_{1M} - 3;\quad I_{1M} = I_1\left(\max\lambda_i = n^{1/2}\right);\quad E = 3kTN_e\end{aligned} \tag{12.18}$$

where n is the average number of statistical segments in a network chain. The invariant I_1 reaches its maximum value I_{1M} when one or more deformation ratios λ_i get equal $n^{1/2}$.

For free swelling, $\lambda_x = \lambda_y = \lambda_z = \lambda = \phi_2^{-1/3}\left(\phi_2^0\right)^{1/3}$, and

$$\Delta\mu_1/RT = \ln\phi_1 + \phi_2 + \chi\phi_2^2 + \nu_e V_{1m}\left(\frac{\phi_2^{1/3}\left(\phi_2^0\right)^{2/3}}{1-\left(\phi_2^0/\phi_2\right)^{2/3}/n} - \frac{2}{f_e}\phi_2\right) = 0 \tag{12.19}$$

We see that the elastic term increases steeply with decreasing n, i.e., with increasing concentration of EANCs until $\left(\phi_2^0/\phi_2\right)^{2/3}/n = 1$, when the elastic term diverges. The swelling degree cannot surpass the limit at which the network chains are fully stretched. The relation between the concentration of EANCs and n reads

$$n = \frac{\rho_2}{\nu_e M_s} \tag{12.20}$$

where M_s is the molecular weight of Kuhn's statistical segment and ρ_2 is the specific gravity of the polymer. If Eq. (12.19) is used for determination of ν_e, we find its value by solution of the following equation:

$$\ln\phi_1 + \phi_2 + \chi\phi_2^2 + \nu_e V_{1m}\left(\frac{\phi_2^{1/3}\left(\phi_2^0\right)^{2/3}}{1-\left(\phi_2^0/\phi_2\right)^{2/3}\nu_e M_s/\rho_2} - \frac{2}{f_e}\phi_2\right) = 0 \tag{12.21}$$

Swelling of Adhering Coating Film

When a cross-linked film deposited on a stiff substrate swells, (a) the film delaminates from the substrate when the swelling stresses exceed the adhesion strength, or (b) only the film thickness increases because the film is prohibited from dilation in plane of the substrate, or (c) the limit of deformation instability is exceeded and surface pattern formation, like buckling or wrinkling, is observed. In case (c), the observed pattern can be used for characterization of elastic properties of the material [14–17].

For case (b): if the film is not detached from the substrate by swelling, it is prevented from changing the planar dimensions (axes x, y) and can expand only vertically (axis, z). Therefore,

$$\lambda_x = \lambda_y = 1; \quad \lambda_z = \lambda = \phi_2^{-1}\phi_2^0 \tag{12.22}$$

and, using Gent's model of a network composed of chains of finite extensibility

$$\begin{aligned}\Delta\mu_1/RT &= \ln\phi_1 + \phi_2 + \chi\phi_2^2 \\ &\quad + \nu_e V_{1m}\left(\frac{\left(\phi_2^0\right)^2\phi_2^{-1}}{1-\left(\left(\phi_2^0/\phi_2\right)^2 - 1\right)/(n-1)} - \frac{2}{f_e}\phi_2\right) \\ &= 0\end{aligned} \tag{12.23}$$

The adhesion effect increases with increasing degree of swelling, i.e., with decreasing cross-link density, contrary to the effect of finite extensibility which is the largest at high cross-link densities.

Effect of Constraints During Film Formation on Swelling of Detached Film

When the coating film is formed by simultaneous cross-linking and solvent evaporation, the EANCs adopt conformations dependent on the volume at which the cross-links are formed. These conformation changes are dependent on relative rates of cross-linking and evaporation. We can relatively easily analyze the extremes—fast evaporation and slow cross-linking or fast cross-linking and slow evaporation. The first case—fast evaporation and slow cross-linking—is easy to handle because almost all cross-linking takes place in the dry state (all network chains are in their state of relaxation). Such a situation exists in baked system. At room temperature cure, the reaction slows down because the film turns glassy.

In the second case—cross-linking takes place first followed by evaporation of major portion of solvent—the network chains oriented vertically (z-axis) are compressed, while the chains oriented in the x,y-plane remain as they were before evaporation, i.e.,

$$\lambda_x = \lambda_y = 1; \quad \lambda_z = \lambda = V_{sw}/V_{dry} = B \tag{12.24}$$

where V_{sw} is the volume of the wet film, V_{dry} is the volume of the dry film, and B is the volume degree of swelling (Eq. 12.3). The dry film is stressed and has a tendency to shrink in the x–y directions when the adhesion is released by detachment of the film from the substrate. The degree to which it shrinks is found by minimizing the Helmholtz energy under the constraint that the volume of the dry film remains constant and that it is equal to $V_{dry} = V/B$. For the Gaussian model, the function to be minimized, y, reads

$$\lambda^2\lambda_z = B \quad y = 2\lambda^2 + \frac{B^2}{\lambda^4} \tag{12.25}$$

where $\lambda = \lambda_x = \lambda_y$.

The minimum of y is obtained when

$$y' = 4\lambda - \frac{4B^2}{\lambda^5} = 0 \tag{12.26}$$

The solution of Eq. (12.26) gives the new set of deformation ratios for the released film equal to

$$\lambda = B^{1/3}, \quad \lambda_z = B^{1/3} \tag{12.27}$$

This result means that the released film is isotropic, and all chains in the dry film have shrunk *to their respective relaxed states (states of ease)* by the same factor $B^{1/3}$. However, with respect to the state of film fixed on substrate, the film shrinks by $100 \times B^{1/3}$ % of its in-plane dimensions and gets thicker by $100 \times (B - B^{1/3})$ % (from B to $B^{1/3}$) when it is removed from the substrate. Care must be taken in the case of films removed from the surface in the glassy state, i.e., in its stressed form. First, the swelling allows the film to relax and experience x,y-shrinkage, which is then followed by an increase in x,y-dimensions with additional uptake of solvent if present. The measured increase of dimensions (in-plane) then does not reflect correctly the level of cross-link density (or interaction parameter).

In most cases, the rates of cross-linking and solvent evaporation are occurring simultaneously, and EANCs are formed at different contents of the solvent. Moreover, gradients in cross-link density and network dilution can develop, and the film can get bent upon swelling. This is a more complicated situation which asks for a more elaborate analysis.

Effect of the Polymer–Solvent Interaction Parameter

The interaction between polymer segments and solvent molecules is an important factor along with cross-link density determining the swelling capacity of the polymer networks. Usually, the pairwise interactions are considered. Their contribution to the degree of swelling is determined by the difference Δw of Gibbs energies, w, of interaction of segment pairs: segment–solvent, segment–segment, and solvent–solvent $\Delta w = w_{12} - (1/2)(w_{11} + w_{22})$. The interaction parameter is a function of this difference $\chi = z\Delta w/RT$, where z is the coordination number of the lattice site (polymer segment). The interaction parameter is linked to the difference between the solubility parameters of the polymer, δ_2, and solvent, δ_1, ($\chi = V_{\mathrm{segm}}(\delta_1 - \delta_2)^2/RT$). This gives only a very rough estimate of χ, especially for systems with hydrogen bonding and other specific interactions. Hansen [18] has considerably improved the estimate of χ, and the method can be helpful for estimating the solvent power of new solvents.

However, for new and newly developed systems, a "calibration" method best serves the purpose. The value of ν_e is obtained from stress–strain dependences for a series of samples differing in the concentration of EANCs, ν_e, measured by an oscillatory shear rheometer, or in extension, or in compression and evaluating them by using theoretical stress–strain dependences based, for instance, on models described by Eq. (12.10) or (12.18). By using these values of ν_e, the interaction parameter and its concentration dependence χ is calculated (Eqs. 12.17, 12.19, and 12.21). The concentration dependence can be expressed as a power series

$$\chi = \chi_0 + \chi_1\phi_2 + \chi_2\phi_2^2 + \dots \tag{12.28}$$

or in a closed form [19]

$$\chi(T, \phi_2) = D(T)B(\phi_2); \quad D(T) = d_0 + d_1/T + d_2 \ln T; \quad B(\phi_2) = \frac{1}{1 - b\phi_2} \tag{12.29}$$

In the case of a Gaussian network, ν_e is proportional to the equilibrium small-strain shear modulus, G_e, or Young modulus, E_e $\left(G_e = RT\nu_e\phi_2^{1/3}\left(\phi_2^0\right)^{2/3};\ E_e \approx 3G_e\right)$. For the models respecting finite extensibility, the stress–strain relations are different and so is the relation between the small-strain modulus and ν_e. Once χ is known

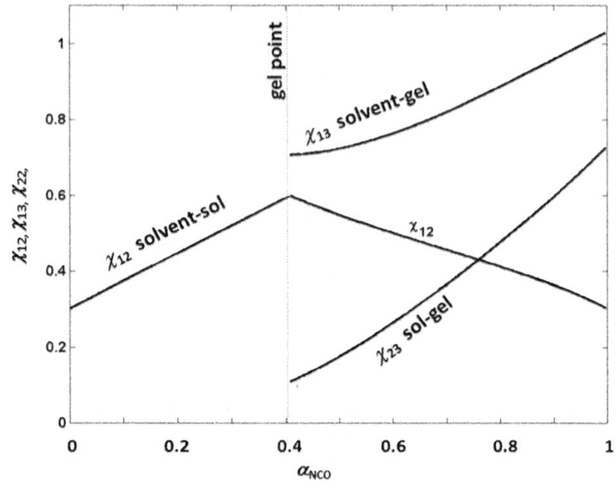

Fig. 12.2 Estimated change of interaction parameters solvent (*1*)–sol (*2*), solvent (*1*)–gel (*3*), and sol (*2*)–gel (*3*) during formation of a network from tetrafunctional polyester star and trimer of 1,6-diisocyanatohexane in the presence of methyl amyl ketone [21]

for a series of samples, the equilibrium swelling method becomes a good tool for determining the variation in ν_e based on the conditions of film formation.

The values of the interaction parameter depend more or less on the progress of the cross-linking reaction because of the changing number and types of interacting sites. For cross-linking of primary chains, the effect is small because the fraction of cross-linked units is usually small [20], but for step reactions, in which many functional groups are transformed into bonds, the change can be large. Such dependence on conversion is illustrated by the change of χ during the formation of polyurethane networks from tetrafunctional polyester star polyol and tri-isocyanate [21]. The situation for a cross-linking system is more complicated due to the transient polydispersity in molecular weight as well as composition of the interacting molecules reaching its maximum at the gel point. A simplification has been used by assuming a pseudoternary system with components: solvent (1), sol (2), and gel (3). The estimated change of the corresponding interaction parameters χ_{12}, χ_{13}, and χ_{23} is shown in Fig. 12.2.

The increase of the interaction parameter χ_{13} is one of the factors causing phase separation during cross-linking in a closed (non-evaporating) system [21]. The change of the interaction parameter should be accounted for, if swelling is used to analyze structural changes of a cross-linking system. When swelling measurements are done with extracted samples, only the interaction parameter χ_{13} is relevant. If the "calibration method" is used (i.e., when ν_e is determined by stress–strain measurement of the extracted sample) to determine the dependence of χ_{13} on conversion, the dependence on changing composition and possible dependence on concentration are merged.

Generally, the conventional solvent systems are such in which the increase of cross-link density as well as worsening polymer–solvent interaction in the course of reaction contribute to the increase of solvent vapor pressure and the possibility of phase separation.

Effect of Structural Parameters of the Network on Swelling and the Swell Test

The main structural parameters of the swelling equation are the concentration of elastically active network chains, ν_e; the polymer–solvent interaction parameter, χ; the volume fraction of network polymer at network formation, ϕ_2^0; the effective functionality of the cross-link, f_e; and the molar volume of the solvent V_{1m}. If the finite extensibility of network chains is considered, its effect is controlled by the number of statistical segments per one EANC, n, which is related to ν_e by Eq. (12.20).

The molar volume of the solvent is available in reference tables. It is important to note that the degree of swelling decreases with increasing size of the solvent molecules if their polarity remains the same. The (average) effective functionality of the cross-link sufficiently past the gel point is close to its chemical functionality; otherwise, it depends on conversion and possible off-stoichiometry of the system and can be calculated using the branching theory. The volume fraction of the polymer at network formation $\phi_2^0 = V_d/V_0$ is equal to the ratio of the dry polymer volume of the extracted sample to the reference volume at which the network chains are in their state of ease; it is assumed that the network chains are at that state when they are formed. Thus, $1 - \phi_2^0$ is the volume fraction of the extractable material, i.e., the added diluent, sol fraction, or other additive not built in the network. At the same cross-link density and other network parameters, the presence of a diluent increases the degree of swelling.

The dependences of the degree of swelling plotted as B or ϕ_2 on the concentration of EANCs for various swelling models (Gaussian network, limited chain extensibility network, adhering network) and for various values of the interaction parameter χ corresponding to solvents ranging from very good to moderately poor ones are displayed in Figs. 12.3a, b, 12.4a, b, and 12.5a, b. Typically, the cross-link densities of coating films are around $10^{-3} - 10^{-2}$ mol/cm^3. A set of four values of the interaction parameters has been chosen and used in the examples (Table 12.1.)

By comparing the plots, one can see that the Gaussian model shows the widest difference in the swelling degree for the selected set of interaction parameters. For models based on the concept of finite chain extensibility of network chains limited by their contour length, the sensitivity of swelling to variation of the interaction parameters is weaker. At low concentration of EANCs, both models give almost the same degree of swelling for a given value of χ, but the difference between the degree of swelling values increases as ν_e increases. At the level of $\nu_e = 0.005$ mol/cm^3, the value of B calculated using the finite extensibility model is only half of that calculated for the Gaussian model. This means that the cross-link density calculated from equilibrium swelling using the Gaussian model is largely overestimated.

When adhering films are swollen, the degree of swelling is considerably lower than that for a free film. The largest difference is found for low cross-link densities,

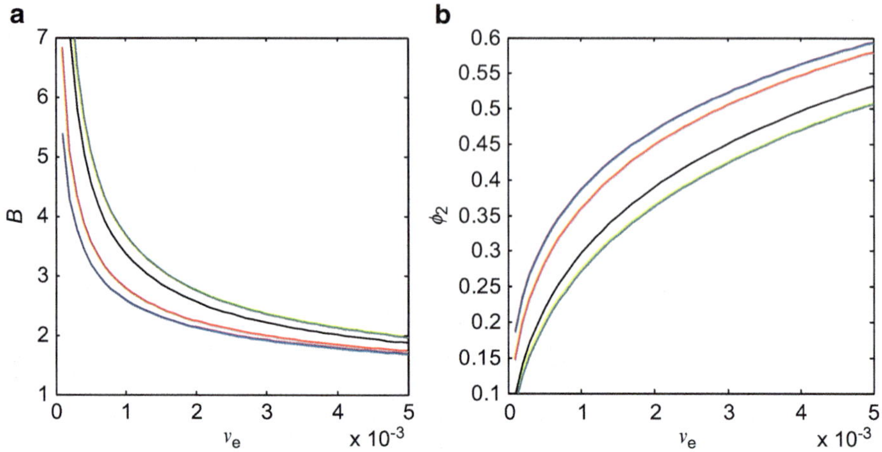

Fig. 12.3 (**a**, **b**) Volume degree of swelling, B, and volume fraction of polymer in the swollen gel, ϕ_2, as a function of concentration of EANCs, ν_e in mol/cm^3, for affine model of Gaussian network (Eq. 12.17), for different values of the interaction parameters. Other details in Table 12.1, curves from top to the bottom, **a**—cases 1, 2, 3, 4; **b**—cases 4, 3, 2, 1

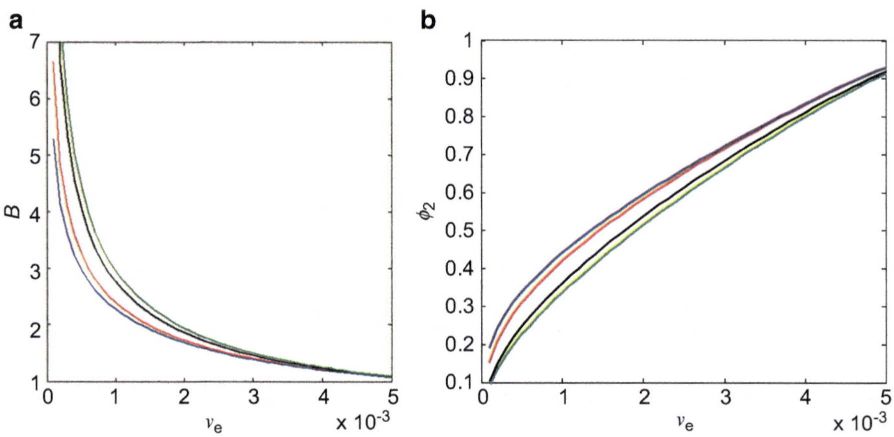

Fig. 12.4 (**a**, **b**) Volume degree of swelling B, and volume fraction of polymer in the swollen gel, ϕ_2, as a function of concentration of EANCs, ν_e in mol/cm^3, for *Gent's model of network with finite extensibility of chains* (Eq. 12.19), for different values of the interaction parameters. Other details in Table 12.1, order of curves as in Fig. 12.3a, b

whereas at high cross-link densities, the degree of swelling is dominated by the presence of cross-links and by the finite extensibility of EANCs.

It would be useful to linearize the dependence of ϕ_2 on ν_e which could make the interpolation of the changes of swelling degree easier. Figure 12.6a, b shows the plot of ϕ_2 vs. $\sqrt{\nu_e}$ as well as the log–log plot. None of these plots are perfectly linear, but the ϕ_2 vs. $\sqrt{\nu_e}$ gives a more compact family of dependences.

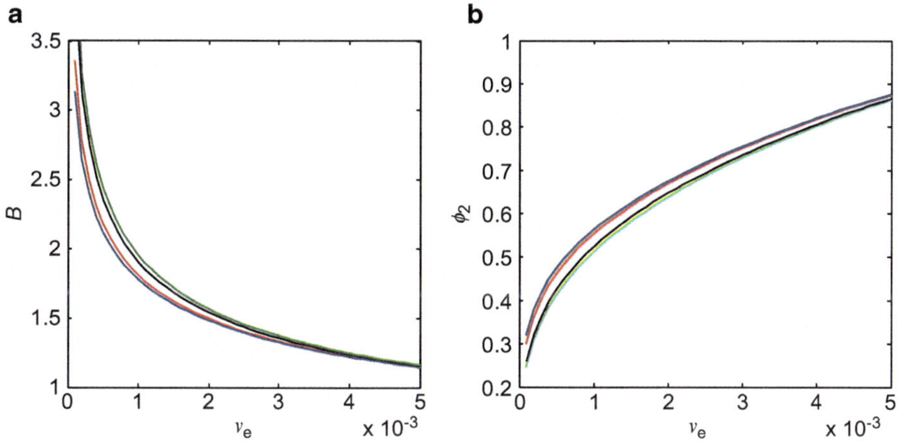

Fig. 12.5 (**a**, **b**) Volume degree of swelling, B, and volume fraction of polymer in the swollen gel, ϕ_2, as a function of concentration of EANCs, ν_e in mol/cm^3, for Gent's model of network with finite extensibility of chains adhering to the substrate (Eq. 12.21), for different values of the interaction parameters. Other details in Table 12.1, order of curves as in Fig. 12.3a, b

Table 12.1 Sets of concentration-independent interaction parameters and other structural parameters used in simulation

Case	a_0	b	Color in figures	Solvent quality
1	0.10	0	Green	(Very) good
2	0.20	0	Black	Good
3	0.40	0.20	Red	Moderately good
4	0.50	0.40	Blue	Moderately poor
	Equation	$\chi = \frac{a_0}{1-b\phi_2}$		
	Other parameters	$\phi_2^0 = 0.8, f_e = 3, V_{1m} = 100$ cm^3/mol, $\rho_2 = 1.2$ g/cm^3, $M_s = 200$ g/mol		

The Swell Test as a Measure of Changes in Cross-Link Density or Solvent Quality

Correlations with Experiment

Typically, the dimensions (area) of a flat piece of film are measured in the dry state, and then the film is immersed into an excess of good solvent (e.g., dichloromethane). After a short time (generally of order of minute(s)), it swells to its maximum, and the dimensions (area) of the swollen film are determined. It is assumed that the film swells isotropically in all three dimensions, and the volume degree of swelling, B, is calculated. The method is very simple and fast, and it has been recommended for use as a diagnostic tool of changes in cross-link density [22–24]. Indeed, changes in cross-link density are the major factor causing changes in

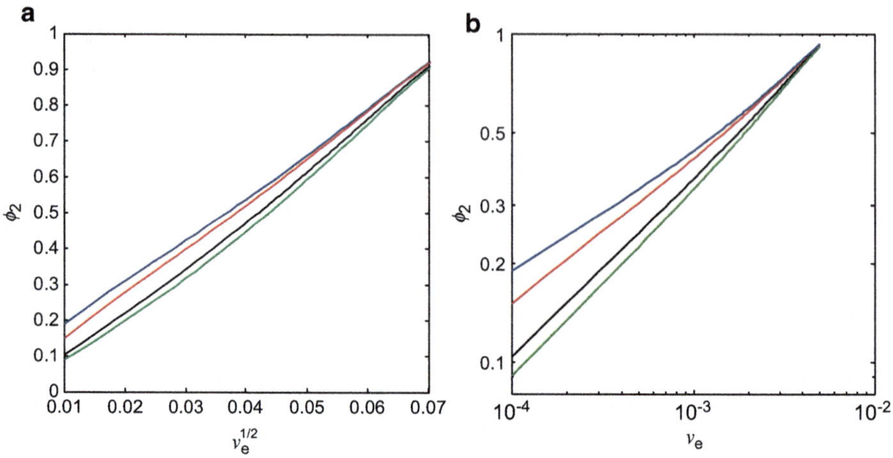

Fig. 12.6 (**a**, **b**) Plots of ϕ_2 against various functional forms of ν_e for Gent's model of network with finite extensibility of chains (Eq. 12.19), for different interaction parameters (Table 12.1, order of curves as in Fig. 12.3b)

ϕ_2 or $B = 1/\phi_2$, if the solvent is good (χ is low) (Gent's model—Fig. 12.6a). At higher values of ν_e, changes in cross-link density dominate the changes in the degree of swelling, especially if the finite extensibility of network chains is taken into account. A comparison of the degree of swelling for two samples of nearly the same composition in a good solvent can be used to determine the relative cross-link density. Quantification of the relation between $\Delta\phi_2$ and $\Delta\nu_e$ is possible, if one knows χ and has reason to believe that it does not change. For the example described in Table 12.1 (case 1, $\chi = 0.1$), $\Delta\nu_e/\Delta\phi_2 \approx 0.0065$ for a wide range of concentrations of EANCs.

The swell test can also be used for characterization of solvent power. In Table 12.2, swell test data (ratio of swollen to dry square areas) for several OH-functional acrylic copolymers that have been cross-linked with the trimer of 1,6-diisocyanatohexane are shown. For the purpose of the present analysis, the exact composition of the copolymers is irrelevant.

Dichloromethane is a (very) good solvent which is used in the swelling test when differences in cross-link density are to be characterized. The structures of OXSOL and PGMEA which are the common solvents are displayed below

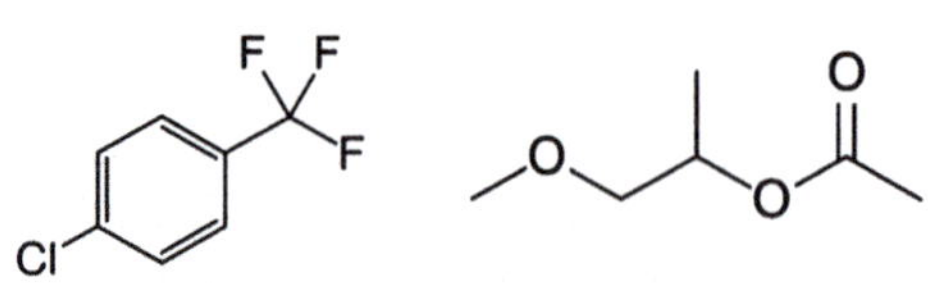

OXSOL PGMEA

Table 12.2 Swell ratios of OH-functional acrylic copolymers 8A–8E cross-linked with trimer of 1,6-diisocyanatohexane in various solvents at 25 °C

	Swell ratio of cross-linked films[a]: *SR* of samples				
Solvent	8A	8B	8C	8D	8E
Dichloromethane	1.62	1.52	1.61	1.67	1.50
p-Chlorobenzotrifluoride (PCBTF, OXSOL)	1.33	1.25	1.39	1.37	1.26
Methyl amyl ketone	1.28	1.25	1.33	1.30	1.27
Amyl acetate	1.21	1.17	1.28	1.33	1.24
Butyl propionate	1.22	1.17	1.25	1.30	1.22
t-Butyl acetate	1.18	1.16	1.23	1.24	1.19
1-Methoxy-2-propanol acetate Propylene glycol methyl ether acetate (PGMEA)	1.28	1.24	1.36	1.26	1.37

[a]The area of swollen film divided by the area of dry film $SR = B^{2/3}$; 8A–8E = various hydroxy-functional acrylic copolymers cross-linked with Desmodur® N3300 tri-isocyanate

Table 12.3 Volume degree of swelling recalculated from swell ratios of Table 12.2

	Volume swelling degree of cross-linked films[a] *B*				
Solvent	8A	8B	8C	8D	8E
Dichloromethane	2.06	1.87	2.04	2.16	1.83
p-Chlorobenzotrifluoride (PCBTF, OXSOL)	1.53	1.40	1.64	1.60	1.41
Methyl amyl ketone	1.44	1.40	1.53	1.48	1.43
Amyl acetate	1.33	1.27	1.44	1.53	1.38
Butyl propionate	1.35	1.27	1.40	1.48	1.35
t-Butyl acetate	1.28	1.25	1.36	1.38	1.30
PGMEA	1.45	1.38	1.59	1.41	1.60

[a]Recalculated from SR

The concentration of EANCs for these samples is not known. From the swell ratio in dichloromethane, one can only conclude that the series of inverse swell ratios indicates the series of increasing concentrations of EANCs, i.e., sample 8D, has the lowest and sample 8E the highest cross-link density. Because "calibration" by mechanical measurements was not done in this case, we can only estimate the values of χ and analyze the outcome of the comparison. For this purpose, the *SR* values of Table 12.2 were recalculated to values of B (Table 12.3).

Based partially on our experience with cross-linked polyurethane networks, we assigned sets of χ values as follows: set 1 (very good solvent) of Table 12.1 to dichloromethane, set 2 (moderately good solvent) to OXSOL, and set 4 (moderately poor solvent) to methyl amyl ketone. Then, for each set and the experimental value of *B* (Table 12.3), ν_e was calculated using Eqs. (12.17) or (12.19). The result is shown in Fig. 12.7a for samples 8D and 8E. The values of ν_e calculated by assuming

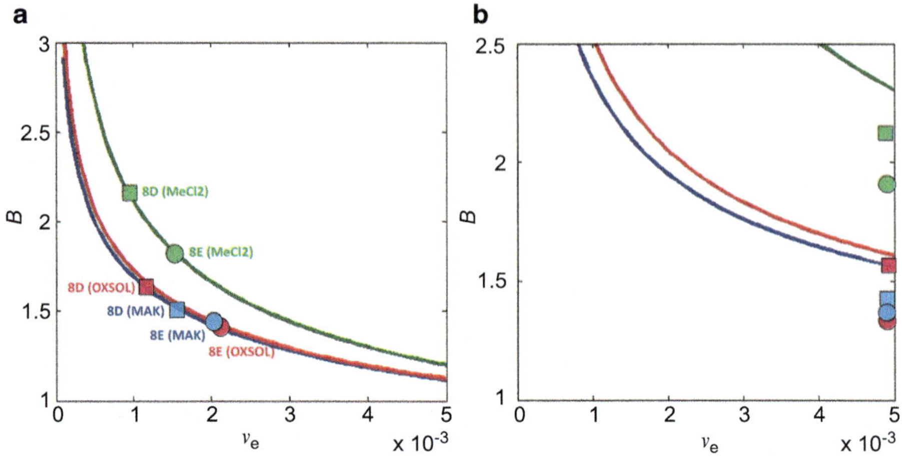

Fig. 12.7 (**a**, **b**) Dependence of the volume swelling degree B on the concentration of EANCs for the sets of interaction parameters of Table 12.1 for the network models of finite extensibility of EANCs (**a**) and affine Gaussian model (**b**). In (**a**), the values of B found for samples 8D and 8E are plotted vs. the corresponding value of ν_e that was calculated assuming certain values of the interaction parameter χ as described in the text

an ideal network with all branch points being elastically active serve as the absolute upper bounds. They vary in the range of $(2.2–4.0)\times10^{-3}$ mol/cm^3. If the choice of χ values is correct and the chosen swelling theory describes the experimental system well, we should get nearly the same values for ν_e from the swelling degree measured for different solvents.

The results in Fig. 12.7a show that the correlation is acceptable when the model of a network with finite extensibility of network chains is applied: for sample 8D, ν_e varies in the range $(1.0–1.5)\times10^{-3}$ mol/cm^3 and for sample 8E $(1.6–2.1) \times 10^{-3}$ mol/cm^3. One should not forget that the interaction parameters used here are only estimated, not determined independently. Better estimates can make these ranges narrower (though not by much). The model of a Gaussian network cannot be used at all, as shown in Fig. 12.7b, because the results are out of bounds for the expected values of ν_e. The calculated values exceed the value 0.005 mol/cm^3, which is unrealistic. This is an important finding. It calls for revision to the evaluation of swelling data, including "calibration" using mechanical measurements, and for finding a proper elasticity theory that respects the finite extensibility of network chains. In addition to Gent's model, for instance, the eight-chain model [25, 26] is worth examining. This model is somewhat "softer" than Gent's model.

Pitfalls of the Swell Test

The advantages of the swell test are both its extreme simplicity and the short time needed to obtain a result. However, the results are not always fully meaningful. Following are some examples of this:

1. It is claimed [24] that a few seconds are sufficient to reach equilibrium with the solvent for coating films. However, there exist exemptions from this general observation. In some solvents, the swelling is much slower than in the majority of solvents although the final (equilibrium) degree of swelling is comparable or even higher. The swelling of the polyurethane networks made with OH-functional precursors is an example. The swelling of all samples in series 8 (Table 12.2) in PCBTF (OXSOL) was very slow, although comparing equilibrium values, PCBTF ranks second right after methylene chloride for most of the samples. Similarly, swelling of cross-linked polystyrene in tetrachloromethane is slower by almost an order of magnitude compared to 1,2-dichloroethane although the equilibrium swelling degree is similar and is also high. We assume that this is due to a slow diffusion of these molecules into the glassy polymer.
2. Extreme care must be taken when the swell test is applied to monitor the changes of ν_e during cross-linking. There are several situations which may occur. (a) The interaction parameter is a function of conversion, which usually causes an increase, but in some cases may cause a decreases of χ, and this change is comparable to the change due to increasing cross-linking. (b) If the sample is in its glassy state and contains unreacted functional groups, the transition to the rubbery state by swelling allows those groups to become mobile and react. Thus, conversion may increase during the test. (c) If sol is still present, the swelling passes through a maximum and decreases again when all sol is extracted. The extraction of sol is a slow process. Thus, the cross-link density can be underestimated because the swelling degree did not reach equilibrium value.

Conclusions

Measurement of the swelling of cross-linked coating films remains one of the key characterization methods offering information on the degree of cross-linking and polymer–solvent interactions. It seems that, indeed, increase of cross-linking density is the major factor in the observed decrease of the degree of swelling, but quantification of this increase of ν_e is impossible. To make the results fully meaningful and quantitative, it is necessary to revise the theoretical basis of swelling. The degree of cross-linking of the coating film is usually relatively high, and the solvents used to swell the sample are good. For this combination, Gaussian theory, although generally used, is inadequate. One should use a theory that takes into account finite extensibility of network chains. The second task is to examine the dependence of the interaction parameter on composition as the latter changes through the progress of reaction, both from a theoretical and an experimental approach. Here, combination with branching theories will be very helpful because those theories can predict variations in the composition of the gel part of the system. A calibration of the swelling equation (of the interaction part) by combining swelling with a mechanical examination (shear, extension, compression, or indentation) is necessary, at least in a certain phase of developing new binder/ cross-linker systems. Also, the new theoretical basis will allow the analysis of equilibrium swelling results for more complicated systems, like adhering (cross-linked) films, or filled systems with a cross-linked matrix, i.e., pigmented films.

References

1. Dušek, K., Dušková-Smrčková, M.: Network structure formation during crosslinking of organic coating systems. Prog. Polym. Sci. **25**, 1215–1260 (2000). doi:10.1016/S0079-6700(00)00028-9
2. Dušek, K., Dušková-Smrčková, M.: Polymer networks. In: Matyjaszewski, K., Gnanou, P., Leibler, L. (eds.) Precise Synthesis, Materials Properties, Applications, Structure-Property Correlation and Characterization Techniques, vol. 3, pp. 1687–1730. Wiley-VCH, Weinheim (2007). ISBN 978-3-527-31446-1
3. Dušek, K.: Network formation in curing of epoxy resins. Adv. Polym. Sci. **78**, 1–59 (1986). doi:10.1007/BFb0035356
4. Dušek, K., Dušková-Smrčková, M., Huybrechts, J., Ďuračková, A.: Polymer networks from preformed precursors having molecular weight and group reactivity distributions. Theory and application. Macromolecules. **46**, 2767–2784 (2013). doi:10.1021/ma3000187
5. Dušková-Smrčková, M., Valentová, H., Ďuračková, A., Dušek, K.: Effect of dilution on structure and properties of polyurethane networks. Pregel and postgel cyclization and phase separation. Macromolecules. **43**, 6450–6462 (2010). doi:10.1021/ma100626d
6. Flory, P.J., Rehner, J.: Statistical mechanics of cross-linked polymer networks. II Swelling. J. Chem. Phys. **11**, 521–526 (1943). doi:10.1063/1.1723792
7. Flory, P.J.: Statistical mechanics of swelling of network structures. J. Chem. Phys. **18**, 108–111 (1950). doi:10.1063/1.1747424
8. Flory, P.J. Principles of Polymer Chemistry. Cornell University Press, Ithaca (1953). ISBN-13 978-0-8014-0134-3
9. Erman, B., Mark, J.E.: Structures and Properties of Rubberlike Networks. Oxford University Press. New York (1997). ISBN: 9780195082371
10. Dušek, K., Choukourov, A., Dušková-Smrčková, M., Biedermann, H.: Constrained swelling of polymer networks: characterization of vapor-deposited cross-linked polymer thin films. Macromolecules. **47**, 4417–4427 (2014). doi:10.1021/ma5006217
11. Dušek, K., Dušková-Smrčková, M., Šomvársky, J.: Effect of constraints on swelling of polymer networks. Macromol. Symp. **358**, 120–127 (2015). doi:10.1002/masy.201500025
12. Edwards, S.F., Vilgis, T.: The effects of entanglements in rubber elasticity. Polymer. **27**, 483–492 (1986). doi:10.1016/0032-3861(86)90231-4
13. Horgan, C.O., Saccomandi, G.: Phenomenological hyperelastic strain-stiffening constitutive models for rubber. Rubber Chem. Technol. **79**, 152–169 (2006). doi:10.5254/1.3547924
14. Ben Amar, M., Goriely, A.: Growth and instability in elastic tissues. J. Mech. Phys. Solids. **53**, 2284–2319 (2005). doi:10.1016/j.jmps.2005.04.008
15. Dervaux, J., Ben Amar, M.: Mechanical instabilities of gels. Annu. Rev. Condens. Matter Phys. **3**, 311–332 (2012). doi:10.1146/annurev-conmatphys-062910-140436

16. Li, B., Cao, Y.-P., Feng, X.-Q., Gao, H.: Mechanics of morphological instabilities and surface wrinkling in soft materials: a review. Soft Matter. **8**, 5728 (2012). doi:10.1039/C2SM00011C
17. Singamaneni, S., Tsukruk, V.V.: Buckling instabilities in periodic composite polymeric materials. Soft Matter. **6**, 5681–5692 (2010). doi:10.1039/C0SM00374C
18. Hansen, C.: Hansen Solubility Parameters: A User's Handbook, 2nd edn. CRC Press, Boca Raton, FL (2007). ISBN 978-0-8493-7248-3
19. Mohite, L.V., Juvekar, V.A.: Quantification of thermodynamics of aqueous solutions of poly (ethylene glycols). Role of calorimetry. Fluid Phase Equilib. **278**, 41–53 (2009). doi:10.1016/j.fluid.2009.01.003
20. Dušek, K.: Quasichemical equilibrium approach to crosslinked polymer solutions. J. Polym. Sci., C, Polym. Symp. **39**, 83–106 (1972)
21. Dušková-Smrčková, M., Dušek, K., Vlasák, P.: Solvent activity changes and phase separation during crosslinking of coating films. Macromol. Symp. **198**, 259–270 (2003). doi:10.1002/masy.200350822
22. Gregorovich, B.V., Adamsons, K., Lin, L.: Scratch and mar resistance and other mechanical properties as a function of chemical nature for automotive refinish coatings. Prog. Org. Coat. **43**, 175–177 (2001). doi:10.1016/S0300-9440(01)00187-4
23. Hill, L.W., Wicks Jr., Z.W.: Design considerations for high solids coatings. Prog. Org. Coat. **10**, 55–89 (1982). doi:10.1016/0300-9440(82)80006-4
24. Hill, L.W.: Structure property relationships of thermoset coatings. J. Coat. Technol. **64**, 28–42 (1992)
25. Arruda, E.M., Boyce, M.C.: A three-dimensional constitutive model for the large stretch behavior of rubber elastic materials. J. Mech. Phys. Solids. **41**, 389–412 (1993). doi:10.1016/0022-5096(93)90013-6
26. Boyce, M.C., Arruda, E.M.: Constitutive models of rubber elasticity: a review. Rubber Chem. Technol. **73**, 504–523 (2000). doi:10.5254/1.3547602

Chapter 13
Chemical Depth Profiling of a Multilayer Coating System Using Slab Microtomy and FTIR-ATR Analysis

Karlis Adamsons and Mei Wen

Background

In automotive coating systems, there are typically several layers, such as clearcoat, basecoat, primer, and electrocoat (e-coat), all serving different purposes. The coatings are typically tested together to document overall system performance with regard to factors such as wet adhesion, impact resistance, and chip resistance. Failures could happen at or near the surface, a certain interface between layers, or within a particular layer.

Different from automotive OEM (Original Equipment Manufacturer) coatings, the basecoats used for automotive refinish coatings contain polymer binders that are generally not crosslinked. However, options are given to body shop sprayers to add a small amount of crosslinker (in a solution form with solvent) to a basecoat to react with the functional groups of the basecoat binders for better system performance. For instance, an isocyanate solution is added to react with the hydroxyl groups of the basecoat binders to form crosslinks. Crosslinking the basecoat can improve its cohesive strength and thus help improve system properties such as wet adhesion, impact resistance, and chip resistance. However, no study has been reported on how adding a crosslinker to a basecoat affects the chemical compositions from the clearcoat to the basecoat, particularly at the clearcoat/basecoat interface and inside the basecoat.

On the other hand, automotive refinish clearcoats typically contain crosslinked polymers. A common clearcoat chemistry is polyurethane based, where polyol resins react with an isocyanate to form a polyurethane network. The isocyanate crosslinker in the clearcoat can penetrate into the underlying basecoat, leading to crosslinking of the basecoat [1]. However, when the clearcoat cures rapidly, there is

K. Adamsons (✉) • M. Wen
Axalta Coating Systems, Coatings Technology Center, Wilmington, DE, USA
e-mail: karl.adamsons@axaltacs.com

M. Wen, K. Dušek (eds.), *Protective Coatings*, DOI 10.1007/978-3-319-51627-1_13

less opportunity for the crosslinker to penetrate into the basecoat [1]. As a result, system properties may suffer.

In this work, a system with a rapidly curing solvent-borne clearcoat and a waterborne basecoat (WBBC) was selected. This coating system has improved system performance when a crosslinker is added to the basecoat. To understand this coating system, it was necessary to address two questions: (1) whether the clearcoat coating components can diffuse into the basecoat to help strengthen the basecoat layer, and (2) what other contributions the addition of crosslinker to the basecoat makes besides crosslinking the basecoat layer, such as interfacial bonding and diffusion of the clearcoat materials into the basecoat. Depth profiling of the chemical composition from the clearcoat to the basecoat, particularly at/near the clearcoat/basecoat interface and inside the basecoat, provides information to answer the questions. In this work, slab microtomy coupled with Fourier transform infrared spectroscopy-attenuated total reflectance (FTIR-ATR) analysis was selected to determine the depth profiles of two coating systems. One system uses a standard, non-crosslinked basecoat, and the other uses a crosslinker-containing basecoat.

Slab microtomy of a coating system is performed to obtain thin slices from top to bottom and coplanar to the surface. These thin slices are then analyzed via FTIR-ATR to yield depth profile information of the chemical composition of the system [2–4]. Slab microtomy can be combined with a number of analytical techniques to determine different properties and composition depth profiles, such as FTIR for photo-oxidation index, component distribution/mixing, and contaminant locus/migration [1, 5, 6], differential scanning calorimetry (DSC) for glass transition temperature (T_g) and thermogram [5], high-performance liquid chromatography (HPLC) for ultraviolet (UV)-screener type and concentration [5], Raman for component distribution [7], UV-Vis for UV-screener type and concentration [2, 3, 5, 8], X-ray photoelectron spectroscopy (XPS) for binding energy spectra of the elements present [5], and X-ray fluorescence (XRF) for elemental composition and distribution [9]. Therefore, this sampling approach provides considerable versatility.

Although this work used slab microtomy for depth-specific sampling, various other techniques such as confocal XRF for elemental composition [10], confocal Raman microscopy for component distributions [11], time-of-flight secondary ion mass spectrometry (ToF-SIMS) for component and contaminant distributions [12], micro-IR on cross sections for degradation and component distributions [13], step-scan photoacoustic IR for component stratification [14], and focused ion beam milling with scanning electron microscopes and energy-dispersive X-ray spectroscopy (SEM/EDS) analyses for elemental distribution [15] have proven to be effective alternatives for chemical depth profiling, depending on the target information required.

In this work, microtomy coupled with FTIR-ATR analysis of cured samples was carried out starting from the top layer of the clearcoat and proceeding to the bottom layer of the basecoat to determine the effect of crosslinker addition to the basecoat on interfacial bonding and basecoat chemical composition. Indentation hardness was also evaluated to gauge the early-stage drying and curing of the clearcoat/basecoat systems to help understand the depth profiling results in terms of diffusion of clearcoat materials into the basecoat.

Experimental

Sample Preparation

In analyzing the depth profiles of clearcoat and basecoat systems, e-coat steel panels were used as the underlying substrates. Two waterborne basecoat solutions were used for preparing the basecoats: a standard one in black that did not have a crosslinker and a crosslinker-containing one. The WBBC solutions were applied to e-coat panels with an automated spray machine. The WBBC was then dried at room temperature for about 40 min. Upon drying of the basecoat, a solvent-borne clearcoat was applied onto the basecoat-coated panels using the same machine. These clearcoat-coated panels were then dried and cured at room temperature. For comparison studies, two types of basecoat-only panels without clearcoat on top were also prepared: non-crosslinked vs. crosslinked. The latter was the basecoat that contains the crosslinker.

The crosslinker-containing basecoat was prepared by adding a solution of water-dispersible polyisocyanate based on hexamethylene diisocyanate (HDI) to the standard basecoat solution. The water-dispersible polyisocyanate reacts with the hydroxyl groups of the basecoat binder to form a urethane linkage. It also reacts with water in the basecoat, which depletes (i.e., evaporates) during the basecoat drying process, to form urea linkages. The solvent-borne clearcoat was a cross-linked polyurea coating, where an amine reacted with an aliphatic polyisocyanate, HDI trimer, to form a crosslinked polyurea.

Coated samples were stored at ambient conditions for 2–3 weeks to allow sufficient drying and curing before slab microtoming was carried out. The thickness of the basecoats was about 10–17 μm, and the thickness of the clearcoats was about 50–61 μm. In summary, two coating systems were examined: System 1 with the clearcoat on a non-crosslinked basecoat, and System 2 with the same clearcoat on a crosslinker-containing basecoat.

Optical Microscope Imaging

Optical microscope imaging was carried out to illustrate the coating structure and how microtoming took place. Coated panels were cut, potted, sanded, and then polished to have a smooth cross-sectional surface for imaging. A small cutaway was potted in a reactive epoxy material, which hardened under vacuum to avoid bubble formation. Then, the sample was sanded in a series of steps using 120, 320, 600, 800, and 1200 grit silicon carbide sandpapers and polished with a slurry of 0.05 μm particle size Gamma Alumina to expose a smooth cross-sectional surface of the coating system. Ecomet™ 6 Variable Speed Grinder-Polisher units were used for all sanding and polishing. The cross-sectional surface was imaged with a Leica® model DM-RXA optical microscope configured with a Zeiss® model AxioCam ICc1 digital camera.

Microtoming

Slab microtoming was done using a Leica® microtome model SM 2500 unit. This is a large-scale, heavy-duty sectioning system, easily handling samples that are in the size range of 1 in. by 1 in. or slightly larger. This has proven to be a convenient size for securing and sectioning the sample coplanar to the surface. Typically, an epoxy support block is used for securing these samples. Prior to attaching a sample, the block is trimmed with the broad carbide blade, providing a smooth surface for sample anchoring. This approach ensures a flat sample to be sectioned coplanar to the surface. As shown in Fig. 13.1, the sample is anchored to the block using a strong, double-sided tape (Microscopical Optical Consulting Inc.), which is slightly porous preventing trapping pockets of air. After the sample is secured, a smooth epoxy block is used to press down on the surface to squeeze out any remaining trapped air.

The carbide blades used in this work have a 50° wedge angle. This angle has proven to work best for relatively hard automotive paint coating systems.

The microtome controller (Fig. 13.1, right side) allows for easy cutting of slices in the 5–10 μm thickness range and for setting up the cutting speed. The individual slices can be cut one at a time in a manual mode or continuously in an automatic mode. The cutting can be stopped after the user has obtained all slices of interest at a given thickness or to change the thickness setting during a series of cuts.

Slice thickness for these studies was approximately 10 μm for the top 40 μm of the clearcoat and reduced to 5 μm for the lower portion of the clearcoat, continuing down into the underlying basecoat and, in some cases, the e-coat layer.

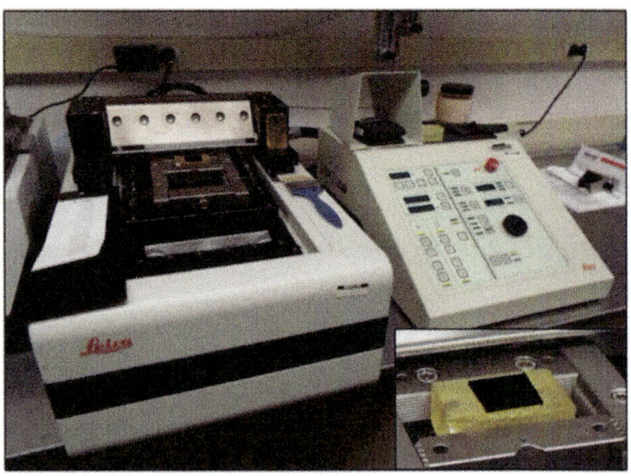

Fig. 13.1 Leica® Slab Microtome model SM 2500 used for cutting a coating system into slices coplanar to the surface. The exploded view (insert) on the right hand corner shows the epoxy polymer anchoring block used for securing the sample for microtoming. This photo is used herein with permission from Leica®

FTIR-ATR

A Nicolet®/Thermo model Nexus 470 FTIR ESP spectrophotometer configured with a Nicolet® Smart DuraSampl*IR*® ATR module was used. FTIR-ATR mode analysis was conducted soon, less than 24 h, after microtoming was carried out. Earlier studies indicate that after 2–3 weeks of aging of the coatings, the residual isocyanate can be neglected. Therefore, not examining the sliced samples right away after microtoming would not affect the compositions of the solids in the samples. Potential evaporation of small amount of residual solvent does not affect the compositions of the solids either. Microtoming was done on the bottom side of each slice of sample. During microtoming, the slices often curled up to form a tight scroll. The scrolls were then flattened and positioned with the outside of the scroll in contact with the internal reflection element (IRE) and held in position with the clutch post. Because the IRE is zinc selenide (ZnSe) with a thin layer of diamond on top (i.e., Diacomp design), the spectrum is a composite (or average) of the upper ~3–5 μm of the slice being analyzed. Spectra obtained were both baseline- and ATR-corrected. Each spectrum recorded in these studies was a co-addition of 32 spectra in order to build up the ratio of signal to noise to obtain high-quality results.

In quantifying the intensity of the individual peaks in the FTIR-ATR spectra, peak deconvolution was carried out by using Gaussian curve fitting with a software package known as Fityk. This is software for data processing and nonlinear curve fitting.

Indentation Hardness

Martens hardness of coating Systems 1 and 2 was measured with a Fischerscope® HM2000 S, instrumented indentation tester configured with a Vickers indenter. A maximum load of 300 mN was used to allow penetration deep into the coating system so that the hardness obtained was overall system hardness instead of just the hardness of the surface coating. A loading time of 50 s was applied. At the maximum load, no creep period was used, because only Martens hardness was used. Then, the load was decreased to zero within 50 s. Martens hardness HM is calculated by the instrument based on the following equation, where $F_{\max}$ is the maximum load and A_s is the surface area of the indenter in contact with the coating at the maximum load:

$$\mathrm{HM} = \frac{F_{\max}}{A_S} \tag{13.1}$$

Martens hardness represents a material's resistance to both elastic and plastic deformation. Martens hardness was evaluated for the two coating systems at 4 h, 1 day, and 7 days after coating application.

Results and Discussion

A typical coating system studied in this work consists of a clearcoat and a basecoat on an e-coat panel. Figure 13.2 shows the cross-sectional optical microscope image of a representative coating system, illustrating the approximate microtoming (or the cutting depth) loci. The film thickness of each layer measured through the microscope imaging method was consistent with the film thickness determined by a film build gauge on single-layer coatings prepared during the spray. Note that due to slight variation of film thickness from point to point, the exact film thickness of microtomed areas cannot be determined with this approach. However, the slab microtome was used to section at every 10 μm on the top 40 μm of the clearcoat and 5 μm at the lower portion of the clearcoat and below.

Figure 13.3 shows the pictures of each slice of a coating system obtained through microtoming. When the basecoat does not have a crosslinker added (System 1), the eighth slice starts to show the black basecoat. The ninth slice is a fragile basecoat

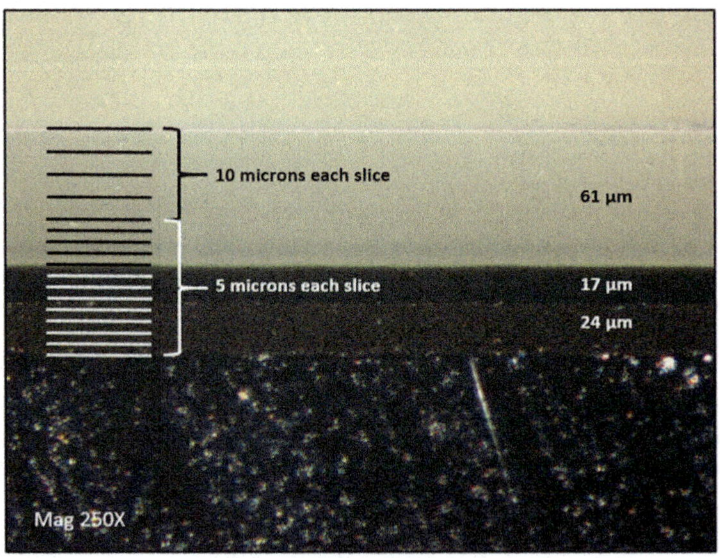

Fig. 13.2 Optical microscope image of the cross section of a multilayer coating system. The top most layer is the clearcoat (~61 μm), next basecoat (~17 μm), and finally e-coat (~24 μm) on stainless steel. The microtoming loci are illustrated. The thickness of each slice changed from 10 μm on the top 40 μm of the clearcoat to 5 μm at the lower portion of the clearcoat and down into the basecoat and e-coat layers. System 2 is shown here

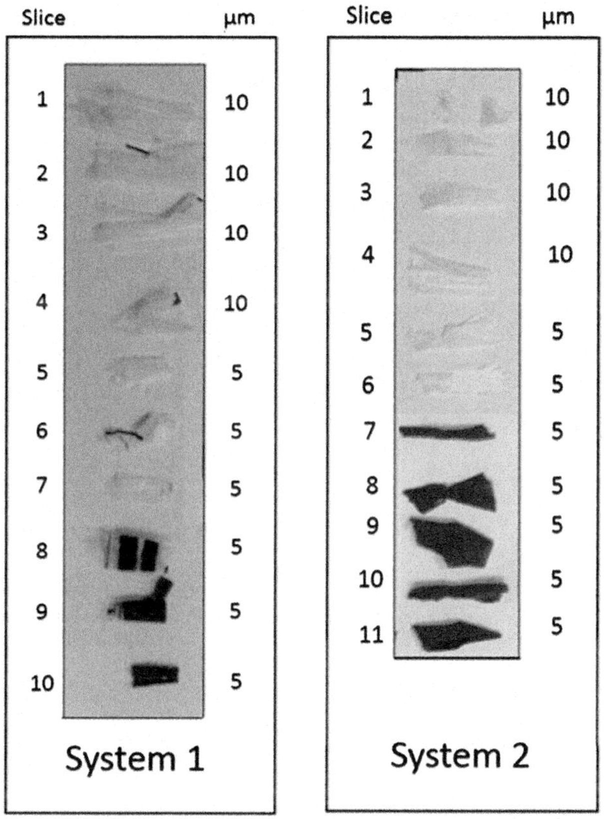

Fig. 13.3 Pictures of microtomed slices obtained from coating Systems 1 and 2

slice which broke into two pieces. It was found that a non-crosslinked basecoat was fragile and tended to break apart in the microtoming process. This fragility indicates low cohesive strength of the non-crosslinked basecoat. The tenth slice is an e-coat layer as indicated from its FTIR spectrum. Note that all FTIR spectrum information will be discussed later in detail. In System 2, the slices #7–9 are the basecoat. Slice #10 starts to transit to the e-coat and slice #11 is an e-coat layer.

Figure 13.4 shows the FTIR-ATR spectra of single-layer coatings, including clearcoat, crosslinked WBBC, non-crosslinked WBBC, and e-coat. The first three coatings were individually sprayed onto e-coat panels and dried. They are often called clearcoat-only and basecoat-only coatings. After microtoming of the clearcoat-only and basecoat-only (both crosslinked and non-crosslinked) coatings, the first slice (top 10 μm) of each individual coating was scanned with the FTIR-ATR system to obtain its spectrum. The spectra are quite different among these samples. The spectra in the 1400–1800 cm^{-1} (wavenumber) range are significantly different among the clearcoat, crosslinked basecoat, and non-crosslinked basecoat. Spectral band assignments are given in Table 13.1. The clearcoat has a carbonyl (C=O) peak from its amine ester resin at 1731 cm^{-1} [17], a C=O peak from the

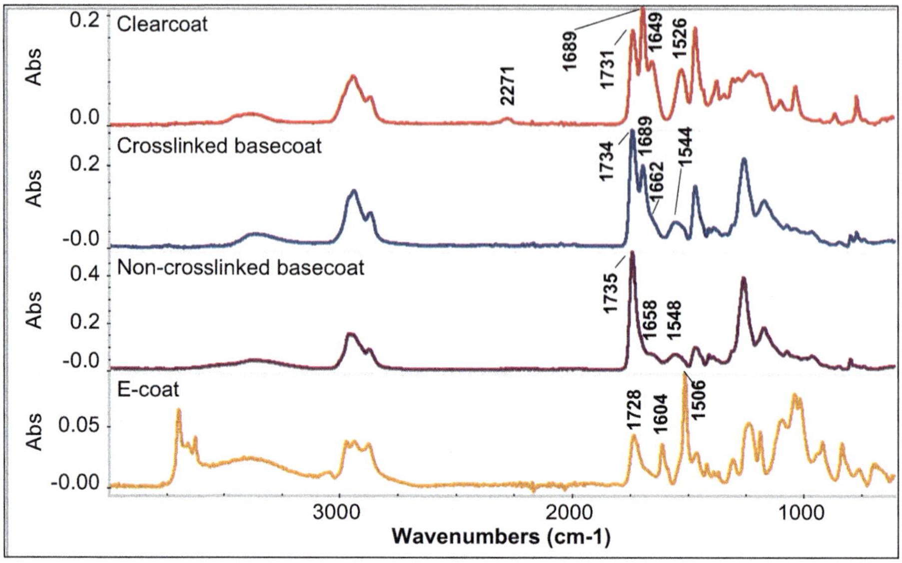

Fig. 13.4 FTIR spectra of model coatings represented in this study, including clearcoat, crosslinked waterborne basecoat (WBBC), non-crosslinked WBBC, and e-coat

Table 13.1 Tentative IR peak assignments for clearcoat, crosslinked WBBC, and non-crosslinked WBBC used in the depth profiling studies

Sub-structure	cm^{-1}	Function type	Reference
Clearcoat			
Ester sub-structure	~1731	Carbonyl	[17]
Isocyanurate ring (HDI trimer)	~1689	Carbonyl	[14]
Urea sub-structure	~1649	Carbonyl	[18]
Urea sub-structure	~1526	Amide II	[16]
Crosslinked basecoat			
Ester sub-structure	~1734	Carbonyl	[17]
Isocyanurate ring	~1689	Carbonyl	[14]
Urea sub-structure	~1662	Carbonyl	[18]
Urea/urethane sub-structures	~1544	Amide II	[18]
Non-crosslinked basecoat			
Ester sub-structure	~1735	Carbonyl	[17]
Urea sub-structure	~1658	Carbonyl	[18]
Urea/urethane sub-structure	~1548	Amide II	[18]

isocyanurate ring structure of HDI trimer at 1689 cm^{-1} [14], a urea C=O peak at 1649 cm^{-1} [18], and an amide II (deformation of N-H and asymmetric stretch of C-N) peak from the urea structure at 1526 cm^{-1} [16]. The NCO peak at 2271 cm^{-1} is mostly gone, indicating near-complete consumption of NCO groups. The crosslinked basecoat has a C=O peak at 1734 cm^{-1}, which is from the ester group component [17], and a C=O peak at 1689 cm^{-1} from the isocyanurate ring from the isocyanate added to the basecoat [14]. It also has a shoulder at 1662 cm^{-1}, which is a peak associated with urea linkage formation [18], and a broad peak at 1544 cm^{-1}, which is an amide II peak composite that contains both urethane and urea groups [18].

Compared to the crosslinked basecoat, the non-crosslinked basecoat does not have a crosslinker added to react with the basecoat binder or the water present in the basecoat solution. The non-crosslinked basecoat also has a C=O peak at 1735 cm^{-1} [17]. In addition, it has a peak from its urea linkage at 1658 cm^{-1} and a peak at 1548 cm^{-1}, which is an amide II peak that contains both urethane and urea linkages [18]. The non-crosslinked basecoat has relatively lower peak intensities at 1658 and 1548 cm^{-1} comparing to the peak intensities at 1662 and 1544 cm^{-1} in the crosslinked basecoat. The e-coat has distinct peaks from the other coating layers. This set of reference spectra helps determine where a microtomed sample originates from, i.e., clearcoat, basecoat, or e-coat layers.

Figure 13.5 overlays full-scale spectra of microtomed slices of System 1 in the ~1475–1800 cm^{-1} range, which is a clearcoat over a non-crosslinked basecoat. The spectra of the first seven slices match the spectrum of the model clearcoat system as shown in Fig. 13.4. Therefore, they all appear to be sections of the clearcoat. Slices #8 and #9 have spectra similar to each other and match the spectral detail of the model basecoat. Slice #10 has spectral detail consistent with an e-coat layer.

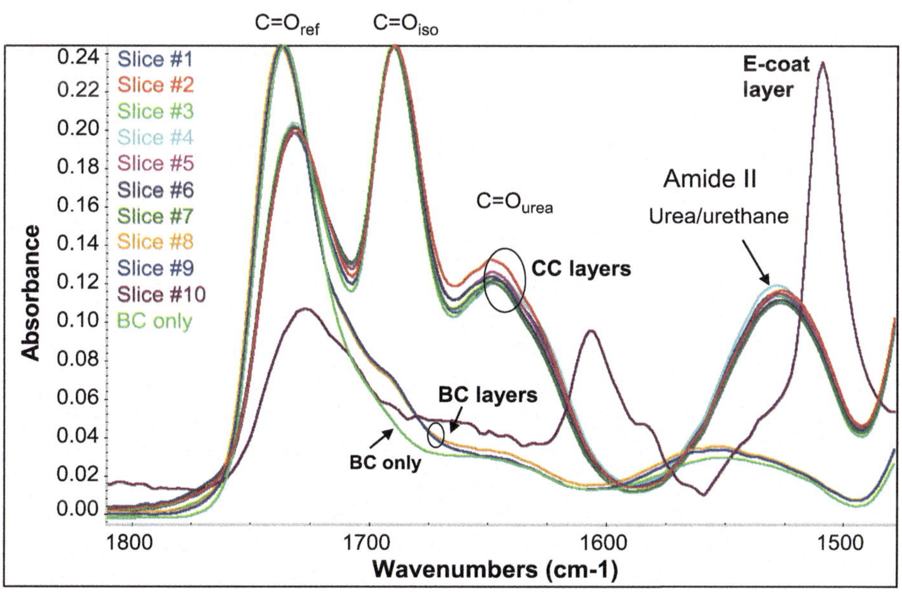

Fig. 13.5 Overlaid, full-scale spectra of ten slices of clearcoat, basecoat, and e-coat of System 1. The ester-type C=O peak at 1731–1734 cm^{-1} is considered as a reference peak and labeled as C=O$_{ref}$. Also, the C=O peaks from the isocyanurate ring, urea structure, and the amide II structures are labeled as C=O$_{iso}$, C=O$_{urea}$, and amide II (urea/urethane), respectively

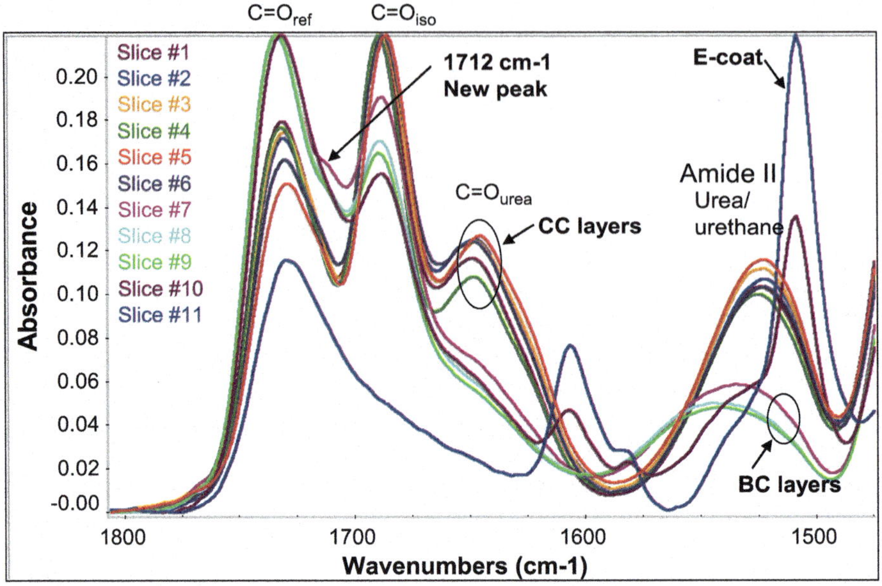

Fig. 13.6 Overlaid, full-scale spectra of 11 slices of clearcoat, basecoat, and e-coat of System 2. The ester-type C=O peak at 1731–1735 cm^{-1} is considered as a reference peak and labeled as $C{=}O_{ref}$. Also, the C=O peaks from the isocyanurate ring, urea C=O, and the amide II structures are labeled as $C{=}O_{iso}$, $C{=}O_{urea}$, and amide II (urea/urethane), respectively

The two distinct sets of spectra between the clearcoat slices and the basecoat slices indicate that interlayer diffusion between the clearcoat and the basecoat is quite limited. What is also shown in Fig. 13.5 is the spectrum of the non-crosslinked basecoat-only sample. The difference of the spectra of the basecoat slices #8 and #9 vs. the spectrum of the basecoat-only sample indicates that there is some isocyanate diffusing from the clearcoat and some limited urethane/urea formation as reflected by the amide II structure peak.

When the crosslinker-containing basecoat is used, the overlaid spectra of the clearcoat and basecoat slices appear quite different as shown in Fig. 13.6. A more gradual change of the urea C=O peak at 1649 cm^{-1} and the C=O peak at 1689 cm^{-1} from the isocyanurate ring structure from the clearcoat layers to basecoat layers is present. Slice #7, which has the basecoat color, shows a spectrum containing some features of the clearcoat and mostly basecoat. Furthermore, the urea C=O peak at 1649 cm^{-1}, the isocyanurate ring C=O peak at 1689 cm^{-1}, and the amide II peak (with slightly varying peak center) gradually drop until slice #9. This indicates that diffusion between the clearcoat and basecoat has occurred. Materials from the clearcoat such as HDI trimer and binder resin appear to diffuse into the basecoat, forming more urea in the basecoat layers. Slice #10 starts to show some e-coat peaks and slice #11 is purely an e-coat layer.

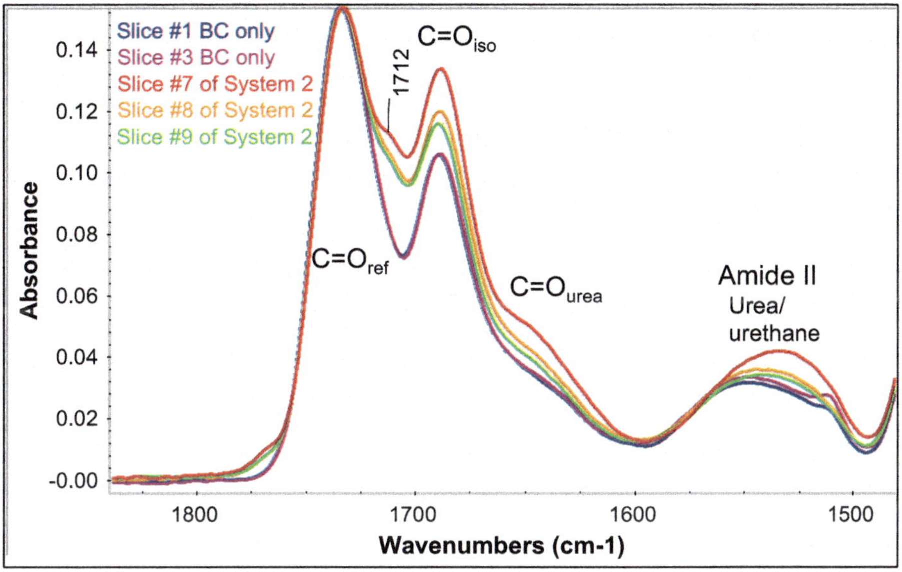

Fig. 13.7 Comparison of the first and third layers of the basecoat spectra of a crosslinked basecoat-only sample and the spectra of slices #7, #8, and #9 of System 2 in the ~1475–1800 cm^{-1} range

A new shoulder appeared at 1712 cm^{-1} in slices #7 to #9, particularly at the first layer of basecoat (slice #7), indicating either an environmental change or a new reaction pathway. This may be due to new functionality or environment.

To better understand the diffusion of the clearcoat materials into the basecoat, the spectra of the first and last slices of a crosslinked basecoat-only sample (coated on e-coat) were compared with that of the seventh, eighth, and ninth slices of System 2. As shown in Fig. 13.7, in the basecoat-only sample, the spectra of the first and third layers are almost identical, indicating there is little variation of chemical composition within the basecoat. In contrast, the spectrum of the top layer basecoat (slice #7) in System 2 where there is a clearcoat above looks quite different from the spectrum of the bottom basecoat layer slice #9. Furthermore, the spectrum of the bottom layer basecoat in this system (slice #9) looks totally different from the spectrum of the bottom basecoat layer in the basecoat-only sample. The higher peak intensity at the isocyanurate ring, C=O urea, and amide II peak positions comparing to the basecoat-only sample indicates migration of isocyanate and clearcoat resin to the basecoat layers to form extra urea in the basecoat.

There is also the formation of the new peak shoulder at 1712 cm^{-1}, particularly in slice #7. This new peak shoulder is not present in the basecoat-only sample.

To quantify the spectrum difference observed above, the ratio of the carbonyl peak from the isocyanurate ring to the carbonyl reference peak is compared. This approach has been used in literature to understand the depth profile of isocyanate in a coating system with urethane clearcoat on a waterborne basecoat [1]. In the clearcoat layers, the carbonyl reference peak is from the ester linkage at

1731 cm^{-1} of the clearcoat resin. This reference peak (ester C=O) then shifts slightly to 1733–1735 cm^{-1} in the basecoat layers, which is considered to come mainly from the acrylic-based ester linkages present. In the clearcoat, the carbonyl peak of the isocyanurate ring structure from the HDI trimer is strong, and the carbonyl reference peak is medium in absorbance strength. Therefore, this ratio is greater than one. In the basecoat that contains a crosslinker, the carbonyl peak from the isocyanate in the crosslinker is weaker than that of the reference peak. Therefore, this ratio is less than one. When there is interlayer diffusion, a gradual change of this peak ratio from high to low from the interface down into the basecoat is expected. When the basecoat does not contain any crosslinker, i.e., the standard, non-crosslinked basecoat, this ratio should be at a minimum level given no diffusion at the clearcoat/basecoat interface.

In quantifying the intensity of the individual peaks, peak deconvolution was carried out by using the Fityk software. Examples of peak deconvolution are shown in Fig. 13.8 for slices #1, #7, and #9 of System 2. Slice #1 represents a clearcoat layer, slice #7 is the first layer of the basecoat, and slice #9 is considered as the last layer of the basecoat. Overall, a good fit was obtained going from the clearcoat to the basecoat layers with minor peak shift observed. The new peak at 1712 cm^{-1} was used in the peak deconvolution of the basecoat layers. In the peak deconvolution of the clearcoat layers, this peak at 1712 cm^{-1} was added for two reasons. First, without this peak, peak deconvolution results in poor fitting. Also, there is literature indicating that free urea, which is not hydrogen-bonded urea, could absorb in this region [19]. Overall, from the clearcoat layers down to the basecoat layers, this 1712 cm^{-1} peak has the highest intensity at slice #7, indicating a preferred reaction product (e.g., possibly non-hydrogen-bonded substituted urea groups) at the interface. Further work is needed to fully understand this.

Figure 13.9 shows the peak height ratio of the carbonyl peak from the isocyanurate ring at 1689 cm^{-1} to the carbonyl reference peak as a function of depth of the two coating systems. The peak ratio of the carbonyl peak from the isocyanurate ring to the carbonyl reference peak (isocyanurate ratio) is about 1.05–1.3 in the clearcoat layers. When the topmost basecoat-containing slice (slice #8 in System 1 and slice #7 in System 2) is characterized, the isocyanurate ratio decreases and keeps dropping as one proceeds lower into the basecoat. The isocyanurate ratio almost plateaus in System 1, showing a minimum gradient of isocyanurate ring structure or the isocyanate in the basecoat. In System 2, the isocyanurate ratio gradually drops to a minimum, indicating a drop of the concentration of the total isocyanate (including both the HDI trimer and the isocyanate from the basecoat crosslinker).

The overall rate of isocyanurate ratio decrease from the last clearcoat slice to the point the isocyanurate ratio becoming almost a constant in the basecoat is much larger in the non-crosslinked basecoat case (System 1—between slice #7 and slice #8) than that in the crosslinker-containing basecoat (System 2, between slice #6 and slice #9). This indicates minimal diffusion of materials from the clearcoat to the basecoat in System 1 and increased diffusion of isocyanate HDI trimer from the clearcoat to the basecoat in System 2. The point where the biggest change of

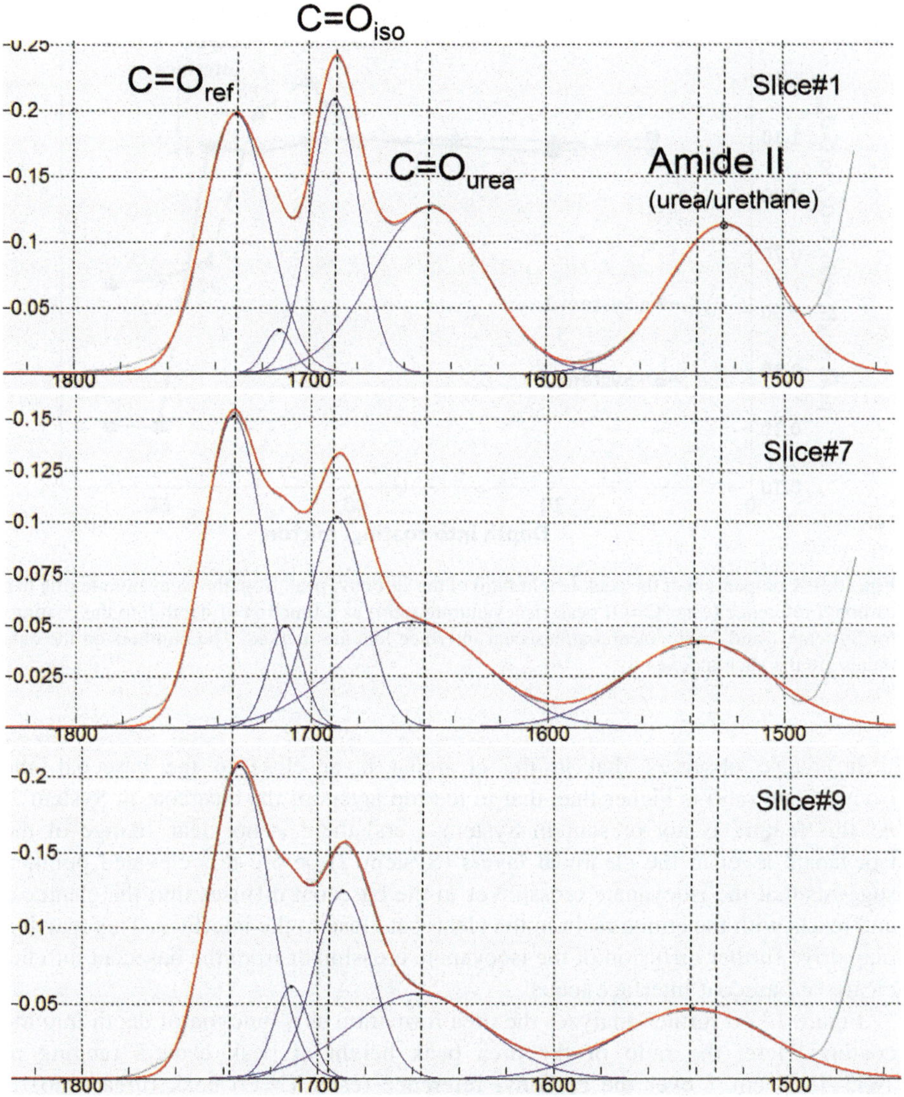

Fig. 13.8 Peak deconvolution examples of slices #1, #7, and #9 of System 2 using Fityk software. The black curves are the original spectrum data, but only the part in the range of 1496.5–1810.9 cm^{-1} was used for curve fitting. The blue curves are the individual peaks obtained from the peak deconvolution, and each red curve is the summation of the individual blue peaks

the isocyanurate ratio occurs for each system is considered to be the clearcoat/basecoat interface as indicated in Fig. 13.9. The CC/BC interface of System 1 is the middle point between slices #7 and #8, and the interface of System 2 is the middle point between slices #6 and #8.

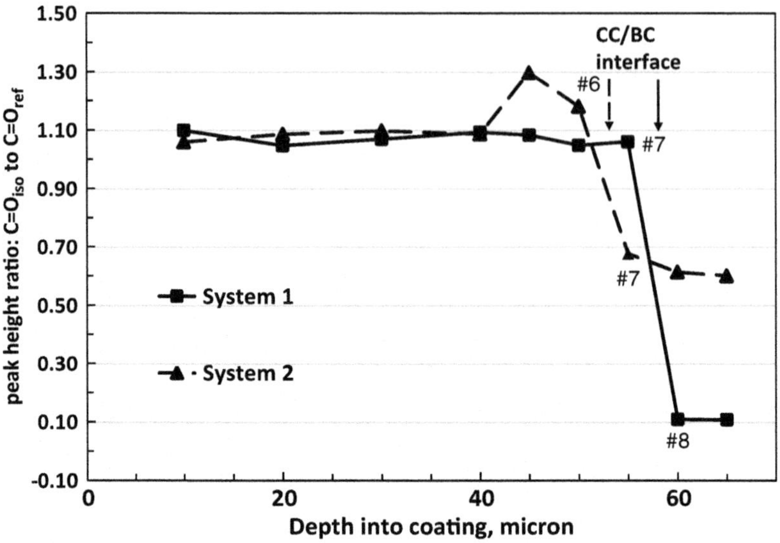

Fig. 13.9 Comparison of the peak height ratio of the carbonyl peak from the isocyanurate ring to a carbonyl reference (ester C=O) peak (isocyanurate ratio) as a function of depth into the coatings for Systems 1 and 2. The clearcoat/basecoat interface loci are marked. The numbers on the data points are the slice indexes

It is also observed that in the clearcoat layer close to the basecoat, the isocyanurate ratio is higher than that in the top layers of the clearcoat in System 2. As this feature is not present in System 1 and there is no clear change of the isocyanate level in the clearcoat layers (System 2) before this elevated point, it suggests that the isocyanate crosslinker in the basecoat diffuses into the clearcoat and reacts with the amine resin in the clearcoat close to the interface. This reaction may drive further diffusion of the isocyanate crosslinker from the basecoat into the clearcoat/basecoat interface locus.

Figure 13.10 further analyzes the urea formation as a function of depth into the coating. Here, the ratio of the urea peak height at peak centers ranging in 1645–1662 cm^{-1} over the carbonyl reference (ester C=O) peak (urea ratio) is used. Note that the urea carbonyl peak becomes weaker, and the reference peak becomes stronger passing from the clearcoat to the basecoat. Therefore, the urea ratio drops from the clearcoat to the basecoat. These depth profiles appear similar to the depth profiles shown in Fig. 13.9. This indicates that not only HDI trimer diffuses from the clearcoat into the basecoat, but also the amine resin probably diffuses into the basecoat assuming that the concentration of water remaining in the WBBC is lower when the clearcoat is applied onto it. There is more urea formation in the crosslinked basecoat than that in the non-crosslinked basecoat.

Similar to Fig. 13.9, Fig. 13.10 also shows an elevated urea level in the clearcoat close to the clearcoat/basecoat interface when the basecoat contains crosslinker.

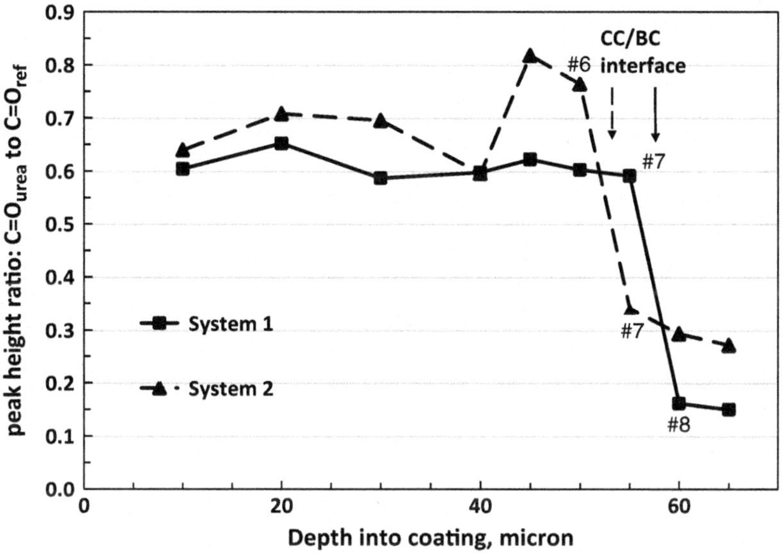

Fig. 13.10 Comparison of the peak height ratio of the carbonyl peak from urea to a carbonyl reference (ester C=O) peak (urea ratio) as a function of depth into the coatings for Systems 1 and 2. The numbers on the data points are the slice indexes

This result is consistent with the earlier hypothesis that the isocyanate crosslinker coming from the basecoat reacts with the amine resin in the clearcoat. Supporting this evidence is the dip of urea formation just before this elevation due to the diffusion of the amine resin to the clearcoat/basecoat interface region.

As urethane formation in the basecoat is also an important reaction to further crosslink the local network in the basecoat, it is helpful to examine urethane formation near the clearcoat/basecoat interface and inside the basecoat. In the non-crosslinked basecoat system, HDI trimer diffused from the clearcoat can react with the hydroxyl groups on the basecoat binder to form urethane. In the crosslinker-containing basecoat system, the isocyanate coming from the crosslinker solution, as well as HDI trimer diffused from the clearcoat, can react with the hydroxyl groups on the basecoat binder to form urethane. Due to the difficulty of separating the urea and urethane peaks, examination of the amide II peak centered at 1524–1550 cm^{-1} was used to determine the total urea and urethane contribution. In the clearcoat, this peak is mainly from the urea group, and the peak is mainly located in 1524–1528 cm^{-1} region. In the basecoat, the amide II peak shifts to higher wavenumbers. In the basecoat layers of System 1, the amide II peak shifts to 1550 cm^{-1} region, whereas in the basecoat layers of System 2, it shifts to 1537–1541 cm^{-1}. The amide II peak is mainly from a combination of urea and urethane groups. Therefore, in the basecoat, its peak intensity can be used as an estimate of the total level of urea and urethane formation.

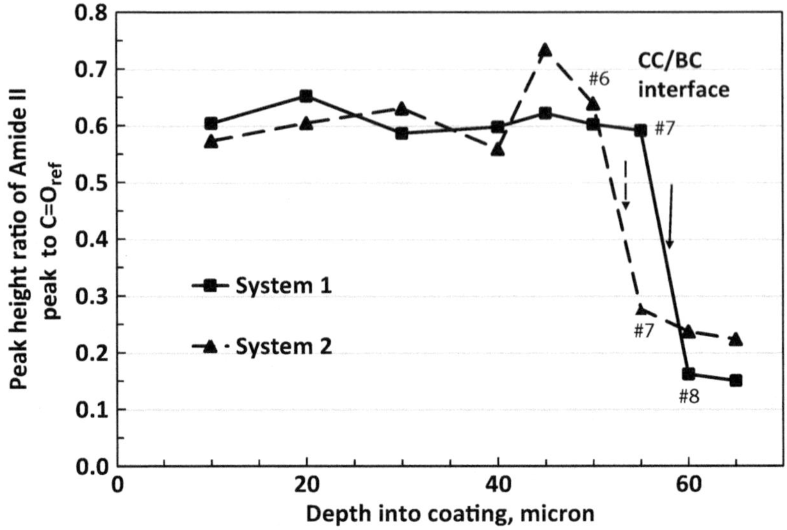

Fig. 13.11 Peak height ratio of the urea/urethane peak to carbonyl reference (ester C=O) peak. The numbers on the data points are the slice indexes

Figure 13.11 shows the peak height ratio of the amide II peak to the ester C=O reference peak (amide ratio). When the basecoat does not contain a crosslinker, there is an abrupt change of the amide ratio from the clearcoat to the basecoat, indicating minimal diffusion between the clearcoat and basecoat. This abrupt change is consistent with the previously discussed results of the isocyanate and urea depth profiles. When the crosslinker-containing basecoat is used, the overall formation of urea and urethane follow a trend consistent with the urea profile reported in Fig. 13.10. It is possible that urea formation is the major contribution of the amide II peak intensity in the basecoat layers of System 2.

The overall urea and urethane formation in the basecoat for both systems can be compared further if we assume the absorptivity of the amide II peak does not change when it shifts its position. Here, if we use the non-crosslinked basecoat-only peak ratio as a base, we can calculate the extra urea and urethane formation by subtracting the peak ratio at each point by this base value for both systems in the basecoats and compare the averages. The non-crosslinked, basecoat-only sample has a relative amide II peak intensity of 0.145 (not shown in the plot). System 1 has on average extra 0.0055 relative amide II (urea and urethane) peak intensity related to the non-crosslinked basecoat-only sample, whereas System 2 has on average an extra 0.0949 relative intensity. In other words, the total extra urea and urethane linkage formed in the basecoat of System 2 is about 17 times of that in System 1. Furthermore, the lowest value in each system can be compared to indicate the weakest linkage. In System 1, the lowest value in its basecoat has extra 0.005 relative amide II peak intensity to the non-crosslinked, basecoat-only sample, whereas System 2 has extra 0.073 relative intensity. Therefore, the minimum extra amount of urea and urethane groups formed in the basecoat of System 2 is about 14.6 times of that in System 1. The difference shown up here gives strong evidence why crosslinker-containing basecoat is much stronger than the standard basecoat.

Figure 13.11 also shows an elevated region of amide II peak signal near the clearcoat/basecoat interface, which is consistent with the extra urea formed as indicated in Fig. 13.10.

Overall, adding a crosslinker to the basecoat leads to a gradual change of the depth profiles from the clearcoat to the basecoat, indicating that it helps diffusion of materials from clearcoat to basecoat and vice versa. As a result, the clearcoat/basecoat interface becomes stronger due to formation of interfacial bonds. Addition of the crosslinker in System 2 also leads to more urea formation in the basecoat compared to that in System 1. The overall urea and urethane linkages in System 2 are also more numerous, leading to higher cohesive strength of the basecoat. The stronger clearcoat/basecoat interface and stronger basecoat help to explain better system properties when a crosslinker is added to the basecoat. From the depth profiles, it is possible that a distinct interphase region forms between the clearcoat and the basecoat, and System 2 has a wider interphase region than that of System 1. Further work to determine the width of the interphase region between the clearcoat and basecoat by using higher-resolution depth profiling techniques would be interesting to investigate.

To understand why adding the crosslinker solution to the basecoat helps diffusion of the clearcoat materials into the basecoat, Martens hardness of the coating Systems 1 and 2 was studied with a micro-hardness tester through an instrumented indentation method. Hardness was measured at 4 h, 1 day, and 7 days after spraying as shown in Table 13.2. A maximum load of 300 mN was applied. At this load, the penetration depth of the indenter goes beyond one tenth of the clearcoat thickness, which is only 5–6 μm. Based on Buckle's rule [20], only a film thickness of ten times the indentation depth is sufficient to neglect the influence of the underneath layer. Therefore, the measured hardness here is not the hardness of the clearcoat alone. Instead, this hardness is an overall hardness of a coating system that includes clearcoat, basecoat, e-coat, and the steel. Note that the total thickness of the clearcoat, basecoat, and e-coat is about 85–100 μm, which is not ten times of the penetration depth values shown in Table 13.2 yet. Therefore, the hardness is even affected by the steel layer under the e-coat based on Buckle's rule. The overall hardness changes when the hardness of any of the layers changes, particularly the basecoat here as it is the only difference of the two coating systems.

As shown in Table 13.2, Martens hardness of the clearcoat/basecoat system with crosslinker-containing basecoat (System 2) is lower at 4 h than that with the

Table 13.2 Martens hardness and maximum penetration depth of Systems 1 and 2

	System 1		System 2	
Time	MH, MPa	Depth, μm	MH, MPa	Depth, μm
4 h	20.3	24.5	18.4	25.8
1 day	33.7	19.0	37.0	18.1
7 day	53.0	15.1	64.6	13.7

standard basecoat (System 1). After that, the hardness catches up and becomes even higher than that of the system with the non-crosslinked basecoat. The initial low system hardness with the crosslinker-containing basecoat is due to the addition of crosslinker solution (i.e., isocyanate in solvent) into the basecoat for spraying. Even though the basecoat is dry to touch before the clearcoat application begins, the reaction of the basecoat crosslinker with the basecoat binder is not complete yet, and likely drying of the solvent coming with the crosslinker is not complete either. The residual isocyanate and solvent probably contribute to the softness of the crosslinked basecoat and further to the overall softness of the clearcoat/basecoat system. As time goes on, crosslinking of the basecoat contributes to the formation of a stronger, harder basecoat, leading to higher system hardness.

The initial softened basecoat created by the addition of crosslinker provides better mobility for the diffusion of clearcoat materials including clearcoat resin and HDI trimer to diffuse into the basecoat. This helps build stronger interfacial bonding. When the crosslinker is not added, even though there is a concentration gradient of the clearcoat materials from the clearcoat to the basecoat, the mobility for diffusion is limited. On the other hand, the clearcoat is a relatively productive clearcoat, which can form a hard, vitrified material in 1–2 h under the air-dry condition. The fast curing of the clearcoat and limited mobility provided by the non-crosslinked basecoat both lead to limited diffusion of the clearcoat materials into the basecoat.

Summary

In this study, slab microtomy and FTIR-ATR analysis proved to be a useful tool to determine the depth profiles of chemical compositions, with particular focus on the isocyanate content, urea formation, and amide II (urea and urethane) formation in a coating system with a polyurea clearcoat and a waterborne basecoat. The analysis illustrates that there is limited isocyanate diffused from the clearcoat to the basecoat and limited urea formation in the standard basecoat. The limited diffusion is likely due to the fast curing of the clearcoat layer and the limited mobility of the clearcoat materials in the dried basecoat.

When a crosslinker solution is added to the basecoat, it leads to a more gradual change of the chemical compositions from the clearcoat to the basecoat. The depth profiles of isocyanate ring peak and urea carbonyl peak indicate that clearcoat resin and HDI trimer diffuse from the clearcoat into the basecoat. Addition of a crosslinker solution into the basecoat facilitates the diffusion of the clearcoat components into the basecoat. This is because unreacted crosslinker and additional residual solvent effectively plasticize the basecoat, enhancing the transport of the HDI trimer and amine resin from the clearcoat into the basecoat. It was also found that an elevated isocyanate and urea levels in the clearcoat close to the clearcoat/basecoat interface when the basecoat contains a crosslinker. This is probably caused by the isocyanate crosslinker coming from the basecoat reacting with the amine resin in the clearcoat.

The comparison of the amide II (urea and urethane) peaks illustrates that a much stronger basecoat was formed in the crosslinker-containing basecoat system than that with the standard basecoat. The stronger basecoat as well as stronger clearcoat/basecoat interface explain why system properties are better when a crosslinker-containing basecoat is used.

Acknowledgments We would like to thank Zuzana Stangova for preparing the coatings and collecting hardness data and Justin Ditchfield for conducting the FTIR-ATR analysis of the slab microtome-sliced films.

References

1. van der Ven, L.G.J., Leijzer, R.T.M., van den Berg, K.J., Ganguli, P., Lagendijk, R.: Interactions between basecoats and clearcoats in car refinish systems. Prog. Org. Coat. **58**, 117–121 (2007)
2. Cliff, N., Kanouni, M., Adamsons, K., Yaneff, P.V.: Migration of reactable UVAs and HALS in automotive plastic coatings. J. Coat. Technol. Res. **1**(3), 201–212 (2004)
3. Adamsons, K.: Chemical depth profiling of automotive coating systems using slab microtome sectioning with IR/UV-Vis spectroscopy and optical microscopy. J. Coat. Technol. **74**(924), 47–54 (2002)
4. Adamsons, K.: A modern analytical toolbox: defect (W&D) studies of automotive coating systems, including depth profiling studies. J. Coat. Technol. Res. **9**(6), 745–756 (2012)
5. Adamsons, K.: Predicting service life performance: our analytical toolbox; Polym. Preprints (American Chemical Society, Division of Polymer Chemistry), **45**(2), 201–202 (2004)
6. Lamers, P.: Migration effects in multilayer paint systems; Abstracts of Papers, 220th ACS National Meeting, Washington, DC, United States, August 20–24, 2000, PMSE-017 (2000)
7. Lambert, D., Muehlethaler, C., Gueissaz, L., Massonnet, G.: Raman analysis of multilayer automotive paints in forensic science: measurement variability and depth profile. J. Raman Spectrosc. **45**(11–12), 1285–1292 (2014)
8. Seubert, C.M., Nichols, M.E., Kucherov, A.V.: Long-term weathering behavior of UV-curable clearcoats: depth profiling of photooxidation, UVA and HALS distributions. J. Coat. Technol. Res. **2**(7), 529–538 (2005)
9. Trojek, T., Trojkova, D.: Several approaches to the investigation of paintings with the use of portable XRF analysis. Radiat. Phys. Chem. **116**, 321–325 (2015)
10. Reiche, I., Mueller, K., Eveno, M., Itie, E., Menu, M.: Depth profiling reveals multiple paint layers of Louvre Renaissance paintings using non-invasive compact confocal micro-X-ray fluorescence. J. Anal. At. Spectrom. **27**(10), 1715–1724 (2012)
11. Vyoerykkae, J., Vuorinen, T., Bousfield, D.W.: Confocal Raman microscopy: a nondestructive method to analyze depth profiles of coated and printed papers. Nordic Pulp Paper Res. J. **19**(2), 218–223 (2004)
12. Lee, Y., Han, S., Yoon, J.-H., Kim, Y.-M., Shon, S.-K., Park, S.-W.: Application of time-of-flight secondary ion mass spectrometry to automobile paint analysis. Anal. Sci. **17**(6), 757–761 (2001)
13. Gu, X., Drzal, P. L., Michaels, C. A., Jasmin, J., Martin, D., Stanley, D., Nguyen, T., Martin, J. W., Mechanical and chemical depth profiling of a UV-degraded polymeric system using nanoindentation and micro-FTIR imaging. 29th Proceedings of the Annual Meeting of the Adhesion Society, pp. 350–352 (2006)

14. Urban, M.W., Allison, C.L., Johnson, G.L., Di Stefano, F.: Stratification of butyl acrylate/polyurethane (BA/PUR) latexes: ATR and step-scan photoacoustic studies. Appl. Spectrosc. **53**(12), 1520–1527 (1999)
15. Chan, D.K., Misquitta, S.F., Ying, J.F., Martner, C.C., Hermsmeier, B.D.: Chemical depth profiling on submicron regions: a combined focused ion beam/scanning electron microscope approach. Surf. Interface Anal. **27**(4), 199–203 (1999)
16. Kaminski, A.M., Urban, M.W.: Interfacial studies of crosslinked polyurethanes; Part I. Quantitative and structural aspects of crosslinking near film-air and film-substrate interfaces in solvent-borne polyurethanes. J. Coat. Tech. **69**(872), 55–66 (1997)
17. Ludwig, B.W., Urban, M.W.: Quantitative determination of isocyanate concentration in crosslinked polyurethane coatings. J. Coat. Tech. **68**(857), 93–97 (1996)
18. Pretsch E., Buhlmann P., Affolter C.: Structure Determination of Organic Compounds, Tables of Spectral Data, pp. 301–302. Springer, New York, NY (2000)
19. de Haseth, J.A., Andrews, J.E., McClusky, J.V., Priester Jr., R.D., Harthcock, M.A., Davis, B. L.: Characterization of polyurethane foams by mid-infrared fiber/FT-IR spectrometry. Appl. Spectrosc. **47**(2), 173–179 (1993)
20. Buckle, H.: In: Westbrook, J.W., Conrad, H. (eds.) The Science of Hardness Testing and Its Research Applications, p. 453. American Society for Materials, Metals Park (1973)

Chapter 14
Characterization of Component Distributions in Acrylic Latex and Paint Films Containing an Alkali-Soluble Resin (ASR)

Wenjun Wu, Dana Garcia, and Steve Severtson

Introduction

Waterborne coatings are complex systems composed primarily of water, polymer, and pigment particles, but they also contain a variety of additives such as dispersants, rheology modifiers, surfactants (e.g., emulsifiers, wetting agents), and coalescing aids. These additives are used in latex synthesis and in paint formulation to facilitate the processing and application of coatings and to manipulate formed film properties. Although coating additives are present at relatively low concentrations, their interactions and distributions in pigmented paint films can have a significant impact on performance. The physical interactions between coating ingredients impact in-can properties such as viscosity, storage, and freeze-thaw stability. The distributions of additives in dry coating films influence properties such as gloss, hiding, color acceptance, tint strength, substrate adhesion, and durability. Of particular interest are the surface-active species. If these components migrate to film surfaces, barrier and adhesion properties as well as the coating appearance can be negatively influenced.

The distribution of water-soluble species and their tendency to enrich at interfaces during the drying of latex paint films depend on the conditions of film formation, as well as the chemical compositions, molecular weights, and rheological properties of both additives and latex components. Surface enrichment of

W. Wu (✉)
Arkema Coating Resins, Cary, NC, USA
e-mail: wenjun.wu@arkema.com

D. Garcia
Arkema Inc., King of Prussia, PA, USA

S. Severtson
Department of Bioproducts and Biosystems Engineering,
University of Minnesota, St. Paul, MN, USA

M. Wen, K. Dušek (eds.), *Protective Coatings*, DOI 10.1007/978-3-319-51627-1_14

surfactants has been reported for many latex films [1–11]. The negative consequences of a nonuniform surfactant distribution on end-use performance properties of adhesives and coatings have also been documented [12–16]. A key parameter in predicting the migration tendency of surfactants is their stability in water. Previously, it was shown that increasing the number of methylene linkages in sodium alcohol sulfates reduced their tendency to enrich at the surfaces of latex-adhesive films [8]. It was also shown that covalently binding a surfactant (polymerizable surfactant) to the backbone of the latex polymers nearly eliminated migration as well as its negative impact on properties when compared with its nonreactive analog [9]. Based on these findings, it has been concluded that structural modifications that promote greater interaction with the polymer phase relative to the aqueous eluent during the film drying process will inhibit surface enrichment. However, further evidence for this hypothesis is required, especially when it involves surface-active, polymeric additives.

Alkali-soluble resin (ASR) refers to polymeric species that possess hydrophobic moieties and carboxylic acid functional groups along their backbone. Upon ionization of acid groups, ASRs are amphiphilic and function as stabilizers and processing aids in the production of emulsion polymers, providing electrosteric stabilization of latex particles [17–24]. This allows reduction or elimination of conventional surfactants when ASRs are present while maintaining or improving latex stability. The ASR-stabilized emulsion polymers have shown performance advantages for a wide range of industrial applications [25–32]. It is conceivable that a high T_g ASR can also enhance hardness-related coating properties when blended with soft latex polymers for low-VOC paints. The ASRs are generally utilized at high concentrations (>20 wt%) to modify the film-forming polymers [18, 19, 33, 34]. The inclusion of ASR at such high levels is expected to alter the film formation process of emulsion polymers. Its influence is likely tied to its distribution in latex films during and after the drying process. Published research on the impact of ASR has examined relatively simple model systems such as poly(n-butyl methacrylate) homopolymer emulsion [35–38]. Disclosures of fundamental research and systematic characterization of ASR's influence on latex paint systems with industrial relevance cannot be found in the open literature.

In this study, a high glass transition temperature (T_g), mid-range molecular weight ASR, poly(styrene/alpha-methylstyrene/acrylic acid), is added to a soft acrylic latex. The model acrylic latex with a practical polymer composition and a low T_g is suitable for low-VOC waterborne coatings. Two microscopy techniques, atomic force microscopy (AFM) and confocal Raman microscopy (CRM), were employed to characterize surface morphology of ASR/acrylic polymer blends and distributions of the ASR and the polymerization surfactant in the acrylic latex and paint films. AFM phase imaging gauges energy dissipation from the tapping frequency phase shift when the AFM tip interacts with film surfaces. These differences are most commonly caused by variations in viscoelasticity, adhesion, and even contact area along the film surface. Given the large difference in T_g values between the acrylic latex and ASR, AFM phase imaging allows for the visualization of the relative spatial distributions of the two components.

CRM combines a Raman spectrometer with a high-resolution confocal microscope. The benefits of confocal scanning can be attributed to the two pinhole apertures, one providing single point illumination to reduce scattered light associated with wide field microscopes and the second pinhole blocking the out-of-focus rays. This spatial filtering feature sharpens the sample analysis and enhances image quality. Samples can be scanned horizontally to map the chemical compositions on XY planes, and the depth profiling can be obtained by changing the focal plane (moving the sample stage in the Z direction). When combined, a 3D compositional map can be constructed. Using the unique Raman spectra of different materials, CRM can be an effective method for visualizing and quantifying the distributions of individual constituents in multicomponent systems [7–9, 39, 40]. Since the acrylic polymer contains no aromatic moieties, Raman peaks associated with the bonds of aromatic components of the ASR can be used to quantify its concentration in the latex and paint films. Distinct chemical differences also exist between the polymerization surfactant and acrylic polymer, allowing for the distribution of the surfactant to be monitored.

This chapter demonstrates that AFM and CRM are a powerful combination for characterizing multicomponent systems with nano- to micron-scale resolution [39, 40]. The spatial distribution of ASR and surface enrichment of polymerization surfactants provide new insights on the transport mechanism and fate of such components in the latex and paint films.

Experimental

Materials

The model acrylic latex (AL) was polymerized from butyl acrylate, methyl methacrylate, methacrylic acid, and an ureido-functional monomer (hydroxyl ethyl ethylene urea methacrylate) by a semicontinuous emulsion polymerization. A C14-C16 alpha-olefin sulfonate was used as the polymerization surfactant to stabilize the latex particles. The pH value of the final latex was adjusted to 8.5 using a 28% ammonium hydroxide solution. The latex solids content was 50%, and the volume- (D_v) and number-averaged (D_n) particle sizes measured via dynamic light scattering were 129 and 113 nm, respectively.

The ASR sample used in this study was poly(styrene/alpha-methylstyrene/acrylic acid) having a number average molecular weight of 6500 and an acid number of 205. It was supplied as an ammonium-neutralized, 32% solution in water. The supplier-reported T_g for the terpolymer is 100 °C. The blends containing 10 wt% (B10) and 20 wt% (B20) ASR were prepared by mixing predetermined amounts of the ASR and the acrylic emulsion under agitation.

Physical properties of the samples used in this study are compiled in Table 14.1. The minimum film formation temperature (MFFT), identified as the point where the

Table 14.1 Physical properties of acrylic latex (AL), ASR, and their blends

Sample ID	ASR (%)	DSC T_g (°C)	MFFT (°C)	D_v (nm)	D_n (nm)
AL	0	−1.7	<0	129	113
B10	10	−1.1	<0	124	106
B20	20	−0.2	<0	123	103
ASR	100	110.5	n/a	n/a	n/a

latex formed a clear and crack-free film, was measured on a rectangular temperature gradient bar. The particle diameters of the neat acrylic latex and its blends with ASR were analyzed by a nanotrac particle size analyzer based on dynamic laser light scattering. The glass transition temperatures were determined by differential scanning calorimetry (DSC) from the dried latex films.

Latex Film Analyses

Water Contact Angle

Latex film samples for contact angle (CA), AFM, and CRM measurements were prepared by making 7-mil drawdowns on Leneta black plastic or white panels. The drawdown films were dried at 25 °C and 50% relative humidity for 7 days. The water contact angles (CAs) on latex films were measured using Rame-Hart Model 250 Standard Goniometer. Measurements were repeated at least 10 times and averaged values were reported.

Atomic Force Microscopy (AFM)

Surface morphologies of polymer films were examined using AFM. AFM images were acquired under ambient conditions using a Molecular Imaging PicoPlus Pico SPM 3000 System (Agilent Technology, Foster City, CA) in tapping mode. The cantilever was operated in the repulsive regime with oscillation amplitudes of 300 nm and an amplitude ratio A/A_0 (feedback amplitude/free-resonance amplitude) in the range of 0.90–0.95. All measurements used integrated silicon cantilevers with tip radii of curvature 5–10 nm, a spring constant in the range of 30–60 N/m, and a measured resonant frequency within 10% of 300 KHz.

Confocal Raman Microscopy (CRM)

Confocal Raman investigation of latex and paint films was conducted on an Almega™ Raman system from ThermoFisher Scientific. The Almega Raman utilized a 532 nm laser (Nd:YVO4 frequency doubled) with a power of ~25 mW.

The combination of 532 nm laser and 50XLWD (long working distance) objective yields a spatial resolution in the order of 300 nm. Spectra were collected from 64 exposures at 1 s each. Spectra data "as collected" and second derivatives (calculated using a Norris filter with a segment of 5 and a gap of 5) were analyzed. Multiple measurements were made to ensure reproducibility. Relative concentrations of the ASR were calculated using the method described by Xu et al. [7].

Depth profiling starts from the surface of the films. To ensure experimental consistency, the zero-depth position was determined under the objective of the Raman microscope. The sample stage was raised until the topographical features of the sample were in optical focus with reflected light, and then the C-H stretching Raman transition was maximized using the 532 nm laser source. This position was defined as the origin, and the depth was gauged relative to this position using a micrometer. The acrylic latex particles used in this study having a volume-averaged particle size of 129 nm and a narrow polydispersity of 1.14 were smaller than the sampling volume of the laser experiment.

Paint Film Analyses

Paint Formulation and Film Preparation

The acrylic latex and the blend containing 20 wt% ASR were evaluated in a simple paint formulation [42] with 35% volume solids (VS) and 31% pigment volume concentration (PVC). The corresponding paints were denoted as Paint 1 and Paint 3, respectively. The formulation ingredients and quantities in grams (per 100 g total batch size) are given in Table 14.2.

Table 14.2 Paint formulations

		Paint 1	Paint 3
Pigment grind	Water	22.00	22.03
	Colloid 226/35	1.36	1.36
	BYK 025	0.04	0.04
	QP4400	0.44	0.44
	Ti-Pure R-706	30.93	30.96
	Ammonia (28%)	0.35	0.35
	Sum	55.13	55.19
Letdown	Acrylic latex (AL)	39.39	31.14
	ASR		12.16
	Acrysol RM-2020	0.35	0.35
	Water	4.95	0.97
	BYK 025	0.18	0.18
	Sum	44.87	44.81
Total		100.00	100.00

Paint films for CRM were prepared on Leneta white panels by using a 7-mil (177.8 μm gap) drawdown bar and dried at 25 °C and 50% relative humidity for 7 days. The thickness of dried coatings was approximately 60 μm.

Results and Discussion

Water Contact Angle on Acrylic Polymer and ASR/AL Blends

Contact angle (CA) is a useful measurement for providing qualitative information on the surface composition of latex films. Figure 14.1 compares the water contact angle values for the AL, B10, and B20 polymer films. Interestingly, water CA does not change significantly with the addition of ASR. The films containing ASR have similar surface hydrophobicity to the neat acrylic polymer. In this study, contact angle measurements were also taken on the films after 50 and 100 cycles of water washing with a sponge. If this treatment removes the surface layer and exposes the underlying regions of the coating, surface enrichment or stratification of ASR in the latex films can then be probed by changes of water CA after scrubbing. The data in Fig. 14.1 shows the surfaces exposed after washing do not possess significantly different contact angles. However, given the relatively high variability of these measurements (indicated by error bars on the graph), the conclusion of uniform vertical distribution of ASR across the thickness of B10 and B20 films needs to be verified by more sensitive techniques. Fortunately, more accurate morphological and compositional information can be provided by AFM and CRM, respectively.

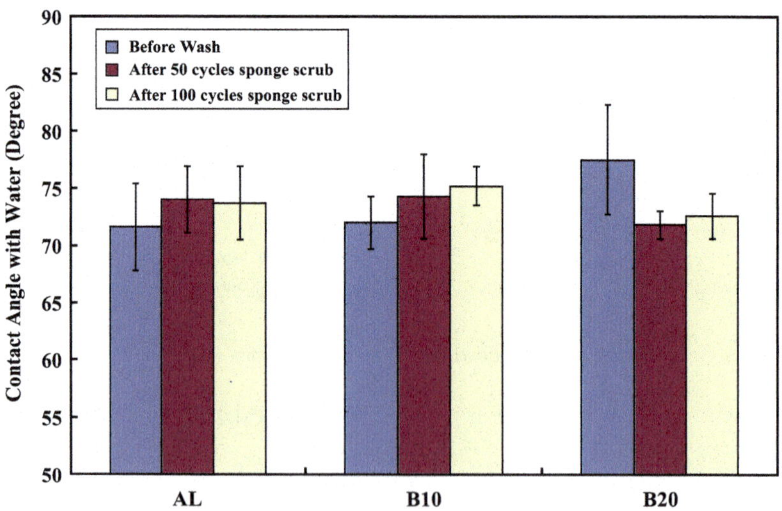

Fig. 14.1 Water contact angles on AL, B10, and B20 films

Film Morphology and ASR Distribution

Surface morphologies of latex films were characterized by AFM. The most commonly used AFM imaging approach for soft polymers is tapping mode in which the cantilever is oscillated at its resonance frequency and brought to the polymer surface. As the cantilever approaches the surface, the amplitude of oscillation is modified (reduced). The resulting phase shift depends on the viscoelastic properties of the sample and the adhesive potential between the sample and cantilever tip. Tapping mode allows examination of soft surfaces by eliminating direct interactions between the AFM tip and the sample surface. Phase composition is inferred from the measured viscoelastic properties, which provides images of surface morphology.

For heterogeneous materials, AFM phase images are often used to characterize differences in the properties of individual components. The phase images outline domains of varying material contrast without requiring prior knowledge of materials' chemical compositions. When tapping mode AFM is operated in repulsive regime, the soft or more energy dissipative material often results in less phase lag, which produces a dark contrast on the phase image while the stiff or less energy dissipative material provides bright contrast. The high T_g ASR as the hard component appears brighter than the soft acrylic polymer in the AFM phase images.

Figure 14.2 shows the AFM phase images of the AL film surface before (2a) and after (2b) being rinsed with deionized (DI) water. The surface of the neat latex film (2a) is covered with long needlelike structures, which are removable with water rinses. Figure 14.2b shows that light rinses with DI water uncover a smooth surface. This is further evidence that surfactants often migrate to the surfaces of latex films. This finding is consistent with the AFM results from other studies involving different polymerization surfactants [7, 10, 11]. Particle boundaries in Fig. 14.2b

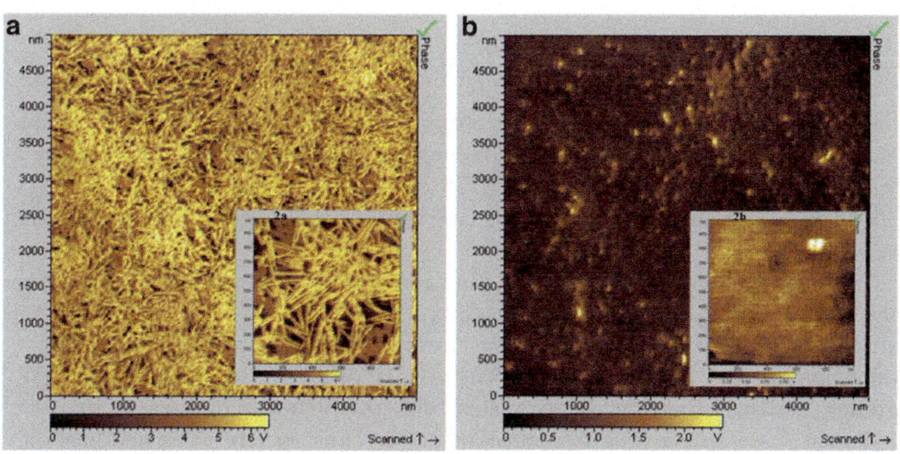

Fig. 14.2 AFM phase images of AL before (**a**) and after (**b**) water rinse

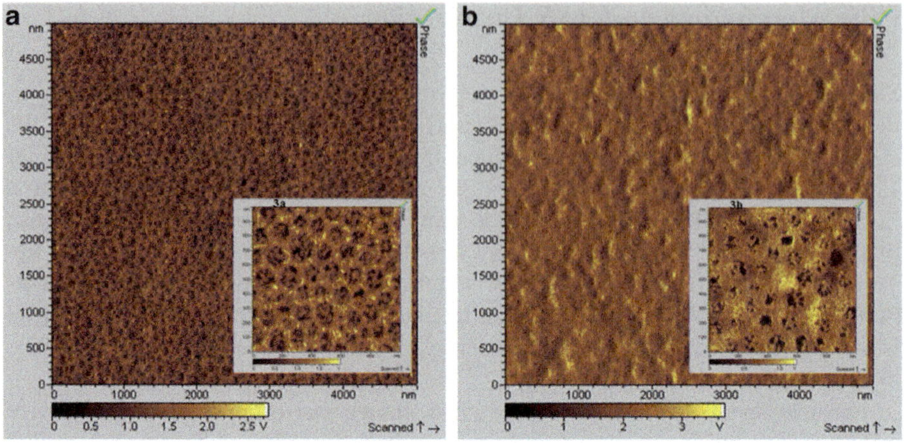

Fig. 14.3 AFM phase images of B10 (**a**) and B20 (**b**) film surfaces

are barely discernible due to particle coalescence, consistent with good film formation expected of the latex having a low MFFT.

Figure 14.3 displays the surface phase images of the latex blended with 10 and 20 wt% ASR, respectively. Both B10 and B20 form coherent films but show the coexistence of hard and soft materials. The surfactant structures shown in Fig. 14.2a are not visible on these two AFM images of film surfaces, indicating that the surfactant is either solubilized or displaced by the polymeric surfactant ASR. Furthermore, a higher magnification (inserts in Fig. 14.3a and Fig. 14.3b) reveals that spacing between dark regions increases when the ASR concentration is increased from 10% (Fig. 14.3a) to 20% (Fig. 14.3b) in the blend.

ASR Distribution in Latex Films

Figure 14.4 displays the Raman spectrum of the ASR with the styrene peaks marked. The ASR has characteristic aromatic C-H bending and ring breathing at 998 cm^{-1}, aromatic ring motion at 1598 cm^{-1}, and aromatic C-H stretching at 3062 cm^{-1}. The bands at 998 and 3062 cm^{-1} are stronger than the Raman transition at 1598 cm^{-1}.

Normalized Raman spectra for the pure components (AL and ASR) and B20 are shown in Fig. 14.5. The 1727 cm^{-1} is the carbonyl (C=O) stretching band associated with acrylic ester on AL. The spectrum of ASR does not exhibit significant transitions in the carbonyl region. The AL sample by design has no aromatic groups, and the sharp bands associated with the phenyl (ph) ring at 3062 and 1598 cm^{-1} are absent. The AL transition at 998 cm^{-1} is very weak and thus only a minor contribution. All signature bands, as expected, are present in the blend

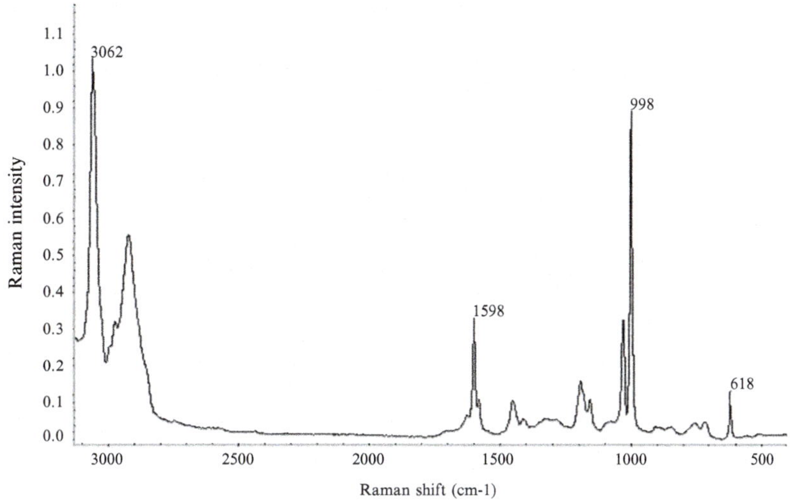

Fig. 14.4 Raman spectrum of ASR (styrene/alpha-methylstyrene/acrylic acid terpolymer)

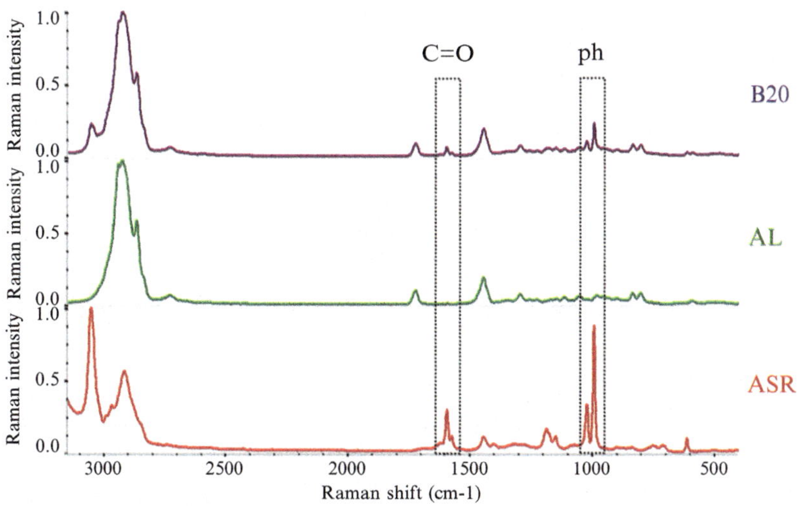

Fig. 14.5 Raman spectra of ASR, AL and B20 samples

sample B20. These bands are employed in quantitative analysis of ASR concentrations. The spectra of pure components can be used as calibration standards for the blends. The peak area ratio $A_{ph}/A_{C=O}$ is proportional to ASR concentrations.

Depth profiling was performed for all the samples. The spectra as a function of depth are shown on a normalized scale for comparison. Figure 14.6 provides spectral examples for the chemical compositions of B20 at the film surface and at various depths down to 40 μm below the film surface.

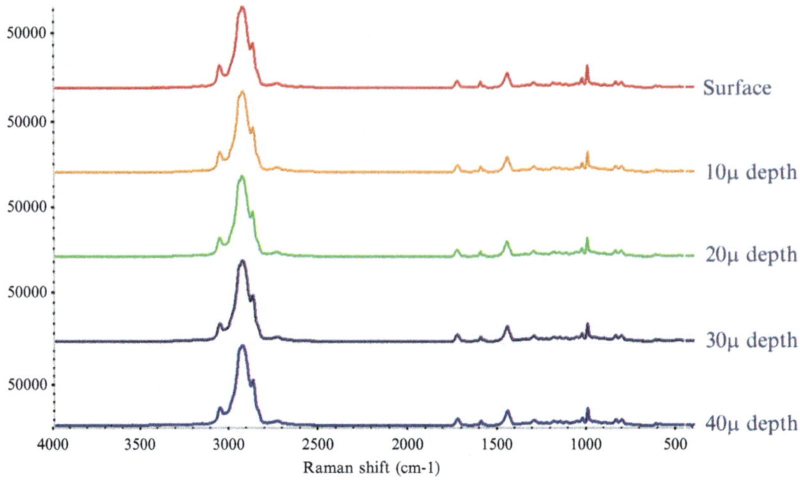

Fig. 14.6 Raman spectra for sample B20 at different depths

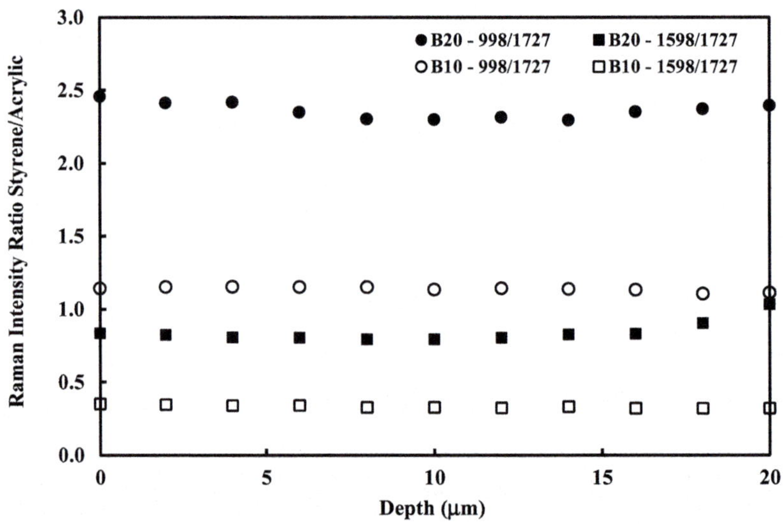

Fig. 14.7 Depth profiling of the relative ASR concentration in B10 and B20 samples

Figure 14.7 presents the Raman intensity ratios for the B10 and B20 films at depths of 20 μm into the latex films. The ratios were calculated using both the 1598 and 998 cm^{-1} peaks of the ASR and plotted in Fig. 14.7. The A_{998}/A_{1727} value is three times the magnitude of the A_{1598}/A_{1727} ratio, a direct reflection of styrene peak intensity difference at 998 and 1598 cm^{-1}, as shown in Fig. 14.4. The two sets of data are self-consistent, and both demonstrate that the ASR concentration is relatively flat as a function of depth, indicating no surface enrichment, stratification, or clustering of the ASR. Additionally, the Raman intensity ratio is doubled for the B20 versus B10. The CRM results therefore confirm a uniform ASR distribution in the B10 and B20 films.

ASR Distribution in Paint Films

Figures 14.8 and 14.9 show the Raman spectra of Paint 1 and 3 films with depth increments of 10 μm up to approximately 50 μm. The presence of rutile titanium oxide (TiO_2) is clearly seen in Figs. 14.8a and 14.9a. The Ti-O stretching and

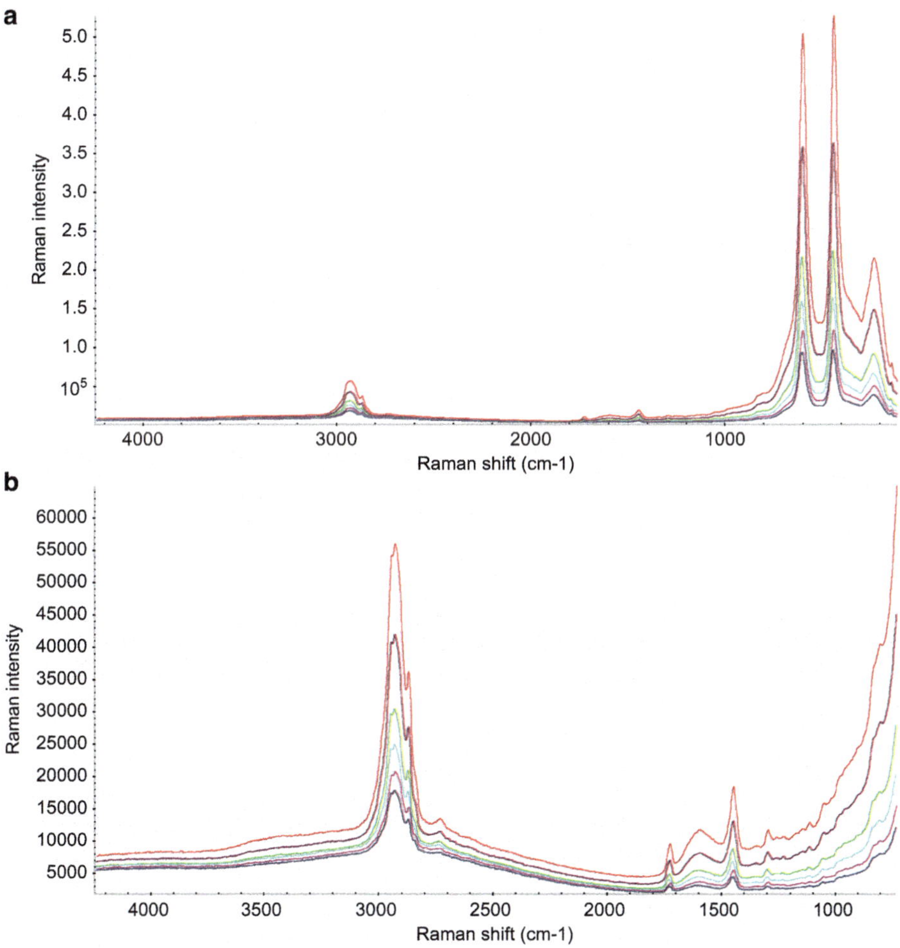

Fig. 14.8 Raman spectrum of Paint 1 film: 4000 to ~150 cm^{-1} (**a**) and 4000 to ~800 cm^{-1} (**b**). The *red* traces represent the top surface, and the subsequent downward spectra are at film depths of 10, 20, 30, 40, and 50 μm

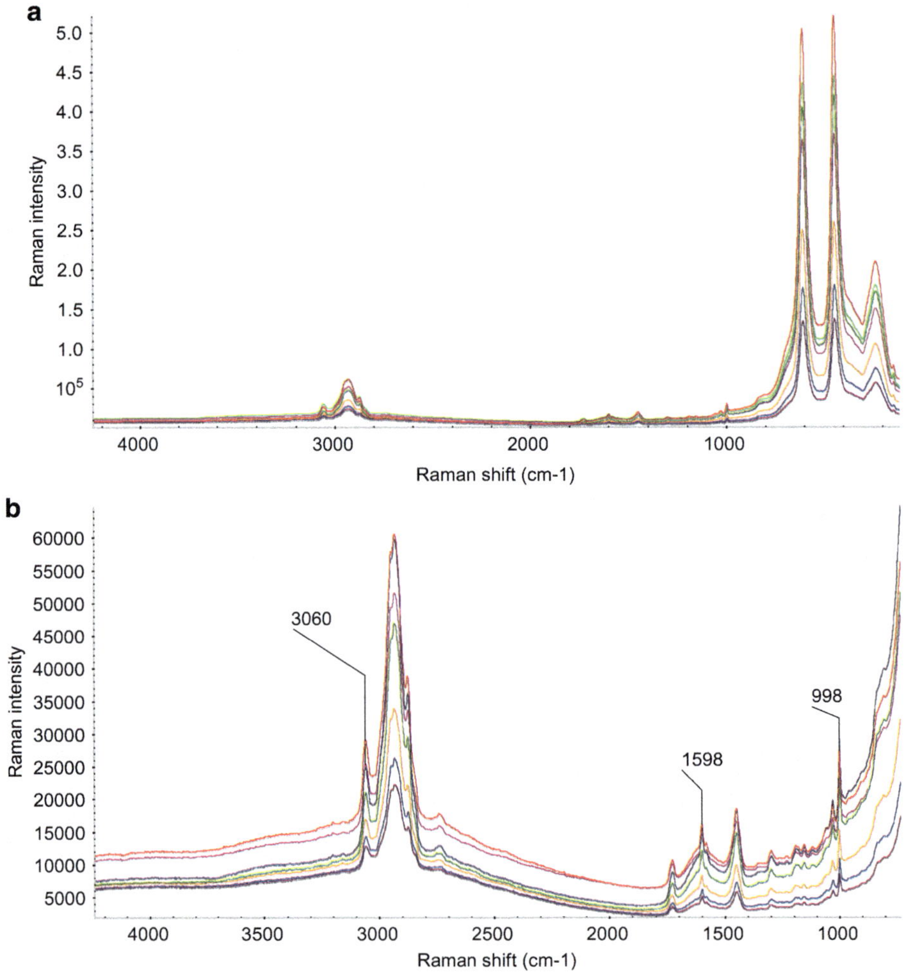

Fig. 14.9 Raman spectrum of Paint 3 film: 4000 to ~150 cm^{-1} (**a**) and 4000 to ~800 cm^{-1} (**b**). The *red* traces represent the top surface, and the subsequent downward spectra are at film depths of 10, 20, 30, 40, and 50 μm

bending transitions are in the spectral range of 650–100 cm^{-1}. However, the strong Raman transitions of TiO_2 at low wavenumbers do not impede the analysis of other spectral features belonging to the organic materials AL and ASR. The Raman peaks of interest for depth profiling of organic paint ingredients from 4000 to ~800 cm^{-1} are shown in Figs. 14.8b and 14.9b.

All spectral intensities in Figs. 14.8 and 14.9 decrease as a function of depth due to attenuation. Compared to the latex films, the attenuation is dramatic for the TiO_2-containing paint samples due to their increased opacity. For quantification of ASR

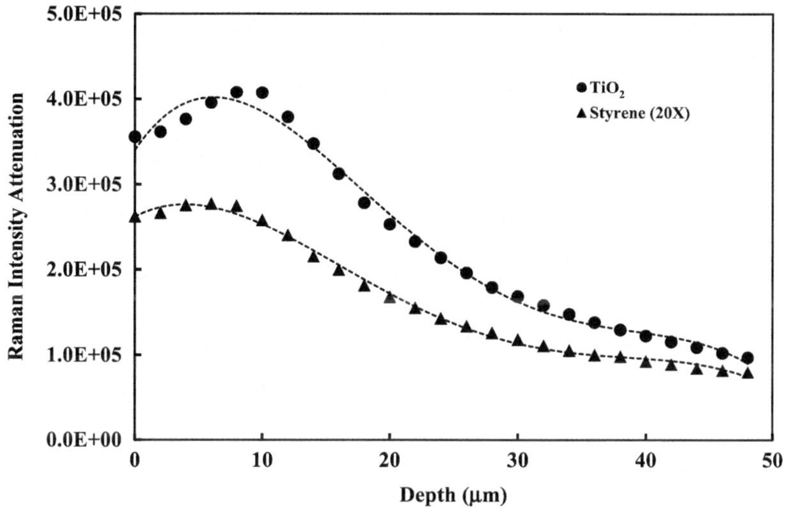

Fig. 14.10 Depth attenuation of TiO_2 and styrene bands (20×) in Paint 3 film

concentration, the intensity ratios of styrene to acrylic signature bands at various depths were used, so that the influence of signal attenuation is minimized.

The ASR component was detected in Paint 3 film. The styrene ring peaks are marked in Fig. 14.9b. No peak frequency shifts were noted. This includes the peak at 3060 cm^{-1}, which does not appear to result from the frequency shift of the 3062 cm^{-1} peak identified previously (Fig. 14.4). Rather, it is a convolution of C-H stretching vibrations from AL and other components of the formulation as well as the presence of pigment.

Figure 14.10 shows the Raman intensities of TiO_2 and the styrene transition at 998 cm^{-1} in Paint 3 as a function of depth. The intensity of the styrene peak at 998 cm^{-1} is weaker than TiO_2; thus a multiplication factor of 20 is used to display the styrene signals on the same graph with TiO_2. Figure 14.10 demonstrates that the depth signal attenuation is similar for TiO_2 and the styrene components, indicating a uniform ASR distribution across the film thickness.

Similar to the latex film study, the distribution of the styrenated ASR in the paint film can be determined by comparing the styrene peaks at either 998 or 1598 cm^{-1} to the C=O peak at 1727 cm^{-1} representing the acrylic constituent. Figure 14.11 shows the Raman intensity ratios of both styrene transitions to the carbonyl stretching. The intensity of the 998 cm^{-1} transition is greater than that of 1598 cm^{-1}, and its styrene to acrylic peak area ratio $A_{ph}/A_{C=O}$ is therefore higher, consistent with the spectral features shown in Figs. 14.4 and 14.9b. Nevertheless, the two sets of data in Fig. 14.11 display little variation as a function of film depth, and both lead to the same conclusion, i.e., the ASR concentration, indicated by the intensity ratios, remains constant across the thickness of the paint film. The CRM experiments yielded an excellent agreement with the previous finding regarding the ASR distribution in the latex films (Fig. 14.7).

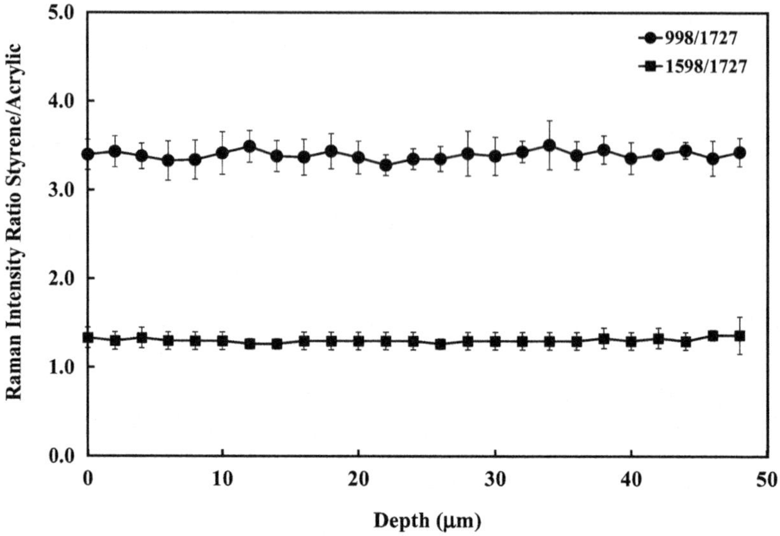

Fig. 14.11 Raman intensity ratios of styrene to acrylic in Paint 3 film

Figure 14.11 represents the averaged intensity ratios from three run profiles at different sample locations. The small error bars displayed in the graph demonstrate reasonably uniform lateral concentrations of ASR as well as good reproducibility of the Raman depth profiling. The results further support the conclusion that the low molecular weight, high T_g ASR is uniformly distributed, not only in the latex film but also in the pigmented paint films.

Surfactant Enrichment on Paint 1 Film

Figure 14.12 compares the second derivatives of Raman spectra in the C-H regions (3000–2800 cm^{-1}) for Paints 1 and 3 film surfaces. The Paint 1 film exhibits compositional inhomogeneities in the first 8 μm of the surface (Fig. 14.12a), which is not observed for Paint 3 film (Fig. 14.12b). These compositional changes are small and can be observed in the second derivatives of the Raman spectrum. The transitions between 2920 and 2925 cm^{-1} correspond to the C-H stretching of methylene linkage $(CH_2)_n$ in long-chain hydrocarbon structures. Higher absolute (negative sign is from the derivative calculation) intensity indicates higher Raman intensity in this spectral region. The differences in Fig. 14.12a thus indicate that a compound rich in hydrocarbon units resides on the top 4 μm of the Paint 1 film, presumably the C14-C16 alpha-olefin sulfonate surfactant used in the polymerization of the acrylic latex (AL). The migration of this surfactant toward air-film interface was observed in the AFM image of the neat acrylic latex film (Fig. 14.2a) and also verified by X-ray photoelectron spectroscopy in an internal

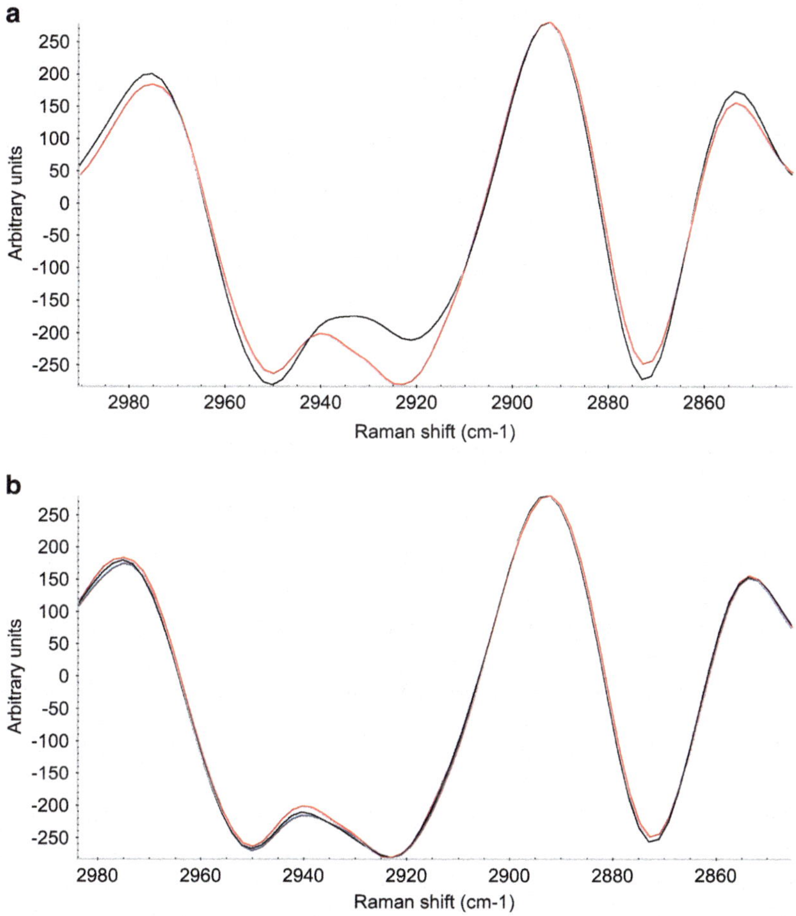

Fig. 14.12 Second derivatives of Raman spectra for the top 8 μm of Paints 1 (**a**) and 3 (**b**) films. The *red* trace represents surface (0–4 μm) and the *green* trace represents depth (4–8 μm)

study. The C-H-rich material is not observed in the Raman spectra of Paint 3, also consistent with the AFM results of its latex film surface (Fig. 14.3b). As shown in Fig. 14.12b, the intensities for the surface and depth of the Paint 3 film are very similar.

In paint formulations, the concentration of polymerization surfactant is significantly reduced but is still substantially higher than its critical micelle concentration (cmc). Figure 14.12 demonstrates that surfactant migration to paint film surface is also apparent.

Figure 14.13 illustrates the second derivatives of Raman spectra for the surface of Paint 1 film after water rinse. The Raman intensities (second derivative format)

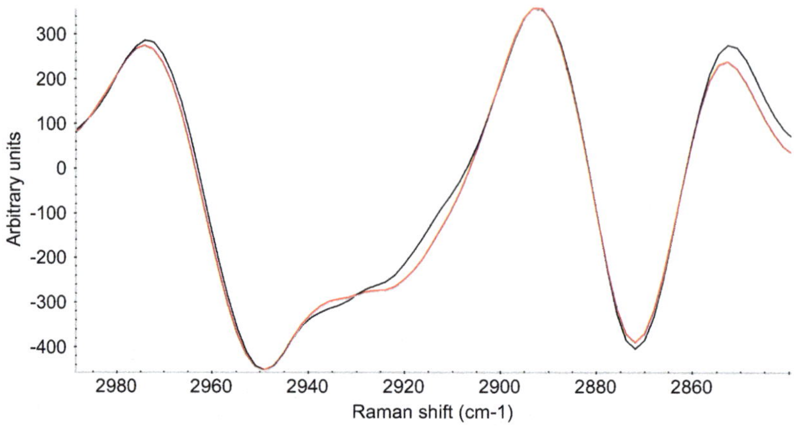

Fig. 14.13 Second derivatives of Raman spectra of water-rinsed Paint 1 film. The *red* trace represents surface (0–4 μm) and the *green* trace represents depth (4–8 μm)

are now much more similar, and the difference is no longer detected. This suggests that a simple rinse of the top surface with water is all that was required to eliminate most of the surface compositional inhomogeneity found in the top 8 μm of Paint 1 film. Any spectral changes in band shapes can be a consequence of the rinse process as the fine surface topography may have been altered. The Raman spectroscopy results should be interpreted in conjunction with AFM and XPS data and provide corroborative support.

The surfactant and ASR distributions have significant implications to the paint performance. The distributions of additives in the paint films depend on many factors including the chemical composition of the additive, latex polymer composition, film formation conditions, type and concentration of other formulation ingredients, and rheological properties of the dispersion particles and formulated paints. There is a large volume of literature documenting the detrimental consequences from nonuniform distribution of surfactants. Negative impact on film properties including surface defects and adhesion failures has been reported [12–15, 41].

The presence of ASR in the paint film can also have a dramatic effect on the paint performance properties. A recent paper has reported that while the inclusion of the high T_g ASR does improve block resistance, it also reduces water resistance, wet adhesion, and scrub resistance of acrylic latexes due to its water-sensitive and alkali-soluble nature of ASR [42].

Conclusions

This chapter reviewed the effects of a high T_g, alkali-soluble resin (ASR), poly (styrene/alpha-methylstyrene/acrylic acid) terpolymer, on the properties of latex and paint films. The film surface morphology and composition of a soft acrylic latex in the absence and presence of the ASR were evaluated using multiple experimental techniques including CA, AFM, and CRM. AFM phase images showed changing film morphology but coherent film formation by the soft acrylic latex with up to 20% ASR in the blend, consistent with the MFFT values. Uniform ASR distribution in the latex film was confirmed by CA measurements, as well as AFM and CRM.

CRM analyses showed evidence of surface enrichment of the polymerization surfactant on the pigmented paint films containing no ASR. The observed inhomogeneity was present in the top 8 μm of the paint film and eliminated through a rinse of the film with water. For ASR-containing latex and paint films, Raman intensity ratios confirmed a constant ASR concentration profile throughout both films and no surface enrichment of the surfactant species. This chapter demonstrates that CRM is a useful technique for characterization of additive distribution. The techniques utilized in this chapter can be applied to characterize component interactions and distributions in different polymer blends as well as other additives in paint formulations.

Acknowledgments The authors acknowledge Helen Xu for collecting the AFM images at University of Minnesota, as well as Arkema Coating Resins (ACR) and Analytical Services and Research (ASR) for supporting the publication of this chapter.

References

1. Aramendia, E., Mallégol, J., Jeynes, C., Barandiaran, M.J., Keddie, J.L., Asuza, J.M.: Distribution of surfactants near acrylic latex film surfaces: a comparison of conventional and reactive surfactants (surfmers). Langmuir. **19**(8), 3212–3221 (2003)
2. Zhao, C.L., Holl, Y., Pith, T., Lambla, M.: FTIR-ATR spectroscopic determination of the distribution of surfactants in latex films. Colloid Polym. Sci. **265**(9), 823–829 (1987)
3. Hellgren, A.-C.: Probing polymer interdiffusion in carboxylated latices with force modulation atomic force microscopy. Prog. Org. Coat. **34**(1–4), 91–99 (1998)
4. Du Chesne, A., Gerharz, B., Liser, G.: The segregation of surfactant upon film formation of latex dispersions: an investigation by energy filtering transmission electron microscopy. Polym. Int. **43**(2), 187–196 (1997)
5. Thorstenson, T.A., Tebeliu, L.K., Urban, M.W.: Surface and interfacial FT-IR spectroscopic studies of latexes. VI. Orientation and spatial distribution of acid functionalities at the film-substrate interface. J. Appl. Polym. Sci. **49**(1), 103–110 (1993)
6. Gundabala, V.R., Zimmerman, W.B., Routh, A.F.: A model for surfactant distribution in latex coatings. Langmuir. **20**(20), 8721–8727 (2004)
7. Xu, G., Dong, J., Zhang, J., Severtson, S.J., Houtman, C.J., Gwin, L.E.: Characterizing the distribution of nonylphenol ethoxylate surfactants in water-based pressure-sensitive adhesive films using atomic-force and confocal Raman microscopy. J. Phys. Chem. B. **112**(38), 11907–11914 (2008)

8. Zhang, J., Severtson, S.J.: Characterizing the distribution of sodium alkyl sulfate surfactant homologues in water-based, acrylic pressure-sensitive adhesive films. J. Phys. Chem. **115**(25), 8138–8144 (2011)
9. Zhang, J., Zhao, Y., Dubay, M.R., Severtson, S.J.: Surface enrichment by conventional and polymerizable sulfated nonylphenol ethoxylate emulsifiers in water-based pressure-sensitive adhesive. Ind. Eng. Chem. Res. **52**(25), 8616–8621 (2013)
10. Mallégol, J., Gorce, J.-P., Dupont, O., Jeynes, C., McDonal, P.J., Keddie, J.L.: Origins and effects of a surfactant excess near the surface of waterborne acrylic pressure sensitive adhesives. Langmuir. **18**(11), 4478–4487 (2002)
11. Mallégol, J., Dupont, O., Keddie, J.L.: Morphology and elasticity of waterborne acrylic pressure-sensitive adhesives investigated with atomic force microscopy. J. Adhes. Sci. Technol. **17**(2), 243–259 (2003)
12. Charmeau, J.Y., Kientz, E., Holl, Y.: Adhesion of latex films; influence of surfactants. Prog. Org. Coat. **27**(1–4), 87–93 (1996)
13. Charmeau, J.Y., Gerin, P.A., Vovelle, L., Schirrer, R., Holl, Y.: Adhesion of latex films. Part II. Surfactant effects at various peel rates. J. Adhes. Sci. Technol. **13**(2), 203–215 (1999)
14. Gerin, P.A., Grohens, Y., Schirrer, R., Holl, Y.: Adhesion of latex films. Part V. Dominating interfacial effect of the surfactant. J. Adhes. Sci. Technol. **13**(2), 217–236 (1999)
15. Charmeau, J.Y., Sartre, A., Vovelle, L., Holl, Y.: Adhesion of latex films. Part II. Loci of failure. J. Adhes. Sci. Technol. **13**(5), 593–614 (1999)
16. Asua, J.M., Schoonbrood, H.A.S.: Reactive surfactants in heterophase polymerization. Acta Polym. **49**(12), 671–686 (1998)
17. Tsaur, S.-L.: Resin-fortified emulsion polymers and methods of preparing the same. US Patent 4,820,762. Accessed 11 Apr 1989
18. Frazee, G.T.: Emulsion polymerization. US Patent 4,879,333. Accessed 7 Nov 1989
19. Frazee, G.T.: Pressure sensitive adhesive compositions. US Patent 4,923,919. Accessed 8 May 1990
20. Lee, D.-Y., Kim, J.-H.: Preparation of small-sized carboxylated latexes by emulsion polymerization using alkali-soluble random copolymer. J. Appl. Polym. Sci. **69**(3), 543–550 (1998)
21. Hwu, H.-D., Lee, D.-Y.: Studies of alkali soluble resin as a surfactant in emulsion polymerization. Polymer. **41**(15), 5695–5705 (2000)
22. do Amaral, M., Asua, J.M.: Synthesis of high solid-content latex using alkali-soluble resin as sole surfactant. Macromol. Rapid Commun. **25**(22), 1883–1888 (2004)
23. Lee, D.-Y., Kim, J.-H., Min, T.-I.: Role of alkali-soluble random copolymer in emulsion polymerization. Colloids Surf. A: Physicochem. Eng. Asp. **153**, 89–97 (1999)
24. Lee, D.-Y., Kim, J.-H.: Emulsion polymerization of styrene using an alkali-soluble random copolymer as polymeric emulsifier. J. Appl. Polym. Sci.: Pt. A: Polym. Chem. **36**(16), 2865–2872 (1998)
25. Ko, K.S., Lee, J.Y., Shim, J.K., Lee, J.Y.: Formulation of emulsion adhesives with removal properties by using alkali soluble resins. J. Nanosci. Nanotechnol. **13**(11), 7467–7471 (2013)
26. Sen, G.: Resins for water-based inks. Am. Ink Maker, **65**(12), 37–38, 40 (1987)
27. Ha, H.-J., OH, S.-H., Kim, S.-H., Park, Y.-J., Jin, Y.-S.: Resin-fortified emulsion polymers for water-borne inks and the method for preparing the same. WO 01/64764A1. Accessed 7 Sept 2001
28. Hu, J., Ma, J., Deng, W.: Synthesis of alkali-soluble copolymer (butyl acrylate/acrylic acid) and its application in leather finishing agent. Eur. Polym. J. **44**(8), 2695–2701 (2008)
29. Lee, D.L., Harry, L.D., Morris, J.D.: Aqueous polymeric dispersions, paper coating compositions and coated paper articles made therewith. US Patent 4,258,104. Accessed 24 Mar 1981
30. Nakamura M., Inoue, T.: Coating material and molded article. US Patent 6,759,473. Accessed 6 Jul 2004
31. Sunada, M., Takanashi, H., Ishikawa, H., Takanashi, A., Yoshino, F.: Composition for water-base paint comprising water-soluble resin and water-dispersible polymer. US Patent 4,179,417. Accessed 19 Dec 1979

32. Hibben, M.J., Wildman, M., Killilea, H., Heiden, W., Johnson, B.: Latex polymer for improved paint washability. WO 2013/134120 A1. Accessed 12 Sept 2013
33. Blank W.J., Layman, R.E.: Surfactant-free polymer emulsion coating composition and method for preparing same. US Patent 4,151,143. Accessed 24 Apr 1979
34. Kiehlbauch, R.A., Tsaur, S-L.: Resin-fortified emulsion polymers and methods of preparing the same. US Patent 4,839,413. Accessed 13 Jun 1989
35. Park, Y.J., Kim, J.H.: AFM study of morphological rearrangement of poly(butyl methacrylate) latex in the presence of alkali soluble resin. J. Ind. Eng. Chem. **3**(4), 305–309 (1997)
36. Park, Y.-J., Lee, D.-Y., Khew, M.-C., Ho, C.-C., Kim, J.-H.: Effects of alkali-soluble resin on latex film morphology of poly(n-butyl methacrylate) studied by atomic force microscopy. Langmuir. **14**(19), 5419–5424 (1998)
37. Lee, D.-Y., Shin, J.-S., Park, Y.-J., Kim, J.-H., Khew, M.-C., Ho, C.-C.: Surface morphology of latex film formed from poly(butyl methacrylate) latex in the presence of alkali-soluble resin. Surf. Interface Anal. **28**(1), 28–35 (1999)
38. Lee, D.-Y., Park, Y.-J., Khew, M.-C., Ho, C.-C., Kim, J.-H.: Mechanistic study on latex film formation in the presence of alkali-soluble resin using atomic force microscopy. Macromol. Symp. **151**, 479–485 (2000)
39. Belaroui, F., Grohens, Y., Boyer, H., Holl, Y.: Polymer. **41**, 7641–7645 (2000)
40. Schmidt, U., Hild, S., Ibach, W., Hollricher, P.: Characterization of thin polymer films on the nanometer scale with confocal Raman AFM. Macromol. Symp. **230**(1), 133–143 (2005)
41. Simmons, J.W., Thornton, R.M., Wachala, R.J.: Defoamers and antifoams. In: Calbo, L.J. (ed.) Handbook of Coating Additives, p. 158. Marcel Dekker (1987)
42. Wu, W., Guo, J., Pu, G., Zhang, J., Severtson, S., Miller, C.M.: Alkali-soluble resins (ASR) and acrylic blends: influence of ASR distribution on latex film and paint properties. Proc. Coatings Tech. Conf., Louisville, Kentucky (2015)

Chapter 15
Advances in NanoScratch Testing of Automotive Clearcoats

Gregory S. Blackman, Michael T. Pottiger, Benjamin W. Foltz, Jing Li, Ted Diehl, and Mei Wen

Introduction

Scratch resistance is one of the highest rated customer concerns for automotive paint systems [1, 2]. Automotive customers would prefer to have pristine, scratch-free finishes for the life of their vehicle. According to Seubert et al. [2], consumers in different geographical regions have different levels of expectation with regard to scratch performance. North American consumers are most concerned about large, fracture scratches. In Europe, where automatic carwashes often utilize stiff, plastic bristles, consumers are more concerned with carwash scratches and the associated loss of gloss. In this field, the terms scratch and mar are both used. Scratch or scratching refers to displacement or removal of material from the surface and mar refers to a change in the perfection or appearance of the surface. Both the mechanical properties and the appearance are important aspects for automotive clearcoats and both will be addressed in this chapter.

There are many different methods to characterize the scratch resistance of a coating, each with different and sometimes unknown statistical capabilities, e.g., taber abraser, crockmeter, wet and dry rub tests, and benchtop carwash. The wear mechanisms in the various performance tests are complex and may not represent the actual damage that occurs to the clearcoat on an automobile. It makes little sense to make subtle changes in coating chemistry, processing conditions, dispersion, etc. and use a wear measurement method that is incapable of discerning a difference. Even if the instrument is statistically capable, the sample itself may have significant variability due to local changes in coating thickness, morphology, distribution of

G.S. Blackman (✉) • M.T. Pottiger • B.W. Foltz • J. Li • T. Diehl
DuPont CR&D, Experimental Station, Wilmington, DE, USA
e-mail: Gregory.S.Blackman@dupont.com

M. Wen
Axalta Coating Systems, Coatings Technology Center, Wilmington, DE, USA

M. Wen, K. Dušek (eds.), *Protective Coatings*, DOI 10.1007/978-3-319-51627-1_15

additives, or even surface active contaminants. A further complication is that the mechanical properties of many paint systems change in the days or weeks after being applied.

Wear depends on energy and sometimes material transfer from one material to another. How this transfer occurs determines which material properties are most important for improving the wear performance. Incomplete understanding of the wear mechanism can lead to focusing on the wrong material property. For a scratch in an automotive clearcoat, the resistance to both plastic deformation and fracture is important as is the recovery over time.

During marring of automotive coatings, grit, sand, or a polymer brush slides across the surface (often in a carwash), leading to the formation of numerous tiny scratches that are microns to tens of microns wide and several microns deep. These are very easy to see in bright sunlight on a dark-colored car and are often described as swirl marks.

There are two components to scratch resistance: damage resistance and scratch visibility [3, 4]. Damage resistance is the ability of the material to resist damage from the deformation typically associated with sharp objects. Scratch visibility is the degree of visual perceptibility of scratch damage by human eyes. In the late 1990s and early 2000s, there was considerable effort focused on developing methods for characterizing the damage resistance of automotive coatings [1, 5–7], which led to ASTM D7187–15[1]. More recently, there has been a lot of work done to understand the relationship between scratch resistance and polymer properties [4, 8–13], which has led to the development of ASTM D7027–13[2].

It is known from the work of Lin and Blackman [5, 6] that there are two primary mechanisms that contribute to damage resistance: plastic plowing and fracture. In ASTM D7187–15, the resistance to fracture damage and the resistance to plastic deformation are extracted from progressive load scratches using a 2 μm radius indenter and loads ranging from near zero to 50 mN. The 2 μm radius indenter was chosen to mimic the asperities on grit or dust particles, resulting in damage that resembles damage encountered in the field. This work also showed that there was a connection between visibility and scratch morphology by performing statistical surveys of appearance [5]. The researchers discovered that when given only a short time to observe scratches on an automotive clearcoat, scratches above the fracture threshold were most easily observed. However, under strong illumination and with plenty of time, survey respondents could see much smaller, plastically deformed scratches. Thirty-five percent of the survey participants were able to see a scratch that was only 6.4 μm wide and 24 nm deep in a 1K melamine clearcoat on a black basecoat.

[1]ASTM D7187–15: Test method for measuring mechanistic aspects of scratch/mar behavior of paint coatings by nanoscratching.

[2]ASTM D7027–13: Test method for evaluation of scratch resistance of polymeric coatings and plastics using an instrumented scratch machine.

In ASTM D7187, the fracture resistance was defined as the load where the first sustained fracture occurred in the coating either by monitoring changes in the force or displacement curves or more commonly by using integrated high-resolution optical microscopy. The plastic resistance parameter was defined as the residual damage, measured 1–2 min after the scratch was produced, at an arbitrary load of 5 mN. However, most polymers have a significant viscoelastic relaxation or recovery after a stress is removed, which leads to an underestimate for the plastic resistance. Recovery continues for hours or days after the scratch is created as shown in a later section of this chapter.

ASTM D7027 recommends using a 500 μm indenter and a load range from 2 to 50 N. This stress causes delamination and severe damage for most automotive clearcoats [14]. Various optical techniques and profilometry are used to assess the resultant damage and define the critical load for polymer failure or delamination. However, the larger load range and indenter size create damage that is orders of magnitude larger than the typical size of swirl marks or scratches from field damage in automotive clearcoats.

In many end use applications, failure is assessed by a visual inspection. The detailed morphology of the scratch, the scattered light intensity, and the contrast difference between the scratch and the undamaged coating nearby all contribute to the visibility of the scratch; see reference [4] and citations therein. A scratch is generally smooth and featureless during plastic deformation, but above the fracture threshold, the scratch morphology can change dramatically (as shown later in this chapter). It is important to determine the correlation between scratch size, shape, and visibility. When the NanoScratch test is coupled to a three-dimensional imaging tool like an atomic force microscope (AFM) or optical profilometer, measurements of scratch morphology and wear rate or wear volume can be calculated as well. These parameters are a direct measure of polymer flow and relaxation under abrasive wear conditions and can be used as a quantitative measure of abrasion resistance.

This chapter discusses an improved methodology with robust statistics for evaluating the scratch resistance of automotive coatings including techniques for evaluating both damage resistance and scratch visibility. Four different types of clearcoats with different crosslink densities were selected to give an overview of how they would perform differently in the test. These clearcoats are crosslinked coatings based on different types of chemistry, including 2K urethane, 1K melamine, and two experimental clearcoats.

Experimental Methodology

Materials

Four different automotive clearcoat formulations with different crosslink densities were selected for this study. The 2K urethane clearcoat (Axalta Coating Systems) is based on acrylic and polyester polyols crosslinking with hexamethylene

Table 15.1 Swell ratios of the clearcoats aged for 1 day and 23 days

Clearcoat	1K melamine	2K urethane	Exp 1	Exp 2
1 day	1.35	1.79	2.16	1.23
23 days	1.35	1.55	1.95	1.20

diisocyanate (HDI) trimer. The 1K melamine clearcoat (Axalta Coating Systems) is based on acrylic and polyester polyols crosslinking with alkoxylated melamine formaldehyde resin. The other two coatings are experimental coatings referred to as Exp 1 and Exp 2. Exp 1 is a crosslinked coating through condensation polymerization with HDI trimer, and Exp 2 is a crosslinked acrylate coating. The 1K melamine and 2K urethane clearcoats are similar to the ones used to develop the original NanoScratch technique that led to ASTM D7187. They are referred to in the standard as Sample A and B, respectively.

The crosslink densities of these coatings were determined by the extent of swelling of the films in dichloromethane. Free-standing films were cut into circular shapes, and the diameters before and after swelling were measured. A swell ratio, determined by the area of the swollen state and the initial area, was reported for each film. Table 15.1 lists the swell ratios of the four clearcoats aged for 1 day and 23 days. The lower the swell ratio, the higher the crosslink density. The ranking of the crosslink densities of these coatings (from high to low) is: Exp 2 > 1K melamine > 2K urethane > Exp 1.

All of the coatings were sprayed with a machine directly on electrocoated steel panels at 23.9 °C and 50% RH. The thickness of the cured clearcoats was about 50 μm.

Test Procedure

The best practice for measuring the damage resistance of polymer coatings is still evolving. The current recommendation is to use a combination of scratch tests to assess damage resistance and optical imaging and Atomic Force Microscopy (AFM) to determine the morphology of the scratch damage and how it heals or recovers over time. Dark field imaging is used as a surrogate for visibility because the contrast is related to scattered light intensity. A progressive load scratch test is used to determine the onset of fracture and a step load scratch test is used for subsequent statistical analysis of plastic deformation resistance and scratch morphology (see Appendix).

The general procedure is to perform a progressive load scratch test using a freshly cleaned 2 μm radius diamond indenter and then to look for any evidence in the force or displacement measurements of a fracture or film rupture event (Fig. 15.1a). Optical microscope images are also used to look for evidence of fracture. If a fracture event occurs, the normal force at that point is recorded (fracture threshold). This process is repeated for several identical progressive

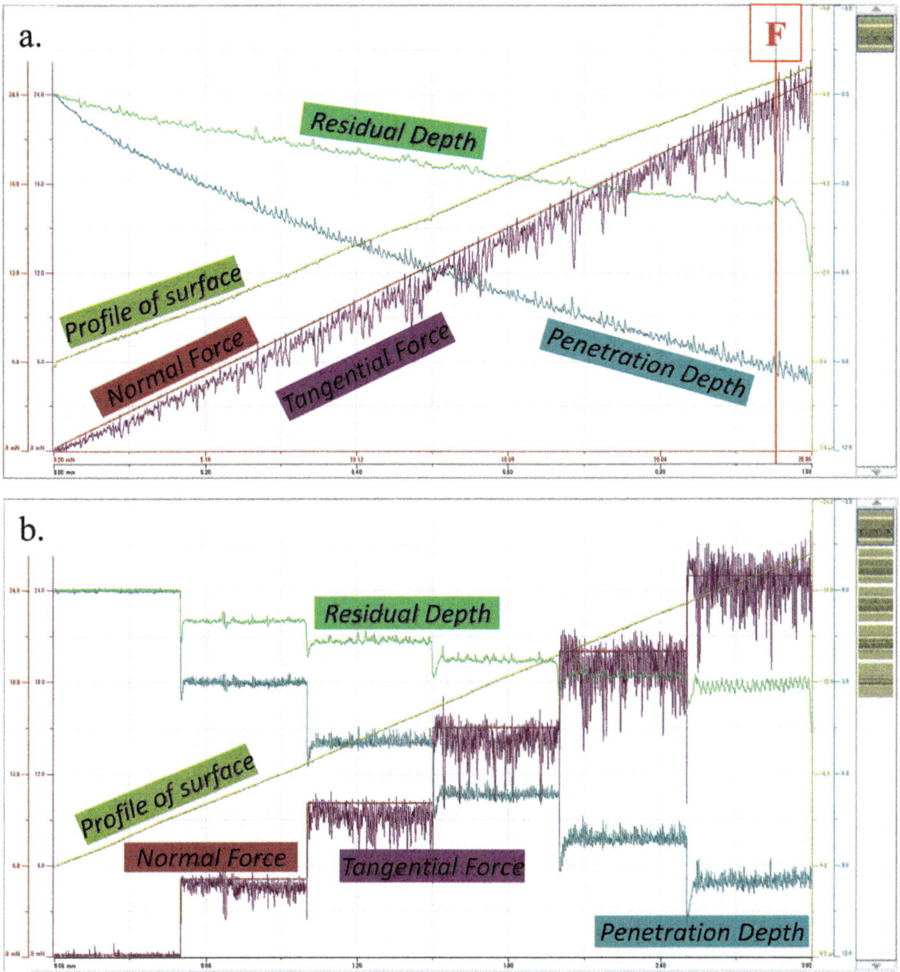

Fig. 15.1 Raw data from NanoScratch tester showing current recommendation for scratch testing of automotive clearcoats. The data in this figure is from the 2K urethane coating aged for 1 day. (**a**) Progressive scratch from 0.2 to 25 mN with the fracture threshold indicated by the "F". (**b**) Step scratch used in statistical analysis. The load steps range from 0.2 to 25 mN in five equal steps

load scratches. Following the progressive load scratch tests, a step load scratch is performed at selected loads above and below the fracture threshold (Fig. 15.1b). Dark field optical micrographs and AFM images are collected at exactly the same locations on the step scratch. The surface is allowed to recover for at least 24 h before AFM imaging. The following section describes the procedure in detail using the 2K urethane as an example. Statistical measurements and standard deviations for all coatings will be discussed in later sections.

Scratch Test Conditions

Scratch tests were performed using a NanoScratch Tester from CSM Instruments (Anton Paar) based on the technology developed by DuPont and Axalta Coating Systems with a 2 μm radius, 90° cone angle diamond indenter. The indenter was cleaned prior to each experiment by sonication for 30 min in reagent grade isopropanol, sonication for 30 min in HPLC grade water, and UV ozone treatment for 10 min. AFM images were collected to qualify the indenter prior to use.

Two types of scratch tests were performed: a progressive load test and a step or incremental load test. In each case, three passes across the surface were performed under a controlled load. The first pass, or pre-scan, records the initial undamaged surface profile of the sample. During the second pass, the load is increased as the scratch is created on the clearcoat surface. The third pass, or post-scan, is performed to record the residual damage. The penetration depth was calculated by subtracting the initial pre-scan profile from the displacement during the scratch. The residual depth was calculated by subtracting the post-scan displacement from the pre-scan profile.

The pre-scan and post-scan were at a constant 0.2 mN load, which for most coatings is below the plastic deformation threshold. The progressive load test was performed over a distance of 1 mm at a speed of 0.5 mm/min. The load was increased linearly from 0.2 mN at the start of the test to 25 mN at the end of the test. The residual depth was recorded 4 min after the damage was produced under these conditions. The step load test was performed over a distance of 3 mm at a speed of 0.5 mm/min. The load was increased in steps from 0.2 to 5, 10, 15, 20, and 25 mN. For the step scratch, the residual damage was recorded 12 min after damage due to the longer scratch. After each pass, the indenter moves back to the beginning of the scratch at the same rate of 0.5 mm/min at a load of 0.2 mN.

Scratching Coefficient of Friction

Sliding friction is the resistance to motion when two objects slide past one another. During a scratch test, the indenter penetrates into the surface and elastic, viscoelastic, and plastic deformation occurs. The lateral force measured in the NanoScratch tester depends on both the surface friction between indenter and clearcoat, but also on the material properties of the displaced material in front of and beneath the indenter. The tangential force is measured during the scratch process and used to calculate the average scratch coefficient of friction at each load step. However, due to the large penetration depths, this parameter is nearly the same for each coating and will not be discussed further in this chapter. At low loads or with larger diameter indenters (less surface damage), the measured friction would likely be more diagnostic.

Scratch Morphology

Optical Microscope

A Nikon® Eclipse® ME600 optical microscope was used to collect dark field images of each scratch. To achieve the best contrast, images were captured in dark field with controlled light level using 10× and 50× objective lenses. The dark field images are used as a surrogate for scratch visibility in this chapter. The gray level in the dark field images ranges from 0 (black) to 256 (full white).

Atomic Force Microscopy

All of the indenters used for this study were examined by AFM to measure their true three-dimensional shape. AFM images of the indenters were collected using a Dimension 5000 V Scanning Probe Microscopy (SPM) (Bruker Corporation). The digital images are 512 × 512 or 512 × 256 pixels. An Icon AFM (Bruker Corporation) was also used to image the scratch morphology in tapping mode. Scanning was carried out in the air using commercially available OTESPA-R3 (Bruker Corporation), nominally 160 μm long, 26 N/m spring constants, and 300 kHz frequency. Quantitative analysis of the area function of the indenters and the scratch morphology is accomplished using SPIP® software (Image Metrology).

Results and Discussion

In a NanoScratch experiment, the penetration depth is the difference between the original surface height profile and the displacement during the creation of the scratch. The residual depth is the difference in displacement between the original profile of the sample and one taken minutes later. The stress distribution around the indenter during a scratch is complex with spatially nonuniform shear, tensile, and compressive elements. However, it is instructive to compare the changes in material behavior during the scratch to a simple stress–strain curve in tension.

Initially, the indenter is applied at low loads and the material responds elastically; the penetration depth depends only on the modulus of the sample and the residual depth would be zero. This corresponds to the Hookean region, or purely elastic portion of a tensile stress–strain curve. As the load increases, the material beneath the indenter starts to deform plastically, analogous to the yield point in a simple tensile stress–strain curve. At this point, the penetration depth is the sum of elastic and plastic deformation, and the residual depth (depth after the load is removed) becomes nonzero. The point on the scratch where the residual depth becomes nonzero is referred to as the elastic/plastic deformation threshold; however, for polymeric coatings, this point can be difficult to quantify.

As the load increases further, more of the total deformation is due to plastic flow of the material. The residual depth would be equivalent to the permanent deformation observed when a force is removed from a material and the resultant change in sample dimension is measured. Eventually, a material limit is reached or a defect in the coating is encountered and film failure or rupture occurs. In a tensile stress–strain curve, this condition is referred to as the elongation to break. In a NanoScratch experiment, this point is defined as the fracture threshold. The goal of the improved method for scratch testing of automotive clearcoats is to provide robust statistics for plastic deformation, fracture, and recovery, so that subtle differences in performance are measureable.

Indenter Characterization

All results from scratch tests depend on the detailed geometry of the indenter used to make the damage, because all such measurements depend on the area of contact at a given load. It is *essential* to monitor the shape and cleanliness of the indenter over time to detect changes. Even diamond indenters wear over time and in rare cases can crack or fracture. Such changes have a profound impact on the NanoScratch results. Debris or contamination on the indenter can also lead to inexplicable variations in the results.

The most sensitive method for characterizing and quantifying the indenter shape, size, and cleanliness is an atomic force microscope (AFM). Figure 15.2a is a three-dimensional AFM image of the nearly perfect, 2 μm radius, 90° cone angle diamond indenter that was used in the present study. The indenter manufacturers typically use optical microscopy at high magnification and examine two side views of the indenter, but occasionally they also use transmission electron microscopy (TEM) to examine an indenter. However, neither technique provides three-dimensional information about the indenter. The optical microscope does not have sufficient resolution to detect small changes in tip radii or shape. Also, neither technique is capable of detecting surface contamination which will often show up in the phase image of a typical AFM. Additionally, some TEM instruments actually contaminate the surface during imaging with a thin layer of carbon generated from the background gases.

Ideally, the indenter should be imaged as received from the vendor and then carefully cleaned and reimaged just prior to each experiment. The three-dimensional image (Fig. 15.2a) is also used to calculate the area function (area as a function of depth) for the indenter as shown in Fig. 15.2b. The vendor specifies the radius of the indenter and the cone angle, but the true radius is seldom the same as the specification, and the indenter apex is almost always contaminated with polishing residue and debris. The calculated area functions for ideal 1 and 2 μm radius indenters are also shown in Fig. 15.2b [15]. The indenter selected for the present study was newly purchased from Anton Paar, and the measured area function is less than an ideal 2 μm indenter. For comparison the area function

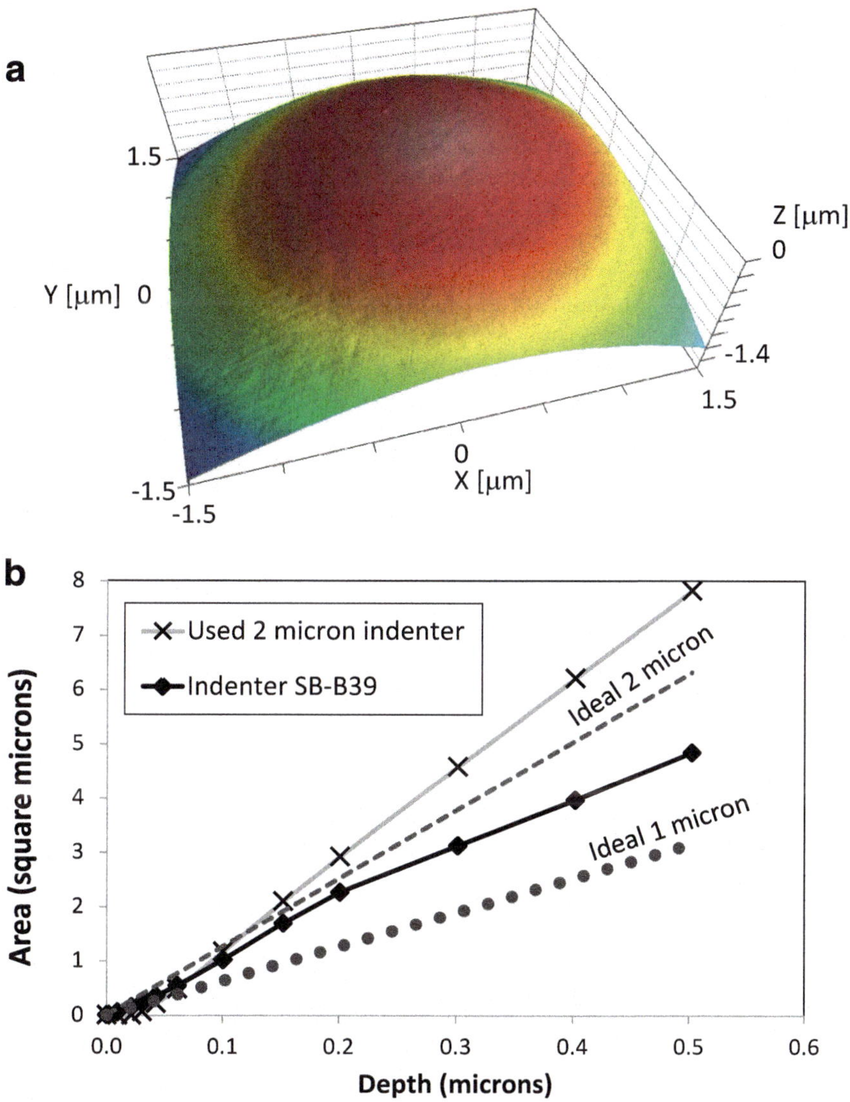

Fig. 15.2 (**a**) Three-dimensional atomic force microscope image of indenter SB-B39 used in this study. (**b**) Indented surface area as a function of penetration depth calculated from the AFM image. The calculated area functions for ideal 1 and 2 μm radius indenters are also included. For comparison the area curve for a used 2 μm indenter is also included

curve for a used 2 μm radius indenter is also shown in Fig. 15.2b. Over time this indenter has become blunt and slightly misshapened. This seemingly small change in indenter radius (1.6–2.6 μm) increases the fracture threshold by 2.8 mN for a typical polymer (unpublished results).

In principle, finite element calculations could be used to determine the differences in stress and stress distributions around the indenter tip using the real geometry, and the NanoScratch test (NST) results could be adjusted accordingly. However, in practice, the indenter is replaced when it is no longer symmetric or when the radius changes too much. For labs that do not have ready access to AFM, changes in fracture threshold for a smooth, flat polycarbonate or PMMA sample could be used to indicate indenter problems. With care the standard deviation on the fracture threshold for such a sample should be less than +/− 0.5 mN. If the fracture threshold varies by more than 1 mN, the indenter should be replaced.

Determination of Fracture Threshold

A typical progressive load scratch for the 2K urethane coating is shown in Fig. 15.1a, with the fracture threshold at 23.8 mN indicated by the "F". There is a slight fluctuation in the tangential force at this point, but it is often difficult to identify the fracture threshold based solely on the output from the NanoScratch tester. In order for the force or displacement curves to be perturbed, there has to be sufficient energy released in each fracture event or a significant change in the local height such that the force feedback loop cannot accurately track the surface. When the height change of the surface is small or the crack width is narrow, the 2 μm radius indenter will not be able to sense the changes in morphology and the displacement curves will show no evidence for fracture. An integrated optical microscope is generally more reliable and is used to inspect the damaged area and determine the load at which sustained fracture occurs.

Three identical progressive load scratches were performed to get a modest level of statistics on the fracture threshold. In the case of the 2K urethane coating aged for 1 day, the average fracture threshold was ~23.4 mN (one of the progressive load scratches ended at 25 mN without evidence of fracture for this sample). Many automotive clearcoats continue to dry and cure for some time after spraying and baking as the crosslinked network develops. Additional NST measurements were performed over time to monitor the changes in surface mechanical properties.

Measurement of Plastic Resistance

In ASTM D7187, a parameter called plastic resistance is extracted from the NST data by recording the residual depth at a 5 mN load. However, relying on a single data point from a progressive load scratch can lead to high standard deviations and the inability to discriminate among samples. Instead, thousands of data points are collected at a constant load over many millimeters of the coating both to obtain sufficient statistics for a valid ANOVA or two-sample t-test and to assess the point to point variability in the surface. The constant load or step scratch also makes it easier to perform tests on other instruments (AFM or dark field imaging) and to monitor the recovery over time or the reflow upon heating. These plots and the subsequent statistical analysis are based on a 3 mm long scratch with over 5600 data points in each 0.5 mm load step (Fig. 15.1b). The processing of the scratch prior to statistical analysis is described in the Appendix. The loads in the step scratch method were selected so that one of the steps would be near a load of 5 mN to compare with historical data, and the range was selected so that at least one step would be above the fracture threshold.

Measurement of Scratch Recovery

Viscoelastic materials are known to recover significantly after a stress is removed. For polymer coatings, there are short and long timescale recovery processes that will reduce the severity of a scratch. The short timescale (e.g., within minutes) recovery can be easily evaluated by the NanoScratch tester. However, once the sample is removed from the stage, the long timescale (e.g., days or longer) recovery is performed using AFM analysis of the scratch.

The scratch morphology and recovery were monitored over time with periodic AFM images of the same area of a step scratch. Figure 15.3 graphically illustrates

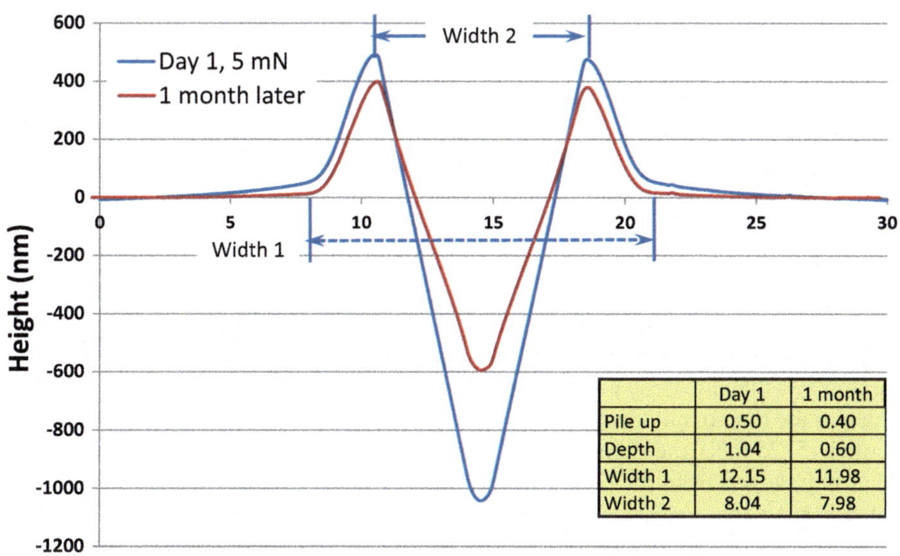

	Day 1	1 month
Pile up	0.50	0.40
Depth	1.04	0.60
Width 1	12.15	11.98
Width 2	8.04	7.98

Fig. 15.3 Average profiles across a 5 mN load step on the 2K urethane coating aged for 1 day and after the scratch has recovered for 1 month. The two curves are of the exact same scratch with 1 month of recovery time between the AFM images

this recovery by comparing the exact same area of a scratch as it recovers over time. The scratch was created 1 day after spraying and imaged immediately after with the AFM. After the scratch has recovered for 1 month, the precise region of the scratch was reimaged with the AFM. The coating in this experiment was the 2K urethane clearcoat. The profiles in Fig. 15.3 are an average of 25 individual AFM scan lines through the exact same region of the 5 mN load step.

The mean residual depth from the NanoScratch tester for this load step was 1.03 μm (recall this is 12 min after the scratch was performed). The average depth from the initial AFM scan is 1.04 μm, which is very close to the residual depth obtained from the NST. However, over the course of 1 month, the scratch recovers to a depth of 0.6 μm. Figure 15.3 also shows that it is the depth that recovers significantly, but the width of the scratch stays the same over time. There are two methods to measure the width of the scratch as indicated in Fig. 15.3. The total damaged area is measured from the widest point on the profile or image where the surface deviates from the undamaged baseline level and is labeled as Width 1 in Fig. 15.3. Width 2 is measured from the maximum pileup on either side of the scratch. The width of the scratch is determined by the width of the indenter at the level it penetrates into the surface, and neither of the measured widths changes appreciably over time.

Short timescale (e.g., within minutes) and long timescale (e.g., days or longer) recovery for the scratches created at different step loads were also calculated. The convention for scratch recovery is to express it as a relative change of the scratch depth after a certain period of recovery over the original depth created during the scratch. The recovery R can be calculated by the following equation:

$$R = \frac{(\mathrm{PD}_0 - \mathrm{RD}_t)}{\mathrm{PD}_0} \times 100\%$$

where PD_0 is the average penetration depth created by NST originally, and RD_t is the average depth after recovery. For the short timescale recovery, RD_t is the average residual depth determined by NST. For the long timescale recovery, RD_t is the average depth determined from the AFM images.

Table 15.2 shows the short timescale (12 min) recovery of the scratches created on the 2K urethane coating aged for 1 day and 23 days. In each load step, the penetration depth and residual depth were higher for the fresher coating. For the coating aged for 1 day, at the 5 mN load step, the indenter penetrated 3.01 μm during the scratch and recovered to 1.03 μm immediately afterwards. The penetration depth during the scratch for the coating aged for 23 days (2.08 μm) was nearly a micron less than the fresh coating, and the residual depth reduced to 0.63 μm. These changes are due to buildup of the network and evaporation of solvent. From the step scratch (5–25 mN), the short timescale recovery of the scratches on the coating aged for 1 day ranged from 65 to 67%. Interestingly, the short timescale recovery of the scratches on the coating aged for 23 days was higher, reaching 70–73%. This indicates that as crosslink density increases, the coating can recover from deformation and damage more easily.

Table 15.2 Comparison of the average penetration and residual depths from the NanoScratch test and short timescale recovery of the 2K urethane coating aged for 1 day and 23 days. Also compared is the average penetration depth and long timescale recovery of the scratch after 9 days of recovery measured by AFM on the coating aged for 23 days. The AFM images are shown in Fig. 15.4

Step load (mN)	5	10	15	20	25
Coating aged for 1 day					
Penetration depth (NST), μm	3.01	4.93	6.64	8.08	9.50
Residual depth (NST), μm	1.03	1.68	2.30	2.81	3.17
Short timescale recovery	66%	66%	65%	65%	67%
Coating aged for 23 days					
Penetration depth (NST), μm	2.08	3.33	4.31	5.23	6.08
Residual depth (NST), μm	0.63	1.01	1.25	1.43	1.71
Short timescale recovery	70%	70%	71%	73%	72%
Depth after 9 days (AFM), μm	0.57	0.86	1.11	1.27	1.34
Long timescale recovery	73%	74%	74%	76%	78%
Width after 9 days (AFM), μm	9.75	13.97	17.03	20.67	22.66

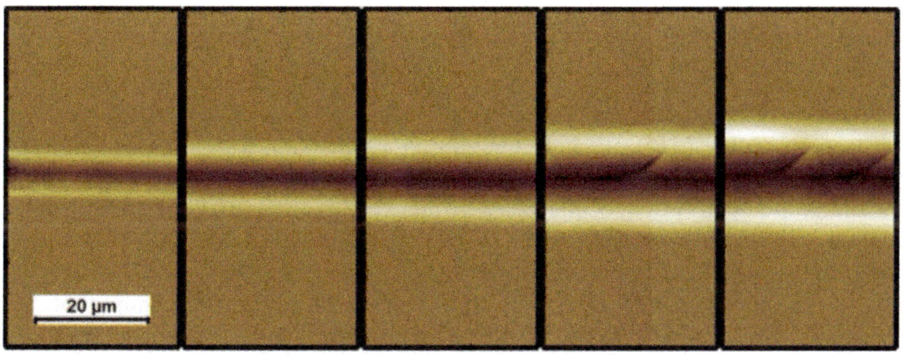

Fig. 15.4 AFM images of scratches 9 days after the scratches were created. The images were obtained from the center of each constant load step for the 2K urethane coating. The original scratches were created on the coating aged for 23 days. From left to right, the step loads used are 5, 10, 15, 20, and 25 mN, respectively

Also compared in Table 15.2 is the long timescale, i.e., 9 days, recovery of the scratches on the coating aged for 23 days. Figure 15.4 shows the AFM images used for determination of the residual depth after 9 days of scratch recovery. The images were obtained from the center of each load step. The long timescale recovery for the scratches ranged from 73 to 78%, which are consistently higher than the short timescale recovery. The healing of scratches over time or with additional heating is of great interest but beyond the scope of this chapter.

The extent of recovery depends on the coating chemistry and on the applied load. The visibility of the scratch depends on the morphology in the fully recovered state. This recovery was explored for various polymers using nanoindentation and AFM by Tweedie et al. [16]. They observed up to 80% recovery of the depth and volume of an indent in a polymer over the first 48 h.

The 2K urethane coating has a fracture threshold of 23.4 mN for a freshly prepared coating (1 day) and 19.1 mN after aging for 23 days (see details later). The 20 and 25 mN images in Fig. 15.4 show the characteristic chevron cracks associated with fracture for the 2K urethane coating. Conventional wisdom suggests that once a material fractures, the recovery or healing will be limited. However, all the materials in the current study recover substantially at all loads. When a scratch is produced, the total deformation is the sum of elastic and plastic components. Some portion of the recovery is due to the elastic and viscoelastic recovery after the indenter passes. This portion recovers even if the material has fractured. It is unlikely that the cracks or crevices generated by fracture will heal even if the depth recovers. Residual strain in the material tends to keep the fracture surfaces open as observed in the various AFM images above the fracture threshold. If the fracture is sufficiently severe, debris is generated and ejected from the surface creating a much larger and more visible scratch.

Scratch Visibility

There is a step change in visibility or perception of a scratch once the material is in sustained fracture [5]. Dark field imaging is a convenient way to determine the fracture threshold, and it is used as a surrogate for scratch visibility. In dark field imaging, the central or specular beam is blocked, and the image is comprised of light that is scattered from the surface. It usually results in a flat black background with enhanced contrast for scratches or other surface irregularities. This is particularly useful for automotive clearcoats where the surface roughness of the undamaged coating is often less than 5 nm RMS (from AFM images of undamaged area), and the scratches produced by the NST are over 1 μm deep.

The fracture threshold for the 2K urethane coating aged for 23 days is 19.1 mN. The AFM images in the center of both the 20 and 25 mN load steps, Fig. 15.4, show the characteristic chevron cracks associated with fracture for this material. Figure 15.5 is a comparison between the dark field image and the AFM image in the center of the 25 mN load step. From the three-dimensional AFM image (Fig. 15.5a), the scratch is 22.9 μm wide and 1.25 μm deep with pileup on either side of the scratch 0.81 μm tall. The crack or crevice in the image is likely deeper than the AFM can penetrate and is the opening to a subsurface pore [6]. In the dark field image (Fig. 15.5b), the mean background gray level is 5 and the maximum peak in the fracture event is 111, although the precise value depends on the microscope and camera parameters.

The scattered light intensity from the dark field image is used as a surrogate for scratch visibility. For a smooth, featureless clearcoat on a black basecoat or e-coat, a feature like a scratch stands out. It is visible because it scatters light, and the more scattered light intensity the more visible the scratch. Figure 15.5c shows the average

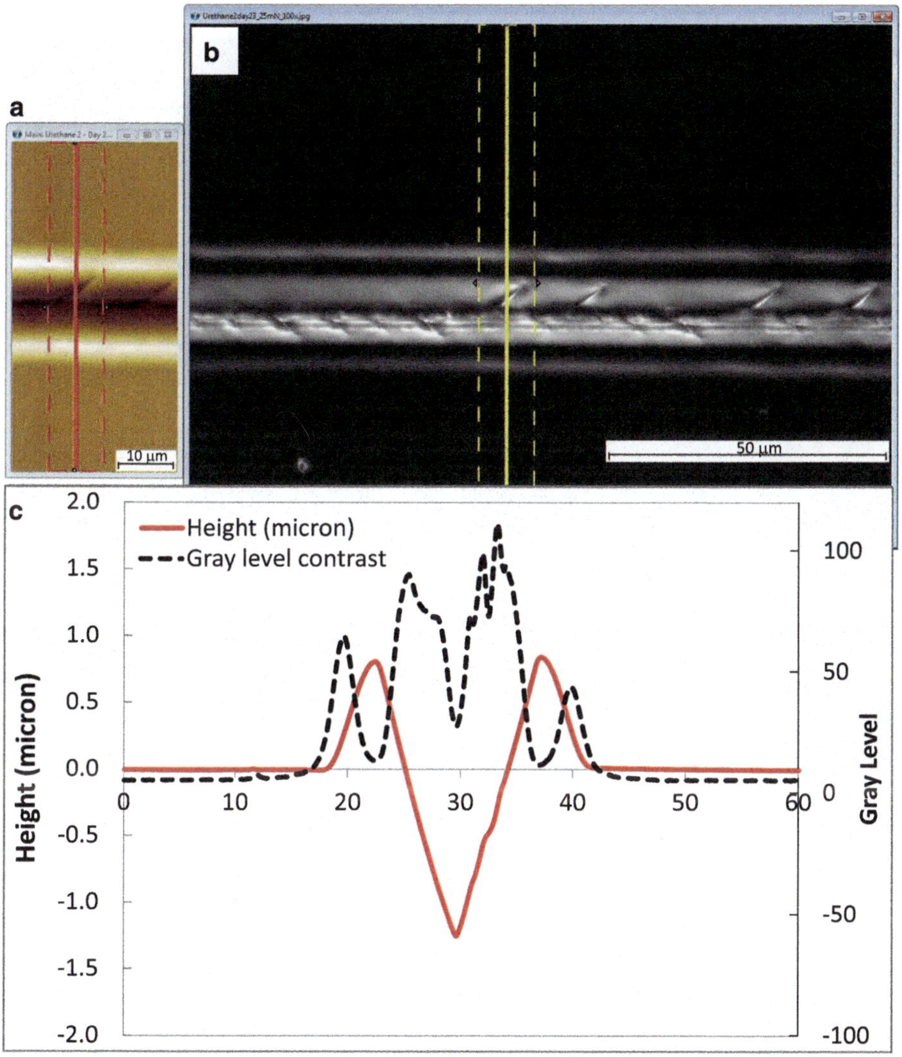

Fig. 15.5 (**a**) Atomic force microcopy image of the center of the scratch created by 25 mN load step and recovered for 9 days for the 2K urethane aged for 23 days. (**b**) Dark field optical image of the same 25 mN region. (**c**) Average profiles through the same fracture event comparing the scratch morphology to the scattered light intensity

profiles through the same portion of a fracture event including both the topographic profile and the light scattering from the dark field image. The most striking feature of the figure is that the pileup on either side of the scratch from plastic flow results in less total scattered light than the fracture event itself. This is the primary reason why the majority of people can easily perceive a scratch that is above the fracture threshold, but it is more difficult to see one that only has plastic deformation [5].

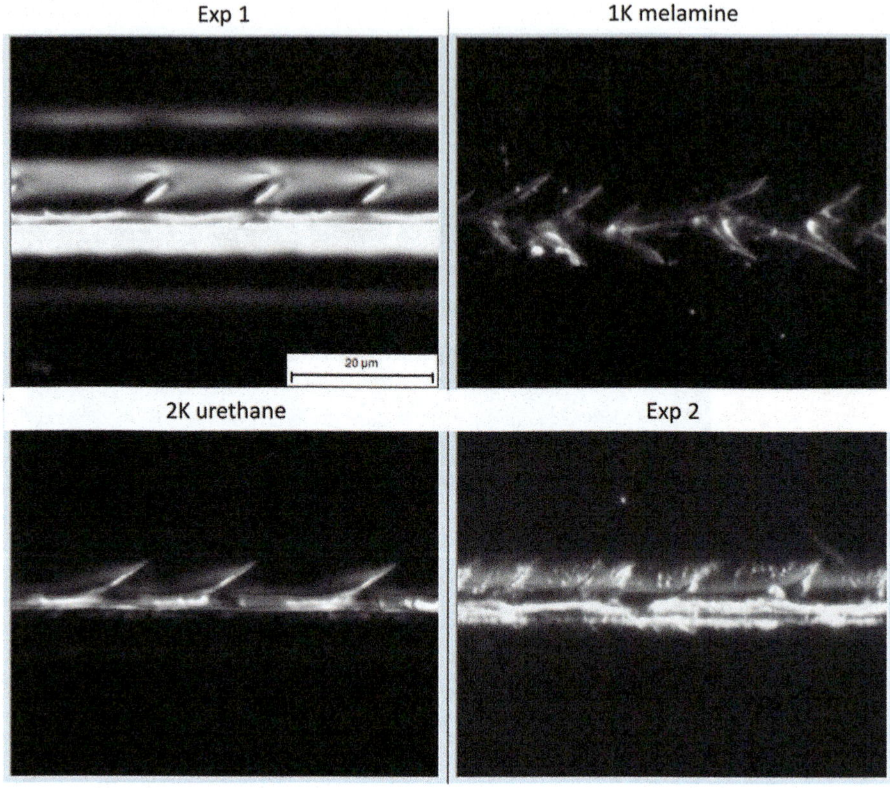

Fig. 15.6 Dark field images from the center of the 25 mN load step for each of the coating types in the present study. The lighting conditions were identical for each of the images. Note the large differences in scattered light intensity, which affects the visibility of the scratches. The scratches were produced on freshly prepared coatings (1 day after spraying), and the dark field images were collected shortly after the scratches were created

The detailed morphology of the scratch above the fracture threshold depends on the properties of the clearcoats and will change from one clearcoat to another (Fig. 15.6). The lighting conditions for each of the images in Fig. 15.6 were identical, but the dark field images are dramatically different. The number and severity of fracture events is different for each coating, which will affect the visibility of the scratches. For example, there are three distinct fracture events in the 60 μm image of the Exp 1 clearcoat, and the contrast between the undamaged coating and the scratch is large. However, on the 1K melamine clearcoat, there are more distinct fracture events, but the contrast is much lower.

To further quantify the scratch visibility based on the contrast in the dark field images, a histogram analysis was used. The images have a background gray level of less than 40 (out of 256 gray levels). Setting the threshold to 40 gets rid of the background, and both the mean gray level and the area of the scratch are calculated

Table 15.3 Visibility measurements of the scratches on different clearcoats in terms of the total scratch area and mean gray level in the scratches based on the 60 μm (scratch length) dark field images in Fig. 15.6

Clearcoat	Exp 1	1K melamine	2K urethane	Exp 2
Scratch area (μm^2)	1267	388	402	678
Mean gray level in scratch	128	73	85	125

based on the pixels that have gray levels greater than 40. The mean gray level of the scratch is a measure of the scattered light intensity which we are using as a surrogate for scratch visibility. The total size (area) of the scratch will also affect the visibility. In general, which coating is the "best" depends on which parameter is most important, the scattered light intensity (contrast) or the size of the scratch.

Table 15.3 summarizes both the mean gray levels of the scratches and the scratch areas for each of the coatings in the present study. Both metrics rank the coatings in the same way: 1K melamine < 2K urethane < Exp 2 < Exp 1. Exp 1 has the lowest crosslink density, and this is consistent with the larger scratch area and minimum side cracks. Exp 2 has the highest crosslink density, and the scratch visibility is similar to Exp 1. This is likely caused by its brittleness and tendency to form cracks when its crosslink density gets too high. The two coatings with medium crosslink densities, 2K urethane and 1K melamine, are the ones with less scratch visibility. Overall, crosslink density appears to play an important role in the scratch visibility. To minimize scratch visibility, crosslink density should not be too low or too high.

Statistical Comparison Between Materials

The fracture threshold varies across coating technologies and can change as the coating ages as shown in Fig. 15.7. By this measure the initial fracture threshold is the highest for the 2K urethane. As the sample ages, the polymer network builds and residual solvent diffuses out, and the fracture threshold drops from 23.4 to 19.1 mN. This lowered fracture threshold indicates that the coating has become more brittle over time. The 1K melamine coating has the lowest fracture threshold at around 8.3 mN. For historical reference, the ASTM D7187 used coatings very similar to the 2K urethane and the 1K melamine, and the fracture thresholds were 18.9 and 8.9 mN for these two materials, respectively. The Exp 1 clearcoat has slightly lower fracture threshold initially than that of the 2K urethane. However, its fracture threshold increases over time and reaches the highest value of 22.8 mN among the clearcoats. The Exp 2 clearcoat has similar fracture threshold initially to that of the Exp 1 clearcoat. The fracture threshold then dropped to 18.2 mN—a phenomenon similar to the 2K urethane clearcoat. The fracture threshold of the coatings aged for 23 days ranks from high to low in the order: Exp 1 > 2K urethane > Exp 2 > 1K melamine. The precise value of the fracture threshold depends on the coating

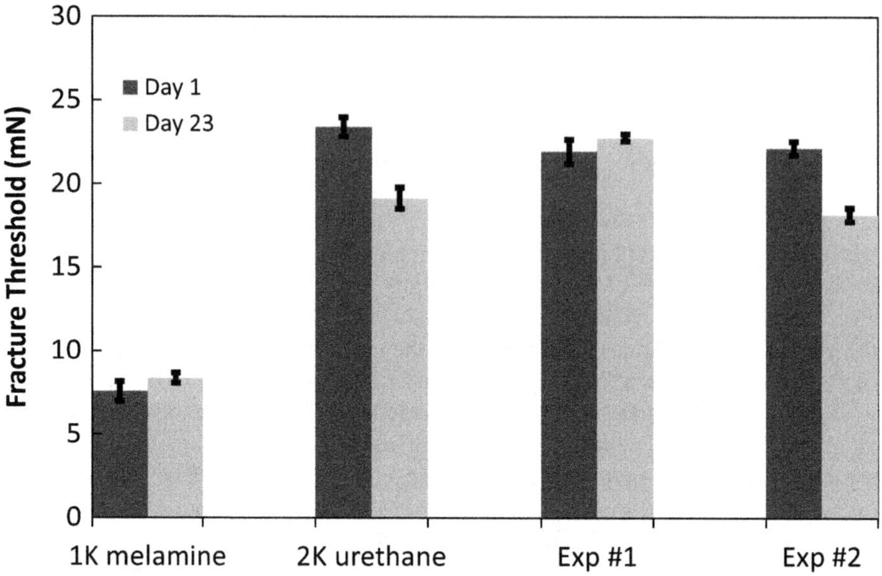

Fig. 15.7 Comparison of fracture thresholds, from progressive load scratches created on clearcoats aged for 1 day and 23 days. The error bars are from the standard deviations of three separate progressive load scratches

chemistry, but also on the radius, sphericity, and roughness of the indenter used to make the scratch as discussed in an earlier section. No clear trend is present in terms of the fracture thresholds and crosslink densities of these coatings.

In a typical step scratch, there are over 280 independent measurements of penetration depth, residual depth, and scratch coefficient of friction at each load step. Detailed processing is described in the Appendix. This amount of data enables robust statistical comparison of samples. Figure 15.8 is a boxplot from an ANOVA calculation performed within Matlab® for the scratches created by the 5 mN load step. The solid portion of each box represents the middle 50% of the data, and each stem is the lower and upper 25th percentile. Statistical flyers are marked by the stars, and there are seldom more than 10–15 outlier points out of the 287 measurements. A general rule of thumb is that if the boxes do not overlap on such a plot, then there is a statistically significant difference between those samples.

It is obvious that the new method with over 280 independent measurements of each parameter has more statistical power than the previous method (ASTM D7187) with only three measurements, but what are the practical consequences of the higher statistical power? If you measure the residual depth at 5 mN from the three progressive load scratches used for the fracture threshold measurement and the data have a standard deviation of 0.1 μm, you can only detect a difference of 0.36 μm between samples. With 280 independent measurements and a standard

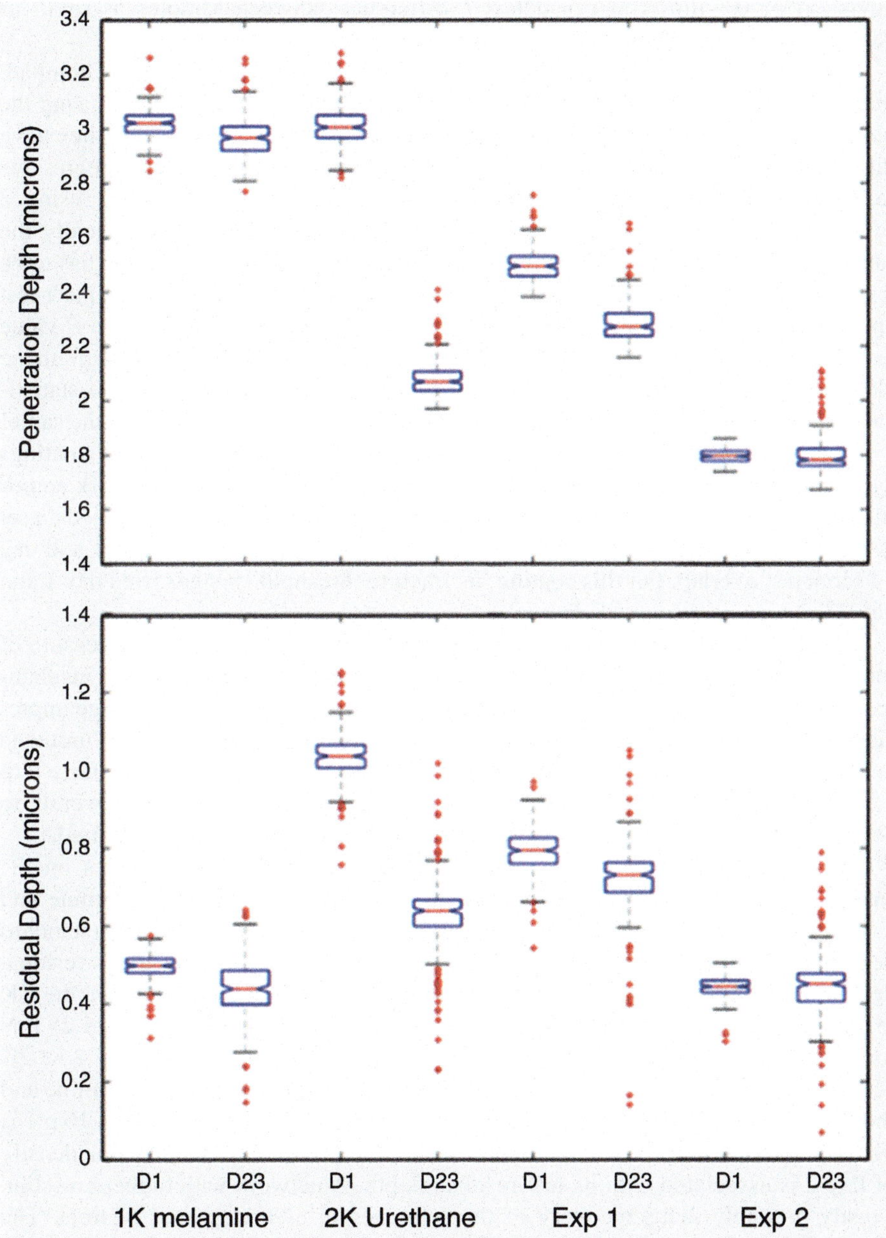

Fig. 15.8 Boxplot output from ANOVA calculation within Matlab for penetration and residual depths at the 5 mN load step tested on the clearcoats aged for 1 day and 23 days

deviation of 0.1 μm, you can detect a difference between samples as small as 0.03 μm in residual depth.

Drawing a specific example from the present study, let us statistically compare the 1K melamine and the 2K urethane clearcoats, both aged for 23 days, using the old and new methods. Previous work on these two coatings suggests that they have different scratch performance. The residual depth at 5 mN from three progressive load scratches was extracted following ASTM D7187 and compared using a two-sample t-test. Under these conditions, the two materials are statistically the same with a *P* value of 0.462 (at the 95% confidence interval, the *P* value has to be less than 0.05 to conclude that the samples are different). The two-sample *t*-test applied to the larger number of data points from the new method yields a *P* value less than 0.001 for the comparison between the 1K melamine and the 2K urethane clearcoats. The new method definitively proves that the two samples are statistically different, while the old method forces you to conclude that they are the same!

In general, the penetration and residual depths are both reduced as the coatings age for a longer period of time. This is likely due to changes in the network and/or further evaporation of solvent over time. The exception in the current 5 mN data set is the Exp 2 coating, where the central 50% of the data between the day 1 and day 23 scratches overlap. For this coating, its fracture threshold changes from day 1 and day 23 though as discussed above.

The residual depth for the scratches on the 23-day aged samples is a measure of the damage resistance of the coating. After aging for 23 days, the polymer mechanical properties are relatively stable, and the performance should not change appreciably. The higher residual depth values associated with Exp 1 and 2K urethane coatings are likely associated with their lower crosslink densities. Particularly Exp 1 has the lowest crosslink density, and its residual depth is the biggest. Overall, to achieve low residual depth, a coating with a high crosslink density is desired.

Figure 15.9 is a plot of the residual depth measured from scratches on the samples aged for 23 days as a function of load. At 5 mN the 1K melamine and Exp 2 clearcoats have identical residual depths within the error bars (standard deviations). The 2K urethane and Exp 1 clearcoats have slightly larger residual depths at the 5 mN level. As the load increases, the residual depth for the 1K melamine and the 2K urethane clearcoats approach each other. At 20 and 25 mN load steps, the ordering for the residual depths from low to high is Exp 2 < 2K urethane < 1K melamine < Exp 1, although the error bars for the 1K melamine and the 2K urethane overlap at these loads. Again the low crosslink density of Exp 1 is likely associated with its high value of residual depth, and the high crosslink density of Exp 2 is associated with its low residual depth. A network with higher crosslink density probably helps recovery of the scratches in a short period of time. The standard deviation of the residual depth is larger once the materials go into fracture as shown most clearly by the data from the Exp 2 clearcoat. The average fracture threshold for the Exp 2 clearcoat aged for 23 days is 18 mN, so both the 20 and 25 mN load steps are above this value.

The traditional approach to assess a coating's scratch and mar resistance is based on the plot of plastic deformation at 5 mN vs. fracture threshold and the area in the

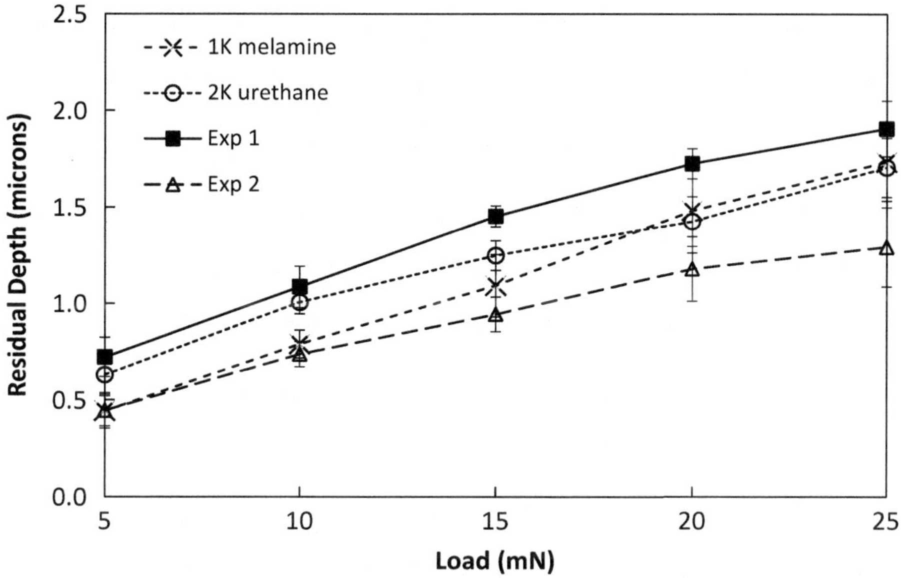

Fig. 15.9 Residual depth from NanoScratch tester on 23-day aged clearcoats as a function of step load. The error bars are the calculated standard deviations of the residual depth over each 0.5 mm long scratch

plot that is associated with the low plastic deformation and high fracture threshold contains coatings with the best scratch and mar resistance [2, 5, 7]. Through this work, it is emphasized that scratch visibility is also important to the scratch resistance. A coating's scratch resistance should be assessed by both its damage resistance and scratch visibility. The damage resistance can be determined by the residual depth and fracture threshold of a coating, and the scratch visibility can be determined by the dark field image analysis.

The coating with the "best" scratch resistance will be a balance of low residual depth (high recovery), high fracture threshold, and low visibility. Figure 15.10 is an attempt to compare all three using a bubble chart. The x and y coordinates are fracture threshold and residual depth (5 mN scratch) and the size of the bubbles is proportional to the mean gray level (scratch visibility). Exp 2 has good damage resistance due to the combination of low residual depth and reasonably high fracture threshold, but its scratch visibility is high. The 1K melamine coating has the lowest fracture threshold, a low residual depth, and the lowest scratch visibility. In contrast, the Exp 1 clearcoat has the highest fracture resistance, the highest residual depth, and the highest scratch visibility. As discussed earlier, crosslink density appears to affect the residual depth and scratch visibility. For low residual depth, a high crosslink density is desired. But for minimum scratch visibility, a medium crosslink density is desired. Further work is necessary to understand how material properties affect the fracture morphology of a coating and the scratch visibility.

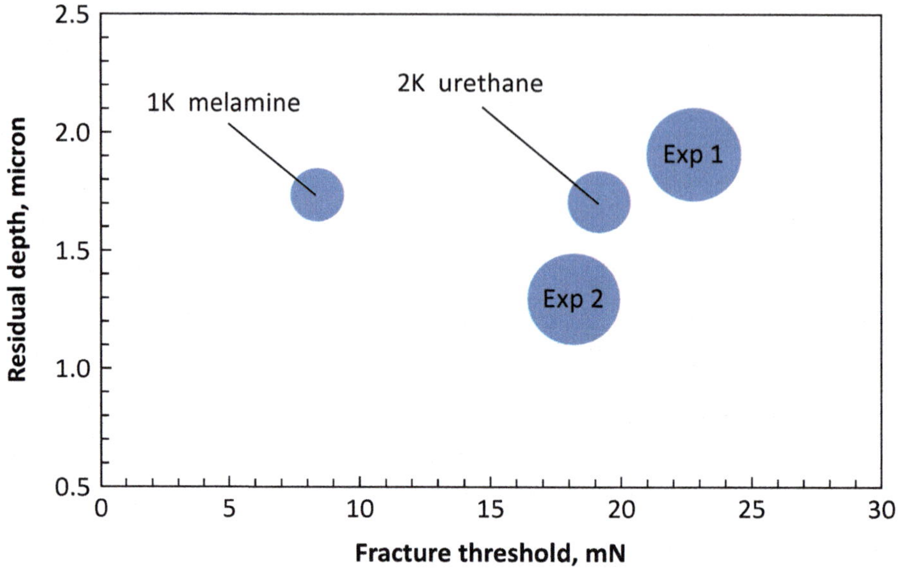

Fig. 15.10 Plot of residual depth at 5 mN vs. fracture threshold and scratch visibility. The data are from NanoScratch tester on coatings aged for 23 days. The bubble diameters are proportional to the mean gray scales from dark field images of each 25 mN scratch in Table 15.3

Conclusion

Advanced methods to quantify scratch and mar damage of automotive clearcoats are described. The new methods offer robust statistics, so that the effect on surface mechanical properties from changes in coating chemistry and coating aging time can be detected. Changes in residual depth as low as 0.03 μm between samples are detectable with the new methods. Analysis of the coatings with different crosslink densities indicates that coatings with higher crosslink densities generally have lower residual depth values. However, there is no correlation of crosslink density with fracture threshold.

Both short and long timescale recovery of scratches are analyzed. The short timescale recovery is measured with the NanoScratch tester, and the long timescale recovery is monitored via sequential AFM images of the scratches. The scratches recover about 70–73% immediately (12 min) after the scratch is made on the 2K urethane coating aged for 23 days. There is further small but consistent recovery days after the scratches are made. Scratch visibility is measured by a combination of dark field imaging and AFM morphology calculations. Dark field imaging offers unique advantages to analyze the visibility of the scratches. Scratch visibility is quantified by calculating the mean gray levels of the scratches and the scratch areas directly from the dark field images. Analysis of the visibility of the scratches created by a 25 mN step load on the coatings aged for 23 days indicates the order of scratch visibility from low to high: 1K melamine < 2K urethane < Exp 2 < Exp 1. Crosslink density appears to play an important role in the scratch visibility. To minimize scratch visibility, crosslink density should not be too low or too high.

There are two components to scratch resistance: damage resistance and scratch visibility. Damage resistance is the ability of the material to resist damage from the deformation typically associated with sharp objects. Scratch visibility is the degree of visual perceptibility of scratch damage by human eyes. This chapter discusses an improved methodology with robust statistics for evaluating both components. Damage resistance can be determined by the residual depth under a given load and the fracture threshold. In general, low residual depth and high fracture threshold give a coating better damage resistance. Scratch visibility depends on the size and scattered light intensity of the scratch, which can be best evaluated with dark field imaging of the scratches made with the NanoScratch tester. A coating with good damage resistance does not necessarily mean it has lower scratch visibility. Depending on the end use application, reduced scratch visibility may be more important than improved damage resistance—"I don't care if the material scratches as long as I cannot see it." Further work is needed to understand how material properties affect the fracture morphology of a coating under scratch conditions. The methods offered in this work open up doors to further understand a coating's scratch resistance.

Acknowledgments The authors would like to thank Li Lin for developing the original NanoScratch technique and for many years of fruitful collaboration; Nick Randall, Bertrand Bellaton, and Richard Consiglio for many helpful discussions before, during, and after the commercialization of their NanoScratch Tester; and Aaron Owens and Melissa Ziegler for advice on statistics.

Appendix *MATLAB*® and *Kornucopia*® *ML*™ Analysis of Scratch Data

The raw data from the NanoScratch test (NST) is exported and processed so that powerful statistical tools such as ANOVA (analysis of variation) and two-sample *t*-tests can be applied and even subtle variations between samples quantified. The amount of raw data to be processed from the NST is extremely large. In addition, the data need to be sliced into subsections for each load step that is applied to a step scratch. Performing this type of data processing using manual cut and paste operations in Excel or similar tools is possible, but it quickly becomes an inefficient process, especially when you need to process tens to hundreds of scratch profiles.

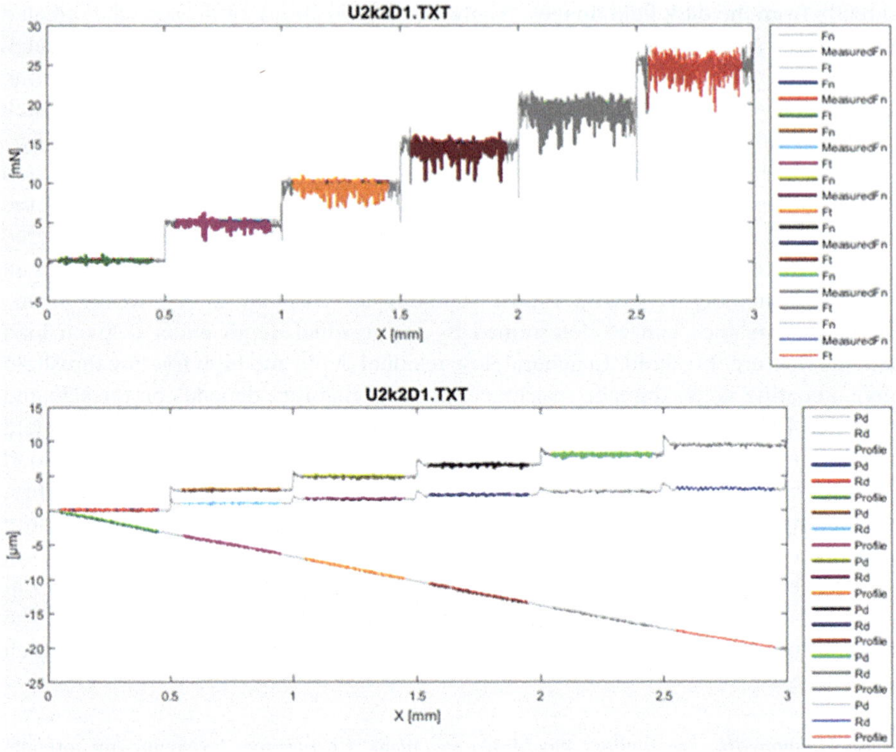

Fig. 15.11 Raw data exported from the NanoScratch tester and imported into Matlab. Kornucopia routines are used to slice the data as indicated by the colored steps in the figure. The top panel contains the normal and tangential force data and the lower panel contains the surface profile and residual depth. The data are for the step scratch for the 2K urethane coating aged for 1 day after spraying

The authors applied a few specialized scripts created using Kornucopia® ML (Bodie Technology, Inc.) and MATLAB® (The Mathworks, Inc.) to address the raw data processing, slicing, filtering, decimation, statistics, and plotting used for much of this work.

Figure 15.11 is one data set from a six-step scratch on a 2K urethane clearcoat. The load transitions are located, and roughly 50 μm at the beginning and end of each 0.5 mm step is removed. This is the region where finite probe size effects, creep, and feedback anomalies most affect the data. The raw data have over 5600 data points in each load step, and after slicing the data and removing the transition regions, there are still over 4500 data points at each load indicated by the different color regions in Fig. 15.11.

The penetration and residual depth measurements are relative to the initial profile of the surface; however, there is often deformation associated with the

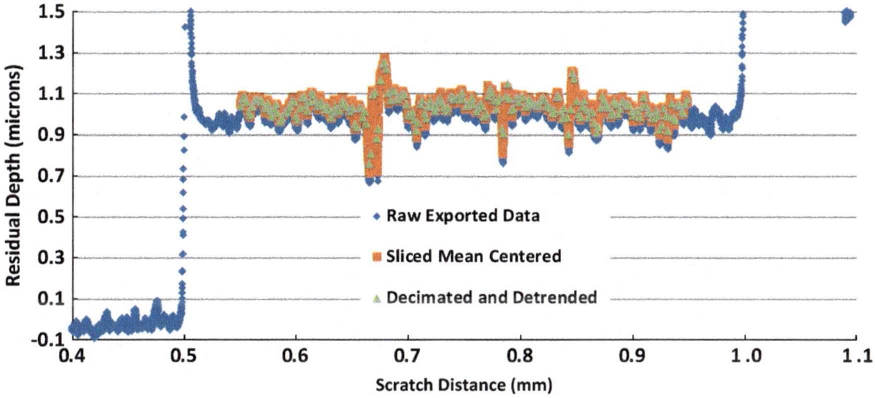

Fig. 15.12 Residual depth data for 5 mN load step for the 2K urethane coating illustrating the various steps in the treatment of the data

pre-scan load. In the NanoScratch test method, the first load step is 0.2 mN (the pre-scan or profiling load). The displacement is corrected by subtracting the mean value in this first load step from each of the other load steps. Figure 15.12 shows the residual depth for a region around the 5 mN load step with the processing steps indicated.

The instrument is capable of collecting data at 100 points per second or higher; however, due to the finite size of the probe, the data is highly correlated. The correlation arises because the spacing between data points is 0.1 μm, and the indenter has a radius of 2 μm. Most statistical analyses require independent measurements; therefore the exported data must be downsampled or decimated prior to the ANOVA calculations (Fig. 15.8). The correlation is significant and is most easily detected in an autocorrelation calculation such as in Fig. 15.13a. The decimated data have very little correlation as shown in Fig. 15.13b.

In the final step, any slope due to creep during the scratch is subtracted from each of the steps. The slope in this region represents a short timescale relaxation of the surface during the scratch. No further analysis was done on the short timescale creep measurements.

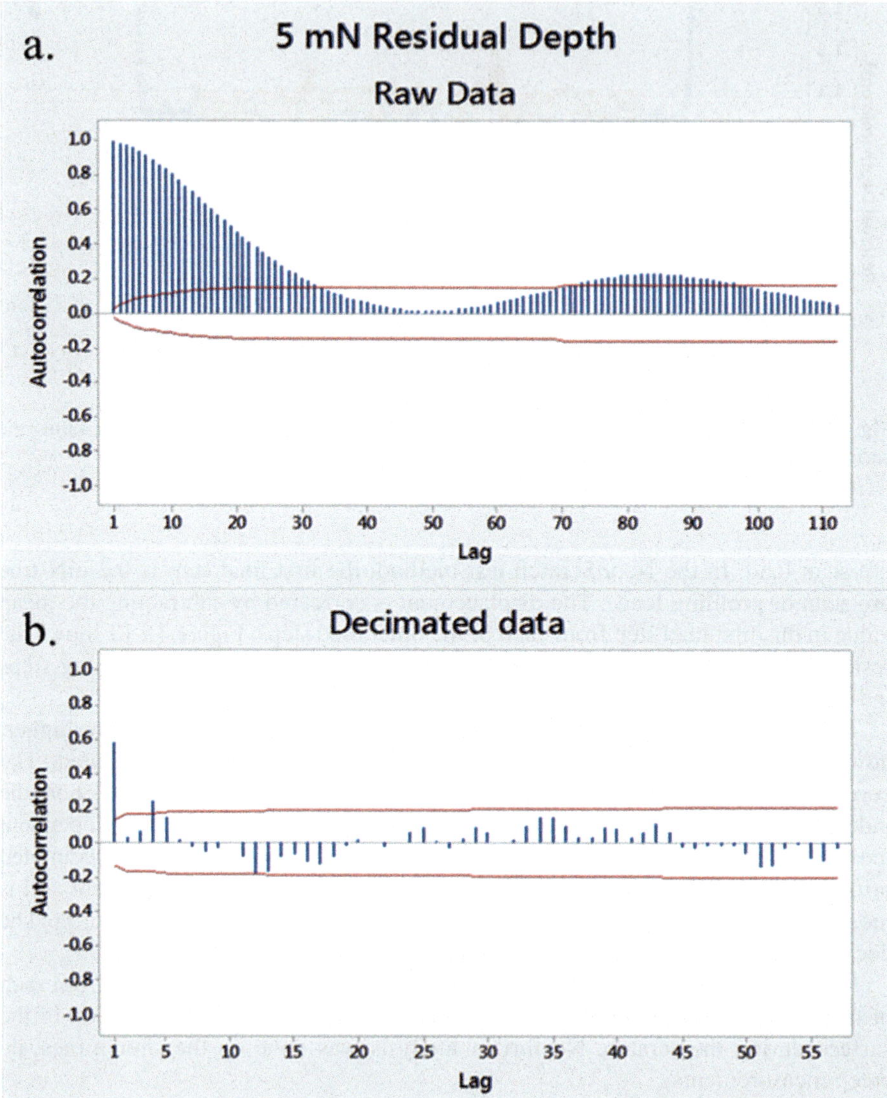

Fig. 15.13 Autocorrelation function performed on residual depth data from 5 mN load step on 2K urethane clearcoat in Fig. 15.12. The *red* lines represent the 5% confidence limits

References

1. Courter, J.L.: Mar resistance of automotive clearcoats: I. Relationship to coating mechanical properties. J. Coat. Technol. **69**, 57–63 (1997)
2. Seubert, C., Nietering, K., Nicholas, M., Wykoff, R., Bollin, S.: An overview of the scratch resistance of automotive coatings: exterior clearcoats and polycarbonate hardcoats. Coatings. **2**, 221–234 (2012)
3. Jiang, H., Lim, G.T., Reddy, J.N., Whitcomb, J.D., Sue, H.J.: Finite element method parametric study on scratch behavior of polymers. J. Polym. Sci. B Polym. Phys. **45**, 1435–1447 (2007)
4. Liu, P., Lee Browning, R., Sue, H.-J., Li, J., Jones, S.: Quantitative scratch visibility assessment of polymers based on Erichsen and ASTM/ISO scratch testing methodologies. Polym. Testing. **30**, 633–640 (2011)
5. Lin, L., Blackman, G.S., Matheson, R.R.: A new approach to characterize scratch and mar resistance of automotive coatings. Prog. Org. Coat. **40**, 85–91 (2000)
6. Blackman, G.S.; Lin, L.; Matheson, R. R.: Micro- and nano-wear of polymeric materials. ACS Symposium Series (Microstructure and Microtribology of Polymer Surfaces), pp. 258–269 (2000)
7. Osterhold, M., Wagner, G.: Methods for characterizing the mar resistance. Prog. Org. Coat. **45**, 365 (2002)
8. Xiang, C., Sue, H.-J., Chu, J., Coleman, B.J.: Scratch behavior and material property relationship in polymers. J. Polym. Sci. B Polym. Phys. **39**, 47–59 (2000)
9. Xiang, C., Sue, H.-J., Chu, J., Masuda, K.: Roles of additives in scratch resistance of high crystallinity polypropylene copolymers. Polym. Eng. Sci. **41**, 23–31 (2001)
10. Browning, R.L., Lim, G.-T., Moyse, A., Sue, H.-J., Chen, H., Earls, J.D.: Quantitative evaluation of scratch resistance of polymeric coatings based on a standardized progressive load scratch test. Surf. Coat. Technol. **201**, 2970–2976 (2006)
11. Bellemare, S.C., Dao, M., Suresh, S.: Mech. Mater. **40**, 206 (2008)
12. Jiang, H., Browning, R., Sue, H.-J.: Understanding of scratch-induced damage mechanisms in polymers. Polymer. **50**, 4056 (2009)
13. Moghbelli, E., Browning, R.L., Boo, W.J., Hahn, S.F., Feick, L.J.E., Sue, H.J.: Effects of molecular weight and thermal history on scratch behavior of polypropylene thin sheets. Tribol. Int. **41**, 425 (2008)
14. Randall, N.X.: Influence of indenter tip radius on the scratch resistance of an automotive clearcoat. CSM Appl. Bull. **25** (2008)
15. Mesa B (Micro Star Technologies) Spherical and rounded cone nano indenters; http://www.microstartech.com/index/ConicalIndenters2.pdf
16. Tweedie, C.A., Van Vliet, K.J.J.: On the indentation recovery and fleeting hardness of polymers. Mater. Res. **21**, 3029–3036 (2006)

Chapter 16
Scratch and Mar Resistance of Automotive Coatings

Jun Lin

Introduction

Scratch and mar resistance is one of the most studied subjects for automotive coatings over the past 20 years [1, 2]. However, due to the complexity of the scratch and mar failure mechanisms and a significant gap between laboratory and field performance [3], development of a true scratch-resistant coating remains a challenge and priority for many paint suppliers.

Scratch and mar resistance of automotive coatings is the surface resistance of the coating to deformation when in frictional contact with hard objects. Since automotive coatings are organic in nature, they are generally soft in comparison to objects they might encounter, such as tree branches, flying stone particulates, shopping carts, car keys, or dirt particles contained in car wash fluids, which are mostly inorganic in nature. As a result, organic coatings are usually scratched when in contact with these environmental objects. "Hard scratching soft" is best represented by Mohs hardness, especially for minerals [4]. For example, diamond is the hardest mineral in nature, and it can scratch any other minerals in nature. For organic coatings, since they are soft by nature, increasing hardness is not necessarily the correct approach to improve scratch resistance as they could still end up significantly softer than the environmental objects of inorganic nature. Understanding the scratch deformation mechanism would be a more promising approach to develop more scratch-resistant coatings. Scratch deformation mechanisms have been discussed in various literature and can be summarized as Fig. 16.1 [5, 6].

The elastic response describes the phenomenon that the coating may recover fully without damage, after the mechanical input is removed. This phenomenon describes the behavior of rubber-like materials such as tires which could survive

J. Lin (✉)
Axalta Coating Systems, Mt Clemens R&D Lab, Mt Clemens, MI, USA
e-mail: Jun.Lin@Axaltacs.Com

M. Wen, K. Dušek (eds.), *Protective Coatings*, DOI 10.1007/978-3-319-51627-1_16

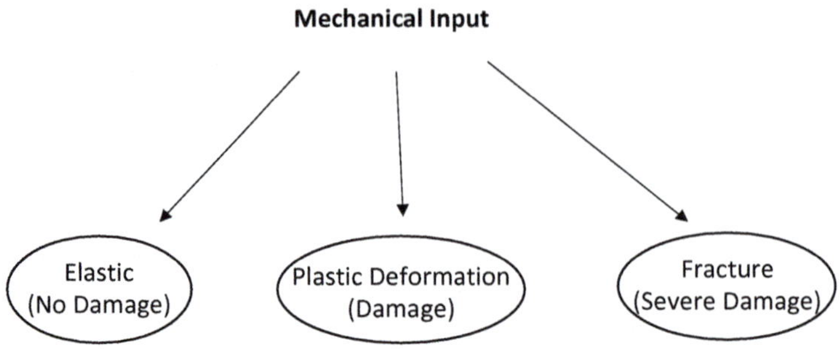

Fig. 16.1 Scratch failure mechanism of coatings

Fig. 16.2 Image of scratches: plastic deformation (*left*); fracture (*right*)

thousands of miles of frictional contacts with concrete or stony roads. However, rubber-like material is impractical to make into automotive coatings due to their ultra-softness, which would create issues for environmental etch resistance [7]. The response of a more practical coating would be more likely viscoelastic or a combination of elastic, plastic deformation, and fracture. Between plastic deformation and fracture, the former is less severe as its deformation is physical, while fracture involves removal of coating material and possible bond-breaking process [2, 8, 9]. Plastic deformation may sometimes be recoverable, especially when the coating is heated above its glass transition temperature (T_g). Fractured deformation is more severe due to its removal of the coating itself and possible cleavage of chemical bonds, usually deep and irregular, and thus more visible. Its higher visibility is a result of light scattering from all possible angles due to the depth and irregularity of these types of scratches (Fig. 16.2).

This study reviews several of the most popular test methods and their working mechanisms, followed by an in-depth discussion on the effect of crosslink density, crosslinking chemistry, chain elasticity, nanoparticles, aging, and weathering on scratch and mar resistance. The failure mechanism of each study will be specifically discussed.

Experimental

For the scratch and mar resistance test, clearcoat (CC) samples of different types of crosslinking chemistries were prepared. Clearcoat samples made for this study normally contain functional polymers, crosslinkers, catalysts, and various additives. The functional polymers were chosen from acrylic polyol or epoxy acrylic. The crosslinkers were chosen from melamine formaldehyde, polyisocyanate, or polyester half acid, depending on the nature of crosslinking chemistry. The catalysts were usually strong organic acid for acrylic melamine, weak acid for two-component (2K) isocyanate crosslinking, and an amine for epoxy acid crosslinking. The samples were usually diluted to applicable viscosities before spraying. For 2K isocyanate, the component A with acrylic polyol was diluted and mixed with the polyisocyanate component B before spraying. These diluted clearcoat samples were then sprayed through a bell applicator onto steel substrate panels which were pre-coated with three layers of different coatings: approximately 25 μm of fully baked electro-coat directly over the steel substrate, around 25 μm of fully baked primer coating over the electro-coat, and 15–20 μm thickness of wet black basecoat over the primer. The coated CC samples were dried at ambient temperature for 30 min before they were baked in a convection oven for a 30 min bake at 140 °C. The resulting CC thickness was approximately 50 μm. The scratch and mar test was usually performed 24 h after the bake.

For scratch and mar weathering studies, the coated panels were exposed on tables at a site located in Jacksonville, Florida. The tables were set horizontally with 0° inclination to maximize the residence time of environmental fallouts. The duration of exposure was typically 14 weeks lasting from the end of May to the end of August each year, or commonly referred to as the Jacksonville exposure period. Panels before and after weathering exposure were tested by Amtec Kistler Car Wash and Crockmeter-9 μm (described in the next section). The percentage of gloss retention for the coating was evaluated, with high gloss retention indicating better scratch and mar resistance.

For the impact of nanoparticles on scratch and mar resistance, colloidal silica of approximately 13 nm in diameter from Nissan Chemical company (IPA-ST) was used for surface treatment in the study. Crockmeter-dry mar and wet mar, Amtec Kistler Car Wash, and Nano Scratch were utilized to test the scratch and mar resistance.

Scratch and Mar Resistance Test Methods

Many different types of scratch and mar test methods have been developed over the years for automotive coatings. These methods include Crockmeter-dry mar (ASTM D 6279–98, using Bon Ami cleaning powder as the abrasive) and wet mar (US6379807, alumina slurry as the abrasive), Crockmeter-2 μm and

a) b)

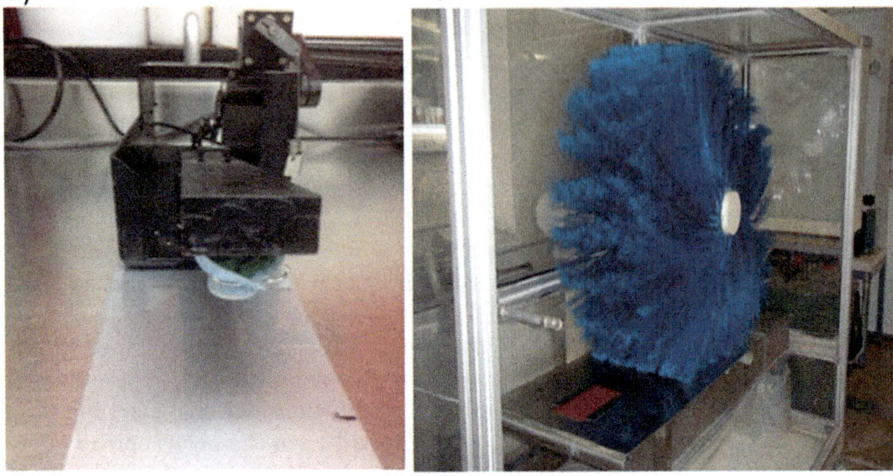

Fig. 16.3 Pictures of (**a**) Crockmeter and (**b**) Amtec Kistler Car Wash

Crockmeter-9 μm with 3M scratch papers, bench top car wash (GM9707P), GM-Opel car wash (GMW14865), Amtec Car Wash (DIN EN ISO 20566), and Nano Scratch (ASTM D7187–05 and Ford Paint Standard Report: PA-0171). With these many different methods, it is not uncommon that conflicting results are produced. This is mainly due to the different failure mechanisms created by each method. The coating industry has evaluated methods to correlate to different types of field failure mechanisms. The current standard methods for testing scratch and mar resistance are on Crockmeter-2 μm and Crockmeter-9 μm, Amtec Car Wash, and Nano Scratch. The focus of the discussion will be on these three methods. Figure 16.3 shows the pictures of a Crockmeter and an Amtec Kistler Car Wash machine.

Crockmeter-2 μm

This method typically runs ten (10) double rubs against the coating surface with the fingertip of the Crockmeter wrapped with a 2-μm 3M scratch paper. The scratch resistance of the coating is determined by measuring the percentage of 20° gloss retention before and after the scratch. This test is considered mild, and the scratch dimension is generally small and concentrated in the surface region. Cross-sectional analysis with an optical microscope on a typical acrylic/ melamine clearcoat suggests the depth of deformation is on the magnitude of 0.2 μm (Fig. 16.4a). The failure mechanism associated with this method is believed to be plastic deformation dominated due to the shallow depth of scratches formed.

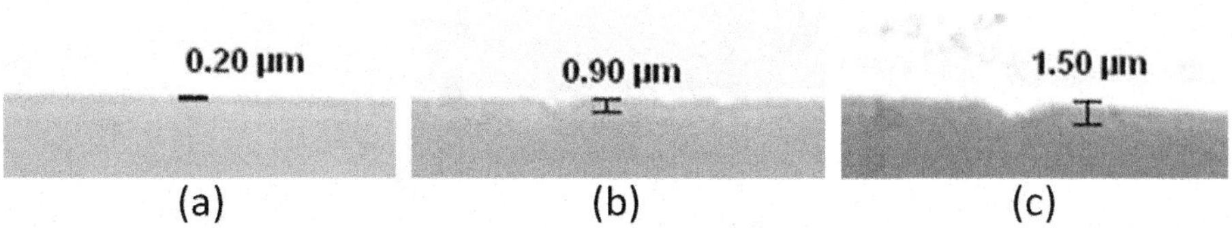

Fig. 16.4 Cross-sectional optical micrograph of representative deformation of acrylic/melamine clearcoat by (**a**) Crockmeter-2 μm, (**b**) Crockmeter-9 μm, and (**c**) Amtec Kistler Car Wash

Crockmeter-9 μm

Different from Crockmeter-2 μm, this method uses a 3M 9-μm scratch paper instead to run the scratch test. Because of the coarser nature of the abrasive particles, the scratch deformation is deeper and may reach as far as 0.9 μm into the coating for the acrylic/melamine-type coating (Fig. 16.4b). The deformation type is considered mainly plastic deformation.

Amtec Kistler Car Wash

This test method simulates the real car wash processes with silica flour slurry (1.5% wt in water) constantly sprayed over a moving test sheet carrier for ten (10) double passes. This method has become an international test standard (DIN EN ISO 20566) along with the Crockmeter-9 μm during the last decade. The distinctive feature of this method is the possibility of producing dominantly fracture type scratches due to high depth of scratch deformation (Fig. 16.4c). Its advantage over a Crockmeter-type test is that it presents an excellent visual display of scratch damage for a coating due to its large scratch area 30×30 cm^2 (Crockmeter generates 1×10 cm^2 of scratch area). Another significant difference of the car wash test is the influence of water medium on the scratch deformation mechanism. Water cools down the coating surface and can dissipate the heat generated by rubbing and friction of the test, which may otherwise help the elastic recovery of the scratch deformation. A further difference with the car wash test is its high speed of scratching, approximately 6.7 m/s (127 rpm with the brush radius around 0.5 m) vs. the approximate speed of 0.3 m/s for a Crockmeter mar, which generates a high strain rate. The combination of high strain rate and decreased temperature promotes a viscoelastic rather than viscoplastic response [10].

Nano Scratch

Unlike other scratch tests, Nano Scratch can quantify the scratch failure mechanism of organic coatings [6]. While more detailed Nano Scratch discussion can be found

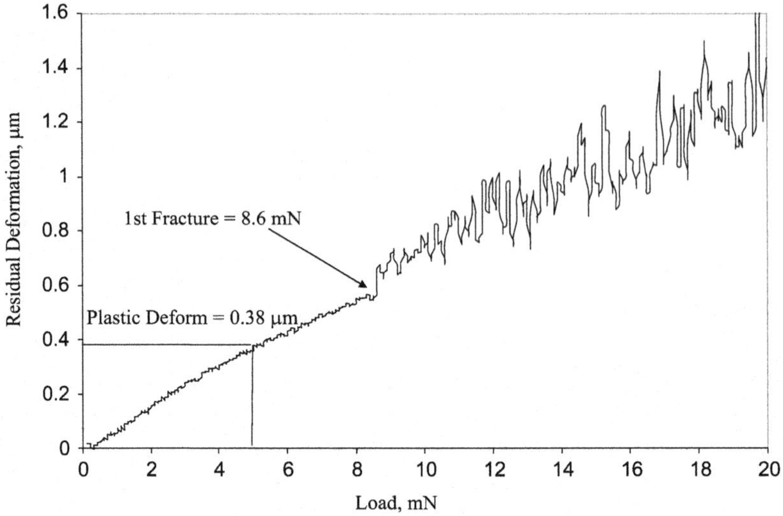

Fig. 16.5 Nano Scratch plot of acrylic/melamine clearcoat

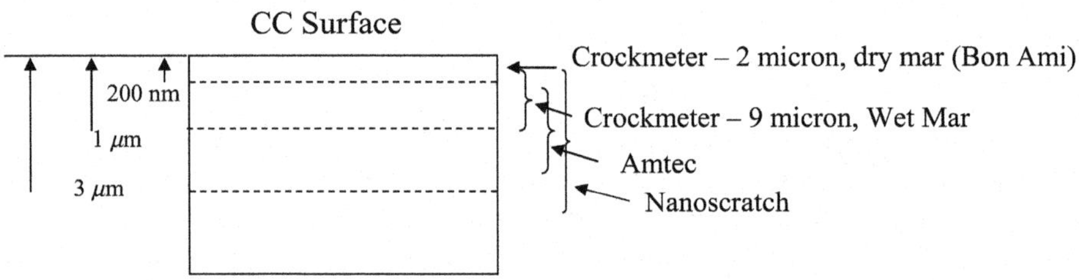

Fig. 16.6 Deformation depth of different scratch test methods

in the preceding chapter [13], the following graph summarizes the basic information needed for a Nano Scratch test. The two features of the graph are the plastic deformation and the minimum load where fracture occurs. The plastic deformation is defined as the residual depth of deformation at 5 mN. This definition takes into the consideration that most automotive coatings have their initial fracture occurring at a load exceeding 5 mN, and thus the deformation at 5 mN would be truly plastic deformation for most coatings. The fracture energy is defined as the load needed to induce fracture in the coating. Figure 16.5 is a plot of residual deformation vs. scratch load for a clearcoat based on acrylic/melamine crosslinking. It shows a significant fluctuation starting at 8.6 mN, which is an indication of irregular morphology or onset of fracture in the coating.

Figure 16.6 summarizes the scratch differences among all different scratch test methods for an acrylic/melamine coating. The depth of deformation is the lowest for the Crockmeter-2 μm test, followed by Crockmeter-9 μm and Amtec Kistler Car Wash. The residual depth of scratch for Nano Scratch is dependent on the scratch load and can reach more than 3 μm when the scratch load is close to 20 mN.

Results and Discussion

Effect of Crosslink Density on Scratch and Mar Resistance

Crosslink density has been one of the key design factors for scratch-resistant coatings, and higher crosslink density generally leads to better scratch and mar resistance. Mechanistically, however, not much has been reported in the literature on how crosslink density impacts the scratch and mar resistance. With the development of novel test methods such as Nano Scratch, understanding the mechanism of crosslink density influence on scratch and mar failure is possible. It is particularly important to recognize that each test method as discussed in the previous section is associated with possibly different types of failure mechanisms. Understanding of the working mechanism can help design the coating to succeed in all different types of tests.

Figure 16.7 compares the structures of crosslinkers popularly used in the current automotive market: acrylic melamine, two-component (2K) isocyanate, and epoxy-acid systems. These three technologies represent vast differences in crosslink densities which are due to the nature of resin design and the type of crosslinking chemistry

As a result, the difference in their coating crosslink densities would be directly related to the type of crosslinker being used, with MF being obviously the most highly crosslinked, followed by 2K isocyanate and then epoxy acid. The differences in crosslink densities are reported in the following Table 16.1, along with the Nano Scratch test results on plastic deformation and fracture energy (force in mN).

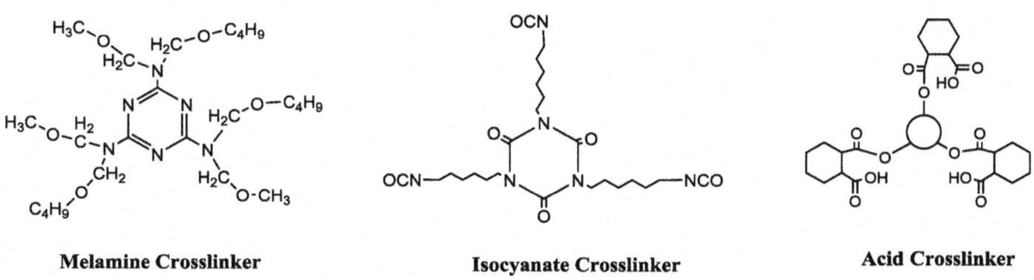

Fig. 16.7 Structures of three different types of crosslinkers commonly used in automotive coatings

Table 16.1 Comparison of scratch and mar resistance between three different types of clearcoats

	Acrylic MF	2K Iso	Epoxy-acid
Crosslink density (mmol/g)	2–2.2	1.5–1.7	1.4–1.6
Plastic deformation (μm)	0.3	0.5	0.7
Fracture threshold (mN)	8	11	8

The test results demonstrate a direct correlation between crosslink density and the plastic deformation measured by Nano Scratch, thus higher crosslink density resulting in better resistance to scratch.

Chain Elasticity vs. Scratch and Mar Resistance

One of the most misguided statements associated with organic coatings is the belief that harder coatings are better for scratch and mar resistance. This statement partially originated from the common understanding that minerals of higher Mohs hardness would be able to scratch minerals of lower hardness. For organic coatings, this statement could also be supported by the fact that coatings baked longer and harder are usually more scratch resistant. This experience, however, could be wrong with regard to scratch and mar failure mechanism, as a function of formulation. Coating hardness is not only related to crosslink density but also highly dependent on resin rigidity and toughness. The rigidity of the reactive resin can inhibit cure and lead to decreased crosslink density which can consequently hurt scratch and mar resistance.

Another issue with increased resin rigidity is the possibility of losing film elasticity. As shown in Table 16.1 of the preceding section, although 2K isocyanate has lower crosslinking density and is generally softer than a 1K acrylic melamine, its Amtec Kistler Car Wash resistance could actually be better than the 1K acrylic melamine. This is related to 2K isocyanate having a higher fracture threshold than the 1K acrylic/melamine coating. Its higher fracture threshold originates from the elastic nature of its 6-C linear hydrocarbon chain linked between the NCO crosslinking site and the central isocyanurate ring as shown in the following diagram.

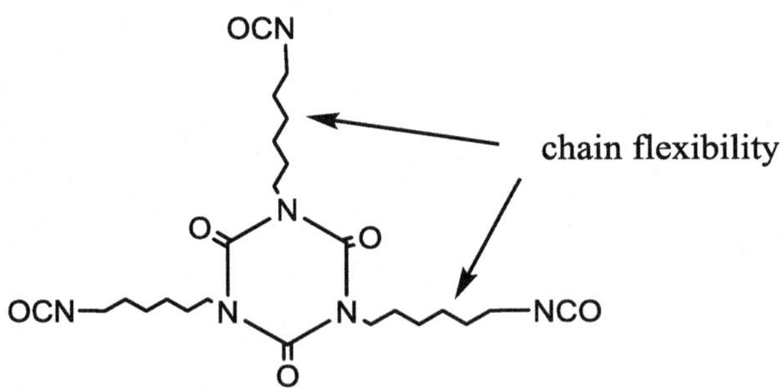

HDI Trimer

The theoretical argument on polymer elasticity is the relationship of the number of carbons with their possible configurations which in turn dictate the system entropy [11]. Thus, the higher the number of atoms present in the jointed chain,

Table 16.2 Influence of crosslinking chain length on scratch resistance of the coating

	Chain length between OH groups (# of bonds)	Fracture threshold (mN)	Plastic deformation (μm)
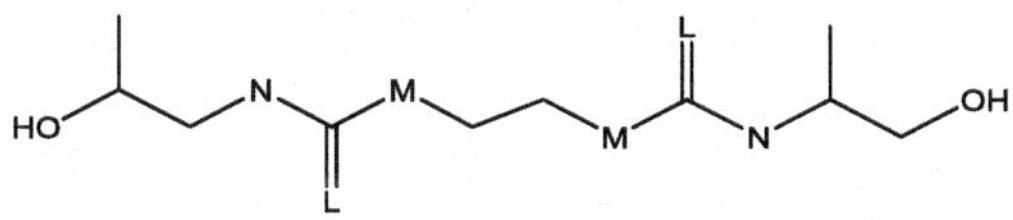	11	15	0.26
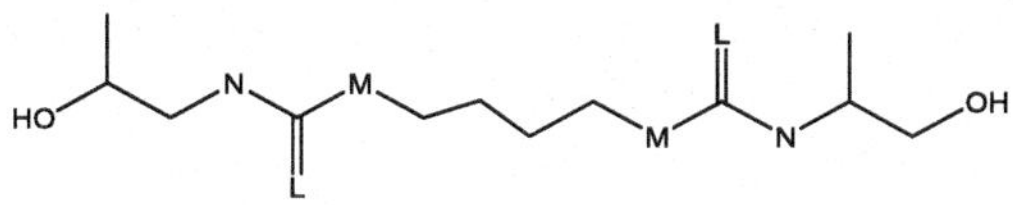	13	16	0.26
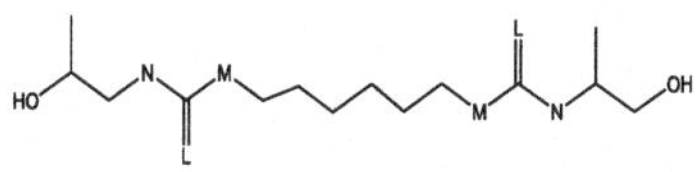	15	18	0.28
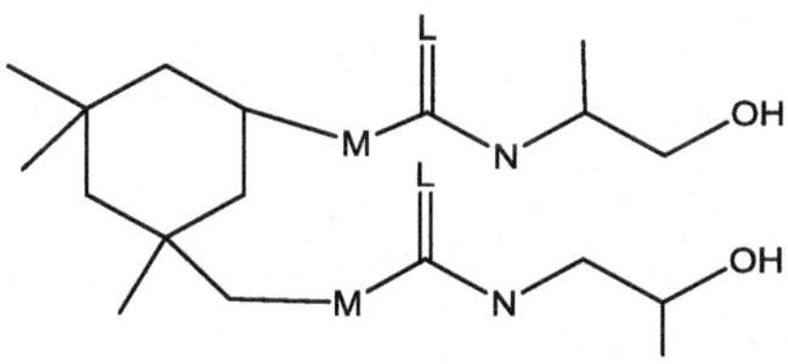	13	13	0.38

the higher the number of configurations and thus the higher the entropy, which in turn means higher fracture energy or elasticity.

To further illustrate the effect of chain configuration on elasticity or fracture energy, Table 16.2 describes a study correlating the number of carbon atoms (chain length) in between the crosslinking sites to the fracture load of the coating measured by the Nano Scratch test. For this study, model hydroxyl functional compounds were prepared to crosslink with the same melamine formaldehyde through the hydroxyl-alkoxylated methylolmelamine reaction. These model compounds were prepared varying the chain length between the two neighboring OH groups. L, M, and N represent different types of atoms of the model compounds. Despite different chain lengths of the model compounds, the crosslink density of the networks was relatively the same as a large excess of melamine was used. One compound also contains cyclic units so that the effect of steric hindrance would also be evaluated.

Table 16.2 demonstrates that the fracture threshold increases with higher number of bonds counted in between the crosslinking sites. The diol that contained a cyclic unit showed decreased fracture threshold. This is expected as steric hindrance inhibits chain rotation and thus reduces the number of configuration and decreases chain elasticity. Plastic deformation, on the other hand, was not correlated to the number of carbon atoms, as they are more correlated to crosslink density.

Aging and Weathering Effect on Scratch and Mar Resistance

Most of the literature discussion on scratch resistance has been focused on tests without consideration of aging or weathering. Although some of the studies showed the effect of weathering on scratching recoveries [12], detailed analysis of failure mechanism due to weathering has not been addressed. In real life, aging and weathering play important roles on scratch and mar resistance. For example, it was found that 2K isocyanate increased scratch resistance 23 days after spraying [13]. This is due to the postcure phenomenon which leads to increased crosslink density in the coating (Fig. 16.8).

Weathering, however, has an effect opposite to the aging effect, where the crosslink density of the coating decreases due to the breakdown of the network as a result of sunlight (UV-light) irradiation. Table 16.3 was a study on the impact of summer exposure in Jacksonville, Florida, on two different types of commercial clearcoats: acrylic melamine vs. 2K acrylic isocyanate.

Table 16.3 demonstrates that both clearcoats had significantly decreased scratch resistance after weathering. However, the isocyanate clearcoat was better than the melamine clearcoat for Amtec Kistler Car Wash resistance after weathering.

To understand the weathering effect, both clearcoats were analyzed with Fourier transform infrared spectroscopy-attenuated total reflectance FTIR-ATR (sampling depth up to 2 μm) before and after Jacksonville summer exposure. Figure 16.9 demonstrates that there was much more significant photooxidation for acrylic melamine than 2K isocyanate, as evident from its more significant increase of OH/NH vs. CH absorption [14]. This suggests that the acrylic/melamine coating

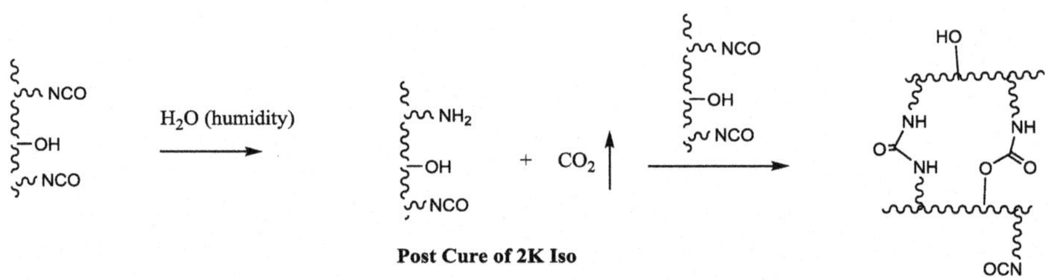

Fig. 16.8 Postcure crosslinking reaction of 2K Isocyanate

Table 16.3 Effect of weathering on scratch and mar resistance of representative clearcoat

	Crockmeter-9 μm[a]		Amtec Kistler Car Wash[a]	
Clearcoat	Before exposure	After exposure	Before exposure	After exposure
Acrylic Melamine	80%	17%	80%	20%
2K Acrylic Isocyanate	25%	13%	80%	40%

[a]Reported as percentage of 20° gloss retention

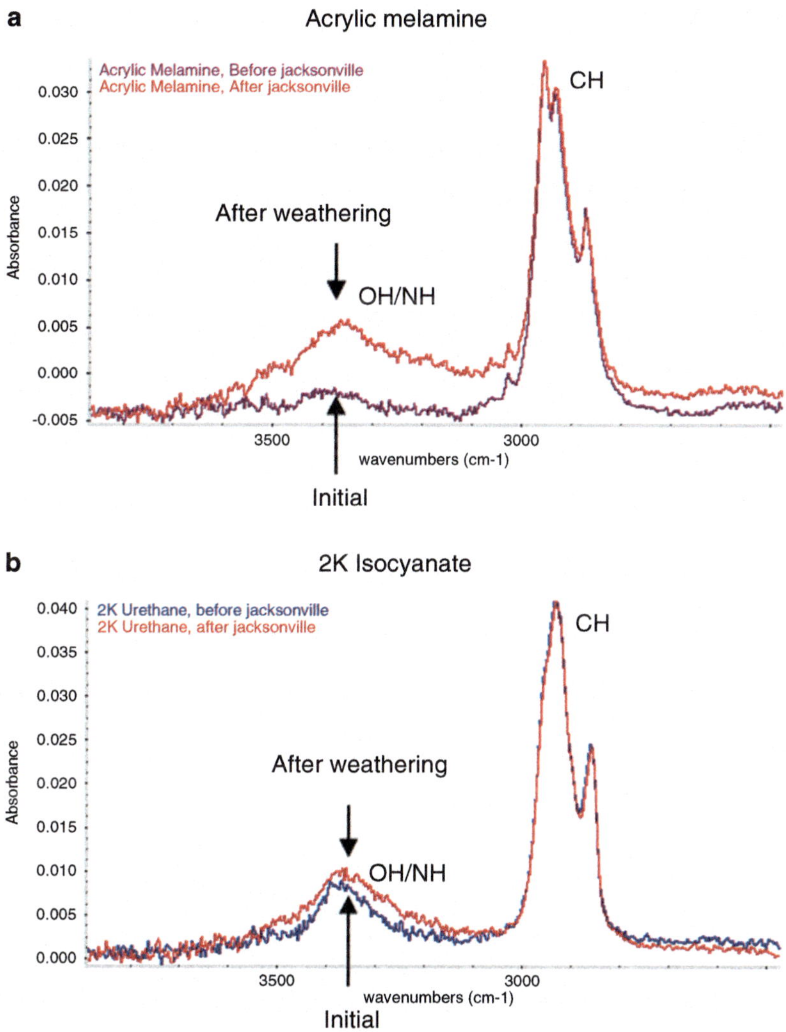

Fig. 16.9 FTIR-ATR spectra of (**a**) 1K acrylic melamine and (**b**) 2K urethane before and after weathering

has a lot more chemical degradation or crosslink network breakdown than the 2K isocyanate coating. Breakdown of the crosslink network decreases the crosslink density, and even more critically it generates defects to become sources for crack initiation and propagation, especially when the coating is subjected to external scratching. The more significant breakdown of acrylic melamine would mean worse resistance to Amtec Kistler Car Wash and Crockmeter tests after weathering.

Nanotechnology

Nanotechnology has been a hot topic in the automotive industry for the last decade. Nano-sized silica particles have been of particular interest due to their transparency, close match of refractive index with the clearcoat, and high Mohs hardness, which makes them ideal candidates to improve scratch and mar resistance in clearcoat. Solvent-based colloidal silica was of particular focus due to the abundant silanol groups on the surface, which makes surface functionalization and paint incorporation feasible. Some of the work was exemplified by recent work in the field (U.S. Patent 6,387,519). Part of the dilemma is that the benefit brought by nanotechnology for improvement in scratch resistance does not justify the amount of effort needed from the automakers to overcome some of the processing challenges such as less effective polishing. The scrutiny on health hazards of nanomaterials also makes many automakers hesitant to adopt them. A further important reason is the general lack of fundamental understanding of how nanomaterial treatment is correlated with performance in the coatings. Below, some of the fundamental work related to nanoparticle surface treatments and how these treatments relate to clearcoat scratch resistance will be discussed.

The following examples represent three types of functionalization on the surface of colloidal silica:

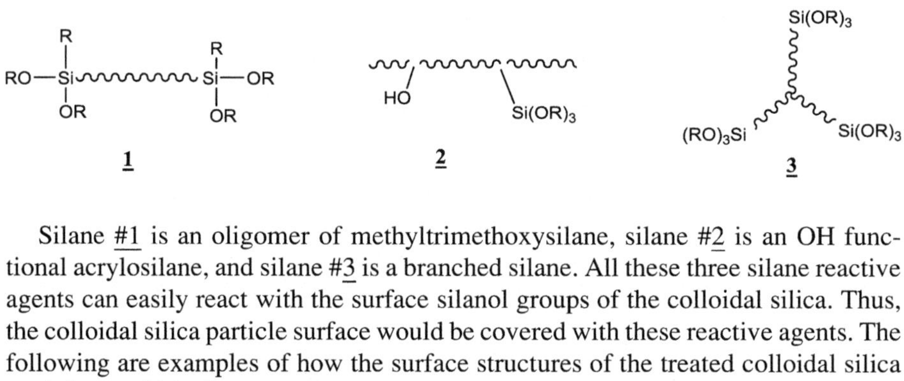

1   2   3

Silane #1 is an oligomer of methyltrimethoxysilane, silane #2 is an OH functional acrylosilane, and silane #3 is a branched silane. All these three silane reactive agents can easily react with the surface silanol groups of the colloidal silica. Thus, the colloidal silica particle surface would be covered with these reactive agents. The following are examples of how the surface structures of the treated colloidal silica particles could look.

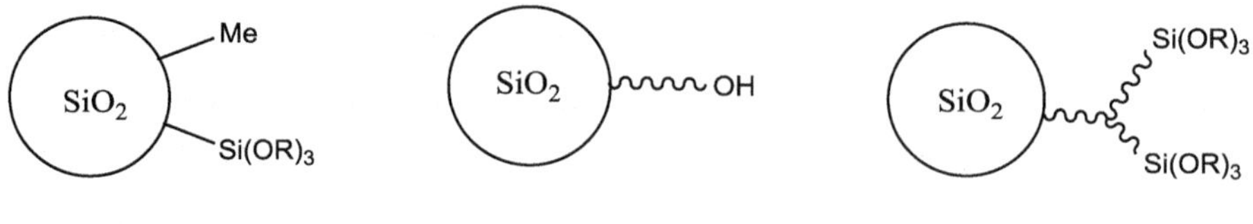

SiO_2 - 1   **SiO_2 - 2**   **SiO_2 - 3**

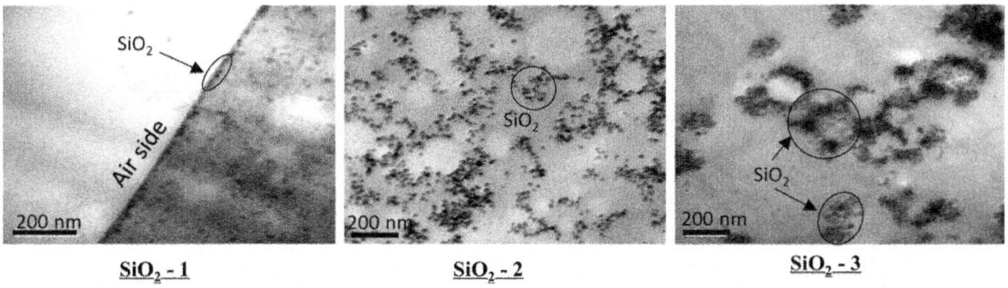

Fig. 16.10 Cross-sectional TEM of acrylic/melamine clearcoat films loaded with three types of surface-treated colloidal silica particles. The clearcoats were baked at 140 °C for 30 min

Table 16.4 Properties of acrylic/melamine clearcoat (CC) loaded with different types of surface-treated nanoparticles

	% Nano	Surface treatment	Crockmeter-dry Mar	Crockmeter-wet Mar	Amtec Car Wash
CC 0	0	None	81	79	70
CC 1	7	Silane #1	96	95	80
CC 2	7	Silane #2	80	78	72
CC 3	7	Silane #3	85	94	80

The above three types of surface-treated particles were incorporated into an acrylic/melamine clearcoat at 7% by weight. A cross-sectional analysis by transmission electron microscopy (TEM) on these three clearcoats demonstrates different particle morphologies as shown below (Fig. 16.10).

The difference among the three images is the degree of particle association. A thin layer of silica particles was observed in the surface region (also called surface stratification) in the colloidal silica treated with silane #1. The treated silica has low surface energy due to the methyl groups on the particle surface, leading to particle migration to the coating surface. In Fig. 16.10, the particles in **2** and **3** are not as highly isolated as **1**, but **2** was only loosely associated, and **3** was tightly linked up clusters. It can be theorized that the tight link-up of **3** is due to the condensation reaction of the active silane groups present on the silica surface.

The three clearcoats as discussed above were tested for Crockmeter-dry mar (working mechanism similar to Crockmeter-9 μm), Crockmeter-wet mar (working mechanism similar to car wash) [10], and Amtec Kistler Car wash. The results are reported in Table 16.4.

It is surprising that only CC 1 shows improved scratch resistance by all three methods. CC 3 had improved Crockmeter-wet mar and Amtec Car Wash but did not have significantly improved dry mar performance. CC 2 did not improve scratch resistance by any methods. The improved scratch resistance in CC 1 is likely due to the formation of surface-stratified silica, whereas the improved wet mar and Amtec Car Wash performance in CC3 is probably due to the densely clustered silica particles in the bulk of this coating. CC 2, which contains only loosely associated silica particles in the bulk, did not show improvement in scratch resistance by

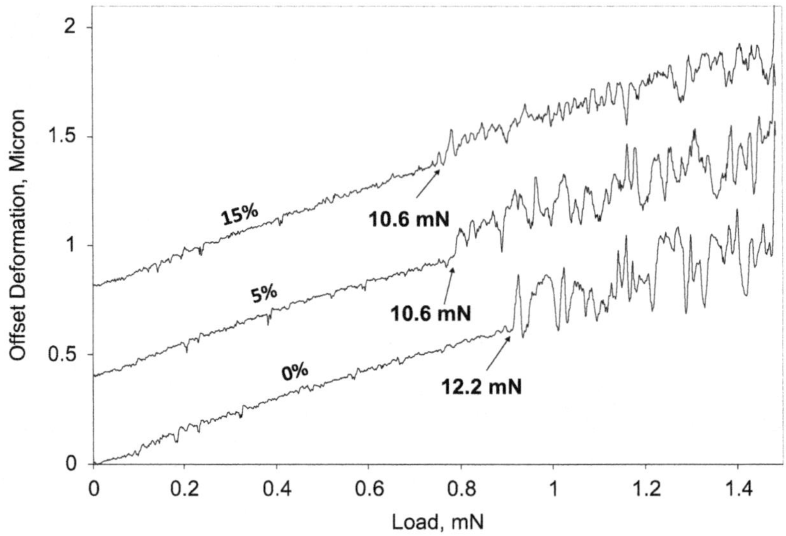

Fig. 16.11 Effect of clustering silica SiO_2-3 on Nano Scratch analysis. Silica are added at 0, 5, and 15% levels (by weight)

crockmeter-wet mar or car wash (penetrating into the bulk interior of the coating) or the crockmeter-dry mar (testing only the surface region).

To understand the mechanism of clustering silica particles on scratch resistance improvement, two levels of clustering silica were added to an acrylic/melamine clearcoat and analyzed by a Nano Scratch tester. The results are shown in Fig. 16.11. Figure 16.11 demonstrates that the presence of clustering silica particles may actually decrease the fracture threshold of the coating. However, the amplitudes of the fracture deformations after the threshold were significantly decreased with 15% loading of clustering silica particles.

It is theorized that tightly clustered silica particles (SiO_2-3) are strong enough to withstand the stress created by film fracture due to the covalent bonding between the clustered silica particles. The silica clusters are also large enough to match the amplitude of the fracture fluctuation, 100–500 nm as shown in Fig. 16.11, and work as a continuous phase to effectively absorb and deflect the energy imposed by the propagating crack fronts (see Fig. 16.12). Figure 16.12 depicts the propagating crack front being arrested by clustered nanoparticles, for which a single nanoparticle or loosely associated aggregates may not be large or strong enough to stop the crack propagation.

For the methyltrimethoxysilane oligomer-treated particles, although surface stratification leads to excellent scratch resistance, potential loss of the surface layer due to photooxidation/degradation of the organic coating may compromise the effectiveness of this approach [14].

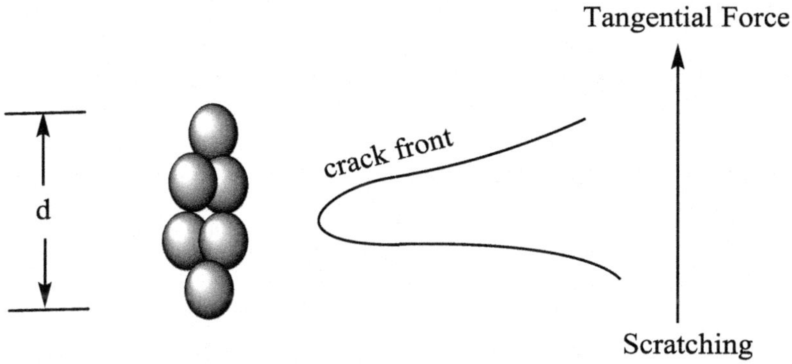

Fig. 16.12 Working mechanism of clustered silica particles to arrest crack propagation during scratching

Conclusions

A few of the popular scratch and mar test methods are discussed, which includes Crockmeter-2 μm, Crockmeter-9 μm, Amtec Kistler Car Wash, and Nano Scratch. Plastic deformation is considered as the primary failure mechanism for the Crockmeter test methods, but Amtec Kistler Car Wash generates more fracture type scratches. Nano Scratch is a method quantitatively measuring plastic deformation and fracture load. From the perspective of coating formulation, crosslink density and chain elasticity are the most important factors for scratch and mar resistance. Higher crosslink density resulted in lower plastic deformation, and higher chain elasticity led to higher fracture load.

Aging and weathering also play an important role for scratch and mar resistance of coatings. Aging of polyisocyanate-based coatings helps scratch and mar resistance due to the increase of crosslink density and postcure reaction from isocyanate groups. Weathering has an opposite effect on scratch and mar resistance due to network degradation. For comparison, 2K isocyanate coating was superior to 1K acrylic melamine for post-weathering scratch resistance, due to its better photo-oxidation resistance.

In terms of addition of nanoparticles, it is demonstrated that surface stratification or clustering of nanoparticles in the bulk can significantly improve the scratch and mar resistance. The mechanism of particle clustering on scratch resistance is likely due to arresting the propagating fronts of cracks initiated during the scratching process.

Scratch and mar resistance will remain to be one of the key attributes for automotive coatings. Challenges still remain on developing coatings with strong scratch resistance after weathering and at plant handling due to their out-of-oven softness and relatively low crosslink density. Still several other factors can play into the work on scratch and mar resistance improvement. For example, continuous drive for improving appearance and decreasing volatile organic compounds (VOC) may lead to development of formulations of lower crosslink density or decreased resin functionality, which compromise scratch and mar resistance. Thus, how to balance all attributes of need while still maintaining or improving the scratch resistance of the coatings could be another focus of future development.

Acknowledgments I would like to thank some of my long-time technicians Robert McEwen and Vince Cianciolo for their significant contribution to this work. I would also like to thank Ford Motor Company for allowing the access to their Nano Scratch equipment for some of the work. Finally, I would like to thank my colleagues and long-time collaborators over the years for many interesting discussions regarding scratch and mar resistance and also continuing friendship, including David Nordstrom, Robert Matheson, Li Lin, Jeffery Johnson, and Marcy Zimmer.

References

1. Courter, J.L.: Mar resistance of automotive clearcoats: I. Relationship to coating mechanical properties. J. Coat. Technol. **69**, 57–63 (1997)
2. Seubert, C., Nietering, K., Nicholas, M., Wykoff, R., Bollin, S.: An overview of the scratch resistance of automotive coatings: exterior clearcoats and polycarbonate hardcoats. Coatings. **2**, 221–234 (2012)
3. Seubert, C., Nicholas, M.: Scaling behavior in the scratching of automotive clearcoats. J. Coat. Technol. Res. **4**, 21–30 (2007)
4. American Federation of Mineralogical Societies: Mohs Scale of Mineral Hardness. http://www.amfed.org/t_mohs.htm
5. Lin, L., Blackman, G.S., Matheson, R.R.: A new approach to characterize scratch and mar resistance of automotive coatings. Prog. Org. Coat. **40**, 85–91 (2000)
6. Lin, L., Blackman, G.S., Matheson, R.R.: Quantitative characterization of scratch and mar behavior of polymer coatings. Mater. Sci. Eng. **A317**, 163–170 (2001)
7. Yaneff, P.V., Adamsons, K., Ryntz, R.A., Britz, D.: Structure/property relationships in flexible silane automotive coatings. Proceedings of Twenty-Eighth International Waterborne, High-Solid, and Powder Coatings Symposium, New Orleans, pp. 109–125 (2001)
8. Ward, I.M., Handley, D.W.: An Introduction to the Mechanical Properties of Solid Polymers, pp. 269–271. Wiley (1993)
9. Sambasivam, M., Klein, A., Sperling, A.H.: The molecular basis of fracture in polystyrene films: role of molecular weight. J. Appl. Polym. Sci. **58**, 357 (1995)
10. Gauthier, C., Lafaya, S., Schirrer, R.: Elastic recovery of a scratch in a polymeric surface: experiments and analysis. Tribol Int. **34**, 469 (2001)
11. Ward, I.M., Handley, D.W.: An Introduction to the Mechanical Properties of Solid Polymers, pp. 29–44. Wiley (1993)
12. Seubert, C., Nichols, M., Henderson, K., Mechtel, M., Klimmasch, T., Pohl, T.: The effect of weathering and thermal treatment on the scratch recovery characteristics of clearcoats. J. Coat. Technol. Res. **7**, 159–166 (2010)
13. Blackman, G.S., Pottiger, M., Foltz, B., Li, J., Diehl, T., Wen, M.: Advances in nanoscratch testing of automotive clearcoats. In: Wen, M., Dusek, K. (eds.) Protective coatings: film formation and properties. Springer, New York (2017)
14. Nichols, M.E., Gerlock, J.L.: Photooxidation induced chain scission and crosslinking in acrylic/melamine coatings. Polym. Mater. Sci. Eng. **79**, 403–404 (1998)

Chapter 17
Appearance of Automotive Coatings

Jun Lin, Jingguo Shen, and Marcy E. Zimmer

Introduction

Appearance is one of the most challenging subjects of automotive coatings due to its subjectivity and complexity. While its subjectivity partially stems from the strong dependence of visual resolution on viewing distance [1], the varying mechanisms through which light interacts with surface structures of different dimensions further complicate the perception of appearance. For non-absorbing materials, light will be scattered in many directions when the size of the object is close to its wavelength [2], but it will be geometrically reflected when the objects or structures are significantly larger [3, 4]. In observing a coating surface, the human eye perceives the larger structures as wavy patterns of light and dark areas known as orange peel in the industry, while the smaller structures detract from the glossiness of a coating. The smaller structures often lead to reflected images having poor definition, known as poor distinctness of image (DOI). Experience suggests that the surface structures of automotive coatings are of multiple length scales and that the formation of those structures is determined by a range of factors including paint formulation, application parameters, curing process, and substrate surface roughness [5–8].

The multiple length scales of the coating surface structures necessitate two distinct yet closely related modes of measurements: focusing on the tendency of the surface to either scatter light from small structures or reflect light from the waviness (or orange peel) of large structures. Instruments that measure the capability of a coating surface to scatter and reflect light include the DOI/Haze meter

J. Lin (✉) • M.E. Zimmer
Mt Clemens R&D Laboratory, Axalta Coating Systems, Mt Clemens, MI, USA
e-mail: jun.lin@axaltacs.com

J. Shen
Coatings Technology Center, Axalta Coating Systems, Wilmington, DE, USA

M. Wen, K. Dušek (eds.), *Protective Coatings*, DOI 10.1007/978-3-319-51627-1_17

from TRICOR systems; Gloss meters, Haze meters, and Wave-Scan DOI from BYK-Gardner; and IQ DOI, Haze, and Gloss meter from Rhopoint Instruments. The nature of light scattering by small surface structures and the refractive indices of the coating materials have strong effects on the measurement results.

The Wave-Scan DOI, first introduced by BYK-Gardner in 2000, represents an advancement in measuring both modes of light reflection: a dullness value, which measures image distortion due to scattering from structures smaller than 0.1 mm and surface waviness/orange peel through an optical profile of the coating surface [3, 4]. DOI is most closely associated with the dullness value. The optical profile, acquired through laser scanning of the coating surface, is processed mathematically into five waves representing surface structure at different length scales: 0.1–0.3 mm for Wa, 0.3–1 mm for Wb, 1–3 mm for Wc, 3–10 mm for Wd, and 10–30 mm for We [3], as illustrated in Fig. 17.1a (courtesy of BYK-Gardner).

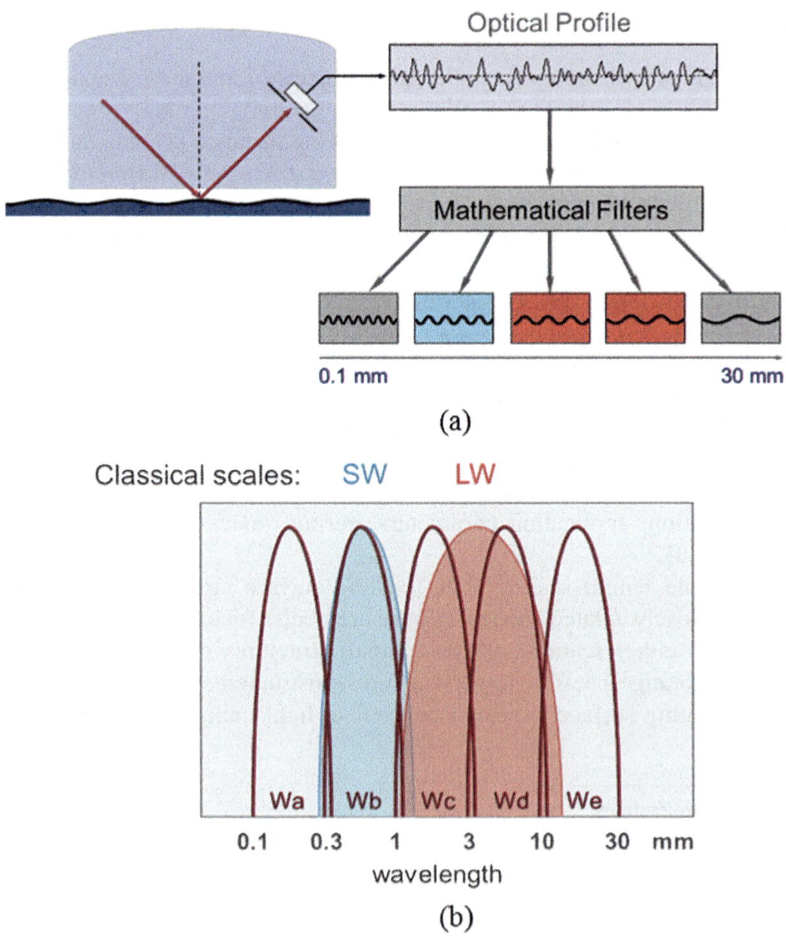

Fig. 17.1 (**a**) Measurement principles of BYK Wave-Scan DOI, (**b**) New wavelength ranges vs. classical scales of orange peel measurement (courtesy of BYK-Gardner)

The amplitude of each wavelength represents the magnitude of light distortion from surface structures in that range and provides an appearance value for that surface structure [3]. Higher amplitudes indicate more pronounced structure or worse appearance. Figure 17.1b compares the ranges of surface structure covered by the five wavelengths, Wa, Wb, Wc, Wd, and We, with the classical scales of long wave (LW) and short wave (SW), as measured by the BYK Wave-Scan Plus, introduced in 1994. The resolution across five size ranges is a significant improvement from the original ranges for SW and LW, as it not only correlates better to the resolution of surface defects by human vision but may also help paint formulators establish a structure–property correlation between paint formulation and process vs. the multiple length scales of structures on coating surfaces.

Despite the Wave-Scan DOI being available since 2000, the subjectivity of visual appearance remains a challenge for paint formulators. Automakers can still be unique when it comes to the measurement and interpretation of appearance. For example, many automakers still set their appearance specifications on the traditional values of SW and LW, and recently, more emphasis has been placed on a DOI specification. Some automakers set a specification on a combined CF value (heavily weighted on dullness or DOI and Wd), while others use a combined R-value, primarily consisting of LW. There are also certain automakers that only value Wa. With these differences, paint formulators must strive to understand the fundamental factors contributing to the surface structures of coatings and develop formulations to meet or exceed expectations from all customers.

There have been relatively few fundamental studies on automotive paint appearance. For example, Basu et al. reviewed the impact of application process and rheology control agents on atomization [5]. Tighe and coworkers evaluated the effect of clearcoat thickness on surface roughness [6]. An interesting study by Peters et al. traced the texture evolution during the curing process [7]. Work has also been carried out to identify the effect of clearcoat and basecoat inter-diffusion on combined system appearance [8]. The effect of substrate on appearance has also been reported [9]. However, most of these studies focused on isolated factors, and comprehensive examination of the interdependent mechanisms influencing coating appearance is still lacking.

This article examines several concepts that are potentially important for the development of surface structure in automotive coatings. These concepts include flow and leveling, substrate telegraphing, gradients of curing and drying, and inter-layer compatibility. Model systems are used to demonstrate and characterize the influences of various factors on appearance, with primary focus being on the appearance as characterized by wave-scan wavelength scales of Wa to Wd.

The model systems consist of both OEM high-bake and refinish low-bake coatings systems. Here, "high-bake" stands for baking at a relatively high temperature such as 140 °C, and "low-bake" stands for baking at a relatively low temperature such as 40–60 °C or air drying. Both systems share similar compositions including functional polymers, reactive crosslinkers, dilution solvents, catalysts, flow additives, etc. but differ in the specific characteristics of the components. Since the OEM system employs a much higher baking temperature, slower solvents and latent catalysts are typically used. The glass transition temperature of the main polymer is also lower to promote a higher extent of cure. Since a refinish system is designed for a lower baking temperature, fast film formation is critical; therefore, fast solvents, highly active catalysts, and resins possessing a higher glass transition temperature are usually selected.

Experimental

Several OEM high-bake and refinish low-bake clearcoat (CC) samples were used as shown in Table 17.1 below. The OEM high-bake samples include three types of clearcoats: 1K acrylic-melamine (acrylic-MF), epoxy-acid, and 2K isocyanate (2K Iso). The refinish low-bake sample uses one type of clearcoat, 2K Iso.

The coating solutions consisted of functional polymers, crosslinkers, catalysts, and various additives. The functional polymers were chosen from acrylic polyols or epoxy acrylics. The crosslinkers were selected from alkoxylated melamine-formaldehyde, poly-isocyanate, or polyester half acid, depending on the nature of crosslinking chemistry and bake conditions chosen. The catalysts were generally a strong organic acid for acrylic-MF crosslinking, a weak acid for high-bake 2K Iso crosslinking, an organotin carboxylate for low-bake 2K Iso crosslinking, or an amine for acid crosslinking of epoxies. The additives included flow additives, such as polyacrylates, for flow and leveling, and UV stabilizers, including benzotriazoles and hindered amine light stabilizers, for paint durability.

For spray application, both the 1K acrylic-MF and epoxy-acid samples were diluted with ethyl 3-ethoxy propionate (EEP) to a measured viscosity of 42 s using a #4 Ford cup. For the high-bake 2K Iso, component A containing the acrylic polyol was diluted to a value of 42 s on a #2 Fisher cup with EEP and then mixed with the poly-isocyanate component B before spraying. The low-bake isocyanate sample was reduced to a Zahn II cup viscosity of about 20 s before application. Samples were sprayed through either a bell applicator (high-bake) or a gravity-fed, air-assisted spray gun (low-bake) onto steel panels that were pre-coated with approximately 25 μm of fully baked electro-coat (E-coat). Depending on the

Table 17.1 List of clearcoat samples used in the study

Category	Clearcoat type	Functional polymers (resins)	Crosslinker	Catalyst type
OEM high-bake CC	1K acrylic-MF	Acrylic polyols	Alkoxylated melamine-formaldehyde	Strong organic acid
	epoxy-acid	Epoxy acrylics	Polyester half acid	Amine
	2K Iso	Acrylic polyols	Poly-isocyanate	Weak acid
Refinish low-bake CC	2K Iso	Acrylic polyols	Poly-isocyanate	Organotin carboxylate

purpose of the evaluation, the CC samples were sprayed either: (1) directly over electro-coat or coil-coated panels, (2) wet-on-wet (2-W) with a basecoat, or (3) wet-on-wet-on-wet (3-W) with a basecoat and a primer.

For the high-bake samples, each wet layer was flash dried at a lower temperature for a brief period before the next layer was applied. After all the layers had been sprayed, the system was fully baked. For the 2-W process, the waterborne basecoat was flash dried for 10 min at 82 °C before the CC was applied. For the 3-W process, a waterborne primer was flash dried at 82 °C for 4 min before being coated with the waterborne basecoat. Then, those combined layers were flash dried at 82 °C for 5 min. Finally, the CC layer was sprayed over the waterborne basecoat, flash dried for 10 min under ambient conditions, and then baked in a convection oven for 30 min at 140 °C. The low-bake samples were baked at 60 °C for 30 min. For appearance evaluations, panels were maintained in either a horizontal or a vertical position throughout spraying and baking.

The various appearance values of the coated panels were obtained with a BYK Wave-Scan DOI. Panel hardness was measured with a Persoz pendulum hardness tester.

For rheological cure profile measurements, an Advanced Rheometric Expansion System instrument (ARES-Melt) was utilized. To run the measurement, about 0.2 grams of a clearcoat sample was pre-dried for 30 min in an oven set at 70 °C to evaporate the solvent. This pre-dried sample was placed between a pair of parallel plates (1-cm diameter) on the instrument, and the gap was adjusted to approximately 0.6 mm for testing. The sample was heated at 4 °C/min to a maximum temperature of 150 °C. For cure analysis, the bottom plate was rotated, deforming the test sample in-between the plates, creating a dynamic shear. A transducer attached to the top plate measured the torque that was produced as a response to this shear strain. The resulting mechanical compliance information, such as storage (G') and loss (G'') moduli, was plotted vs. time and plate temperature.

Results and Discussion

Film Formation: Low- vs. High-Bake Coatings

In automotive coatings markets, global paint suppliers often supply two distinctively different types of paint products. Refinish paint, which is utilized at body repair shops to repair damaged vehicles, is typically cured at ambient to low temperature, while the paint used by OEM automakers to paint car bodies is heated for an extended time at a much higher temperature (e.g., at 140 °C for 30 min).

Figure 17.2 conceptually depicts the film formation process for an OEM coating going through the stages of a high-temperature bake as a function of baking time. The left Y-axis represents film formation, which is characterized by a viscosity

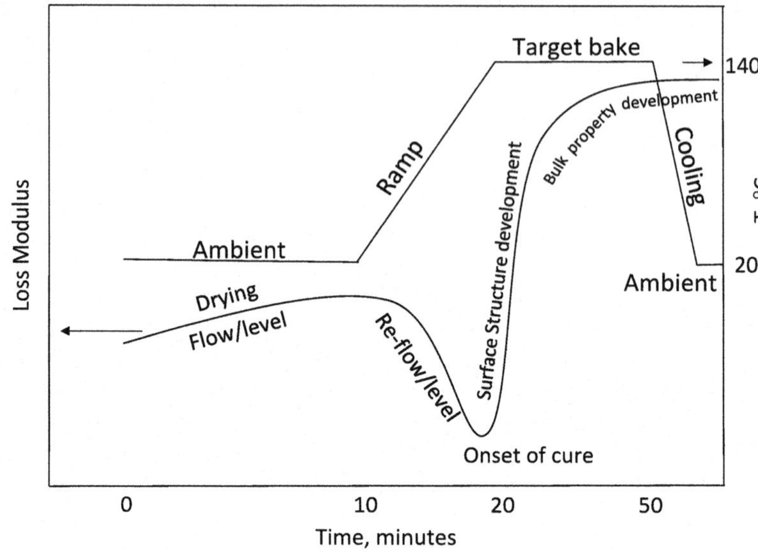

Fig. 17.2 Film formation for OEM high-bake coatings

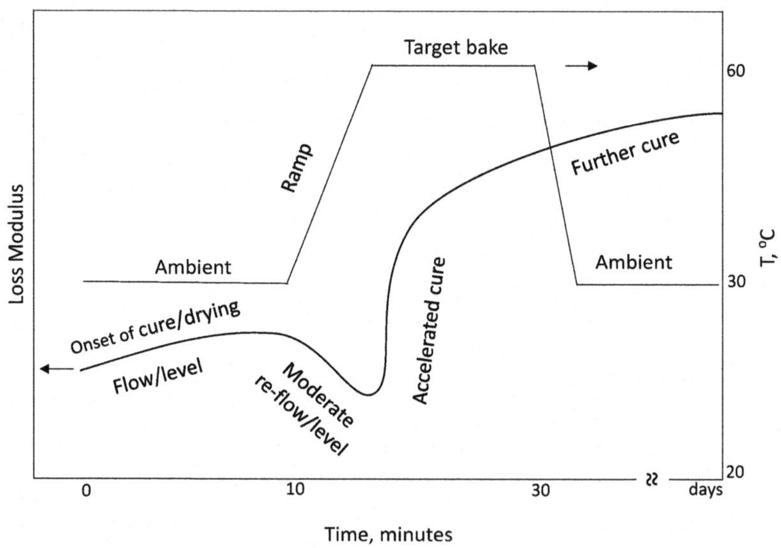

Fig. 17.3 Film formation for refinish low-bake coatings

change or modulus development. The right Y-axis represents panel temperature which ramps up from ambient conditions to the target baking temperature. Immediately after being sprayed, the paint begins to flow and level; however, this process slows down as the viscosity increases due to loss of solvent. As the temperature ramps up, the paint experiences a significant drop in viscosity, leading to significant reflow and leveling until the paint temperature reaches the onset of cure. The start of cure is when the crosslinking reaction begins, and the elastic and loss moduli develop characteristics of solid film formation. The surface structure develops first, within just a few minutes, due to heat transfer from hot air contacting the surface of the coating and initiating cure reactions there. The bulk film properties develop as the cure continues. The cooling process can also be important for the development of surface structure due to thermal shrinkage of the coating film or a possible thermal expansion mismatch with the substrate [7].

Conversely, Fig. 17.3 represents the film formation process of a refinish low-bake coating. Comparing Figs. 17.2 and 17.3, the film formation processes are significantly different between the OEM and refinish coatings. For the OEM coatings, flow and leveling dominate from the start of the drying process until the onset of cure, which is when crosslinking chemical reactions initiate. For the refinish coatings, the chemical reactions (cure) begin much earlier, theoretically right after the polyol component A is mixed with the poly-isocyanate component B in the paint cup. At repair shops, the baking temperature can be as high as 60 °C. As a result, the refinish coatings may struggle to achieve the same extent of flow and leveling as that of the OEM coatings. Another significant difference between the two types of paint products is in structure development. For the OEM coatings, the timeframe for the development of surface structure is minutes, while a refinish coating may take hours or even days to develop its structure. As a result, flow and leveling can be a more easily controlled factor for the final appearance of an OEM coating than a refinish coating. On the contrary, both surface structure and appearance development of a refinish coating can initiate as soon as the paint is sprayed and last much longer. A more detailed discussion on structure development and gradients of cure and drying will be provided in a later section.

Flow and Leveling

Flow and leveling is a process for a paint to develop a smooth surface, therefore exhibiting good appearance. The paint surface may start out "rough" after paint application, but surface tension drives the surface to be as smooth as possible in order to minimize the surface area or energy. An equation was derived by Orchard to describe flow and leveling based on observations of brush-applied paint (Eq. 17.1 and Fig. 17.4) [10–12]:

$$\ln\frac{a_0}{a_t} = \frac{16\,\pi^4}{3} \cdot \frac{\gamma x^3}{\lambda^4} \cdot \int \frac{\mathrm{d}\,\mathrm{t}}{\eta} \qquad (17.1)$$

This equation demonstrates that the most significant factors for flow and leveling are coating film thickness and the structure wavelength, as they are of cubic and quartic order, respectively. Time and viscosity are also important, but at lower

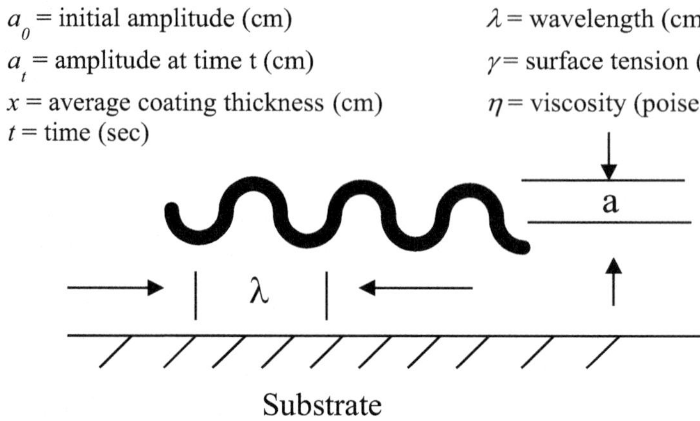

Fig. 17.4 Illustration of Orchard's flow and leveling equation

orders. Since this equation assumes constant surface tension during the leveling process, its applicability to air-drying systems has been challenged due to the development of a surface tension gradient as a result of solvent evaporation from the surface [13]. Models for the leveling of thixotropic coatings (coatings deviating from Newtonian rheology) have also been proposed [7, 14]. However, since solvent composition and rheological behaviors are generally similar for the same classes of automotive coatings, the validation of even the orders of magnitude between factors such as film thickness, structure wavelength, and leveling as suggested by Eq. (17.1) is still beneficial.

To evaluate the relative importance of and interactions between film thickness, structure wavelength, and leveling, a model 1K acrylic-MF clearcoat was sprayed to achieve a gradual increase of film thickness, ranging from 10 to 100 μm, over black basecoat (BC) on a smooth coil-coated substrate. Via profilometry, the arithmetic average height of roughness of the coil-coated substrate, R_a, is equal to 0.1 μm. The panel was dried for 10 min at ambient conditions to allow for solvent evaporation, followed by a bake at 140 °C for 30 min, all while being kept in a horizontal position. Appearance was evaluated at four of the wavelength factors as a function of clearcoat film thickness (Fig. 17.5).

Dullness with a wavelength range of less than 0.1 mm is not considered in this study as the origins of such structures are usually not associated with flow and leveling. Also, We, covering a wavelength range of 10–30 mm, is not discussed. Surface structures of automotive paint systems with this length scale take too long to achieve leveling and do not reach a plateau value even at 100 μm coating thickness. The arrows in Fig. 17.5 point to the critical coating thicknesses beyond which the decreasing wave-scan factors reach their plateau values or the completion of leveling. More specifically, as indicated in the graph, the critical coating thicknesses of Wa, Wb, Wc, and Wd are 16, 21, 52, and 74 μm, respectively. The results

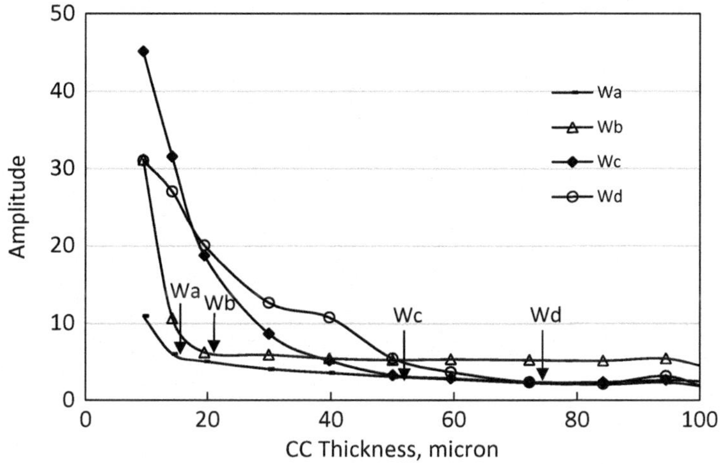

Fig. 17.5 Leveling of clearcoat over black BC dried and baked horizontally

are nominally consistent with Eq. (17.1); coating structures of longer wavelengths require higher thicknesses to achieve leveling at a given time. However, the observed values of critical coating thickness do not exactly fit the equation. For example, the ratio of average wavelength between Wb and Wd is 10. Since the amplitudes for Wb and Wd were similar at both low and high film thicknesses, it is reasonable to assume that the respective values of ln (a_0/a_t) would be similar at a given time. Therefore, a factor of 10 in wavelength ratio would be equivalent to a thickness ratio of 22 according to Eq. (17.1), which is considerably higher than the 3.5 thickness ratio measured for Wd vs. Wb (74/21) upon leveling. Potential explanations for these differences include:

1. The critical coating film thicknesses that identify the point of leveling were extrapolated from fitted data points, which contributes to systematic error.
2. The time factor was not considered. All of the critical film thicknesses representing leveling were identified at a fixed time (after the panels were taken out of the oven). There is a possibility that Wb reached the plateau of leveling much earlier than Wd during the bake cycle. This could lead to considerable error in the calculation of the ratios.
3. Film formation of automotive paint is a complicated process with multiple stages of drying and curing, as suggested by Fig. 17.2. The structure development after the start of cure caused by volume shrinkage may also influence the final appearance.

In conclusion, Eq. (17.1) is at least qualitatively valid with regard to the influence of film thickness and structure wavelength on leveling for automotive paint that is dried and baked in a horizontal position. For paint that is dried and baked in a vertical position, this relationship needs to be reevaluated, due to the potential effect of gravity on flow and leveling [7].

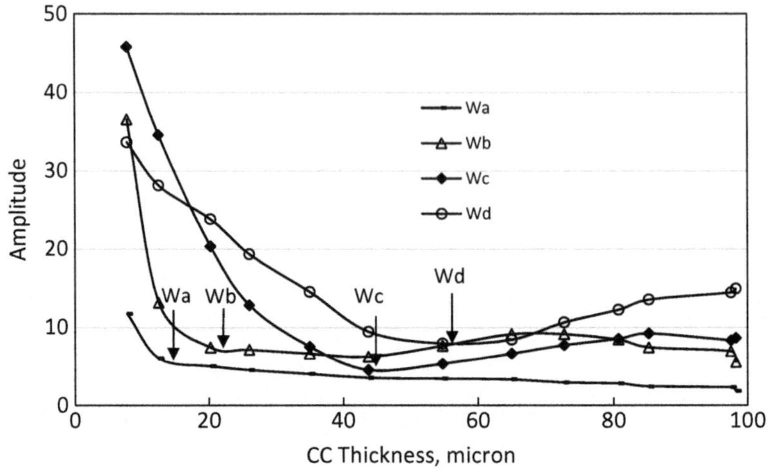

Fig. 17.6 Leveling of acrylic-MF clearcoat over black basecoat dried and baked vertically

The same model acrylic-MF clearcoat used to generate the data in Fig. 17.5 was applied in the same manner over a black basecoat (BC) on a coil-coated substrate, but flashed and baked vertically. The appearance as a function of clearcoat film thicknesses was characterized using the same four wave-scan factors (Fig. 17.6).

A similar trend of leveling vs. film thickness was observed for the vertically baked system: critical coating thickness for Wd > Wc > Wb > Wa, with a few important differences:

1. The film thickness where leveling occurs on the vertical panels tends to be lower than that on the horizontal ones, especially at longer wavelength ranges. For example, leveling for Wc is at 52 μm for the horizontal panel, but 46 μm for the vertical panel; for Wd, leveling of horizontal panel is at 72 μm, but 54 μm for the vertical.
2. The amplitudes on the vertical panels at leveling are generally higher than on the horizontal panels, especially at longer wavelength ranges. For example, the amplitude of Wd is 8 for the vertical panel but around 2 for the horizontal panel at the critical coating thickness.
3. The amplitudes of the vertical panels increase after reaching a minimum as the film thickness further increased, with the exception of Wa.

To highlight the differences between the appearance of the vertical and horizontal panels, the data were plotted in Fig. 17.7 comparing vertical vs. horizontal at each of the four wave-scan factors.

Figure 17.7 demonstrates that although significant leveling differences between horizontal and vertical panels are observed at longer wavelength ranges (i.e., Wc and Wd), the initial decrease of the amplitude against increasing film thickness is very similar. It is only after reaching the critical coating thickness that the values for the two orientations began to diverge: the amplitudes of Wc and Wd for the vertical

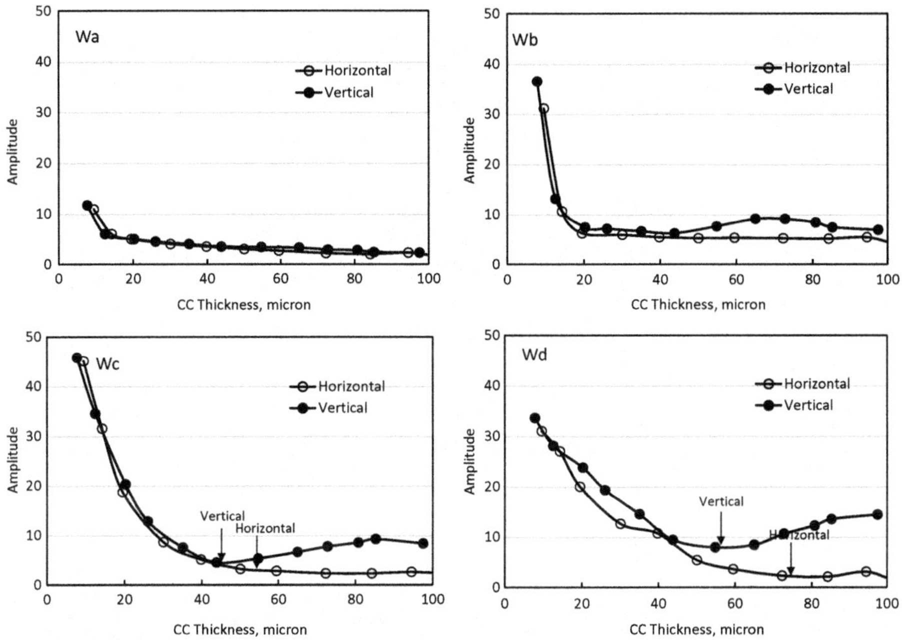

Fig. 17.7 Horizontal vs. vertical for leveling of Wa-Wd structures

panels rise with increasing film thickness, while the amplitude reduction continues for the horizontal panels. It is arguable that the difference observed in critical coating thickness is due to the higher minimum amplitude, or texture, for the vertical panels. The question is why vertical orientation generally leads to higher wave-scan amplitudes than the horizontal orientation.

To help explain the difference in appearance between the vertical and horizontal panels, the concept of sagging must be considered. Sagging is defined as the volume of paint being pulled through a given cross-section due to gravity, as indicated by the shaded slices in Fig. 17.8 [15].

The total volume of paint that sags over a given cross-section of the film (1.0X) during time t can be calculated from the equation expressed in Eq. (17.2) [15]:

$$V_s = \frac{X^3 \rho g t}{3\eta} \tag{17.2}$$

where V_s is total sagged volume, X is total film thickness, ρ is coating density, g is acceleration of gravity, t is time, and η is viscosity in poise.

Equation (17.2) suggests that the total sagged volume of a coating is strongly dependent on the coating thickness given its cubic order. This relationship is similar to the impact of thickness on leveling as expressed in Eq. (17.1), which means a film of higher thickness will create a significantly higher volume of sagging than a film

Fig. 17.8 Representation of paint sagging past a given cross-section of the film

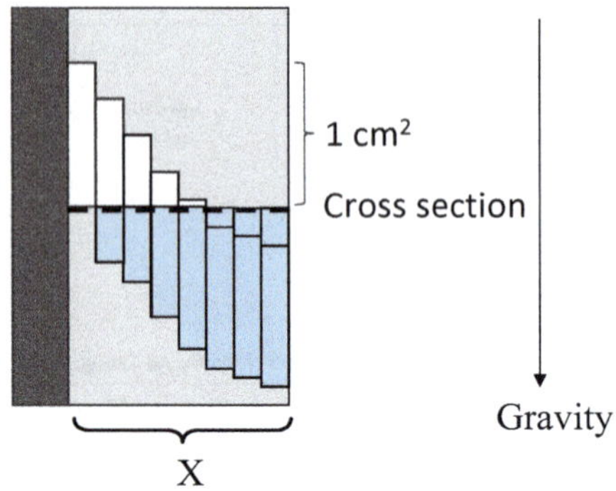

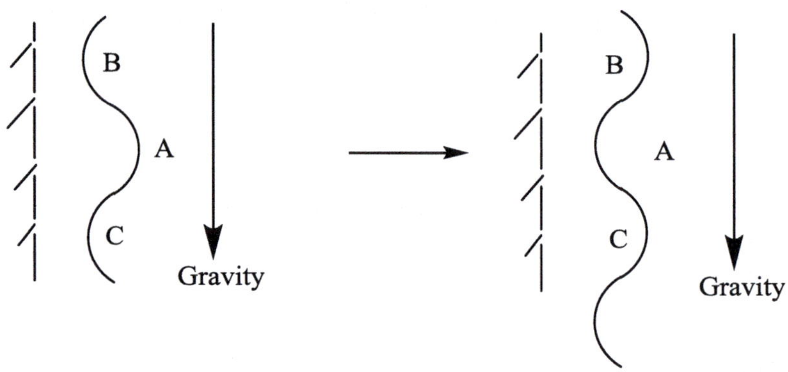

Fig. 17.9 Representation of gravity-driven material movement

of lower thickness. The dependence of sagging on film thickness can create a non-ending wave movement on a vertical surface as demonstrated in Fig. 17.9.

The point A in Fig. 17.9 corresponds to a thicker cross-sectional area, which produces a higher volume of sagging than point B or C, according to Eq. (17.2). As a result, at time *t*, the volume of the sagging material under A will move to the lower cross-sectional point C, creating a new high amplitude. By contrast, cross-sectional area B will produce less volume of sagging to replace the material that moved away from A. Thus at time t, a new surface wave is formed, with B and C being thicker and A thinner. This sequence of material movement creates a wave movement on the paint surface, and the surface structure is maintained. The competition between surface tension-driven leveling and gravity-driven wave movement leads to a higher amplitude at final leveling, as demonstrated in Fig. 17.9. This is one explanation for why vertical panels are always worse than horizontal panels for appearance.

A potentially more suitable theoretical model to describe gravity-driven surface wave movement of liquids is found in the field of fluid dynamics. The "shallow-liquid wave propagation" model specifically studies the movement of fluids with thicknesses significantly smaller than the wavelength of the surface waves [16]. Since an automotive clearcoat has a typical film thickness of 40–50 μm and the wave-scan wavelength ranges are on the order of millimeters, the small thickness to wavelength ratio fits the definition of a "shallow-liquid" well (thickness $\leq$ 1/20 of wavelength). The "shallow-liquid" theory predicts that if a gravity-driven arbitrary wave is traveling along the surface of a shallow-liquid layer with a given velocity, it will continue to travel with the same velocity, and the shape of the wave will not change as it moves along the surface [16]. Thus, this theory predicts that the vertical surface waves of paint will remain undamped due to the continuous surface wave movement. This partially explains why vertical appearance is generally worse than the horizontal one, and the characteristic of film shallowness becomes even more appropriate in the longer wavelength ranges, such as Wc or Wd. In conclusion, gravity-driven paint flow that propagates in the form of surface waves retains surface structure and therefore resists leveling.

At shorter wavelength ranges, such as Wa, where the wavelength to paint thickness ratio (assuming 50 μm thickness) is in the range of 2–6, the "shallow-liquid wave propagation" theory may not apply. Thus, surface tension-driven flow and leveling will dominate, which is one of the reasons that a significant difference is not observed for Wa between vertical and horizontal orientations. When the wavelength to thickness ratio exceeds 6–20 (at Wb or longer wavelength ranges for a 50 μm liquid), gravity-driven "shallow-liquid wave propagation" takes effect, explaining why the amplitude of vertical panels increases at higher thickness for Wb to Wd. Interestingly, the coating industry occasionally uses a terminology of "micro-sagging" to describe a structured appearance due to local sagging [17]. Micro-sagging refers to a vertical appearance when the amplitudes of long waves stop to decrease with an increase of coating thickness. This phenomenon can be interpreted as a result of "shallow-liquid wave propagation."

Substrate Telegraphing

It is well known in the automotive industry that poor metal quality or high surface roughness of the substrate can "telegraph" to the paint surface and thus, leading to poor appearance [9], especially for processes where wet paint is sprayed directly over the rough substrates. For example, electro-coat (E-coat) surface roughness is a critical control characteristic for 3-W (primer, BC and CC sprayed wet-on-wet-on-wet) and primerless (BC and CC are sprayed wet-on-wet) processes. Figure 17.10 shows a schematic representation of the substrate telegraphing effect on the surface texture of a high-bake OEM coating utilizing either an addition or a condensation type of crosslinking chemistry.

The idealized scheme shows that atomized paint droplets deposited on a rough substrate will form a surface that telegraphs the structure of the substrate due to flow and leveling and solvent loss. More specifically, flow and leveling can lead to the initial formation of a fully leveled surface, which creates film thickness variations across the substrate. The thicker regions (corresponding to the valley areas on the substrate) will shrink more due to a higher absolute volume of solvent content, while the thinner regions (corresponding to the peak areas on the substrate) will

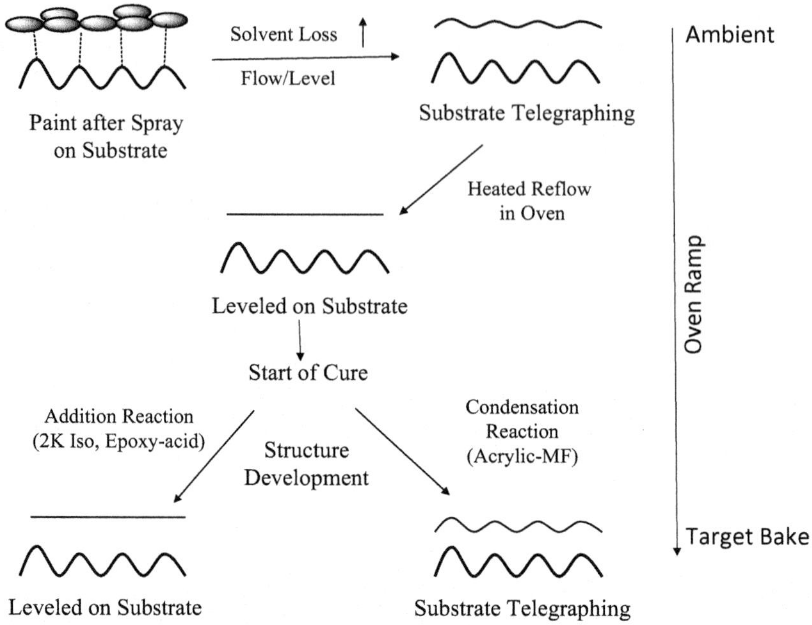

Fig. 17.10 Flow and leveling and crosslinking chemistry on substrate telegraphing

shrink less. Therefore, the evaporation of solvent will lead to the formation of a wet/uncured paint surface reproducing the substrate topography. For an OEM coating, the structured surface has another opportunity to flow and level when the paint is subjected to a high-temperature bake, where paint viscosity becomes significantly lower. Then, as the oven temperature further increases, cure initiates and different scenarios can occur, depending on the type of crosslinking reactions. The onset of cure leads not only to the inhibition of flow and leveling due to network formation but also to the possibility of film shrinkage. The acrylic-melamine crosslinking reaction is a condensation reaction as shown in Fig. 17.11. The amount of alcohol released can be as high as 15% weight lost and a corresponding volume shrinkage, depending on the level of melamine and acrylic polyol. In contrast, the epoxy-acid and 2K isocyanate (2K Iso) chemistries are addition reactions. No further weight is lost beyond the solvent, and only minimum film shrinkage is expected.

As a result, coatings using a condensation reaction such as acrylic-MF clearcoat can form a structured surface that telegraphs the substrate. In comparison, coatings with addition reactions, such as 2K Iso and epoxy-acid, can form a paint surface of significantly less structure. It is also important to note that the structure development process for OEM coatings can be very short. For example, the start of cure for a typical acrylic-MF coating occurs around 100 °C, and the structure development process can end by 120 °C, due to a significant level of crosslinking, which locks up the surface structure. For an oven ramp-up rate of 10 °C/min, this translates to a structure development time of only 2 min, which is drastically different from

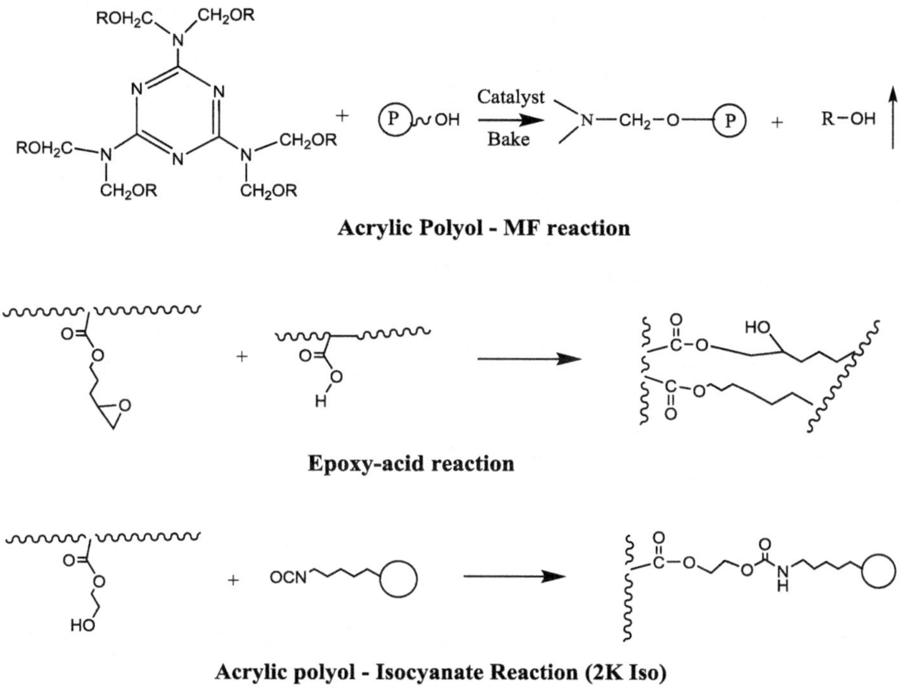

Fig. 17.11 Crosslinking of acrylic-MF condensation reaction vs. epoxy-acid and 2K Iso addition reaction

the structure development process for low-bake systems to be discussed in a later section.

To experimentally evaluate the effect of crosslinking chemistry on substrate telegraphing, clearcoats with acrylic-MF and epoxy-acid crosslinking chemistries were directly sprayed over both a rough E-coat substrate (surface roughness R_a ~0.5 μm) and a smooth coil-coated substrate (R_a ~0.1 μm) without the use of primer or basecoat. From this comparison, the ability of a given clearcoat to cover the E-coat roughness can be determined. The horizontal drying and baking protocol was used to minimize any complications from gravity-induced sagging.

As shown in Fig. 17.12a, over the rough E-coat substrate, the epoxy-acid clearcoat provides a significantly lower SW amplitude than the acrylic-melamine clearcoat. Also interestingly, the acrylic-MF clearcoat appears to have higher wave-scan amplitude with increasing clear thickness. This can be due to its higher absolute volume release of alcohol combined with a more significant gradient of curing at a higher thickness, together creating a "dieback" type of appearance defect which is more commonly seen in a low-bake system. This type of appearance defect will be discussed in detail in the next section. Notably, even the epoxy-acid possesses some level of surface texture over E-coat after bake. An experiment that measured weight loss during bake demonstrated that even the epoxy-acid coating can lose 4–5% of coating weight from the bake, which helps explain why

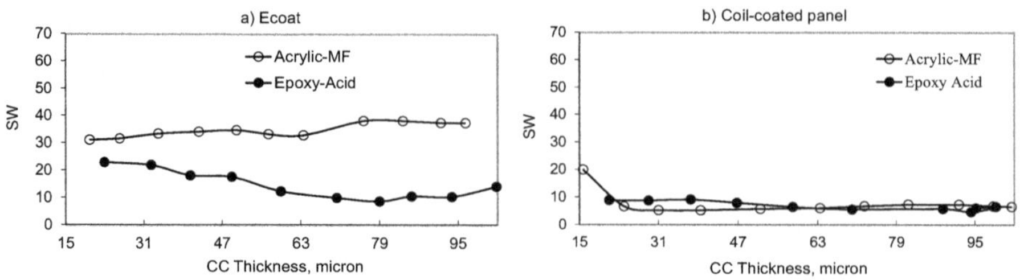

Fig. 17.12 SW vs. clearcoat thickness for acrylic-melamine and epoxy-acid clearcoats over (**a**) E-coat substrate and (**b**) coil-coated substrate, dried and baked horizontally

epoxy-acid may also possess some degree of substrate telegraphing. The volume shrinkage observed in epoxy-acid could be due to the presence of low Mw portions of both the epoxy and acid polymer distributions, which might become volatile at the high-temperature baking conditions. Other low-molecular-weight components of the coating formulation, such as UV stabilizers, catalysts, and slow solvents, could all contribute to the weight loss or volume shrinkage in the coating. The solidification of the coating films (liquid state for the systems studied in this work) due to crosslinking may increase the film density and thus contributes to the volume shrinkage as well.

When the same clearcoats were sprayed over the much smoother coil-coated substrate (surface roughness $R_a = 0.1$ μm), very different results were obtained, as shown in Fig. 17.12b. The acrylic-MF clearcoat over a smooth substrate gives comparable SW values to that of the epoxy-acid clearcoat. In the low-thickness region, the acrylic-MF clearcoat was even slightly better than the epoxy-acid clearcoat. The difference is probably due to better flow and leveling of the melamine crosslinker, which has a lower molecular weight than the acid counterpart in the epoxy-acid clearcoat. In summary, volume shrinkage has less of an effect on surface texture when the substrate is smoother.

Gradients of Drying and Curing

The gradients of drying and curing refer to the variance of both solvent retention rate and crosslinking rate as a function of the depth in coating through the curing process. More specifically, nearly all coatings tend to cure and dry faster at the surface than in the bulk portion of the coating. Further, as a coating vitrifies from the surface down, a crosslink density gradient can be established. Experimental evidence for the existence of such depth-wise drying and curing gradients includes an indentation hardness profile of 2K low-bake isocyanate clearcoat vs. cure time

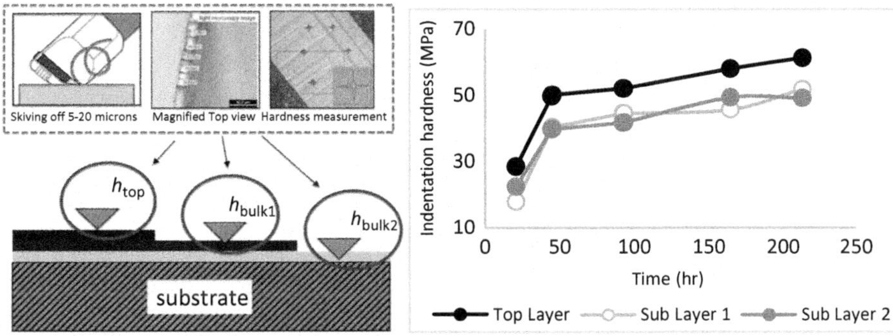

Fig. 17.13 Indentation hardness gradient determined via skiving techniques

obtained from skiving techniques [18], as shown in Fig. 17.13. The results show that the top layer (skin-containing layer) has a higher hardness as a function of cure time, which is likely due to lower solvent content and consequently higher T_g [19].

Many factors, including the nature of the crosslinking reactions, solvent–resin interactions, solvent concentration, and external conditions, such as temperature, pressure, and humidity, potentially control the drying and curing gradients via complicated mechanisms [20, 21]. For example, the solvent diffusion rate is a limiting factor that can create drying gradients as the cure state of the film changes. More specifically, the evaporation of solvent at the skin layer promotes cure and system T_g by raising the concentration of reactants, but as solvent evaporates and the viscosity of the surface layer increases, reactant mobility is suppressed. As a result, the cure rate will decrease and the film will begin to vitrify, which in turn controls the solvent diffusion rate. Therefore, gradients of drying and curing can result from the interplay of solvent retention, crosslinking kinetics, and glass transitions/vitrification of the curing film.

The influence of drying and curing gradients on coating appearance typically manifests through a phenomenon known as "dieback," where surface textures develop late in the cure cycle. For low-bake coatings, the timing can range from hours to days. One key mechanism proposed to explain this delayed structure development is wrinkling caused by gradients of drying and curing due to trapped solvents or unreacted low-molecular-weight reactants underneath the surface skin layer, migrating to the semi-cured surface and reswelling the constrained network [21]. Dieback is often facilitated by relatively faster drying and curing of the coating at the surface layer of the coating film vs. the bulk coating, or faster "skin formation." In theory, faster skin formation would entrap more solvent in the bulk coating, causing steeper depth-wise drying and curing gradients, and in turn, more entrapped solvent would generate more swelling of the semi-cured surface over time, leading to poorer appearance. Therefore, in general, less steep or controlled gradients of drying and curing result in less dieback, hence better appearance. For example, in theory excellent appearance in a high-bake coating could be achieved by ramping up the oven temperature very slowly. It is worth noting that

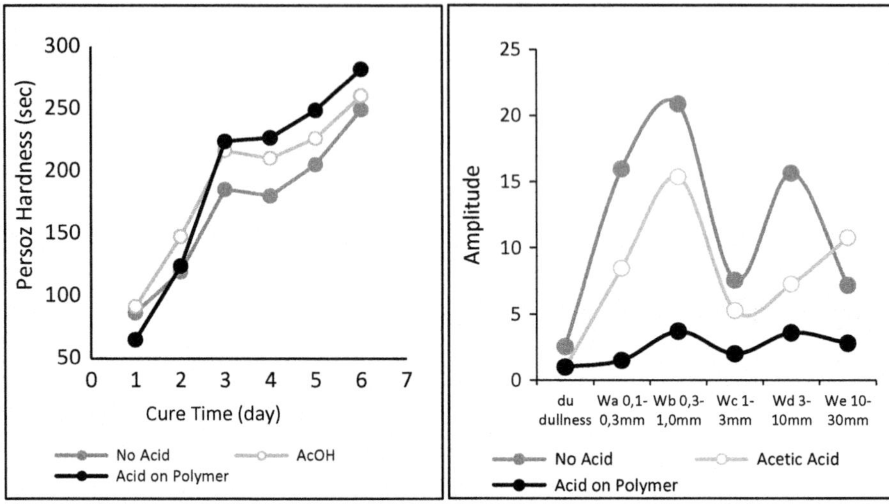

Fig. 17.14 Persoz hardness as a function of cure time and wave-scan appearance data at 7 days of cure time for a 2K isocyanate clearcoat system with different reaction inhibitor options

through this mechanism, the gradients of drying and curing inherently have fundamental effects on the balance of appearance and hardness development in the coating film.

Formulation factors can be utilized to adjust the gradients of drying and curing, including resin properties, solvent choices, and catalyst systems, among others. For example, in a low-bake 2K isocyanate coating, the cure gradient can be minimized by the use of a crosslinking reaction inhibitor—carboxylic acids. Figure 17.14 shows typical appearance and hardness development data from a model clearcoat system with various types of carboxylic acids as inhibitors of the hydroxyl–isocyanate reaction. By design, the group of inhibitors varies in terms of both volatility and mobility, in the order of Acid on Polymer < Acetic Acid < No Acid, assuming a mobility of infinity for No Acid. Conversely, the order of the rate of film skin formation (deduced from increases in early Persoz/pendulum hardness values) was Acid on Polymer < Acetic Acid < No Acid, also likely to be the order of lowering carboxylic acid concentration in the skin layer. Therefore, it is reasonable to assume that steeper cure gradients and more dieback will be present in systems with more volatile and mobile inhibitors, which agrees well with the observed trend in appearance: No Acid < Acetic Acid < Acid on Polymer.

Multi-Layer Compatibility

Discussions thus far have focused on factors that affect the surface structure of a single layer. However, coating systems are multi-layer, and any incompatibility among different layers can disrupt interfaces and create structures that telegraph to

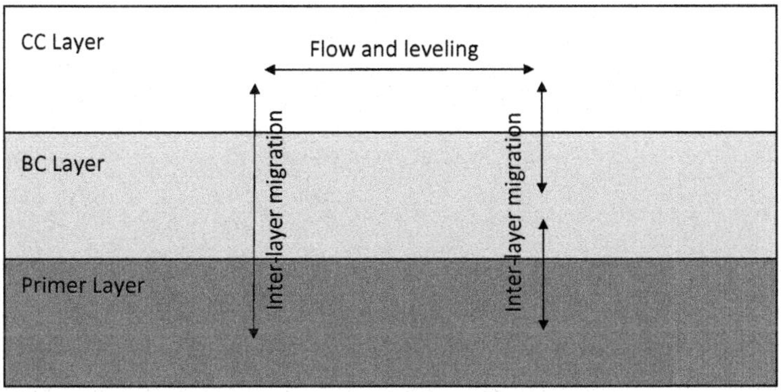

Fig. 17.15 Effect of inter-layer migration on flow and leveling of the topmost level in a multi-layer 3-W system

the surface of the top layer. One such scenario for multi-layer systems is the potential for ingredients to migrate from one layer into a different layer. Concentration differences between layers can drive ingredients of high mobility to migrate across the interface and potentially disrupt its smoothness (Fig. 17.15).

Another scenario for multi-layer systems is the potential for a cure rate difference between neighboring layers, which can disrupt the interface and create structures. A significant mismatch of cure rate between two neighboring layers can create heterogeneous structures across the interfacial region and subsequently telegraph to the surface of the top layer and hurt the overall system appearance.

To test these hypothesized scenarios, acrylic polyols of three different weight average molecular weights (M_w) were prepared: 3000, 4000, and 5000 g/mol. Each resin was formulated into an acrylic-MF clearcoat. The glass transition temperature (T_g) of the polyols was varied from −10 to 15 °C, as calculated with the Fox equation [22]. Resin T_g was chosen as a design factor because a lower T_g is expected to promote mobility, especially at elevated temperatures. Finally, to create a cure rate mismatch between the basecoat and clearcoat, portions of the OH groups in the acrylic polyols were converted into urethane groups that are less reactive with melamine-formaldehyde (MF), via reaction with an alcohol half-blocked isophorone diisocyanate (IPDI). The catalyst concentration was kept constant because it also affects other properties than just the reaction rate. Clearcoat samples formulated with these polyols and MF have slower cure rates than the basecoat, as the cure rate of the basecoat is similar to the reaction rate of MF with the polyols prior to OH conversion. Table 17.2 describes the design of experiment (DOE) used to study the impact of the three factors (resin M_w, T_g, and % OH conversion) on appearance of the combined basecoat–clearcoat system.

In Table 17.2, HEMA is hydroxyethyl methacrylate, and % OH conversion indicates the weight percentage of the OH monomers (HEMA) that have been converted into urethane groups through reaction with the half-blocked IPDI.

Table 17.2 Full factorial DOE for BC–CC interaction study

Run	Base acrylic			% OH conversion	Spray group
	% HEMA in resin (Wt)	T_g, °C Polyol	M_w		
1	38	2.4	5000	50	Day 1, 2
2	38	2.4	5000	50	Day 1
3	38	15	5000	50	Day 1
4	38	−10	5000	50	Day 1
5	38	2.4	4000	50	Day 1
6	38	2.4	3000	50	Day 1
7	38	2.4	5000	40	Day 1
8	38	15	5000	40	Day 1
9	38	−10	5000	40	Day 2
10	38	2.4	4000	40	Day 2
11	38	2.4	3000	40	Day 2
12	38	2.4	5000	30	Day 2
13	38	15	5000	30	Day 2
14	38	−10	5000	30	Day 2
15	38	2.4	4000	30	Day 2
16	38	2.4	3000	30	Day 2

Clearcoat samples made with the resins in Table 17.2 were sprayed over a black solvent-borne basecoat that had been applied directly on E-coat substrate. The coated panels remained in a horizontal position during flash and bake. Both the BC and CC were sprayed to their respective target thicknesses (approximately 20 μm for the BC and 50 μm for the clearcoat). The appearance of the clearcoats was measured with the BYK Wave-Scan DOI. Because of the large number of panels to be sprayed, the experiment was split into 2 days. The clearcoat sample made with the Run 1 resin was included on both days as a control. The appearance values of the clearcoat samples sprayed on different days were normalized to the results of the control clearcoat sprayed on the same day to minimize the effect of day-to-day variation on the analysis. A statistical analysis tool was used to analyze the effects of the three variables on appearance. Figure 17.16 shows the main effect plots for T_g, M_w, and % OH conversion of the acrylic polyols on short wave (SW) and long wave (LW) magnitudes after normalization.

Figure 17.16 shows that lower T_g resins are worse for short wave performance but better for long wave. This is consistent with the hypothesis that higher mobility resins can facilitate flow and leveling but disrupt interfacial structure through inter-layer migration. Interestingly, molecular weight does not demonstrate a significant effect on either SW or LW. This might indicate that the M_w range of 3000–5000 g/mol is not sufficiently large enough for the resins to be differentiated statistically. The effective size of a polymer molecule is proportional to the root square of the chain length [23], which makes the "effective size" difference even smaller. To demonstrate a significant M_w effect, the M_w at the lower limit may need to be 1000 g/mol or less. The effect of % OH conversion on system appearance is

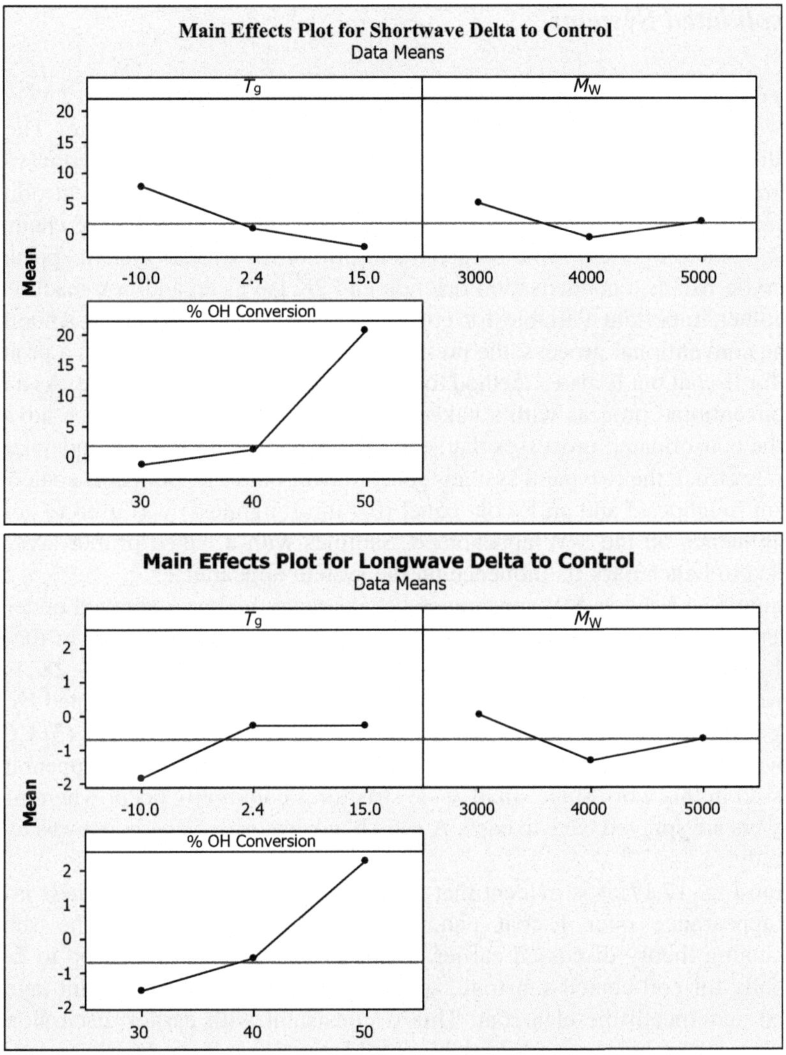

Fig. 17.16 Effects of T_g, M_w, and % OH conversion of the acrylic polyols on SW and LW of clearcoat over solvent-borne black basecoat. "Delta to control" means the amplitudes of the short wave and long wave values were subtracted by that of the control sample sprayed on the same day

significant. Earlier discussion suggested that an increase of % OH conversion would slow down the CC cure rate and lead to a larger cure rate differential with the BC. The large effect of % OH conversion on both SW and LW appearance supports the hypothesis that a cure rate differential between neighboring layers can disrupt the interface and telegraph structure to the top layer.

Consolidated Systems

Conventional OEM painting processes spray BC and CC wet-on-wet (2-W) over a primer that has been pre-coated and baked on an E-coat substrate. The 3-W consolidated process applies the primer, basecoat, and clearcoat wet-on-wet-on-wet directly over an E-coat substrate. To help understand the factors controlling the appearance of 3-W process, two clearcoats with different crosslinking chemistries were evaluated over the same basecoat and primer for overall system appearance: 1K acrylic-MF as a condensation reaction and 2K Iso as an addition reaction.

Another important variable for consolidated systems is substrate smoothness. For the conventional process, the pre-baked primer not only serves as a protective layer for E-coat but is also a method to smooth out the substrate texture. As a result, the conventional process with a baked primer layer has an appearance advantage over the consolidated processes that spray directly over E-coat. To understand the substrate effect, the two paint systems sprayed over both a smooth coil-coated panel (0.1 μm roughness) and an E-coat panel (0.4 μm roughness) was used to compare their influence on the 3-W appearance. Samples with a baked primer layer were included to benchmark its influence on the system appearance.

Figure 17.17 shows SW appearance vs. clearcoat thickness sprayed under eight different process conditions. These plots provide more understanding of the effect of substrate (coil-coat vs. E-coat) and CC type (1K acrylic-MF vs. 2K Iso) on appearance. The potential for inter-layer interactions can be determined by comparing results for coating systems with (1) CC only, (2) CC over BC, (3) CC over BC over a baked primer, and (4) a full 3-W system. Only SW appearance is discussed in this work since small-scale structures commonly occur when multiple wet layers are sprayed back to back. A solid black waterborne basecoat was used for this study.

From Fig. 17.17, it is evident that appearance over coil-coated panels is better than appearance over E-coat panels. This is consistent with the substrate telegraphing theory discussed earlier. Comparing the graphs from top to bottom, especially for coil-coated substrate, appearance degrades as more paint layers are applied underneath the clearcoat. This is consistent with earlier discussions that appearance can be compromised by inter-layer ingredient migration and/or a gradient of cure. With more layers being sprayed and baked together, there are more opportunities for ingredients in a given layer to migrate to other layers. It is also more likely that the cure differential across the different interfaces can amplify the interface disruption and thus, hurt the system appearance. Further, since more wet layers are underneath the clearcoat layer, it is possible that the structure development process takes longer, approaching the mechanisms of low-bake cure systems for surface structure development. As a result, wrinkling may also become a potential appearance defect for the 3-W system.

For the clearcoat comparison, the 2K Iso produces superior appearance vs. the 1K acrylic-MF; the improvement is even more significant when the clearcoat is applied over additional paint layers and rougher underneath layers or substrates.

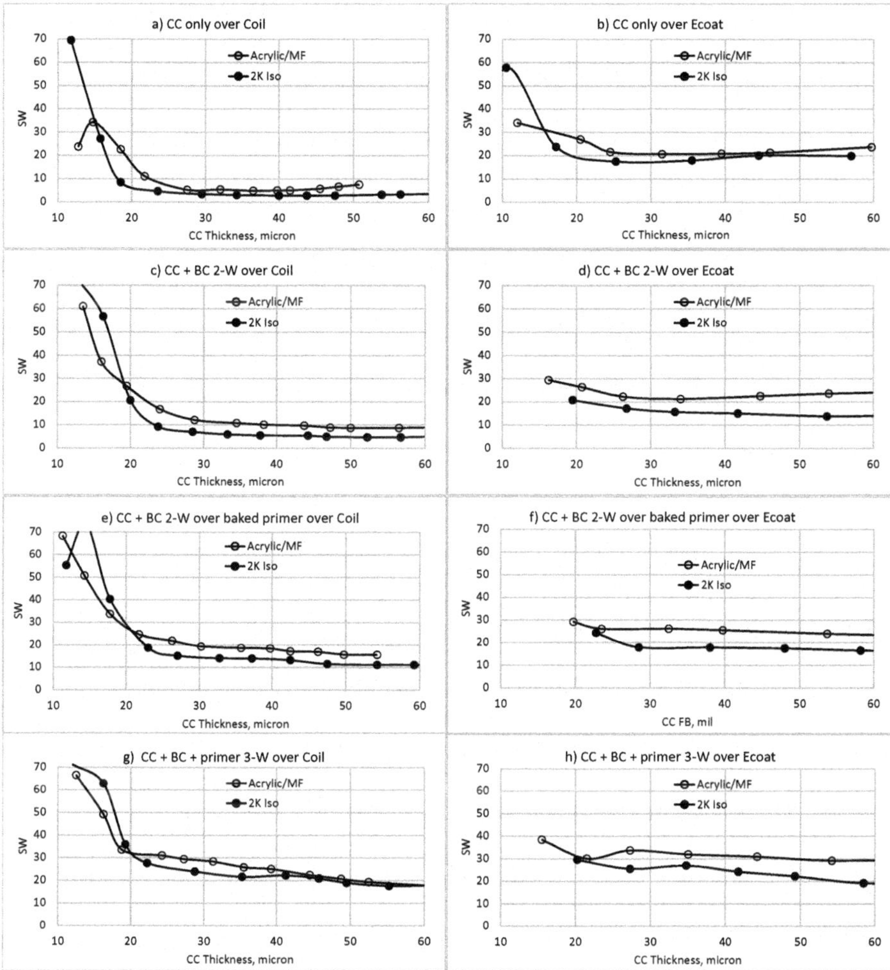

Fig. 17.17 SW Appearance of Acrylic-MF vs. 2K Iso coating systems sprayed under eight different layering conditions described in the individual graphs. 2-W represents two layers sprayed wet-on-wet with CC over BC. 3-W represents three layers sprayed wet-on-wet-on-wet with CC over BC over primer

There are three possible mechanisms contributing to the superior appearance of the 2K Iso. The most important one is probably its addition-reaction type of crosslinking chemistry. With significantly less volume shrinkage in the 2K Iso clearcoat during cure, the structure development of this clearcoat is expected to be significantly less than that of the 1K acrylic-MF, as discussed in the earlier sections.

As another advantageous mechanism, the 2K Iso clearcoat has a significantly slower cure rate than the 1K acrylic-MF as demonstrated in Fig. 17.18. This figure

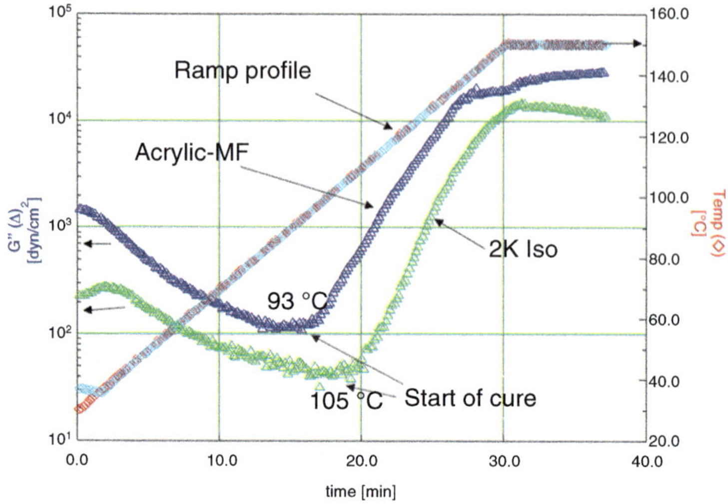

Fig. 17.18 Rheological cure profile of the 1K acrylic-MF vs. 2K Iso

compares the rheological cure profiles of the 1K acrylic-MF and the 2K Iso, generated with an Advanced Rheometric Expansion System instrument. As Fig. 17.18 demonstrates, the onset of cure for the 2K Iso clearcoat occurs later (19 vs. 16 min) at a higher temperature (105 vs. 93 °C) than for the 1K acrylic-MF, meaning that the 2K Iso has more time to flow and level than the 1K acrylic-MF. The lower viscosity of the 2K Iso also contributes to more efficient flow and leveling. The slower cure rate and lower viscosity may also allow the clearcoat to level out any disrupting structural information telegraphed from a lower interface (i.e., CC–BC and BC–primer interfaces), especially when combined with its lower level of volume shrinkage during cure.

The final advantage of the 2K Iso clearcoat is the potential for the isocyanate crosslinker to react with any primary or secondary amines that migrate up from the BC and primer layers. The 1K acrylic-MF clearcoat contains an acid catalyst, and the migration of amines from the sublayers could neutralize the acid catalyst in the clearcoat region close to the BC–CC interface. Lower catalytic activity near the BC would create a gradient of cure through the clearcoat, with the exterior surface curing faster. This type of cure gradient can lead to wrinkling of the clearcoat surface, an appearance defect discussed earlier for the low-bake systems.

Overall, achieving excellent appearance for consolidated processes is a challenging goal. To achieve optimal system appearance, all of the factors discussed need to be taken into consideration: flow and leveling, substrate telegraphing, gradient of cure, and inter-layer compatibility.

Conclusions

Four basic concepts relating to the development of surface structure in automotive coatings have been discussed. These concepts include flow and leveling, substrate telegraphing, inter-layer compatibility, and gradients of drying and curing. Good flow and leveling is essential to achieve excellent coating appearance. However, when coatings are oriented vertically during spraying and baking, gravity can drive sagging and propagate surface waves that contribute to a retention of surface texture in the coating. Volume shrinkage is a driving force for telegraphing substrate structure. Under high-temperature baking conditions, crosslinking chemistries employing addition reactions are more effective at minimizing telegraphing than those using condensation reactions, which produce volatiles and contribute directly to volume shrinkage. Inter-layer compatibility is critical to the system appearance. Ingredient migration and mismatch of cure rates between neighboring layers can both disrupt the interface structure and telegraph such disrupted structure to the clearcoat surface. Although the result is poor appearance at the clearcoat surface, the source is the system of mismatched layers. More specific for the low-bake systems, gradients of drying and curing can lead to surface wrinkling that is a major source of surface structure formation. Eliminating or minimizing the gradients is critical for achieving excellent appearance in these systems.

For automotive paint manufacturers, improved appearance will continue to be one of the most important goals to satisfy the demands of the automotive paint market. The influences of paint processing and chemistry on appearance remain as areas that demand a greater depth of mechanistic understanding. Interpreting data from instrument measurements and establishing correlations to visual perception will require automakers and paint suppliers to work together, ultimately leading to optimized specifications and better final appearance for consumers. However, excellent paint appearance cannot be pursued at the expense of other performance attributes, and research efforts should combine both goals.

Acknowledgments The authors would like to thank the following technicians who contributed to the experimental work of this study: Christopher Abbott, Robert McEwen, Vince Cianciolo, Jeff Reiff from Mt. Clemens R&D Laboratory of Axalta Coating Systems at Michigan, and David Frempong and Sherell Malcolm from Coatings Technology Center of Axalta Coating Systems at Delaware. Gabi Kigle-Böckler of BYK-Gardner is also acknowledged for discussion on BYK Wave-Scan DOI.

References

1. BYK Additives & Instruments, QC Solutions for Coatings and Plastics, 2012/2013
2. Hulst, H. C. V. D.: Light Scattering By Small Particles, pp. 131–171. Wiley, New York (1957)
3. Kigle-Bockler, G.: Surface quality control on high and medium gloss surfaces: wave-scan dual, 4th Wave-Scan User Meeting, 2006

4. Konieczny, J., Meyer, G.: Computer rendering and visual detection of orange peel. J. Coat. Technol. Res. **9**, 297 (2012)
5. Basu, S.K., Zhou, J., Harding, A., Moncier, J., Williams, C., Baker, L., McCreight, K.: Effect of atomization and rheology control additives on particle size and appearance of automotive coatings. Proc. Int. Waterborne High-Solids Powder Coat. Symp. **38**, 234 (2011)
6. Tighe, C., Giroux, J., Henshaw, P.: Effect of clearcoat thickness on surface roughness. JCT Coatings Tech. **5**(6), 24 (2008)
7. Peters, C.A., Nichols, M.E., Ellwood, K.R.: The evolution of surface texture in automotive coatings. J. Coat. Technol. Res. **8**, 469 (2011)
8. Amookht, S., Kandi, S.G., Mahdavian, M., Moradian, S.: The effect of clear coat and basecoat interdiffusion on the appearance of automotive coating system. Prog. Org. Coat. **76**(10), 1325 (2013)
9. Osterhold, M.: Influence of substrate structure on coating appearance and the relevant characterization methods. Materialwissenschaft Werkstofftechnik. **29**(3), 131 (1998)
10. Orchard, S.E.: On surface levelling in viscous liquids and gels. Appl. Sci. Res. A. **11**, 451 (1962)
11. Smith, N.D.P., Orchard, S.E., Rhind-Tutt, A.J.: The physics of brush marks. JOCCA. **44**, 618–633 (1961)
12. Patton, T.C.: Paint Flow and Pigment Dispersion, 2nd edn., pp. 554–555. Wiley, New York (1979
13. Overdiep, W.: The levelling of paints. Prog. Org. Coat. **14**, 159–175 (1986)
14. Cohu, O., Magnin, A.: The levelling of thixotropic coatings. Prog. Org. Coat. **28**, 89–96 (1996)
15. Patton, T.C.: Paint Flow and Pigment Dispersion, 2nd edn., pp. 570–574. Wiley, New York (1979)
16. Currie, I. G.: Fundamental Mechanics of Fluids, 3rd edn., pp. 201–248. Marcel Dekker, New York (2003)
17. Gidcumb, Y., Kubish, S., Templeman, C., Richards, G., Ikeura, Y.: Development of robust anti-chipping performance primer for various application process conditions. SAE Technical Paper 2016-01-0536 (2016)
18. Duskova, M., Dusek, K., Shen, J., Wen, M., Douglas, C. B., Barsotti, R.: Solvent entrapment and film formation in polyurethane coatings. 248th ACS National Meeting, San Francisco, CA (2014)
19. Duskova-Smrckova, M., Dusek, K.: Processes and states during polymer film formation by simultaneous crosslinking and solvent evaporation. J. Mater. Sci. **37**, 4733–4741 (2002). doi:10.1023/A:1020843020379
20. Basu, S.K., Chung, J.A., Francis, L.F., McCormick, A.V., Scriven, L.E.: Modeling the depthwise gradient in curing and skin formation in wrinkling coatings. Ind. Eng. Chem. Res. **46**, 3358 (2007)
21. Basu, S.K., Scriven, L.E., Francis, L.F., McCormick, A.V.: Mechanism of wrinkle formation in curing coatings. Prog. Org. Coat. **53**, 1 (2005)
22. Hiemenz, P., Lodge, T.: Polymer Chemistry. CRC Press, Boca Raton, FL (2007)
23. Flory, P. J.: Principles of Polymer Chemistry, pp. 390–431. Cornell University Press, New York (1953)

Chapter 18
Craters and Other Coatings Defects: Mechanisms and Analysis

Clifford K. Schoff

Introduction

Organic coatings provide decoration and protection for countless structures, objects, and assemblies. Unfortunately, it also is possible for them to suffer a variety of defects that can undermine the properties for which they are designed. Examples include craters, dewetting, telegraphing, soak-in, pinholes, orange peel, sagging, air entrapment, and color problems. Baked coatings may experience additional defects such as popping, gassing, and wrinkling. The reader will note that the defects listed all occur during application or soon afterwards. This chapter is limited to such problems and does not deal with field failures such as loss of gloss, cracking, and blistering. The descriptions of the defects and the comments on their mechanisms stem from observations by the author during more than 40 years of lab and customer plant experience.

Surface Tension-Driven Defects

Surface tension has beneficial effects in that it works to smooth out spray droplets, roller patterns, or brush marks in wet coatings in order to minimize the surface area and thereby produce excellent appearance. However, the existence of contaminants, temperature differences, or composition variations across the surface of the coating can cause localized differences in surface tension and lead to undesirable flow and the formation of defects. Surface flow occurs from a region of low surface tension to one of higher surface tensions. For example, away from a contaminant and towards

C.K. Schoff (✉)
Schoff Associates, Allison Park, PA, USA
e-mail: c_schoff40@msn.com

M. Wen, K. Dušek (eds.), *Protective Coatings*, DOI 10.1007/978-3-319-51627-1_18

an uncontaminated area, resulting in defects such as craters, dewetting, and telegraphing or from a warmer area to a cooler one to produce picture framing (fat edges).

Craters

General

The best known and most troublesome type of surface flow is that involved in the formation of craters, which are small bowl-shaped depressions that occur all too often in paint films. Some of them have raised rims, but many do not (see Figs. 18.1 and 18.2). There may or may not be any material in the center of the crater. Craters are caused by low surface tension contamination in the coating, under it, or that has just fallen on it. They usually occur during or immediately after paint application.

Fig. 18.1 Top view of a crater with a rim

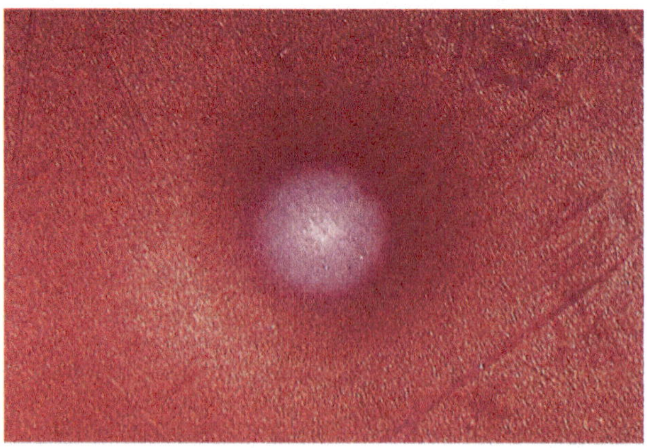

Fig. 18.2 Top view of a crater without a rim

Some small, deep craters look like pinholes and are difficult to distinguish from air entrapment, gassing, or solvent popping

Surface Flow

Surface flow such as in the formation of craters has been reviewed and analyzed by Fink-Jensen [1]:

The velocity of a wave front at the edge of a crater or other surface tension-driven flow is

$$\nu = (\partial\gamma/\partial x)(h/\eta)$$

where $\partial\gamma/\partial x$ is the surface tension gradient (the driving force for flow), which is the difference in surface tension ($\partial\gamma$) over the distance (∂x) between the regions of different surface tension, h is the wet film thickness, and η is the viscosity. For a crater, the equation becomes

$$\nu = (\partial\gamma/\partial r)(h/\eta)$$

where $\partial\gamma/\partial r$ is the difference in surface tension along the radius of the crater, ∂r.

A parameter that is useful in the analysis of surface flow is called flux. It defines the material that flows away or is pushed away in terms of a cross-section of the affected area. Therefore, it is two-dimensional as shown in Fig. 18.3.

$$\text{Flux} = J = 1/2(\partial\gamma/\partial x)\left(h^2/\eta\right)$$

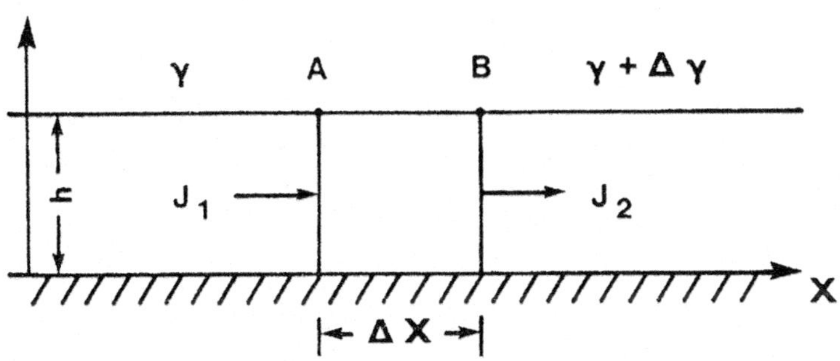

Fig. 18.3 Flux diagram: $\Delta\gamma$ is the difference in surface tension between areas A and B. ΔX is the distance between them. The letter h is the thickness of the coating. J_1 and J_2 are fluxes. If the surface tension at B is higher than the one at A, there will be flow from A to B; J_1 will be larger than J_2. The film thickness at A will decrease, leaving a depression, and the thickness at B will increase. The result will be a crater-like structure, possibly with a raised rim

The rate of change of wet film thickness is

$$\partial h/\partial t = -\partial J/\partial x = -\left(h^2/2\eta\right)\left(\partial^2\gamma/\partial x^2\right)$$

where $\partial^2\gamma/\partial x^2$ = the change in surface tension gradient with distance (e.g., as moving away from the point of contamination in the crater).

Is it possible to calculate the velocity of the crater edge? Assume an initial wet film thickness, h, (100 μm) and an initial viscosity, η, (0.5 Pa·s) for an automotive clearcoat after application. The problem is lack of knowledge or even a good estimate for $\partial\gamma/\partial x$, the surface tension gradient. The difference in surface tension probably ranges between 1 and 10 mN/m. The distance over which this difference occurs may be as little as 0.1 mm (perhaps even smaller) or as large as a few mm. That gives a value for the gradient of perhaps 0.1 to 100 Pa. If the gradient is 0.1 Pa, then using a thickness of 100 μm and viscosity of 0.5 Pa·s, velocity is = (0.1 Pa) × (100 × 10^{-6} m)/(0.5 Pa·s) = 20 × 10^{-6} m/s = 20 μm/s, which is very slow. The coating probably would dry before the crater would be noticeable. However, if the gradient is 100 Pa, then velocity = (100 Pa) × (100 × 10^{-6} m)/(0.5 Pa·s) = 2 cm/s, which seems more reasonable.

It turns out that there are data in the literature on the velocity of spreading of one liquid over another and the velocity of craters. Isobutanol on water spreads very rapidly with a maximum velocity of 48 cm/s. The velocity drops as water is added to the butanol or butanol is added to the water (thereby increasing the solubility of one liquid for the other), but the maximum still is greater than 20 cm/s [2, 3]. Spreading velocities of surfactants on water range from 1 to 30 cm/s. The spreading speed was inversely proportional to the molecular weight of insoluble surfactants [4]. Weh and Linde [5] studied craters produced by drops of silicone oil on a 75% glycerol/25% water film and on an alkyd-amino varnish. They showed that the rate of increase in radius of a crater formed by placing a 0.5 mm diameter drop of silicone oil on the glycerol/water film was ~ 8 mm/s at 0.25 s and ~ 5.5 mm/s at 0.5 s. The rate of increase in radius of a crater formed by placing a 110 μm diameter drop of silicone oil on the varnish was ~ 1.6 mm/s at 0.5 s and ~ 1.0 mm/s at 1 s.

Several years ago, a colleague took a "movie" of the formation of a crater in an automotive acrylic clear using a computer as a recorder [6]. The crater was initiated by placing a drop of a spray lubricant on a wet film of automotive clear. The crater that formed had a complicated shape with two rims or crests. The velocity of one of the crests was measured at different times as follows:

Time	Approximate velocity
0.1 s	12 mm/s
0.25 s	2 mm/s
0.5 s	1 mm/s
1.0 s	0.6 mm/s
2.0 s	0.3 mm/s

The values are low compared to those of surfactants and other materials on water as noted above, but very comparable to those for silicone oil on a varnish. The lower velocities are not surprising considering that paints and varnishes are considerably more viscous than water. We can safely say that the initial velocity of a crater wave is high. This points to the existence of large initial surface tension gradients (forces for flow), presumably due to the operation of moderately large surface tension differences over very short distances, such as 10 mN/m over 0.1 mm. In that case, $\partial\gamma/\partial x = (10\ \text{mN/m})/(10^{-4}\ \text{m}) = 10^5\ \text{mN/m}^2 = 100\ \text{Pa}$, which is a considerable force.

Crater Formation Time

Initial crater front velocities are high, but how long does it take to form a crater? Growth is rapid initially and then slows down. Weh and Linde's varnish took 1 s to reach 1 mm in radius, and more than three additional seconds to reach 2 mm. My colleague's crater reached 1 mm in radius after 0.1 s, but took an additional 0.9 s to reach 2 mm and another 1.5 s to reach 2.5 mm. A crater may keep growing for several seconds, but it will be noticeable in a very short time. Zero time is when a contaminant reaches the surface, either by falling on it or by rising up to it from the bulk of the film (although, with the latter, spreading may well be different).

This means that there will be very little time (tenths of a second to perhaps a second) for viscosity increases after application to occur to reduce surface flow. Many if not most pigmented coatings build viscosity quickly, which undoubtedly is why craters are less common in them than clearcoats, most of which show relatively slow increases. I would expect that adding additional silica or microgel has little or no effect on crater resistance in clears unless added at high levels. Personal experience at several auto plants has showed that raising silica levels in clears (then reducing the paint to normal cup viscosity) hurt flow-out and leveling, but did not reduce cratering. The key is to have a higher viscosity in the paint as it hits the car body. Applying paint with less reduction, i.e., a higher cup viscosity will do this, but usually hurts appearance. Having a higher application viscosity has helped reduce craters in several cases, but has had no effect in others (the viscosity probably was not high enough).

In another crater scenario [7, 8], it is possible to have a crater that is not initiated by a surface tension gradient. If a drop of an insoluble low surface tension contaminant such as a silicone falls on a wet clearcoat, it will spread very rapidly and form a fluid lens. The inertia of the underlying paint causes it to pile up in front of the lens, thereby forming a rim. Viscous drag causes a thinning of the underlying paint. On baking, the silicone will disappear, but a crater will remain. In practice, there probably will be some interaction between the silicone and the wet paint, some interdissolution or intermixing, in a narrow interphase. This will cause a surface tension gradient, but the surface tension difference will occur over such a small distance that there will be a very large gradient. Presumably, the lower the solubility of the silicone and the wet paint in each other, the smaller the interphase, and the larger the gradient. There will be a very strong driving force for flow at the edge of the lens. This will act ahead of the bulldozing effect of the spreading lens.

There are indications that the more the lens and the paint intermix, the lower the maximum velocity of the front and the sooner that the surface tension gradient drops off. Rapid mixing should prevent formation of a noticeable crater.

Crater Size and Contaminant Volume

It is interesting to calculate the size of a droplet of contaminant necessary to cause a crater of a certain size. Let us consider a crater of diameter 2 mm and assume that the lens that formed it also was 2 mm in diameter and was a constant 1 μm thick (it could be as thin as a monolayer). The volume of this cylinder is

$$\pi r^2 h = \pi(0.1\,\mathrm{cm})^2 \times \left(10^{-4}\,\mathrm{cm}\right) = 3.1 \times 10^{-6}\,\mathrm{cm}^3 = 3.1\ \mathrm{nL}$$

This is equivalent to a spherical drop of diameter 0.18 mm, the volume of which is (4/3) $\pi r^3 = 4.18 \times (0.09 \times 10^{-1}\ \mathrm{cm})^3 = 3.1 \times 10^{-6}\ \mathrm{cm}^3 = 3.1\ \mathrm{nL}$

However, since a crater front can move ahead of the lens itself, a droplet of this size may well produce a crater larger than 2 mm or, conversely, it may take a smaller droplet than 3.1 nL to give a 2 mm crater.

Why does a crater stop? If there is no mixing with the paint and the lens viscosity and paint viscosity do not change, spreading will continue until there is a monolayer of the contaminant. In practice, there will be mixing and the viscosities will increase (the paint through loss of solvent, the lens through taking in paint), and the surface tensions will begin to converge. The surface tension gradient will decrease, and the resistance to the movement of the lens will increase until the crater stops.

Causes of Craters

A wide range of low surface tension materials can end up on, in, or under a wet paint and cause craters. Common ones include (1) Foreign materials that contact the surface of the wet coating—oil droplets (from air lines, machinery), oven aerosols and drips, chain lubricants, overspray, dirt, fibers, antiperspirants, and other personal care products; (2) Substrate properties or contamination—oil, grease, fingerprints, dirt, and mold release agents; (3) Paint components or contamination—additives (especially when poorly dissolved or dispersed), gel particles, contaminated raw materials, contaminated containers, etc. The great number of possible causes makes the identification of the root cause or causes of cratering difficult. Much detective work usually is needed.

It should be pointed out that not just any contaminant will cause a crater. Several factors are important: (1) Surface tension—contaminant surface tension must be lower than that of the wet paint film; (2) Solubility—most crater-causing contaminants are at least partially soluble in the paint; (3) Rate of dissolution of the contaminant—if rapid, no crater should occur, if moderate to slow, then a crater is likely; (4) State—most craters are caused by liquids (although they may be very viscous or adsorbed on a solid).

Crater Prevention

Crater prevention begins with the development of paint formulas that contain components such as surface active additives and solvents that give the wet paint a surface that is homogeneous with low surface tension. With such a surface, any surface tension gradients that do occur will be small. Also, the surface is likely to interact with low surface tension contaminants to give interdissolution, possibly rapidly enough that no crater will form, or if one does begin, it will be small and/or flow out. Silicone surfactants are particularly effective in accomplishing the above scenario, but they must be used at very low levels or repair or recoatability may not be possible. Silicone oils (linear dimethyl siloxanes and polydimethyl siloxanes) are so highly surface active and mobile that they never should be used in paint or allowed to enter a paint or auto plant. They are notorious for causing plant and equipment contamination as well as adhesion and surface defect problems. Polyether or polyester modified polydimethyl or polymethyl alkyl siloxane surfactants that are added at 0.1–0.3% on total paint and are well dispersed are much less likely to cause problems.

Another strategy is to raise the low shear viscosity of the paint by increasing the pigment level or adding a thickener or thixotrope. This makes the coating more crater resistant by making it more difficult for the paint to flow away from a contaminant. Paints that are sag resistant are likely to be crater resistant. However, they may show more orange peel and not have the required smoothness. A combination of surface active additives and rheology modifiers often has been used to improve crater resistance without hurting appearance.

Good formulation practices must be followed by the production of clean, well-filtered paints that are filled into absolutely clean cans, pails, drums, totes, or tank wagons. Contaminated shipping containers have caused many crater problems. Once the paint reaches the customer, it must be kept in clean tanks or other containers and applied under clean conditions. Many root causes of cratering have been traced to bad customer-operating procedures, poor housekeeping, and lack of proper training for workers. In a factory, the paint application area should be separated from the rest of the plant and must be kept clean. The air must be filtered to remove dirt and contaminants. In spray applications, the compressed air must be filtered and kept free of water and oil. Water by itself does not cause craters, but it can act as a carrier for oil and other contaminants.

Additional information on craters and cratering may be found in references [9–13].

Wetting and Dewetting

Craters are not the only surface defects driven by surface tension forces. Normal application methods usually insure initial spreading of the coating over substrates or undercoats. However, during or shortly after application, retraction or dewetting of the coating may be observed resulting in crawling, pull-away, cratering, picture framing, pinholes and holidays (discontinuities), and roughness and bumpiness. These defects involve one form or another of dewetting. Flow occurs away from low surface tension contamination on a substrate (see Fig. 18.4) as in crater formation. Dewetting also may occur at an edge or hole in a part. This involves somewhat different behavior and is discussed in the next section.

Wetting is a process by which a liquid interacts with a solid. From the standpoint of coatings, wetting involves bringing a liquid paint into contact with a substrate or a dispersant into contact with a pigment, displacing air and moisture from the surface and adsorbing onto the surface. Wetting is critical for film formation, good appearance, and adhesion. Wetting and wettability usually are described in terms of a sessile or resting drop of paint or another liquid (see Fig. 18.5).

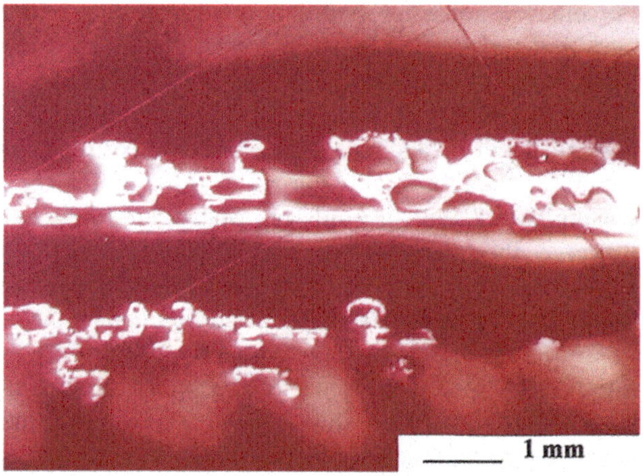

Fig. 18.4 An example of a red industrial topcoat dewetting over a white primer

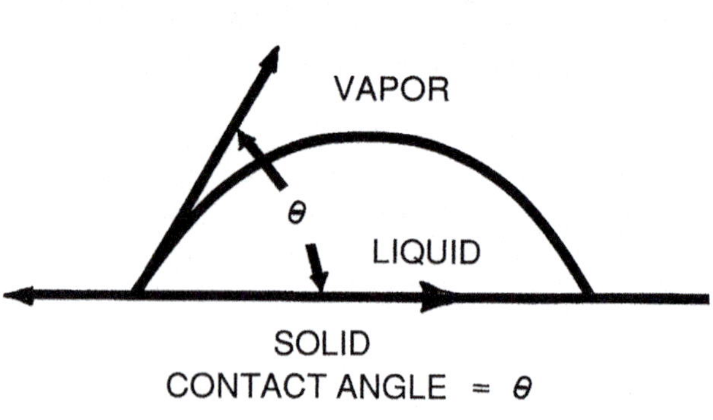

Fig. 18.5 Sessile drop showing the contact angle of the liquid on the solid surface (from Ref. [13], with permission)

The contact angle (θ) between the drop and the surface is a measure of wettability. A drop of liquid that spreads and forms a low contact angle (0–20°) with the surface is said to wet well, and the surface is said to be highly wettable by that liquid. There is an intermediate situation where the drop neither spreads nor pulls away and gives a moderate contact angle (20–45°). This intermediate state is common with paints and typical substrates and sometimes is called non-dewetting. Dewetting is when the drop pulls back, beads up (sometimes looks like it is trying to climb off the surface), and gives a contact angle greater than 45 (often much greater). We say that the surface is not wettable by that liquid. Wettability is determined by the competition between the paint-substrate adhesive forces and the cohesive forces within the wet paint [14]. Adhesive forces cause the paint to spread. Cohesive forces cause the drop to bead up. The contact angle is determined by the balance between the two forces. Wetting occurs when the liquid paint "likes" the substrate, i.e., is compatible with it, and beading occurs when the substrate is not "liked."

Cleanliness or lack of it can have a great influence on the wettability of any surface. Dirty surfaces are difficult to wet and may give adhesion problems or surface defects even if initial wetting seems to be adequate. Good cleaning can dramatically improve wetting and is the reason that there are rigorous cleaning steps before the painting of automobiles and many other factory-coated objects and assemblies. It is equally important to make certain that architectural surfaces and those to which maintenance paints are applied are clean. Unfortunately, dust, dirt, and other contaminants may soil previously clean surfaces outdoors or in a factory before they are painted. It is important to clean surfaces properly and then keep them clean. Additional cleaning before paint application may be necessary.

Additional information on dewetting and wetting may be found in references [1, 7, 8, 10–16].

Picture Framing (Fat Edges)

A common coating defect is a relatively thick bead or band along the edges of a painted panel, car door, or other part or assembly. This defect is called fat edge or picture framing [12, 13]. Fat edges are common in flow and dip coating where gravity-driven flow causes paint to be thicker on lower areas of the coated part, particularly at the bottom edge. However, the defect is much more widespread and can occur with other applications such as spraying, roll coating, and even electrodeposition. With these coatings, surface tension gradients during baking or drying cause the defect. Often, solvents evaporate more rapidly from edges. Those regions become richer in higher surface tension coating vehicle (and may experience evaporative cooling which further raises surface tension) and a surface tension gradient is set up which causes flow of material to the edges. Sometimes, the bead is back from the edge. In this case, the edge probably has heated up faster on baking than the rest of the part or panel. This lowers the surface tension and gives a gradient that causes flow away from the edge, producing a bead offset from the edge. However, rapid loss of solvent may reverse the flow. Addition of silicones or other surfactants may lessen problems by reducing surface tension gradients. A higher pigment to binder ratio or addition of a thixotrope decreases the rate and amount of flow to or from the edge.

A related defect is dewetting from sharp edges. A paint may initially coat all areas equally, but on baking, a reduction in viscosity allows the coating to flow easily and surface forces work to minimize the surface area. At edges, this manifests itself as flow to produce a smooth cross-sectional curve and less coating at the edge (sometimes no coverage at all). Additional flow from the edge is likely to occur on baking if the edge heats up more rapidly, causing the surface tension to drop in that area, thereby producing a surface tension gradient. Poor coverage tends to cause decreased resistance to weathering and corrosion. The term blue edge is often used to describe metal show-through. Poor edge coverage can be a difficult problem because addition of surfactants does not stop the driving force to minimize surface area. Higher pigment loadings and the addition of thixotropes are more effective because flow away from the edge is prevented or reduced. However, this can result in excessive orange peel.

Telegraphing

This defect involves the reproduction of surface features on a substrate by the coating applied over it. The coating makes fingerprints, wipe marks, detergent residues, or sand scratches (see Fig. 18.6) more obvious instead of covering them up. The main cause of this defect is flow due to surface tension: flow away from residues such as finger oils and sanding dirt or flow away from sharp edges of sand scratches.

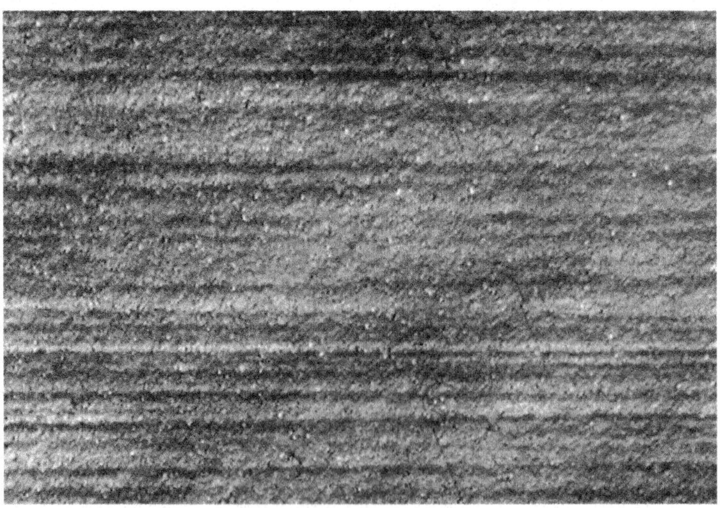

Fig. 18.6 Telegraphing by an automotive basecoat of sanding scratches in a primer

Aluminum flakes in automotive metallic coatings have been known to line up along sand scratches or wipe marks, leaving silvery streaks noticeable from a considerable distance. Sanding telegraphing (also called sand mar) is where soak-in of paint into sanded areas causes changes in gloss, color, or flake orientation in coatings with aluminum or mica flakes. Wetting and flow on sanded areas are affected by the roughness, porosity, and the increased surface area and wettability of the abraded surface. Paint also can soak into wood to a greater or lesser extent depending on grain, hardness, degree of weathering, etc.

There is another form of telegraphing called bond line readout [17]. It involves the showing of underlying structural features of a large, flat part or assembly such as a car hood after it has been coated. It is mainly seen over plastic parts with bonded reinforcements, but can occur over metal (e.g., in panels baked on racks). The term bond line comes from the fact that the reinforcement backing normally is bonded to the part with adhesive. Either reinforcement telegraphing or temperature gradient telegraphing would be a better name for this defect, which is caused by surface temperature differences between backed and unbacked areas. The backing acts as a heat sink, so heat-up and cool-down of backed and non-backed areas are different. The defect can occur with air dry coatings or during a prebake flash or early in a bake due to different degrees of cooling by solvent evaporation. The temperature differences cause temperature gradients, which in turn lead to surface tension gradients (lower temperature means higher surface tension), which cause localized flow leading to the formation of steps (up to several μm) and ridges. These defects are noticeable to the eye, especially if aluminum flakes in the coating cause dark-light boundaries along the step or ridge. There may be color differences as well.

Bond line effects have been seen immediately on application (when any evaporative cooling was only just beginning). It is not known whether this was due to carryover temperature differences from the primer bake or because of electrostatic effects. Bond line readout is greatly affected by the application process (type of gun or bell, number of passes), flash time (longer flash reduces the effect), the thickness of any underlying coatings, the design of the part, and the temperature of the oven (higher temperature has been known to diminish or prevent the defect). Silicone surfactants often are good for reducing surface tension gradients and higher viscosity increases resistance to flow, so a surfactant–thixotrope combination should prevent or reduce the defect.

Convection Flow Defects

During the drying and baking of paints, convection may occur within the wet film causing circular flow patterns roughly perpendicular to the plane of the film [12, 13, 18]. Flow moves from the coating substrate interface, rises up to the surface, and then drops back down. A cross-section might look like Fig. 18.7. The driving forces are surface tension and density differences between the substrate interface and the surface of the coating. The defects are made worse by low viscosity, rapid solvent evaporation, and pigment flocculation. Pigments rising with the flow may be

Fig. 18.7 Diagram of convective flow showing the up-and-down pattern of motion (from Ref. [12], with permission)

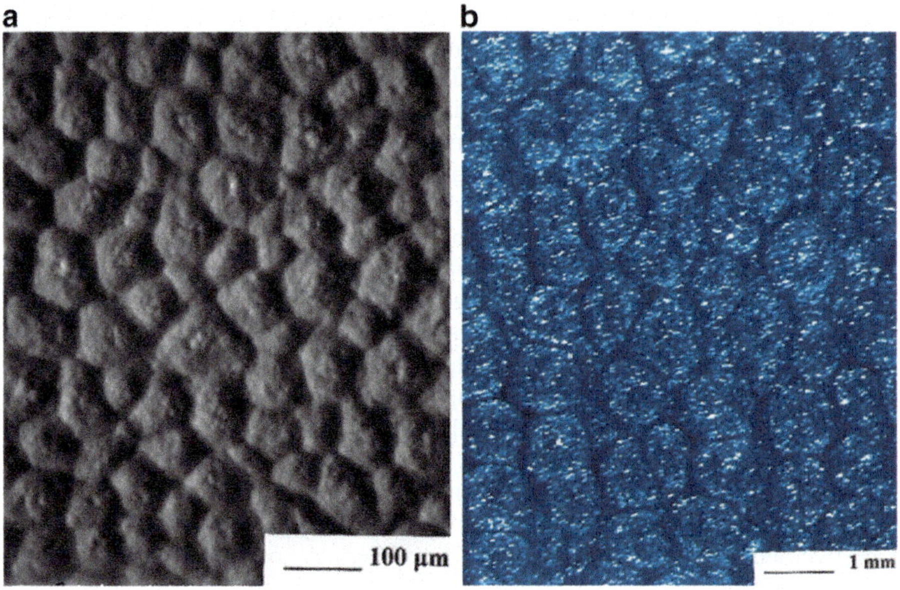

Fig. 18.8 (a) Bénard cells in a white can coating. (b) Bénard cells in a blue automotive metallic topcoat

trapped at the surface producing the defects called flooding and floating. If only one or just some of the pigments come to the surface, the color will be changed. Although the cause is convection flow, it can be worse if one pigment has very small particles, which tend to float to the surface. In automotive coatings, aluminum flakes are supposed to layer themselves throughout the coating, but with convection, they may completely cover the surface. When this happens, we call the defect flooding. The entire surface is covered with a single pigment. Floating usually produces streaks or mottled areas, rather than affecting the whole surface. Flocculation of one of the pigments can contribute to either problem. Rubbing the wet paint with a finger (known as rub-up) will show that the surface is a different color from the interior of the layer.

Bénard cells are cell-like structures on the surface of the paint film. In theory, they should be hexagonal, but often are four or five sided or even roughly circular. It is common to have a mixture of shapes as shown in Fig. 18.8a, b, which show two different sizes of cells in two different kinds of coatings. Bénard cells have very noticeable centers and are produced by convection circulation patterns induced by solvent evaporation. They can occur during air drying as well as baking. The whole cell is in motion with currents streaming up the center and flowing down at the cell edges. An easy way to visualize this motion is to take the lid off a can of paint containing aluminum pigment and wait for a few minutes as solvent begins to evaporate. Pouring a little of the paint into a Petri dish produces the same effect. The flow patterns lead to segregation and deposition of the different coating components in different parts of the cell. The cells occur in networks that cause color irregularities in coatings with multiple pigments and can produce surface effects such as roughness, low gloss, and waviness in virtually any coating, including clears. It is rare for clears to show the hexagonal surface pattern, but I am convinced that unexpected waviness in clearcoats and the occasional waves induced in automotive basecoats by clearcoats both are due to convection flow. Bénard cells may be quite large (several centimeters across) or so small that a microscope is needed to see them. In the latter case, the cells are likely to give a dull or matte appearance to the paint film without the observer understanding why. This has been seen in coil coatings and can coatings. Figure 18.8a shows an example of the latter. It should be pointed out that defects like the one shown in Fig. 18.8b are very rare in automotive topcoats. This example was not a line failure, but was induced in the lab for teaching purposes. However, Bénard cells, flooding, and floating have been seen in automotive coatings as well as many other products.

Convection flow can be prevented by lowering the film thickness, increasing the viscosity of the coating, minimizing surface tension gradients across the film surface, and preventing or reducing flocculation. Thin multiple coats are less likely to suffer convection flow than a single thick paint film. Raising the viscosity will slow down the circulation pattern. Introducing a surfactant or a low energy polymer or oligomer that comes to the coating surface prevents surface tension gradients that arise from solvent evaporation and uneven heating. Reducing the rate of evaporation by reformulating with slower solvents has helped in some cases, but can make convection worse if the coating does not dry rapidly enough. Flocculation can be prevented or reduced through better dispersion practices and the use of wetting agents.

Wrinkling

This phenomenon may be a defect or a desired property (the latter in certain special effect coatings). Wrinkling in a coating is the state of having surface folds or corrugations as seen in Fig. 18.9. The topography may be coarse and very obvious or fine with microscopic corrugations that give low gloss. Wrinkling is most likely to happen during cure and can occur in powder coatings as well as liquid products. Curing normally occurs from the top down and results in a crosslinked elastic skin over a viscous liquid lower layer. Basu and coworkers [19] describe a mechanism in which liquid from below diffuses up into the skin, which then swells, but cannot

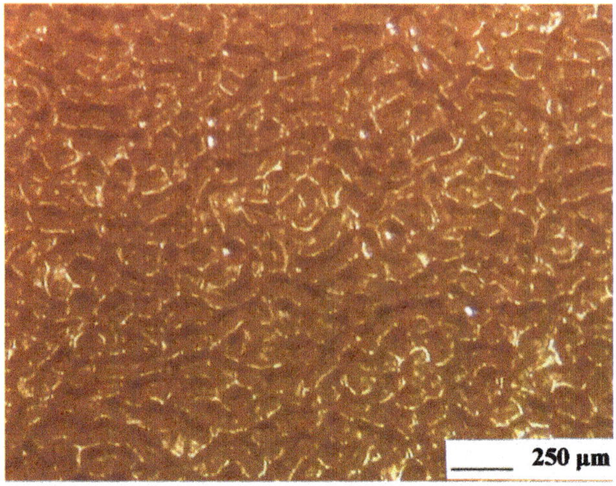

Fig. 18.9 Wrinkling in an industrial coating

expand in its plane because it is attached to the layer below. This causes compressive stress to develop in the upper layer. The stress is relieved by buckling that produces wrinkles. The remainder of the coating then cures and solidifies, freezing in the wrinkles. The wavelength and amplitude of the wrinkle depend on the thickness of the coating, increasing as the coating thickness increases.

Wrinkling also may be seen when paint is applied over an uncured or partially cured coating. If the undercoat takes in solvent and swells, a compressive stress will develop causing the coating to buckle in a manner similar to the skin described above. The process involves top-down diffusion rather than bottom up, but the mechanism should be the same. Solvent alone can cause wrinkling in the same way. Probably, the most common example of this is the action of paint remover, which wrinkles a film before it lifts it. In applications where wet paint is applied over another coat that is wet or tacky, such as automotive base/clear systems, wrinkling is likely to involve a differential degree or rate of cure between the clear and the basecoat beneath it. A rapidly curing clearcoat may act as a skin or a slowly curing one may swell the basecoat.

Wrinkling can be reduced or prevented by recoating or topcoating only after a film has cured properly, applying thin coats and drying (and, if possible curing) in between, employing slow solvents to keep the film open and reduce skinning, avoiding strong solvents in topcoats over uncured or partially cured undercoats and adjusting catalyst levels in base and clear coats so that they react at similar rates.

Volatile-Related Defects

Surface tension is involved in a great many coatings defects, but there are other causes as well. Most paints contain volatile solvents that must evaporate during air drying or baking. Other gases such as air may be incorporated during stirring,

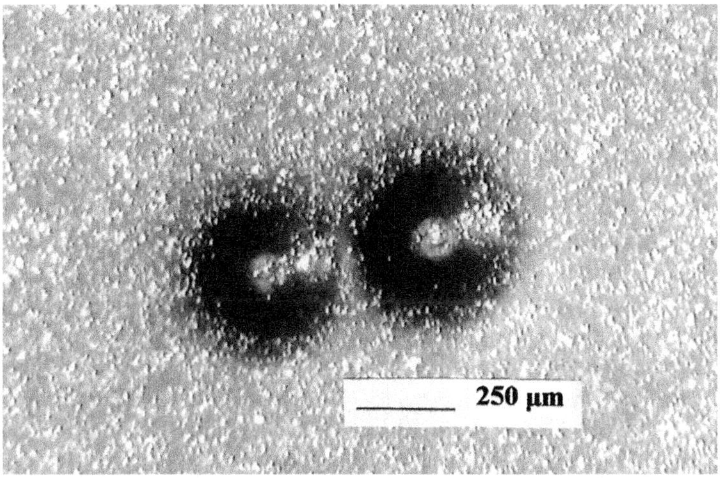

Fig. 18.10 Solvent pops in an automotive base-clear combination. These originated in the primer-surfacer, the coating below the silver metallic basecoat

pumping, or spraying. In addition, volatile materials may come from cure reactions or porous substrates.

Popping

Popping involves the formation of defects by the often-violent expulsion of trapped solvent or other volatiles during baking. The result may be miniature volcanoes (see Fig. 18.10), pinholes, craters, dimples, bubbles, or a few of each. The defect is often called solvent popping, but may be due to volatiles from the substrate (also called gassing) or air entrapment. It can be very difficult to distinguish between these mechanisms, although true solvent popping is more likely to involve several or all of the manifestations listed above. In addition, dehydration of waterborne basecoats can lead to popping-type defects in clearcoats applied over them. Too much dehydration (high bake temperature) produces a porous basecoat, which takes in solvent from the clear, which later blows out resulting in pinholes or pops. Too little dehydration (low bake temperature) leaves water in the basecoat, which then is expelled through the clear producing pinholes or pops. Sloppy paint application can have an effect on popping. Overspray and spits can trap solvent, blow out on baking, or absorb solvent from subsequent coats, and then blow out from underneath the second coat. Irregular spraying or roll coating can give thick spots that pop as does electrostatic spray wrap, fat edges, or sags.

With so many possible mechanisms, how do we identify the root cause? The best tools for characterizing pops are light microscopes. A low power (5–50×) stereo microscope is good for initial viewing. However, it is often necessary to cut a coupon from a part or panel and cross-section it to identify whether the defect indeed is due to volatiles and is not a crater, also to determine the originating layer, i.e., substrate, primer, and basecoat. A higher power research or metallurgical microscope (100× to perhaps 300× or as much as 500×) is necessary for looking at cross-sections. The specimens are prepared by cutting the coupon close to, but not right at the defect, and then sanding and polishing until the cut is at or near the center of the defect. See references [13, 20] for examples of cross-sections.

Crosslinking reactions of polyols and other reactive resins with melamines and other amino resins give off volatiles such as water, ammonia, and alcohols. As long as the surface of the coating stays open, their emissions will hardly be noticed, but if the film sets up before the condensation products or other volatiles are released, the result can be popping. Monofunctional alcohols (usually the same one used in synthesizing the melamine) are added to many formulations that contain melamine resins in order to improve package stability. However, these and other fast formulation solvents should not stay around long enough to cause problems. Volatile release from amino resin cure reactions is aided by thin films, keeping the surface open with slow tail-end solvents and slowing the cure reaction.

Moisture-curable urethane coatings crosslink after application by reaction of the isocyanates with water to form amines, which then react with yet another isocyanate to form substituted urea crosslinks. Cure rates depend on the amount of moisture in the vicinity of the coating. At high humidity and temperature, cure is rapid, but the carbon dioxide released by the reaction of isocyanate with water can be trapped as bubbles and, possibly, blown out as pops, especially in thick films. Again, keeping the film open and controlling the cure rate are keys to preventing the bubbles and pops. All waterborne coatings can be affected by the rate of water release after application. However, 2K waterborne urethanes are even more sensitive because water is a reactant as well as a carrier solvent. If water is lost slowly, there will be a larger extent of reaction of isocyanate with water, inhibiting full crosslinking of the coreactant groups. Since loss of water is affected by film thickness, thicker films tend to increase the relative reaction with water. For ambient cure coatings, relative humidity affects rate of water loss. Above about 70% RH, bubble formation caused by increased evolution of carbon dioxide may result in reduction of gloss or even noticeable blisters, bubbles, and pops.

Powder coatings have a tendency to trap air and other volatiles, resulting in bubbles and pops. Many powder coatings contain 0.1–1% benzoin as an additive to improve the appearance of films and act as an anti-pinholing agent and degassing aid. Benzoin definitely helps reduce popping, but it also causes yellowing. The mechanism still is not known. One study showed that benzoin plasticizes the melt and increases the flow window of polyester-glycoluril formulations [21], which should allow better release of volatiles. Another investigation has shown that benzoin evaporates from the applied film during the first 6–8 min of cure and that entrapped bubbles escape in the first 5–6 min of cure or they do not escape at all. The authors suggest that as the benzoin leaves the film, voids left in the film permit rapid diffusion of other volatiles [22]. Another study showed that derivatives of benzoin such as *p*-chlorobenzoin are effective in degassing and lead to less yellowing of films than benzoin [23].

Gassing

Gassing refers to eruptions, bubbles, or pinholes due to volatiles that come from the substrate during baking. The defects look like solvent pops and the initial conclusions is that there is a popping problem. Much effort may be spent on solving the wrong problem. Gassing may occur in electrodeposition coatings (and subsequent topcoats) due to electrical discharge and hydrogen evolution during deposition (pinhole gassing or rupture) or to volatiles coming from the zinc layer of electrogalvanized steel (galvanized gassing) or in coatings for plastics due to volatiles from the plastic substrate (see Fig. 18.11). Gassing defects resemble solvent popping, but are caused by volatiles that originate from the substrate rather than the paint layer. The volatiles force their way up into the wet paint layer as it is baked, resulting in pinholes, volcanoes, and/or bubbles. Subsequent coats rarely seal these defects, particularly if the area has been sanded. Defects occur in the new layer above the original defects. A cross-section would show a series of connected voids through the layers.

Gassing is most common over plastic substrates and zinc-coated steel, although it also can occur over porous aluminum castings. Gassing over plastics usually is

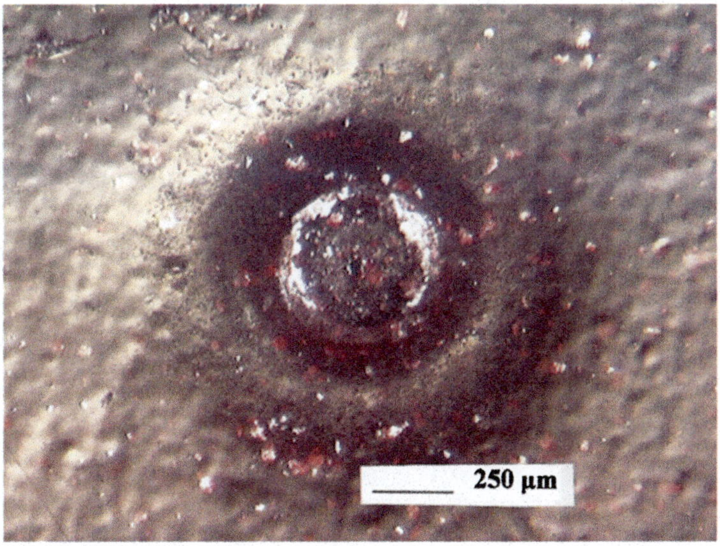

Fig. 18.11 Gassing in a coating over a plastic part. A cross-section showed that the hole in the center went all the way down to the surface of the plastic. There was a void in the plastic near the origin (from Ref. [20], with permission)

due to air or moisture blowing out of voids near the surface of the plastic as the paint is baked. Figure 18.11 shows a magnified view of such a defect. Many plastics have a thin, but relatively dense "skin" at the surface, but if this skin is missing, has been removed by sanding, or itself has voids, there is nothing to prevent volatiles from traveling up into the paint. Some primers are more effective than others at sealing plastics, but the best remedy is rigorous quality control of the plastic parts so that they do not have voids and pinholes at or just below the surface.

Gassing has been known to occur over electrogalvanized steel and, occasionally, over hot-dipped galvanized. It is due to the release of hydrogen and, possibly, methane from flaps, blisters, or other defects in or under the zinc layer. In a number of cases, the problem has been shown to be due to inadequate cleaning or contamination of the base steel before the zinc layer is applied. However, at other times, no reason or source of the defects has been found. Galvanized gassing depends entirely on the zinc-coated steel, but changing electrodeposition parameters such as solvent level, bath temperature, time, and voltage may help. Another defect, similar in appearance, called pinhole gassing or rupture, can occur in electrodeposition coatings at high deposition voltages [24, 25]. With anodic electrodeposition, the cause of the defects is a process called *Joule heating* where water in the deposited paint is heated to the point of boiling by the deposition process itself. The vapor blows out through the paint film resulting in pinholes. Voltage reduction usually solves the problem. With cationic electrodeposition, the defects are caused by a different mechanism that involves electrical discharge followed by hydrogen evolution and water vaporization. Figure 18.12 shows pinhole gassing in a cationic

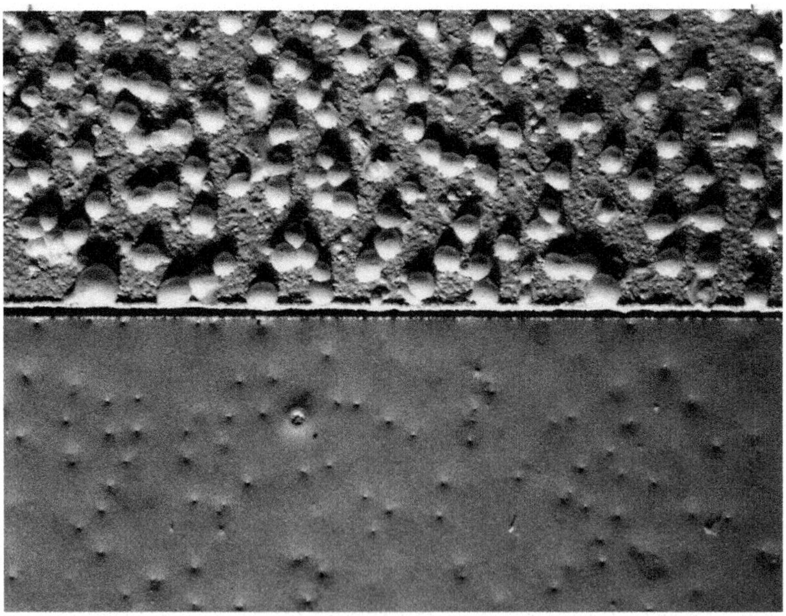

Fig. 18.12 Cationic electrodeposition pinhole cratering over zinc-iron alloy steel before (upper image) and after baking (lower image)

electrodeposition coating. Electrical discharge is more likely to occur over zinc-coated steels, especially zinc-iron alloys, but it can occur over cold-rolled steel. The defects can be prevented or reduced by lowering and/or ramping or stepping voltage, adding solvent and raising the bath temperature.

Air Entrapment

This term refers to the trapping of air bubbles in a coating, which can result in bubbles, pinholes, and crater-like defects. The size of the defects can be large and very obvious or very small and give the appearance of haze or scum. Air entrapment often is mistaken for other volatile-related defects, so it is important not to jump to conclusions when holes or bubbles are seen. Most tests for air entrapment involve rapid stirring or shearing (usually with a kitchen blender) of the paint in question as well as a control that does not show the defect and comparing the results. The evaluation can be by eye, weight per gallon measurements, or after spray-outs, but the idea is to see which paint picks up more air. One difference that has been noted between solvent popping and air entrapment is that spraying thinner and thinner coatings will eventually get rid of popping, but often makes air entrapment more noticeable.

A coating does not need to be baked to have the defects occur. In fact, baking sometimes allows release of the air bubbles. Bubbles are a possibility wherever there is a stirring or shearing action that can lead to vortexing, turbulence, or cavitation. This can occur in manufacturing and during application. Spraying provides many opportunities for bubble formation. Air may dissolve under high pressure (as in airless spraying) in the circulation system and be released in the spray droplets or in the film. Atomization may cause bubbles, particularly with worn or damaged gun tips or chipped bells, or produce doughnut or cup-shaped droplets that trap air when they hit the substrate. Even normal spherical spray particles can trap air as they splash onto the substrate. Bubbles also may form because of a tendency of the paint to foam or because air is trapped in a porous substrate.

The best way to avoid air entrapment is to prevent the formation of bubbles in the first place. Proper dispersion and mixing practice can reduce air entrainment during manufacture. Correct matching of the batch size with the size of the manufacturing equipment is essential. Proper choice, maintenance, operation, and adjustment of paint application equipment can prevent the trapping of air during application. Airless spray has caused air entrapment problems with a number of waterborne coatings. If bubbles cannot be prevented completely, then keeping the coating surface open longer via slower solvents may allow air to be released before the film sets up.

Defects Related to Flow or Lack of It

Some defects producing poor coating appearance either are caused by flows not related to surface tension or only partly involving it. Other defects are caused by insufficient flow. The former include sagging and related defects. Examples of the latter are poor flow-out and leveling and orange peel.

Sagging

Sagging, running, curtaining, and slumping describe unsightly gravity-driven flows on vertical surfaces. As is common with paint terms, there is not a consensus as to exactly what is meant by each term. They can be very obvious such as tears or fingers of paint running down a wall or the side of a cabinet (sags or runs). They also can be subtle such as a slightly irregular line or curtain of basecoat or clear along style lines or window sills of a car or truck. Sometimes, the sag would not be noticeable except that fine popping occurs in the thick area. Slumping tends to be thought of more in connection with plastering and cement work where a whole area drops down, but it can occur with paints, especially thickly applied ones such as bridge and maintenance paints. All of these defects are caused by too much flow (too low a viscosity). This excess flow can occur on application (cold sag) or, with baked coatings, in the oven (hot sag).

Sagging can be reduced or prevented by raising the low shear rate viscosity of the paint and/or applying thinner coats of paint. The film viscosity can be raised by using faster solvents or by introducing thickeners or thixotropes. The latter are additives such as fumed silica, treated clays, microgels, and castor oil derivatives that form physical networks. Associative thickeners in waterborne paints serve a similar purpose. Structures form after application to reduce flow and also modify viscosity–temperature behavior to prevent sag in the oven. Unfortunately, they also can prevent flow-out and leveling on horizontal surfaces, so amounts of the additives must be chosen carefully. Hot sag also can be reduced with the addition of materials such as waxes, certain castor oil derivatives, and acrylic microgels that prevent the sharp viscosity drop that usually occurs on heating. It should be pointed out that it is not necessary or desirable to completely stop gravity-driven flow. It just needs to be slowed down enough so that it does not become noticeable before the paint sets up.

Poor Leveling and Orange Peel

Poor leveling and flow-out of spray droplets, brush marks, and roller patterns tend to produce poor appearance. Readers who are DIY painters undoubtedly have

Fig. 18.13 Orange peel (at moderate magnification) in an industrial coating

experience with ugly brush marks that refuse to flow out. A common defect on many industrial and automotive coatings is a bumpy or rough surface called orange peel because it resembles the peel of an orange (see Fig. 18.13). Depending on the degree of bumpiness and the application, this may be objectionable or not. Automobile companies demand that orange peel be minimized and most horizontal surfaces are quite smooth. However, examination of vertical surfaces on most cars will show that smoothness does not happen everywhere. A regular orange peel pattern is considered to be pleasing with appliances and is less apt to show fingerprints. Orange peel is acceptable for many other industrial coatings. One thing appreciated about it is that it covers up metal irregularities and other underlying defects. In addition, orange peel and the viscosity behavior that causes it can prevent or cover up craters as well as preventing sagging. It is possible to have both leveling and sag resistance by building in thixotropy (time-dependent behavior) to produce low viscosity immediately after application and then to raise viscosity rapidly to a high level (usually by building a three-dimensional structure) [13].

Conclusions

Many different surface defects may occur during the application of organic coatings. Some of the common ones are craters, dewetting, telegraphing, picture framing, sagging, and orange peel. Additional defects may happen when a coating is baked, particularly popping and gassing. Such defects damage the appearance of coatings, but also the protective aspects. Solving defect problems is difficult and takes much detective work. A good optical microscope is a big help in defect identification and root cause analysis, but more powerful tools such as Fourier transform IR (FTIR) and scanning electron microscopy (SEM) often are needed, especially for crater analysis. Defect prevention begins with development of formulations that provide control of rheology and surface tension during application and on drying or baking. Beyond that, many defects are affected by cleanliness in the paint plant and at the user site.

References

1. Fink-Jensen, P.: Surface disturbances in fluid paint films, Färg Lack, **8**, 5–14, 39–42 (1962); Farbe Lack, **68**, 155–162 (1962)
2. Suciu, D.G., Smigelschi, O., Ruckenstein, E.: The spreading of liquids on liquids. J. Colloid Interface Sci. **33**, 520–528 (1970)
3. Ruckenstein, E., Suciu, D.G., Smigelschi, O.: Spreading on liquids: effect of surface tension sinks on the behavior of stagnant liquid layers. In: Schrader, M.E., Loeb, G. (eds.) Modern Approaches to Wettability: Theory and Applications, pp. 397–422. Plenum Press, New York (1992)
4. O'Brien, R.N., Feher, A.I., Leja, J.: Spreading of monolayers at the air-water interface. J. Colloid Interface Sci. **56**, 474–482 (1976)
5. Weh, L., Linde, H.: Kraterförmige Oberflächenstreifen in Anstrichfilmen, verursacht durch Silikonzusatz. Plaste Kautsch. **20**, 849–860 (1973)
6. Stiger, R.M.: PPG Industries, Inc., unpublished results (2001)
7. Bierwagen, G.: Surface dynamics of defect formation in paint films. Prog. Org. Coat. **3**, 101–113 (1975)
8. Bierwagen, G.: Surface defects and surface flows in coatings. Prog. Org. Coat. **19**, 59–68 (1991)
9. Hahn, R.J.: Cratering and related phenomena. J. Paint Technol. **43**(562), 58–62 (1971)
10. Hansen, C.M., Pierce, P.E.: Surface effects in coatings processes. Ind. Eng. Chem. Prod. Res. Dev. **13**, 218–225 (1974)
11. Kornum, L.O., Rasschou Nielsen, H.K.: Surface defects in drying paint films. Prog. Org. Coat. **8**, 275–324 (1980)
12. Pierce, P.E., Schoff, C.K.: Coating Film Defects, rev edn. Federation of Societies for Coatings Technology, Blue Bell, PA (1994)
13. Schoff, C.K.: Coatings defects: diagnosis and cure. J. Coat. Technol. **71**(888), 56–73 (1999)
14. Johnson Jr., R.E., Dettre, R.H.: Wetting of low energy surfaces. In: Berg, J.C. (ed.) Wettability, Surfactant Science Series, vol. 49, pp. 1–73. Marcel Dekker, New York (1993)
15. Hansen, C.M.: Surface dewetting and coatings performance. J. Paint Technol. **44**(570), 57–60 (1972)
16. Schoff, C.K.: Wettability phenomena and coatings. In: Schrader, M.E., Loeb, G. (eds.) Modern Approaches to Wettability: Theory and Applications, pp. 375–395. Plenum Press, New York (1992)
17. Blunk, R.H.J., Wilkes, J.O.: Surface-tension-driven flows of coatings: bondline readout formation. J. Coat. Technol. **73**(918), 63–71 (2001)
18. Hansen, C.M., Pierce, P.E.: Cellular convection in polymer coatings-an assessment. Ind. Eng. Chem., Prod. Res. Dev. **12**, 67–70 (1973)
19. Basu, S.K., Scriven, L.E., Francis, L.E.: Mechanism of wrinkle formation in organic coatings. Prog. Org. Coat. **53**, 1–16 (2005)
20. Schoff, C.K.: Optical microscopy. JCT Coatings Tech. **3**(2), 36–43 (2006)

21. Jacobs, W., Foster, D., Sansur, S., Lees, R.G.: Durable glossy, matte and wrinkle finish powder coatings crosslinked with tetramethoxy methyl glycoluril. Prog. Org. Coat. **29**, 127–138 (1996)
22. Maxwell, B.E., Wilson, R.C., Taylor, H.A., Williams, D.E., Farnham, W., Tria, J.: Understanding benzoin's mode of action in powder coatings. Prog. Org. Coat. **43**, 158 (2001)
23. Jahromi, S., Mostert, B., Derks, A., Koldijk, F.: Mechanism of action of benzoin as a degassing agent in powder coatings. Prog. Org. Coat. **48**, 183–193 (2003)
24. Schoff, C.K.: Electropaint–substrate interactions. J. Coat. Technol. **62**(789), 115–123 (1990)
25. Schoff, C.K.: Electrodeposition of coatings V: surface defects. JCT Coatings Tech. **10**(8), 44 (2013)

Chapter 19
Degradation of Polymer Coatings in Service: How Properties Deteriorate Due to Stochastic Damage

S.G. Croll

Introduction

All materials are fallible to some degree because they are imperfect to start with and are then assaulted by the environment. Many biological systems can repair themselves since they have mechanisms for absorbing material and energy from the environment that can create more material, but most engineering materials have no means of self-repair although it is a highly sought goal in materials science. Unfortunately, the environment damages almost all materials. Crosslinked polymers, in particular, are an extremely useful and versatile class of materials, but they are susceptible to attack from the environment. Such polymers are usually attacked by the more energetic photons in the solar spectrum and commonly suffer oxidation and hydrolysis.

It is comparatively easy to measure the chemical changes that occur in service, and it is very easy to quantify the performance of a coating because its service life is usually determined by a macroscopic physical property and how that deteriorates. However, at present there is no theory or model that connects chemical changes with service life. In fact, commercial products are probably too complex for any attempt to predict the failure mechanism to be practical, but a descriptive model that uses material parameters as variables would be very useful in quantifying relative performance. In many branches of engineering or technology, including coatings, it is very useful and common to employ extreme value statistics to correlate failure data and extrapolate the reliability distribution function [1–5]. This approach can be readily correlated a posteriori to material properties or variables, but there does not yet seem to be any connection established a priori with material properties. It is

S.G. Croll (✉)
Department of Coatings and Polymeric Materials, North Dakota State University, Fargo, ND, USA
e-mail: stuart.croll@ndsu.edu

M. Wen, K. Dušek (eds.), *Protective Coatings*, DOI 10.1007/978-3-319-51627-1_19

possible to use analytical tools to measure and compare rates of chemical change and then choose the material that changes the least or slowest, but that approach is not predictive of the end-use behavior. It will always be advantageous to have an approach for quantifying the degradation where the end-use property of interest is linked to the fundamental changes in the material(s), even if that approach is descriptive rather than predictive.

Degradation may cause several properties to change. For example, corrosion protection may change as well as the appearance of the coating. Color and gloss changes are visually obvious and easily measured in a nondestructive way. Protective properties, e.g., toughness or corrosion barrier protection, are much more difficult to measure. It is common to use simple or obvious visual changes and assume that other more complicated properties are deteriorating in an analogous way. This assumes that the failure process is closely linked for all the properties and that their progress towards failure follows the same kinetics. A fundamental understanding of how the properties deteriorate and what determines the rate of degradation would clarify the situation. Some properties are measured as an average over a significant area or volume of material, e.g., gloss or color, but properties such as toughness and corrosion protection are compromised by individual defects that are not assessed in the same way.

Ideas for linking chemical or microscopic changes to changes in product performance will inevitably become more specific and intricate as they become more accurate. Here, the search is for a description that is more broadly applicable and descriptive although not necessarily predictive. As a starting point, it is a good principle to explain phenomena by the simplest hypothesis possible [6] and one that is true in most circumstances. This is that degradation is due to a huge number of aggressive events, via aggressive photons or molecules, etc., which impinge upon the material at many random times and across its surface at random. In normal circumstances, this must be correct, and it will be shown here that it allows the deterioration in physical properties to be understood and described in terms of the size of a characteristic damaging event linked to very simple principles of physics and chemistry. One might then add details for a chosen material or situation and provide a more specific description. The objective is to provide a comprehensible framework without so much complexity that the underlying phenomena cannot be distinguished [7].

Background

Opportunities for Degradation

As a first step, we can gauge the number of possibilities for causing degradation. In the spectrum of sunlight at the Earth's surface, the most damaging photons are those in the more energetic ultraviolet and blue end of the spectrum. Figure 19.1 shows

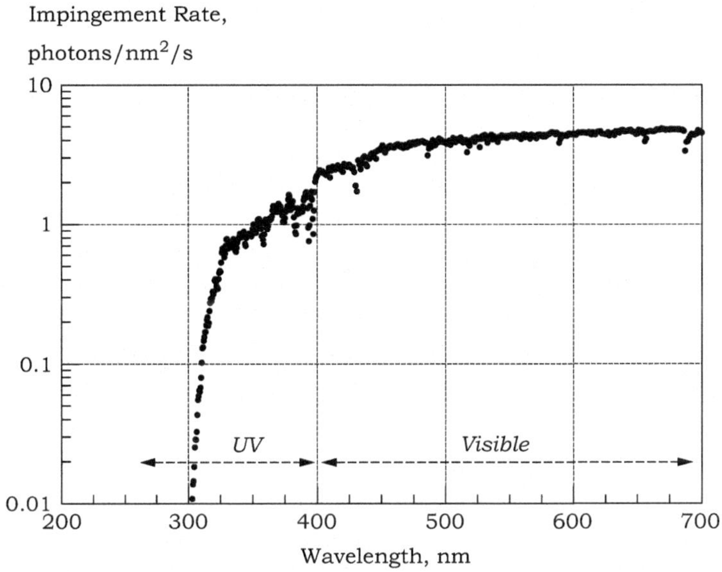

Fig. 19.1 Impingement rate at the Earth's surface of solar photons in the UV-visible spectrum. From the global tilt solar spectrum [8], by converting the energy values, E, using $E = hc/\lambda$: h = Planck's constant, c = velocity of light in vacuo, λ = wavelength

the number distribution of photons (at 1 nm intervals) deduced from the ASTM G173 "global tilt" spectrum [8].

One can see that in the middle of the UV-visible region, around 350 nm, approximately 1 photon is arriving every second, on average, on each square nanometer. Thus, it is no surprise that photodegradation occurs readily. In a similar vein, one can easily find values of parameters [9] to show that air at sea level contains ~ 8.5 mol/m^3 oxygen (~ 5 × 10^{24} mol/m^3) and ~ 0.66 mol/m^3 of water (~ 4 × 10^{23} mol/m^3) at 50% relative humidity and 25 °C. There are more than enough molecules of oxygen and water to participate in oxidation, hydrolysis, and related activities.

When a coating fails to prevent corrosion or fails in some other duty, we might become alarmed when rust spots or blisters occur at a frequency of approximately 1 fault/cm^2. Organic polymers are composed of carbon, hydrogen, oxygen, nitrogen, and a few other elements. If one imagines that each atom or bond has a characteristic size of 1 angstrom (Å), then a square centimeter of organic polymer has 10^{16} atoms or bonds confronting the environment. Even if only one bond in a 100 is susceptible, every square nanometer contains at least one opportunity for degradation, and there are 10^{14} plausible starting points across a square centimeter where failure in protection might start. There are huge numbers of aggressive species in the normal, natural world and a huge number of opportunities for them to attack the surface of any material.

Statistics of Random Events

Many textbooks contain details on the justification for the use of the Gaussian function to describe the statistics of random chance. A few relevant points are reiterated in a simple fashion. A binomial experiment is one where there are many (n) identical, independent events (e.g., photons bombarding a polymer surface) where there is one of two outcomes, either success or failure. Here, we are interested in tracking the increasing number of successful photons or molecules that cause a bond or chemical group to change. If the probability of success in one event is p, then the probability of failure (bond or group survival) is $q = 1-p$. The probability, P, of k successes within n events (couched in the binomial probability distribution) is given by

$$P(k) = \frac{n!}{k!(n-k)!}p^k q^{n-k}$$

The numbers above indicate that the number of assaults on a typical polymer causing degradation readily falls within the definition of large numbers, and a continuum representation would be feasible. In the continuum limit, the binomial probability distribution can be approximated well by a Gaussian function,

$$P(k) = \frac{1}{\sqrt{2\pi npq}}\exp\left[-\frac{(k-np)^2}{2npq}\right]$$

In fact, the central limit theorem says that, in fairly general circumstances, mean values taken from random samples of a large population tend to fall on a "normal" distribution. The number of photons or aggressive molecules that impinge on the material is the number of events, n, and the number that has been absorbed and done damage is np and is where the probability density $P(k)$ has its maximum and represents the mean state of the material. In an example of the use here, n would be given by the number of photons that had impinged on the material, which increases with exposure time, and p would be obtained from measurements of the fraction of photons absorbed multiplied by the quantum efficiency for the relevant degradation chemistry. In a more familiar form,

$$P(x) = \frac{1}{\sqrt{2\pi\sigma^2}}\exp\left[-\frac{(x-\mu)^2}{2\sigma^2}\right] \qquad (19.1)$$

The variable used now is x, and the mean, $\mu = np$, and the standard deviation, $\sigma = \sqrt{npq}$, are readily deduced by comparing the equations. In the circumstances here, the probability of individual success (damage) is considerably less than one so the probability of failure (no damage), q, is almost identical to one, and it is easy to appreciate that the standard deviation is given by the square root of the mean, to a very good approximation.

Failure caused by degradation occurs when the damage accumulated in service reaches a value where the change in properties has reached an unacceptable level, e.g., gloss loss, or the change precipitates a sudden different failure, for example, a rust spot appears. Damage will be caused by the continual exposure to the myriad of photons or aggressive molecules that impinge upon the material and eventually are successful in creating damage. However, the starting polymer material may not be perfect but incorporate some defects that inevitably occur during the formation of the polymer. Defects, and how they arise, during the formation of a crosslinked polymer are discussed in more detail in another chapter [10], but some of the salient points are reiterated below.

Intrinsic Defects in Crosslinked Networks

Crosslinked coatings are formed from two or more reactive precursors that are usually small molecules dissolved in solvent. The initial mixing process randomly disperses the ingredients, but there will be localities where a concentration fluctuation occurs. As the system starts to react, while it is still fluid, i.e., containing solvent and below the gel point, material may diffuse and find suitable opportunities to react so the effect of this initial inhomogeneity probably diminishes. However, crosslinking reactions produce larger molecules that are much less mobile and isolated from possible co-reactants. Thus, an unreacted end might remain unreacted or react, but with the crosslink junction where its other end is attached, so a loop forms, or unreacted pendants remain, at random. Thus, the network has defects whose size would be on the order of the reactive groups or perhaps somewhat larger, in other words, a few atoms in size. Thus depending on the defect concentration, one can see that aggressive molecules from the environment, which have a similar or smaller size, e.g., water or corrosive ions, would find it much easier to penetrate and cause harm, especially when the environment has contributed more defects due to degradation.

One can see [10] that a highly functional network, especially if it is cured rapidly at a high temperature, has many loops and an appreciable concentration of unreacted pendant network strands. One might expect that a network derived from high functionality crosslinkers, and crosslinked at a high temperature, might have a high resistance to permeability, but as many as 75% of the network junctions in a six-functional network may have a loop attached and thus have a defect in their neighborhood, as a result of the stochastic nature of network formation, regardless of any specific chemistry.

As the earlier chapter [10] goes on to explain, a crosslinked network comprises macromolecules that initially grew separately, like clusters in diffusion-limited nucleation, and is very similar to the growth of dendritic polymers, crystal growth or the preparation of particles in solution [11, 12]. These macromolecules (gel particles) comprise more or less reacted material where the surface has only limited remaining functionality with which to join a neighboring gel. Thus, there are intergranular fissures within the final crosslinked mass, which are not spanned frequently by chemical bonds, where the gel particle surfaces came together. Spatially, neighboring chemical groups may not be bonded together directly, where the gel clusters came together, but be rather indirectly connected by longer bond chains around the network.

Thus, the stochastic nature of the crosslinking process means that the resultant network cannot be ideal and must contain faults. At one end of the spectrum, the faults are approximately the size of a small molecule (loops and ends), and possibly approximately 1 nm or so, but the other end of the spectrum of imperfections are fissures along former intercluster boundaries where some features may be at least an order of magnitude greater in extent. Inevitably, these regions of low crosslinking are a mechanical weakness and may be opened up further when only a few bonds are broken. Also, if a coating is required to be a preventative barrier to corrosion, it is evident that any imperfection in a crosslinked network that allows water and corrosive salts to pass more easily will shorten the service life because exposure degrades the coating and adds/completes pathways through the polymer coating. These defects arise, not always due to poor mixing or application but due to the stochastic nature of the reactions between huge numbers of reactive molecules [10].

Degradation Kinetics of Physical Properties

Customarily, criteria of durability, or failure, are almost always determined by the deterioration in a macroscopic physical property. Different polymers are affected differently by ultraviolet photons, heat, and so on, but if one can identify a description of the phenomena that are common to all materials, then one might find a scheme that is broadly useful and then build upon that with the necessary specifics for a given situation. In concert with Ptolemy and Occam's Razor, the approach assumes that useful progress can be made with the most basic, general ideas. The focus will be on long-term performance, i.e., the "wear-out" regime. We know that a material may be bombarded by huge numbers of photons, water molecules, oxygen molecules, and possibly acid pollutant molecules. These aggressive species arrive frequently, at random locations and at random times across a surface. Many will penetrate and affect bulk properties. In either case, we can estimate how surface erosion or bulk degradation accumulates and can be linked to changes in physical properties [13, 14].

Monte Carlo computer techniques are well suited to model the consequences of random processes. It is fairly easy to imagine how accumulated damage caused by random, repeated events can be simulated provided that the requisite understanding of the physical or chemical processes is available. Such a Monte Carlo approach could deal with coatings that are made of more than one material, or suffer from multiple degradation processes, and thus ultimately might be useful in particular, well-characterized instances. Unfortunately, no such commercial software exists, and developing a simulation for a particular case might not be cost- or time-effective. The essence of a statistical model containing simple equations [14] is repeated briefly here. The basis is that degradation accumulates from random results and that simple average values or standard deviations can be used in well-known models of physical properties. It assumes that the central limit theorem applies and the Gaussian representation of random processes is adequate. The following invokes only one degradation mechanism in each case, but the resultant equations are simple enough that one can see that adding other mechanisms to a model would be possible if required, but the point of diminishing returns might be quickly reached.

If we characterize the environment by the flux of aggressive species and the material by the fraction of photons absorbed and the quantum yield for damage, we are seeking the most probable outcome although we know that there will be variability in all of these quantities and that there will be variation in the actual outcome. Values for absorption cross-section and quantum yield include the statistical likelihood of a photon being absorbed and its subsequent chances of causing degradation. It is accepted that degradation starts on the exterior surface of a material and progresses deeper [15]. At the surface of a material that is being degraded:

$$\begin{aligned}\Delta N(t) &= \text{Average number of fragments removed (or other damage)} \\ &= \text{Flux of photons} \times \text{Area concentration of labile bonds} \\ &\quad \times \text{Absorption cross-section} \times \text{Quantum efficiency for bond scission} \\ &\quad \times \text{Time}(t) \\ &= \phi \cdot N(0) \cdot a \cdot q \cdot t \\ &= K(=\text{constant}) \times t\end{aligned} \tag{19.2}$$

The symbols are defined by the parameters in the order that they occur in Eq. (19.2). The single, simple parameters for the flux and the absorption cross-section disguise the fact that radiation penetrates. A more complete expression would include factors from Beer's law or the Kubelka-Munk model to account for the penetration of radiation and how it might depend on the concentration of absorbing species or any pigmentation. However, it is doubtful that the presentation here would gain from the inclusion of additional parameters and complexity.

In normal composites, such as coatings and components such as pigments and fillers, they may be inert and control the concentration of the polymer, but not otherwise participate. The simple dependence on time supposes that the exposure period is much longer than any diurnal or other fluctuations. It would be simple to express exposure in terms of radiation dosage, ϕt, instead of time. These kinetics assume that the degradation immediately affects the whole surface, which is an aspect that will be dealt with later. Photon flux could be replaced by water molecules and the absorption cross-section by its equivalent. Equation (19.2) assumes that if a fragment leaves, then the material underneath is similarly assaulted and so there is no change in the degradation rate caused by a change in

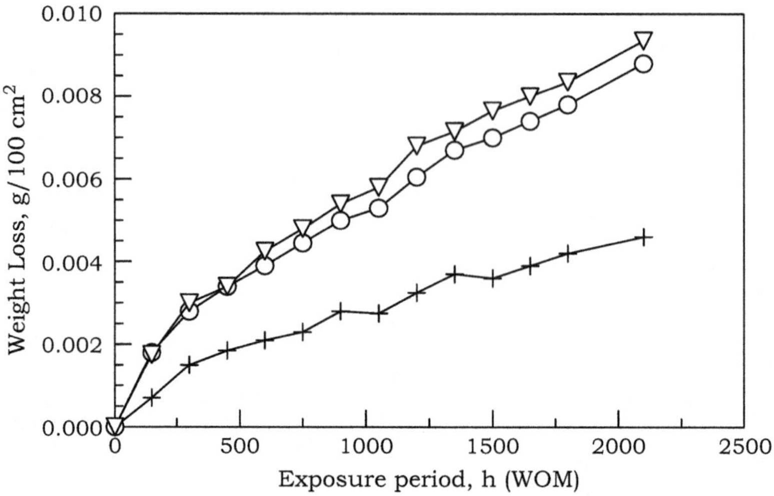

Fig. 19.2 Examples of weight loss with exposure time. Examples from three different alkyd housepaints exposed in a Weather-Ometer®

concentration. This follows zero-order chemical kinetics and thus obeys reciprocity [16]. The photon flux can be measured for a given environment, the absorption cross-section could be related to the material's spectroscopic absorption coefficient and the concentration of absorbing species, and the quantum efficiency depends on how many of the absorbed photons break bonds rather than be dissipated in other processes. All parameters could be measured. The product of the absorption cross-section and the quantum efficiency gives the probability of a bond being broken and is very much less than one.

Using this approach, mass loss or thickness reduction as a result of photo-oxidation, etc. is predicted to be a linear function of exposure time, which is a good approximation to the behavior in Fig. 19.2 and in other investigations [17–22] even when detailed kinetics of polymer chain scission is investigated [23]. It is common in pigmented films for the initial external surface to be rich in polymer so the initial rate of weight loss is often slightly higher. Equation (19.2) can also be applied in descriptions of how other properties might change. In any case, weight loss, or its equivalent in thickness loss, would be a simple measure of durability but unfortunately is not as common an approach as measuring appearance properties.

Gloss

Gloss is the specular reflectance of a surface expressed with respect to the reflectance of a standard glass surface. Since it is determined by the reflection at the exterior surface, and it is very important in the quality of appearance, gloss is a common parameter used to assess the durability of a material. Reduction in gloss is caused principally by an increase in surface roughness due to degradation. Roughness is often expressed as the root mean square deviation in the height of the surface, so the roughness would be the standard deviation in the average number of polymer segments removed at a location (after sufficient exposure has occurred to randomize the surface).

The properties of the Gaussian representation of a stochastic process, Eq. (19.1), show that the standard deviation (roughness), as a function of exposure time, would be given by the square root of the average number of fragments removed, and we can use the result in Eq. (19.2).

If ε is the characteristic (linear dimension) of the fragment size removed at each degradation event,

$$\mathrm{Roughness}(t) = \sigma_{\mathrm{height}}(t) = \varepsilon\sqrt{\Delta N(t)} = \varepsilon\sqrt{Kt} = \sqrt{kt} \tag{19.3}$$

The new parameter, k, lumps K and ε together for simplicity. Thus, roughness increases as exposure continues. A simple, and robust, physical model of how specular reflectance is affected by random surface roughness is that of Bennett and Porteus [24–27]. If the roughness is smaller than the wavelength of light,

$$\frac{R_S}{R_0} = \exp\left[-\left(\frac{4\pi\ \cos\varphi\ \sigma_{\mathrm{height}}}{\lambda}\right)^2\right] \tag{19.4}$$

where

R_s = specular reflection of the rough surface
R_0 = reflection of a perfectly smooth surface of the same material given by Fresnel's equations
φ = "specular" angle, i.e., gloss measurement angle
λ = wavelength of the illumination

The Bennett and Porteus model, B&P, is the first term in the more exact Kirchhoff model of the reflectance from randomly rough surfaces which includes diffuse reflectance into the specular angle [26, 28]. If the determination of gloss uses an illumination spectrum, its distribution can be easily included, but using a single value of wavelength from the middle of the spectrum models results very well and captures the underlying behavior well. The shadowing and masking effect of features larger than the wavelength of light may be included [29], but using B&P gives illustrative results that often model results well at high gloss angles (20°) [13].

The earlier chapter [10] on defects caused in the crosslinking process shows that it is likely they are present in the unweathered material and that there would be distribution in their sizes. Thus, there may well be an initial surface roughness, $\sigma_{\mathrm{initial}}$, due to application or intrinsic material properties. If we assume that it is also random, then it can be combined with the incremental roughness due to degradation:

$$\sigma^2_{\text{height}}(t) = \sigma^2_{\text{initial}} + \Delta\sigma^2(t) = \sigma^2_{\text{initial}} + kt \tag{19.5}$$

Equation (19.5) can be easily inserted into Eq. (19.4),

$$\begin{aligned} R_s(t) &= R_0 \exp\left[-\left(\tfrac{4\pi\cos\varphi}{\lambda}\right)^2\left(\sigma_{\text{initial}}{}^2 + kt\right)\right] \\ &= R_0 \exp\left[-\left(\tfrac{4\pi\ \cos\varphi\ \ \sigma_{\text{initial}}}{\lambda}\right)^2\right]\exp\left[-\left(\tfrac{4\pi\ \cos\varphi}{\lambda}\right)^2 kt\right] \end{aligned}$$

The rate of decay of gloss depends on the relative sizes of the initial roughness compared to the size and rate of the degradation events. For simplicity, σ_{initial} can be removed from consideration by calculating the relative gloss with respect to the initial value, $R_s(0)$:

$$R_s(t)/R_s(0) = \exp\left[-\left(\frac{4\pi\cos\varphi}{\lambda}\right)^2 kt\right] \tag{19.6}$$

Thus, the relative reduction in gloss due to weathering is predicted to follow a monotonic, exponential decay with exposure time, which is shown in the data plotted in Fig. 19.3 and found in many other examples [22, 30–33], although a more exact treatment of how gloss depends on surface texture that includes diffuse reflectance contributions and shadowing and masking may be necessary to model seemingly more complicated data at lower gloss angles (60°) [34]. Inevitably, gloss eventually diminishes to very low values where further degradation does not

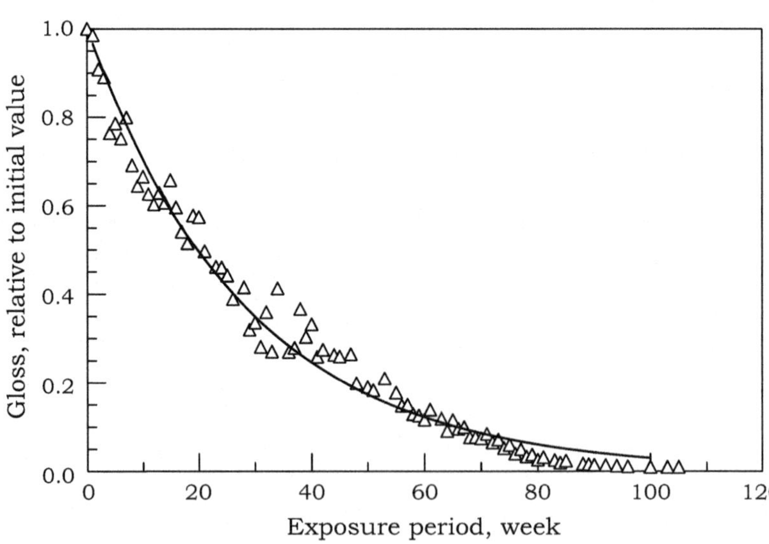

Fig. 19.3 Reduction in gloss with respect to the initial gloss (measured at 20°) of a pigmented polyurethane coating exposed in accelerated weathering (QUV-A)

produce any further perceptible change although the material may still be eroded, thus losing its protective capabilities.

So far, the discussion has covered appearance or similar properties that are measured as an average across an area or perhaps throughout a volume. In fact, Eq. (19.2) has also been used in the Wentzel equation for the contact angle on a rough surface to show how surface wettability increases with degradation [13, 14]. However, other important properties reach a crux in a different way. For example, mechanical strength or resistance to repeated mechanical stresses may be what determines the service lifetime of a material. It is well known that the toughness of a material, i.e., resistance to fracture, is linked to the size of the largest crack that exists in the material. In this case, failure is determined by when the damage causes a defect large enough to provoke cracking or adhesion loss. Another example is corrosion protection where the first instance of rust under the coating determines that failure has occurred. Although we would prefer to disregard the first few rust spots, they indicate that corrosion has happened, is certainly spreading, and the damage is large enough to manifest itself through the paint film(s). Corrosion protection failure may be when the coating has cracked, or it may be a more complicated situation where hydrophilic regions and inherent defects together with the damage caused by outside agencies first combine to form a pathway somewhere through the coating for water and corrosive species.

Toughness

The model here of deterioration in toughness and how a coating might fare upon exposure assumes again that the material is homogenous and that the failure is due to the accumulated degradation over an area of the material. There is a huge literature that expresses material toughness through the Griffith equation for the strength of materials, as we shall, but it is worth pointing out that the link between fracture strength and how the distribution of flaws within a material lead to the eventual failure is still the subject of research [35]. However, the simple approach here is an attempt to capture the principal cause for diminishing toughness and show how using the stochastic approach leads to an expression that can be used to describe the trend.

The Griffith equation for the stress at failure for a brittle material is

$$\text{Stress}_{\text{Griffith}} = \sqrt{\frac{2EG_c}{\pi a}} \simeq \text{Stress}_{\text{External}} + \text{Stress}_{\text{Internal}} \tag{19.7}$$

where E = tensile modulus; G_c = energy required to form unit area of crack surface; a = linear size of an existing crack (the factor π shows that a circular crack is assumed, for simplicity). Any internal stress due to curing conditions supplements the effect of externally applied stress, but that will be ignored here. If it is assumed that the modulus and the crack energy are material constants that do not change during exposure (although, if known, those variations should be inserted), the focus is restricted to the size, a, of the crack that becomes large enough to provoke failure. If the damage grows from the exterior, as it often does [15], it is the increasing surface roughness that is the source of the flaws that determine toughness. Equation (19.5) that relates the standard deviation in height (roughness) to exposure period can be used as a measure of crack size with exposure time,

$$a = m\sqrt{\sigma_{\text{initial}}^2 + \Delta\sigma^2(t)} = m\sqrt{\sigma_{\text{initial}}^2 + kt} \tag{19.8}$$

Cracking toughness is determined by the first defect that becomes large enough to provoke brittle failure, and thus instead of using the value of the standard deviation as a measure of the linear size (crack radius) to model the damage, a multiple (m) of that is used so that the value better represents the first time that such a size is achieved, rather than the average time

$$\text{Stress}_{\text{Griffth}}(t) = \sqrt{\frac{2EG}{\pi m\sqrt{\sigma_{\text{initial}}^2 + kt}}} \tag{19.9}$$

Choosing to represent an extreme value by this simple approach is suitable for this discussion because it is simple, and it yields a description of the trend with exposure time, being related back to the loss of material in Eq. (19.2) without involving more complicated analyses using the statistics of extreme values. The use of extreme value theory seems attractive [36] but has some intrinsic difficulties, especially in providing a description related to the properties of the materials or a likely distribution of defects [5, 37].

In this approach of linking the stochastic accumulation to damage with the Griffith equation, it would be simple to generate another model if we believed that the reduction in toughness is due to bulk defects caused by leaching of degraded material from inside the polymer, or if the leaching of material creates internal stress, which could be included in Eq. (19.7). However, assuming the defects occur as a result of external degradation is the simplest case fits data reasonably well and creates an example for discussion.

After some simple algebra, Eq. (19.9) can be rearranged to eliminate the need to know explicitly (or guess) the parameters m, E, or G_c, or how they change during exposure, and to couch the trend in terms of the relative reduction in strength:

$$\frac{\text{Stress}_{\text{Griffith}}(t)}{\text{Stress}_{\text{Griffith}}(t=0)} = \left(1 + kt/\sigma_{\text{initial}}^2\right)^{-1/4} \tag{19.10}$$

In this case, the initial defects cannot be eliminated from the expression for relative deterioration in strength, in contrast to how the relative gloss could be represented. Few examples of how toughness of coatings diminishes with exposure

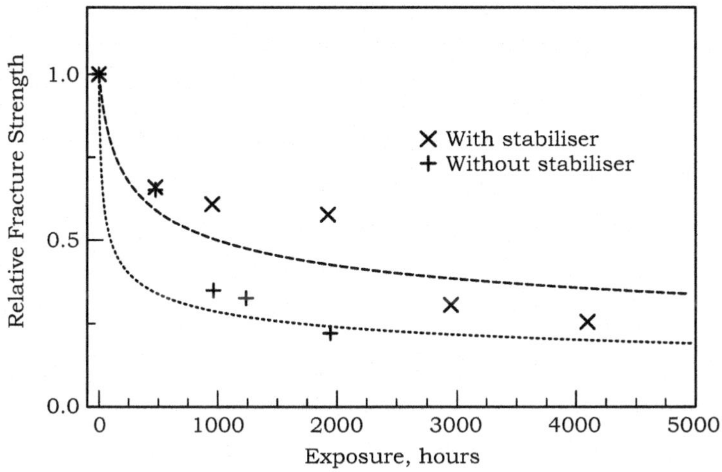

Fig. 19.4 Reduction in fracture strength of automotive coatings, with exposure. Comparing the trend predicted by Eq. (19.10) with data from Nicols et al., with permission from M. E. Nichols [37]

exist [37, 38], but this trend gives a reasonable representation of some published data [37], see Fig. 19.4 below.

The model follows the reduction in the fracture strength reasonably well considering the simplicity of the underlying model. The loss in toughness is modeled here via an increasing defect size due to degradation, rather than any changes in material parameters. However, embrittlement of coatings via changes in G_c, due to degradation, is also dealt with elsewhere in this book.

The model in Eq. (19.10) was constructed from the assumption that damage is stochastic, occurs at the surface, and that the venerable Griffith equation describes the tendency to crack. It portrays what is obvious; that if the material starts with significant flaws ($\sigma_{initial}$) (of a scale that provides stress concentrations rather than smoother wavy features), then its life will be shortened. In addition, by taking this surface roughness to be the source of the flaws leading to failure, there is a link between the reduction in gloss and the reduction in toughness. It seems reasonable that a coating that loses gloss will have become less tough. This approach may provide a means of linking, for example, a property that can be determined in a nondestructive manner, gloss, with a property that can only (so far) be determined in a destructive way and with a considerable degree of difficulty.

The equations use reliable and well-tested physical models to describe how properties deteriorate in service. They are all very simple, but they yield a different dependence on exposure time that can easily be traced back to the physical description of the property. We naturally tend to link changes in appearance (they are easily detected and nondestructively quantified) to changes in protective qualities (difficult and possibly destructive to measure), but trying to predict one from another will almost certainly be unsuccessful because, if nothing else, the rate of decay is certain to be different. Initial gloss depends on the roughness and consistency of the initial surface and so depends on the defects occurring in the application of the coating, but the relative change thereafter does not. The calculation for the relative fracture stress yields no such convenience because the initial level of defects is also convoluted into the rate of decay of the toughness, and so even the very simple zero-order kinetics in Eq. (19.2) is manifested in a more complicated way.

Other Degradation Kinetics

Degradation is not necessarily limited to simple, zero-order kinetics, as used in Eq. (19.2). Other kinetics are useful in the models of physical properties [39]. Spectroscopic assessments of degradation are another example of how this approach might be applied where the size of a spectroscopic absorption peak is used to measure how the concentration of a chemical group either increases or decreases. The rate of loss, or change, depends on the concentration, $N(t)$, of material remaining to be changed. In this case, the analogy of Eq. (19.2) would be

$$\Delta N(t) = -K'N(t) \cdot \Delta t \tag{19.11}$$

The use of K' here indicates that its components, as before in Eq. (19.2), i.e., absorption cross-section and quantum efficiency, may well be different when dealing with chemical change rather than ablation, even in the same material. The solution of Eq. (19.11) is straightforward:

$$N(t) = N(0) \cdot \exp(-K't) \tag{19.12}$$

Spectroscopic examination of the chemistry of degradation can measure an absorption band that gradually diminishes as that functionality is degraded, and these usually decay in an approximately exponential fashion as Eq. (19.12) would suggest [40–43]. Other degradation measurements measure a quantity that increases with degradation, for example, yellowing experiments, Fig. 19.5, or other measurements of oxidation have trends where the effect increases and eventually reaches a plateau [42, 44–46] such that the measurement follows

$$N_{\mathrm{Changed}}(t) = N(0) - N(t) = N(0)[1 - \exp(-K't)] \tag{19.13}$$

In this case, the rate of change eventually stops when there is nothing left to change (to the depth that radiation or penetrant reaches).

The description of degradation must be tailored according to what is known about the system of interest. In fact, if Eq. (19.11), instead of Eq. (19.2), is a more suitable representation of the kinetics that controls the increase in roughness, and thus gloss (or toughness or contact angle), then we should use it instead [32].

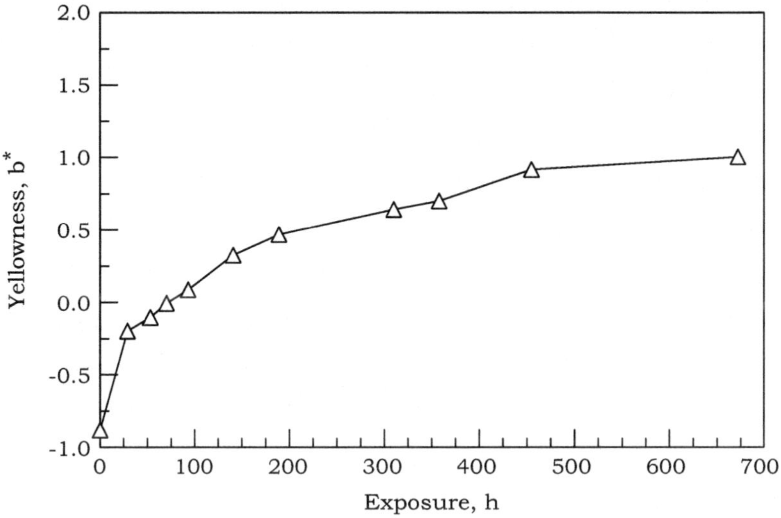

Fig. 19.5 Yellowing (b*, CIELAB) of a polyurethane coating in an accelerated exposure test (QUV-A)

Other kinetics might be suitable when, for example, there are reactions that are accelerated as degradation proceeds. For example, if the product of degradation causes the material to absorb more radiation and thus accelerate the degradation. The example below is drawn from the degradation of a polyester–urethane that yellows [18]. There is increasing absorption at the short wavelength end of the solar spectrum as exposure continues, Fig. 19.6a.

Thus, the effective dosage [47] increases, as quantified from the changing absorption spectrum, $A(\lambda,t)$, and the irradiation spectrum, $E_0(\lambda,t)$, (solar spectrum here),

$$d_{\text{eff}}(\lambda,t) = \int_{\lambda_{\min}}^{\lambda_{\max}} E_0(\lambda,t)\left(1 - 10^{-A(\lambda,t)}\right)\phi(\lambda,t)\,\mathrm{d}\lambda \tag{19.14}$$

where

d_{eff} = effective dosage at a given time [W/m^2]
$\lambda_{\max}$, $\lambda_{\min}$ = maximum and minimum photolytically effective wavelengths
$\phi(\lambda,t)$ = quantum yield of interest
t = exposure time
λ = wavelength

In this case, the degradation kinetics could be represented by a simple exponential, autocatalytic growth. However, it might be more physically reasonable to represent it by the logistic equation [48] which models accelerating, exponential growth, but with a limit, so the overall form is sigmoidal. The logistic equation can

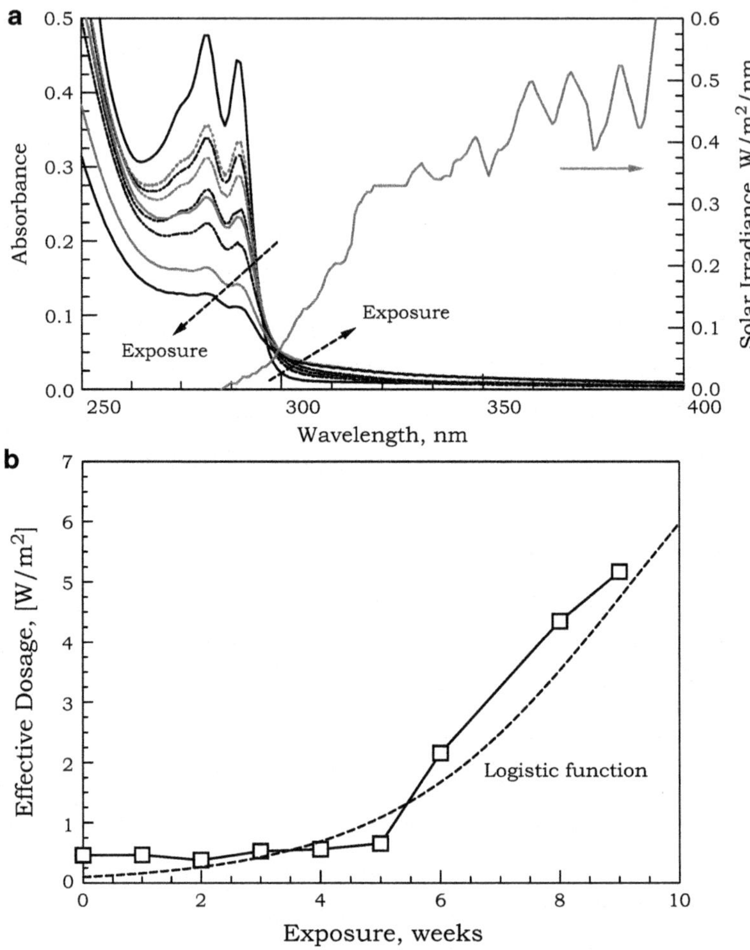

Fig. 19.6 Example of how yellowing may change absorbance. (**a**) Shows the change in absorbance of a polyester–urethane with increasing exposure. (**b**) The integrated effective dose (Eq. 19.14) increases as the absorption within the solar spectrum increases with degradation, with an example of how a Logistic function could model an autocatalytic increase

either approximate first-order kinetics that reach a plateau after a brief initial acceleration period or could be used to model accelerating (exponential) growth over a substantial period if the eventual limit is very large, so this might be a very versatile, physically plausible form to adopt for modeling the kinetics of degradation.

In any case, there should be a causal link between the phenomena involved and the model chosen, rather than a desire to merely codify the data with a smooth curve.

Nonsimultaneous Degradation

If degradation affects everywhere quickly, the trends should follow approximately the equations given earlier. However, often there is an apparent induction period where the measured rate is initially slower but accelerates, as in the example given in Fig. 19.6. Data with this trend are common in gloss, yellowing, or other spectroscopic studies [49–54] and have been attributed to chemical processes that change as the degradation proceeds, i.e., effectively autocatalytic either due to some specific possibilities in the chemistry or as shown in Fig. 19.6. That may be true, but there is also a simpler explanation, which was offered a number of years ago, that not all the material in the sample starts to degrade at the same time [2].

The model offered here provides a means to demonstrate the simpler explanation. It is quite possible in the early stages of degradation, in a material for which the probability for degradation is very low, that some fraction of the material in the sample is not affected [2] within the period of the first few measurements. In the analysis above, the assumption was explicit that degradation immediately affects the whole surface (or volume) and so average values or standard deviations would properly represent the state of the sample. Some very simple equations resulted that conveyed the essence of stochastic degradation employed in well-known models of physical properties. If one measures a property, $A(t)$, averaged over an area (or volume), e.g., gloss, in the early stage, the measurement gives a result where the undamaged fraction retains its original value. The damaged fraction is deteriorating, but each part of the surface (or bulk) has a property value that depends on when it started to degrade.

If the degradation progresses in a random fashion, then a Gaussian representation can again be used. The cumulative Gaussian distribution function, $\Phi(x)$, is the integral of the Gaussian function Eq. (19.1),

$$\Phi(x) = \frac{1}{2}\left[1 + \mathrm{erf}\left(\frac{x-\mu}{\sigma\sqrt{2}}\right)\right]$$

The part of the sample that is not affected will shrink with time but retain the original value of the property, $A_{\mathrm{unaffected}}$. Exposure time, t, will be used as the continuous variable now, but more properly it might be the number of aggressive photons or aggressive molecules that test the surface. The area of the affected portion increases stochastically, and the property will change according to when that part of the sample started to degrade. If changes in the degraded area do not spread by some other means to the unaffected material, the property being monitored is given by the simple sum,

$$A(t) = [1 - \Phi(t)] \cdot A_{\mathrm{unaffected}}(0) + \int_0^t \frac{1}{\sqrt{2\pi\sigma_{\mathrm{T}}^2}} \exp\left[-\frac{(t' - \mu_{\mathrm{T}})^2}{2\sigma_{\mathrm{T}}^2}\right] \cdot A_{\mathrm{affected}}(t - t')\ \mathrm{d}t' \qquad (19.15)$$

The first term on the right-hand side gives the diminishing contribution from the undegraded fraction, and the second term gives the increasing contribution as the degraded area increases in a stochastic way, and the degradation within that area proceeds. Values of μ_T, the mean time after which the whole surface is likely to have been affected, and the standard deviation, σ_T, would be needed and would be derived from the parameters used in Eq. (19.2), etc. Examination of Eq. (19.2) shows that there would be a time after which the number of fragments will equal the area concentration of labile bonds, $\Delta N(t)/N(0)$. This would be the characteristic mean time, μ_T, after which the whole surface would have been affected and can be seen to be $1/(\phi a q)$ in Eq. (19.2). The standard deviation can be determined by the square root of this quantity. The application of this relation depends on whether the property of interest is measured using a technique where this applies. For example, in measuring weight loss, gloss, or color, one measures a value averaged over an area and similarly one would expect this idea to apply in some spectroscopic techniques. Other techniques may explicitly examine only degraded locations and so one would only need to consider the kinetics of degradation without the effect of its spreading.

The model discussed in this chapter provides some simple functional forms for $A_{\text{Affected}}(t)$ in Eqs. (19.2), (19.6), and (19.13) so that Eq. (19.15) can be readily evaluated to demonstrate the effect. For example, if this approach is used to model gloss, Eq. (19.6) gives an expression for the affected property:

$$\begin{aligned} A_{\text{Affected}}(t) &= \frac{\text{Gloss}(t)}{\text{Gloss}(0)} = \frac{R_s(t)}{R_s(0)} \\ &= \exp\left[-\left(\tfrac{4\pi \quad \cos\varphi}{\lambda}\right)^2 kt\right] = \exp[-\alpha t] \end{aligned} \tag{19.16}$$

Here, α has been used to group parameters together for ease of representation. Thus, a complete expression for how gloss changes with respect to its initial value is given by:

$$A(t) = \frac{\text{Gloss}(t)}{\text{Gloss}(0)} = [1 - \Phi(t)] + \int_0^t \frac{1}{\sqrt{2\pi\sigma_T^2}} \exp\left[-\frac{(t' - \mu_T)^2}{2\sigma_T^2}\right] \exp[-\alpha(t - t')] \mathrm{d}t'$$

where

$$\Phi(t) = \frac{1}{2}\left[1 + \operatorname{erf}\left(\frac{t - \mu_T}{\sigma_T\sqrt{2}}\right)\right]$$

Fortunately, the speed of modern computers makes this model easy to evaluate. Unfortunately, there are several parameters in these equations that must either be measured or possibly determined by curve-fitting, and so only illustrative diagrams of the effect are presented using arbitrary units for the exposure time and choices for the other parameters, Fig. 19.7.

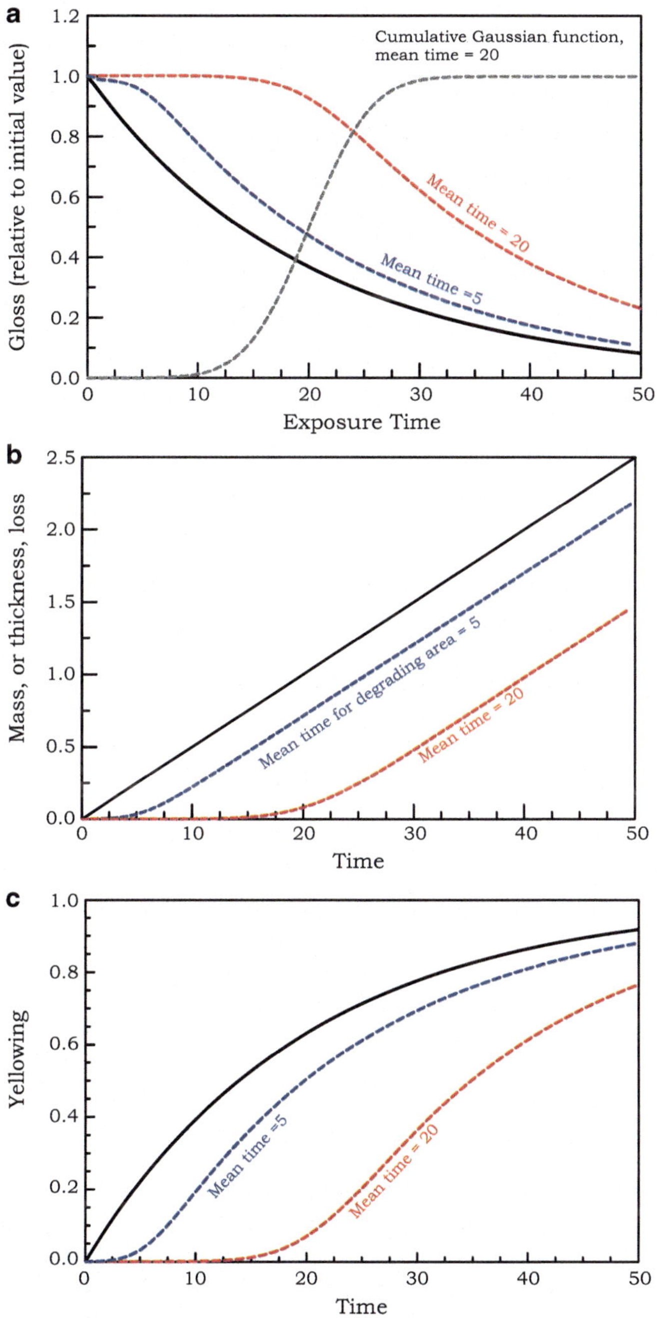

Fig. 19.7 How a significant time for degradation might produce an induction-like period. All units are arbitrary. The underlying trend if the whole sample starts degrading immediately is the solid black line. (**a**) Gloss with a characteristic mean time, μ_T, of either 5 or 20 time units. A cumulative Gaussian function for a 20 unit characteristic mean time is included for illustration. (**b**) Linear weight loss or thickness loss with a characteristic mean time of 5 or 20 time units, (**c**) Yellowing or other spectroscopic measurement (assuming first-order chemical kinetics, Eq. (19.13))

In Fig. 19.7a, the value of α, the decay constant for the trend for the gloss in Fig. 19.7a was chosen to be the reciprocal of μ_T, i.e., 1/5 or 1/20. This was done not only for simplicity but also to illustrate the inverse connection since a stable material would have its whole area affected after a long time, related to $1/(\phi aq)$, and its underlying deterioration rate would be small, related to (ϕaq). The same connection was exemplified for the case of yellowing, Fig. 19.7c, where the exponential constant ($1/K'$ in Eq. (19.13)) for the underlying yellowing was also taken as the reciprocal of μ_T for the two examples. In Fig. 19.7b, the underlying rate of loss of mass is 1/20 in the units used on the diagram. In each case, the standard deviation, σ_T, is calculated as the square root of the characteristic mean time.

In each of the examples, the value of μ_T is apparent from the graphs, if one extrapolates back to the point of zero deterioration. Thus, it should be possible to deduce its approximate value and thus a value for $1/(\phi aq)$ which also underlies the decay in the properties in this approach. If the apparent induction period is not due to another identifiable cause, then it indicates something about durability. A material is one that degrades quickly if there is only a brief period before the whole sample is affected. However, if the property decays only slowly, then the amount of damage done in each event must be small. If there is a long period before the deterioration becomes apparent, then the material must be quite stable, but if the subsequent decay is rapid, it would mean that each successful damage event causes large damage (large ε) although it happens infrequently. Such observations might lead to more detailed understanding of the material and its likely weak links.

An analogous situation is where the material is stabilized against degradation by an ultraviolet absorber, antioxidant, or other ingredient that combats the degradation process. Such ingredients themselves have a finite lifetime, so the protection would diminish as the stabilizer degrades, presumably in a stochastic process, with a characteristic mean lifetime, μ_{Stab}, that has a standard deviation, σ_{Stab}. Not all of the stabilizer would fail after the same time, some would last longer and protect that region of the sample, whereas other areas might lose protection earlier. Thus, the measured property of the material would similarly be the property averaged over those areas where the protection continued and those regions where degradation had started.

Summary

It is possible to build an approach, from very basic principles, that can estimate how the physical properties that usually determine service life deteriorate during exposure. Environmental conditions may be localized and variable, so the approach here examines the deterioration of coating properties and not on other definitions of failure, e.g., blistering or rusting, that include a process dependent on the substrate and probably a particular location. Instead of dealing with the whole, formidable problem of linking composition directly to service performance, the approach here breaks the problem into stages.

First, we use that degradation processes are random, involving vast numbers of aggressive incidents. This is very basic and general and does not require any specific chemical or physical knowledge of the processes. The absorption of a photon or aggressive chemical molecule happens according to a statistical chance, so we can represent the state of a material according to the mean number of aggressive events and measurements of the likelihood of absorption, e.g., spectroscopic absorbance. Typical exposure conditions and the response of materials are consistent with the central limit theorem and thus permit the use of the normal distribution. The likelihood of damage occurring after the arrival of a single photon or molecule is small, so we can calculate the standard deviation from the mean. Thus, an estimate of the mean, i.e., most typical response, and its variance are possible. Then, the materials information about the coating and its environment can be incorporated applying very well-known and reliable theories that connect physical changes, such as roughness or flaw size to end-use properties such as gloss or toughness. Despite most materials being complicated composites in a variety of changing environments, this provides a basis for understanding the kinetics of degradation and models that can be used to quantify and compare how materials deteriorate which can be made more exact or specific with added information. In all cases, there is a reciprocal relationship between rate of deterioration, measured at any time (most usefully at the start) and likely service. An advantage to incorporating mechanistic knowledge is that the approach can be used to design new and improved materials. The scheme here assumes that the service life is determined by the accumulation of damage superimposed on the flaws generated in the formation process. Depending on the property of interest, it may not be the weakest link that causes failure, and even if the eventual failure is sudden, it will be a result of the environmental damage added to the built-in flaws, so it is questionable whether the statistics of extreme values are applicable without further insight. In any case, it seems difficult, at present, to link the parameters of extreme value functions with material parameters in a way that would be useful in the design of a material.

In a simple case, degradation follows zero-order kinetics, and a simple, reciprocal relationship exists between rate of degradation and service life. Deterioration in other properties can be modeled, such as gloss via its dependence on surface roughness or toughness, by using an estimate of the largest flaw caused by degradation. The model here shows how the deterioration in various properties may be linked. One can appreciate which properties may be connected and, perhaps, use a nondestructive test to assess how other properties might change. However, it also demonstrates that different properties are sensitive in different ways to the degradation process and decay with a different rate, even though the basic mechanism of deterioration may be identical. One cannot expect simple correlations between properties as a material degrades.

The model provides simple equations that can be used to demonstrate the effect when the initial degradation across the area or volume of a sample of material starts at random times instead of simultaneously. Again, this can be quantified using Gaussian statistics. There is an apparent induction time which may indicate something about durability without being due to autocatalytic chemistry.

This statistical approach to describing durability uses the very basic notion that the environment assaults a material many times, at random intervals and at random locations. One cannot say that this approach is the best or only possibility, but there are grounds for retaining optimism in understanding how durability is related to physical and chemical mechanisms. Further progress is possible. Mechanistic approaches can be put together that enable viable insights into how durability depends on the initial structure of the coating, the environment, and the criterion for failure, so that robust product design or selection is possible.

Acknowledgments The author would like to thank his colleagues, students, and friends over many years for ideas, data, and guidance.

References

1. Guseva, O., Brunner, S., Richner, P.: Service life prediction for aircraft coatings. Polym. Degrad. Stab. **82**, 1–13 (2003)
2. Celina, M., George, G.A.: A heterogeneous model for the thermal oxidation of solid polypropylene from chemiluminescence analysis. Polym. Degrad. Stab. **40**, 323–335 (1993)
3. Martine, J.G., Oliverio, S., Rodrigue, F., Santiago Sanchez, L., Eduardo Ramirez, V., Allen, N. S.: Prediction of photoageing stability of plasticised PVC films containing UV-stabilisers. Polym. Degrad. Stab. **54**, 49–55 (1996)
4. Danzer, R., Supancic, P., Pascual, J., Lube, T.: Fracture statistics of ceramics—Weibull statistics and deviations from Weibull statistics. Eng. Fract. Mech. **74**, 2919–2932 (2007)
5. Manzato, C., Shekhawat, A., Nukala, P.K.V.V., Alava, M.J., Sethna, J.P., Zapperi, S.: Fracture strength of disordered media: universality, interactions, and tail asymptotics. PRL. **108**, 065504 (2012)
6. Ptolemy C.: attributed in The Science of Conjecture: Evidence and Probability before Pascal, J. Franklin, Johns Hopkins University Press, Baltimore (2001)
7. Transtrum, M.K., Machta, B.B., Brown, K.S., Daniels, B.C., Myers, C.R., Sethna, J.P.: Perspective: sloppiness and emergent theories in physics, biology, and beyond. J. Chem. Phys. **143**, 010901 (2015)
8. ASTM G173—03.: Standard Tables for Reference Solar Spectral Irradiances: Direct Normal and Hemispherical on 37° Tilted Surface, Subcommittee: G03.99 (2012)
9. Haynes, W.M. (editor).: CRC Handbook of Chemistry and Physics, 95th edn. CRC Press, Boca Raton (2014)
10. Kroll, D.M., Croll, S.G.: Heterogeneity in crosslinked polymer networks: molecular dynamics simulations. In: Wen, M., Dusek, K. (eds.) Protective coatings: film formation and properties. Springer, New York (2017)
11. Meakin, P.: Formation of fractal clusters and networks by irreversible diffusion-limited aggregation. Phys. Rev. Lett. **51**(13), 1119–1122 (1983)
12. Thanh, N.T.K., Maclean, N., Mahiddine, S.: Mechanisms of nucleation and growth of nanoparticles in solution. Chem. Rev. **114**, 7610–7630 (2014)
13. Croll, S.G., Hinderliter, B.R.: Estimating service lifetimes in weathering: an optimistic view. J. Coat. Technol. Res. **4**(3), 217–230 (2007)
14. Croll, S.G., Hinderliter, B.R.: Monte Carlo simulation applied to coating deterioration during weathering, service life prediction, challenging the status quo. In: Martin J. W., Ryntz R. A., Dickie R. A. (eds), pp. 107–122. Federation of Societies for Coatings Technology (2005)

15. Fechine, G.J.M., Rabello, M.S., Souto Maior, R.M., Catalani, L.H.: Surface characterization of photodegraded poly(ethylene terephthalate). The effect of ultraviolet absorbers. Polymer. **45**, 2303–2308 (2004)
16. Martin, J.W., Chin, J.W., Nguyen, T.: Reciprocity law experiments in polymeric photodegradation: a critical review. Prog. Org. Coat. **47**, 292–231 (2003)
17. Bauer, D.R.: Predicting in-service weatherability of automotive coatings: a new approach. J. Coat. Technol. **69**(864), 85–96 (1996)
18. Croll, S.G., Skaja, A.D.: Spectroscopic adsorption and effective dosage on accelerated weathering of a polyester-urethane coating. J. Mater. Sci. **37**, 4889 (2002)
19. Gu, X., Raghavan, D., Nguyen, T., VanLandingham, M.R., Yebassa, D.: Characterization of polyester degradation using tapping mode atomic force microscopy: exposure to alkaline solution at room temperature. Polym. Degrad. Stab. **74**, 139 (2001)
20. Wypych, G.: Handbook of Material Weathering. ChemTec, Toronto, Canada (1995)
21. Hollande, S., Laurent, J.-L.: Degradation process of an industrial thermoplastic elastomer polyurethane-coated fabric in artificial weathering conditions. J. Appl. Polym. Sci. **73**, 2525–2534 (1999)
22. Wood, K.A., de Robien, S.: A quantitative model for weathering-induced mass loss in thermoplastic paints. In: Martin, J.W., Ryntz, R.A., Chin, J., Dickie, R.A. (eds.) Service Life Prediction of Polymeric Materials, Global Perspectives, pp. 457–474. Springer, Berlin (2009)
23. Reddy, S.K., Anseth, K.S., Bowman, C.N.: Modeling of network degradation in mixed step-chain growth polymerizations. Polymer. **46**, 4212–4222 (2005)
24. Bennett, H.E., Porteus, J.O.: Relation between surface roughness and specular reflectance at normal incidence. J. Opt. Soc. Am. **51**(2), 123–129 (1961)
25. Toporets, A.S.: Specular reflection from a rough surface. Opt. Spectroscopy. **16**, 54–58 (1964)
26. Beckmann, P., Spizzichino, A.: The Scattering of Electromagnetic Waves from Rough Surfaces. Artec House, Norwood, MA (1987)
27. Hunt, F.Y., Galler, M.A., Martin, J.W.: Microstructure of weathered paint and its relation to gloss loss: computer simulation and modeling. J. Coat. Technol. **70**(880), 45–53 (1998)
28. Bendler, J.T., Feldman, S.F., Hatti, H., Hobbs, S.Y.: Approximate model of diffuse reflectance from rough polymer surfaces. J. Appl. Phys. **83**(2), 998–1004 (1998)
29. Bourlier, C., Berginc, G., Saillard, J.: One- and two-dimensional shadowing functions for any height and slope stationary uncorrelated surface in the monostatic and bistatic configurations. IEEE Trans. Antennas Prop. **50**(3), 312–324 (2002)
30. Fischer, R.M.: Accelerated weathering test development with fluorescent UV-condensation devices. SAE Technical Paper 841022 (1984)
31. Wernståhl, K.M., Carlsson, B.: Durability assessment of automotive coatings—design and evaluation of accelerated tests. J. Coat. Technol. **69**(865), 69–75 (1997)
32. Hinderliter, B.R., Croll, S.G.: Predicting coating failure using the central limit theorem and physical modelling. ECS Trans. **24**(1), 1–26 (2010)
33. Croll, S.G., Hinderliter, B.R., Liu, S.: Statistical approaches for predicting weathering degradation and service life. Prog. Org. Coat. **55**, 75–87 (2006)
34. Croll, S.G.: Application, and limitations, of current understanding to model failure modes in coatings. J. Coat. Technol. Res. **10**(1), 15–27 (2013)
35. Alava, M.J., Nukala, P.K.V.V., Zapperi, S.: Statistical models of fracture. Adv. Phys. **55**(3–4), 349–476 (2006)
36. Fortin, J.-Y., Clusel, M.: Applications of extreme value statistics in physics. J. Phys. A: Math. Theor. **48**, 183001 (2015)
37. Nichols, M.E., Peters, C.A.: The effect of weathering on the fracture energy of hardcoat over polycarbonate. Polym. Degrad. Stab. **75**, 439–446 (2002)
38. Seubert, C.M., Nichols, M.E., Cooper, V.A., Gerlock, J.L.: The long-term weathering behavior of UV curable clearcoats I. Bulk chemical and physical analysis. Polym. Degrad. Stab. **81**, 103–115 (2003)

39. El-Mazry, C., Correc, O., Colin, X.: A new kinetic model for predicting polyamide 6-6 hydrolysis and its mechanical embrittlement. Polym. Degrad. Stab. **97**, 1049–1059 (2012)
40. Martin, J.W., Nguyen, T., Byrd E., Dickens B., Embree N.: Laboratory apparatus and cumulative damage model for linking field and laboratory exposure results. In: Martin, J.W., Bauer, D.R. (eds.) Service Life Prediction Methodology and Metrologies, ACS Symposium Series, vol. 805, pp. 119–143 (2002)
41. Allen, N.S., Robinson, P.J., White, N.J., Swales, D.W.: Photo-oxidative stability of electron beam and UV cured acrylated epoxy and urethane acrylate resin films. Polym. Degrad. Stab. **19**, 147–160 (1987)
42. Rivaton, A., Serre, F., Gardette, J.-L.: Oxidative and photooxidative degradations of PP/PBT blends. Polym. Degrad. Stab. **62**, 127–143 (1998)
43. Mailhot, B., Morlat-Thérias, S., Bussière, P.-O., Gardette, J.-L.: Study of the degradation of an epoxy/amine resin, 2 kinetics and depth-profiles. Macromol. Chem. Phys. **206**, 585–591 (2005)
44. Allen, N.S., Edge, M., Rodriguez, M., Liauw, C.M., Fontan, E.: Aspects of the thermal oxidation, yellowing and stabilisation of ethylene vinyl acetate copolymer. Polym. Degrad. Stab. **71**, 1–14 (2001)
45. Davidson, R.S.: The photodegradation of some naturally occurring polymers. J. Photochem. Photobiol. B, Biol. **33**, 3–25 (1996)
46. Malešič, J., Kolar, J., Strlič, M., Kočar, D., Fromageot, D., Lemaire, J., Haillant, O.: Photo-induced degradation of cellulose. Polym. Degrad. Stab. **89**, 64–69 (2005)
47. Martin, J.W.: Quantitative characterization of spectral ultraviolet radiation-induced photodegradation in coating systems exposed in the laboratory and the field. Prog. Org. Coat. **23**, 49–70 (1993)
48. Verhulst, P.-F.: Recherches mathématiques sur la loi d'accroissement de la population. Nouveaux Mémoires de l'Académie Royale des Sciences et Belles-Lettres de Bruxelles. **18**, 1–42 (1845)
49. Diepens, M., Gijsman, P.: Photodegradation of bisphenol A polycarbonate. Polym. Degrad. Stab. **92**, 397–406 (2007)
50. Feller, R.L., Curran, M., Colaluca, V., Bogaard, J., Bailie, C.: Photochemical deterioration of poly(vinylbutyral) in the range of wavelengths from middle ultraviolet to the visible. Polym. Degrad. Stab. **92**, 920–931 (2007)
51. Larché, J.F., Bussière, P.-O., Gardette, J.L.: Characterisation of accelerated ageing devices for prediction of the service life of acrylic-melamine/urethane thermosets. Polym. Degrad. Stab. **96**, 1530–1536 (2011)
52. Larché, J.F., Bussière, P.-O., Gardette, J.L.: Photo-oxidation of acrylic-urethane thermoset networks. Relating materials properties to changes of chemical structure. Polym. Degrad. Stab. **96**, 1438–1444 (2011)
53. Pospíšil, J., Nešpurek, S.: Photostabilization of coatings. Mechanisms and performance. Prog. Polym. Sci. **25**, 1261–1335 (2000)
54. Monney, L., Dubois, C., Chambaudet, A.: Evolution of the thin photo-oxidation layer of an epoxy matrix during artificial photo-ageing. Angewandte Makromolekulare Chemie. **273**, 6–11 (1999)

Chapter 20
Long-Term Mechanical Durability of Coatings

Mark E. Nichols

Introduction

The decorative and protective functions of a coating are only preserved if the coating remains intact and attached to the substrate during the coated object's service life. Thus, the mechanical properties of the coating, how they change over time, and the mechanical and chemical interaction between the coating and the substrate are critical to the long-term performance of the complete material system.

While clearly important, the mechanical properties of coatings have traditionally been probed via simple, application-specific test methods. As one example, the hardness of a coating can be quantified via indentation (Knoop, Vickers), scratch (pencil hardness), or mechanical damping (pendulum hardness). All provide a different answer, none of which is directly relatable to a more fundamental property, such as crosslink density, modulus, or strength of the coating. Other specific tests for coatings include scrub resistance, blocking resistance, and abrasion resistance. All are useful, as coating formulators use the results of these tests to improve their products, but none measure any one intrinsic property, as each test confounds several different fundamental material properties. Abrasion resistance, for example, is a function of the coating's coefficient of friction, fracture toughness, and resistance to plastic deformation.

Fundamental mechanical properties of coatings, however, are difficult to measure. Coatings, by definition, are attached to a substrate—often with excellent adhesion. Traditional materials test methods often require the preparation of free films. For some properties, this can be perfectly acceptable. However, the use of low surface energy substrates to ease the release of films can change the nature or rate of cure and the disposition of additives in the coating. In addition, low surface

M.E. Nichols (✉)
Ford Motor Company, Dearborn, MI, USA
e-mail: mnichols@ford.com

M. Wen, K. Dušek (eds.), *Protective Coatings*, DOI 10.1007/978-3-319-51627-1_20

energy substrates are difficult to wet and uniformly coat, and the handling of thin free films can be difficult due to their fragile nature and susceptibility to small flaws.

The measurement of mechanical properties of films attached to substrates also has its own challenges but clearly has unique advantages as well. If material properties as a function of exposure time are of interest, then coatings attached to substrates are the only possible source of samples, and a wealth of specimens often exist, as paint systems are routinely exposed for outdoor and accelerated durability testing. Progress has been made in the use and interpretation of nano-indentation [1] and scratch measurements [2] (see also Chaps. 15 and 16 in this monograph). These methods are now a useful means for determining some mechanical properties. Techniques to measure stresses in coatings [3–5] (see also Chaps. 10 and 11 in this monograph), as well as the chemical composition and the distribution of additives [6] (see also Chaps. 13 and 14 in this monograph), are most informative when the coatings are firmly attached to the intended substrate.

For many coating systems, the resistance to fracture is an important material property that can be a useful probe of both network structure and an indicator of in-service performance. In particular, the fracture toughness or fracture energy of a coating is directly related to its propensity to crack under the application of mechanical loads. Many coatings undergo cracking during their lifetime due to a combination of increasing brittleness and increasing internal tensile stresses. The appearance of cracks in coatings and thin-layered structures has been observed by people for centuries [7, 8]. The distinct cellular nature of the cracking patterns has attracted the attention of geologists, art historians, and electronic materials scientists (Fig. 20.1). Geologists, in particular, have noted and studied these cracking patterns [9]. However, apart from art historians, little attention has been paid to this phenomenon in the organic coatings community [10, 11].

In this chapter, the relationship between stresses, fracture properties, weathering and the patterns of crack formation in coatings is reviewed. Methods for assessing the fracture resistance of ductile and brittle films are discussed, and the intrinsic and extrinsic coating properties that influence how cracks form over time in coatings is demonstrated for specific coating systems.

Fracture Mechanics

Crack Initiation

While the majority of the discussion in this manuscript focuses on crack propagation, a brief discussion of crack initiation in coatings is warranted. At the macroscopic scale, most coatings when applied ideally are free of large defects. The rupture strength of these pristine materials is significantly higher than materials that contain even vanishingly small flaws. To initiate a crack in a pristine material would require the nucleation and growth of a defect due to fatigue or some other

Fig. 20.1 Cracking patterns observed in different coatings/materials. Clockwise from upper left: laminated film outdoor sign, seventeenth century Dutch painting, asphalt coating on road, and automotive coating

heterogeneous process. However, application conditions are rarely ideal, and most coatings contain defects ranging in size from submicron to millimeter scale. Defects can be contaminants, pigment agglomerates, voids, craters, or any region within the coating whose mechanical properties are discontinuous with the bulk coating.

Even in nominally defect-free coatings, external stresses often lead to the formation of defects early in a coating's lifetime. Certainly, macroscopic damage can be induced by, for example, stones impacting [12] the surface of an automotive coating, abrasion between beverage cans transported by railway, and the scraping of a shoe over a coated wood or vinyl floor. The defects produced are often macroscopic and well above the necessary threshold to act as initiators of crack propagation when the coating is subjected to even moderate stresses. In addition, many coatings will degrade heterogeneously during photodegradation. Research by workers at NIST has shown that local nanometer scale voids are produced during photodegradation [13]. These voids can grow as time progresses leading to supercritical defects in the coating. Croll has modeled such nonhomogeneous degradation and finds that locally higher rates of photooxidation can drive void formation [14].

Crack Propagation

Almost all mechanical failures in coatings are due to the propagation of cracks—either within the coating or between the coating and the substrate. The fracture mechanics that describes the propagation of cracks in a coating or at the interface between the coating and substrate are fundamentally the same as those that describe the propagation of cracks in the bulk of a material. When the strain energy release rate, G, exceeds the critical strain energy release rate, G_c (also called the fracture energy), a crack will advance. The units of G and G_c are energy/area (J/m^2), and the release rate refers to the energy required to overcome the creation of two new surfaces per unit area. They do not describe time-dependent phenomena, as the term release *rate* implies. For thin films on an elastic substrate, such as an organic coating on a metal or rigid polymer, the strain energy release rate can be written as [15]

$$G = \frac{\pi h \sigma^2 (1 - \upsilon^2) g(\alpha, \beta)}{2E} \tag{20.1}$$

where h is the coating thickness, E is the coating modulus, σ is the stress in the coating, ν is Poisson's ratio of the coating, and $g(\alpha,\beta)$ is a constant related to the mismatch between the modulus of the coating and substrate. For organic coatings on metallic substrates where the crack extends through the full thickness of the coating to the substrate (fully cracked geometry), $g(\alpha,\beta)$ is approximately 0.72 [16]. For coatings with no elastic mismatch with the underlying layer, which is a good approximation for multilayer coating systems where the crack does not reach the substrate, $g(\alpha,\beta)$ is approximately 1.3.

As the crack advances, the material behind the crack tip relaxes as the stress is relieved. For systems that are uniaxially loaded, cracks will propagate parallel to one another and perpendicular to the direction of applied stress (Fig. 20.2). The average equilibrium crack spacing, λ, is related to the fracture energy, the coating thickness, and the applied stress by [17]

$$\lambda \cong 5.6 \sqrt{\frac{G_c h E}{(1 - \upsilon^2)\sigma^2}} \tag{20.2}$$

For more complex loading, for example, equal biaxial loading, the growth of cracks is more complex, as the local stress fields are affected by crack intersections and nonparallel neighbors. Unlike the uniaxial loading condition, no simple analytical solution exists to predict the crack spacing under biaxial or more complex loading. Several researchers have attempted to model the growth of cracks subjected to biaxial loading [18, 19]. However, these studies have been mostly restricted to the growth of cracks in brittle, inorganic films on substrates, such as semiconducting materials on computer chip supports. Those cases are more straightforward than the case of organic coatings on various substrates, as the

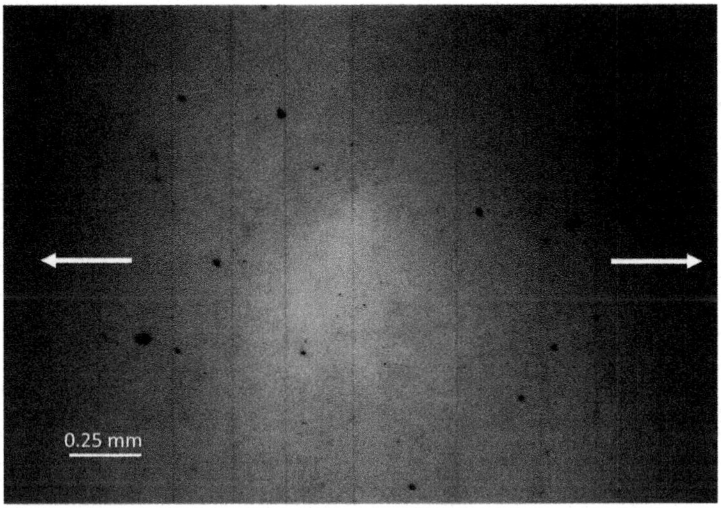

Fig. 20.2 Cracks in a polysiloxane hardcoat on polycarbonate. Cracks grew during a tensile test of the PC/hardcoat system. *Arrows* indicate the direction of the applied load

stresses in inorganic films are strictly related to residual stresses due to lattice constant mismatches and to thermal stresses due to the operating environment. The more complex stresses in organic paint systems are discussed presently.

Stresses in Organic Coatings

The stresses that develop in organic coatings play a key role in the determination of a coating's useful life. In general, the stresses arise because the coating, by definition, is attached to a substrate and is thus constrained from changing its dimensions when the surrounding environment changes or when the coating itself changes. For example, a common source of stress in coatings is due to changes in the temperature of the environment. Essentially, all materials have a non-zero coefficient of thermal expansion (CTE). When the temperature changes, materials expand or contract in response to the temperature change. Most organic materials have a CTE approximately at 100×10^{-6} 1/°C, and that of most metals is approximately an order of magnitude less. The change in the stress in the coating can be calculated from

$$\Delta\sigma = E\Delta T(\alpha_c - \alpha_s) \tag{20.3}$$

where α_s and α_c are the CTEs of the substrate and coating, respectively, and ΔT is the change in temperature imposed upon the system. When the temperature is decreased, a tensile (positive) stress in the coating is induced. Two *caveats* are worth mentioning. First, the modulus of organic coatings is a function of temperature, particularly near the glass transition temperature (T_g) of the binder. Therefore, the modulus at the final temperature should be used to calculate the stress. Secondly, and related to the first *caveat*, the modulus of organic coatings is a function of time, and therefore the stresses in the coating will begin to relax after they are generated. For crosslinked coatings, the stresses will progress towards an equilibrium value when modulus reaches an equilibrium level, which is related to the crosslink density of the coating. However, at very long times, secondary processes such as physical aging or crystallization may prevent true equilibrium from being reached. For lacquers, in the absence of any long-term chemical changes, the stresses may relax towards zero as viscous flow begins to dominate. While some very long time scale data exist for selected polymers, little data are available for binders of particular interest to coatings formulators. Of course, over very long time scales, some chemical changes are likely to occur, as most polymers are prone to react with oxygen, UV light, or other environmental constituents.

Analogous to the temperature-induced stresses; stresses are induced or can be relaxed in a coating when the coating and/or substrate absorbs moisture. When a coating absorbs moisture, it will want to expand. In addition, the coating will become plasticized by the water, and its modulus will decrease. Water can also induce chemical changes, which result in the formation or breaking of crosslinks or main chain bonds. When the water is removed from the coating, the coating contracts and stiffens. Thus, assuming the coating is affected more than the substrate, drying will induce tensile stresses in the coating. The thermal- and moisture-induced stresses in coatings are reversible and largely track the diurnal cycle. These cyclic stresses may drive fatigue failures in coatings, but no work has yet been published that demonstrates true fatigue crack growth in organic coatings.

Longer term stresses can arise in coatings due to both physical and chemical changes that take place over the course of weeks or years. Perera [20, 21] has quantified the stresses in coatings that arise due to physical aging, which is the process by which coatings proceed towards their equilibrium thermodynamic state when held below their T_g. As this process leads to densification, the constrained coating experiences a net tensile stress as physical aging proceeds [22, 23]. Similarly, continued chemical composition changes can lead to further densification. Most polymerization and crosslinking reactions result in a decrease in specific volume. As these reactions can continue for some time after a coating is applied and solidified, the tensile stress in the coating can increase. From a formulation perspective, coatings that show little shrinkage as cure progresses should have lower internal stresses. Ideally, small molecule products of crosslinking reactions should diffuse out of the coating during cure. However, in practice, this is not the case, and the continued evolution of these molecules can lead to stress buildup due to both volume shrinkage and decreased plasticization (modulus increase). Internal stress buildup in high-solids coatings is particularly important as more chemistry occurs post-application for these coatings. Polymerization reactions that are known to be volume neutral or volume positive could mitigate these curing stresses. The opening of cyclic structures during cure is known to be volume positive. In particular, the work of Bailey and Endo on expanding monomers based on spiro-orthocarbonates may be worth revisiting as one means to reduce stresses in high-solids coatings [22, 23].

When summed, the total stresses in coatings can clearly build up to levels that induce cracking. For example, clearcoats in automotive paint systems can experience stresses approaching 20 MPa as weathering progresses [24]. Stresses in the primer layer can be even larger, as the high pigment loading in primers increases the modulus of the primer. Stresses in architectural coatings are substantially lower, as these coatings typically have lower T_gs, and moduli and cannot support the high stresses unlike high T_g automotive coatings.

In summary, the stresses in coatings tend to increase over time and are largely tensile. In addition to the slowly increasing tensile stresses, dynamic stresses due to temperature and moisture changes in the environment are superimposed upon the specimen.

Long-Term Resistance to Cracking in Coatings

The remainder of this discussion will focus on cracking in the coating, but it can easily be translated to cracking at an interface via the appropriate modifications in the fracture mechanics.

The difficulty in fully assessing the fundamental resistance of coatings to fracture lies in Eqs. (20.1) and (20.2). The driving force for cracking, i.e., the strain energy release rate, is proportional to the stress squared and linearly proportional to thickness. While the thickness of many industrially applied coatings is relatively uniform over reasonably macroscopic dimensions, the paint thickness of works of art or architectural paints applied by DIY painters varies greatly and leads to spatially varying driving forces.

The stress in organic coatings generally increases over time, as discussed in the preceding section. In addition, most organic coatings embrittle over time. As such, observation of or even testing the fracture resistance is confounded. Did cracks appear during service because the internal stress increased (G increased) or because the fracture energy (G_c) decreased? Either will impact the failure criterion ($G>G_c$). Observation and testing of the *apparent* fracture energy cannot separate these two important differences. The crack spacing in a tensile test cannot answer these questions either, as both the fracture energy and stress are on the right side of the crack spacing equation (Eq. 20.2). To probe the mechanism behind cracking, independent measurements of either the stress or the toughness is required.

The stress in coatings can be directly measured by a number of methods. Most commonly, a coating can be applied to a very thin, flat substrate. Upon curing, the internal stress in the coating causes the bi-material structure to curve in the direction of the material that undergoes a tensile stress, as first noted by Corcoran [25]. Measurement of the curvature of the sample allows for the calculation of the stress in the coating if the thickness of the coating (h) and substrate (x) and the modulus of the substrate (E_s) are known [21],

$$\sigma = \frac{4dE_s t^3}{3l^2 h(x+h)(1-\upsilon)} \tag{20.4}$$

where d is the midpoint deflection of the sample supported on two knife-edges, l is the distance between the two knife-edges, and t is the thickness of the substrate. Alternatively, other methods to measure the stress in thin films have been developed that use stress-sensitive tracer molecules or small changes in the IR spectrum [26].

The measurement of the fracture energy independent of the internal stress is more challenging. The production of stress-free coatings is problematic due to both thermal and curing stresses, which are difficult to avoid. If the coating can be removed from the substrate, the stresses will typically relax and the fracture energy can be assessed on the free film. Many techniques and geometries have been used to measure the fracture energy of free films. The most commonly used geometries are the single edge notch (SEN) or double edge notch (DEN) style. For these tests, a pre-crack of known length is inserted into one (or two for the DEN test) side of the specimen [27]. The force required to propagate the crack can then be measured as a function of the ligament length (the length of the uncracked width of the specimen). The work (area under the force displacement curve) can then be plotted as a function of ligament length. The y-intercept is then the essential work of fracture required to propagate a crack in the film. While relatively straightforward, the usual difficulties in both preparing and handling free films still apply. The procurement and preparation of free films after weathering is even more challenging; as it is difficult to remove coatings from a substrate for which they have been designed to adhere.

If one can measure or estimate either the internal stress or the fracture energy, one has powerful tool to probe the behavior of a coating and its response to stress and weathering. For example, thicker coatings are more prone to cracking due to the higher driving force that is placed upon thick coatings (Eq. 20.1). Failure envelopes for coatings of various thicknesses can therefore be created to understand the impact of process variables (thickness) on long-term performance.

The rate of embrittlement over time can be measured for a coating by assessing its fracture energy as a function of exposure time [28]. The embrittlement of durable coatings has been found to be mainly due to photooxidation of the binder resin [29]. Hydrolysis plays a less important role in durable coatings, but can certainly be significant in alkyds and other less durable resins. Photooxidation can be quantified readily via IR spectroscopy and other means [30]. The effect of photooxidation on the network has been shown to depend on the presence of certain stabilizer additives, and both increases and decreases of crosslink density have been found [31].

As an example, the fracture energy of an acrylic-melamine automotive clearcoat as a function of accelerated weathering time is shown in Fig. 20.3. The data is

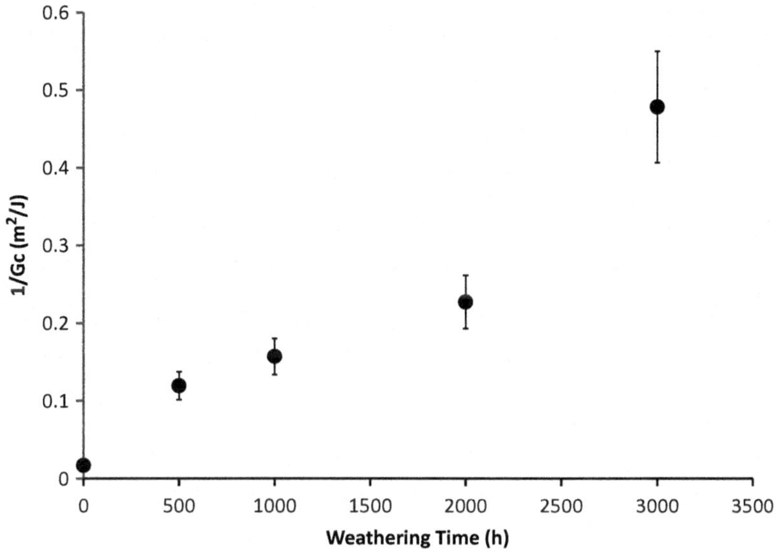

Fig. 20.3 Inverse of the fracture energy of an automotive clearcoat as a function of accelerated weathering time. Weathering protocol was SAE J1960 with borosilicate inner and outer filters

plotted as $1/G_c$ vs. time, as most coatings show an initial rapid embrittlement followed by a slow, long-term embrittlement, and the resulting behavior is approximately linear when plotted in this manner. Data for this clearcoat was collected on full paint systems firmly attached to the substrate, as described in previous work [15]. Crack channeling occurred in the clearcoat alone and was arrested at the top of the basecoat. Independently, the weathering-induced stress in this system was measured via the shim-bending method described above and found to be on the order of 5–6 MPa. The clearcoat whose data are shown in Fig. 20.3 is known to embrittle an unacceptable amount, and will eventually crack during weathering, as the fracture energy of the clearcoat approaches that of inorganic glass after 2000 h of accelerated weathering.

Failure envelopes for this coating can be generated using the data from Fig. 20.3, the thickness of the clearcoat, and the stress that may arise in the clearcoat. The failure envelope, shown in Fig. 20.4, describes the combination of thickness and stress that this coating can tolerate after a given amount of weathering. Each curve represents a particular stress state in the coating, which arises due to the factors previously enumerated. In general, the stress levels will increase with exposure time. Early in the coating's lifetime, the coating is tough and can withstand significant stress before cracking. However, as weathering progresses, the coating embrittles and the stress increases. If one estimates the maximum stress that the coating will experience during its service life, then a maximum film thickness can be calculated, below which the coating will survive. Conversely, if the coating thickness is well controlled, then the lifetime can be predicted, given an estimate of

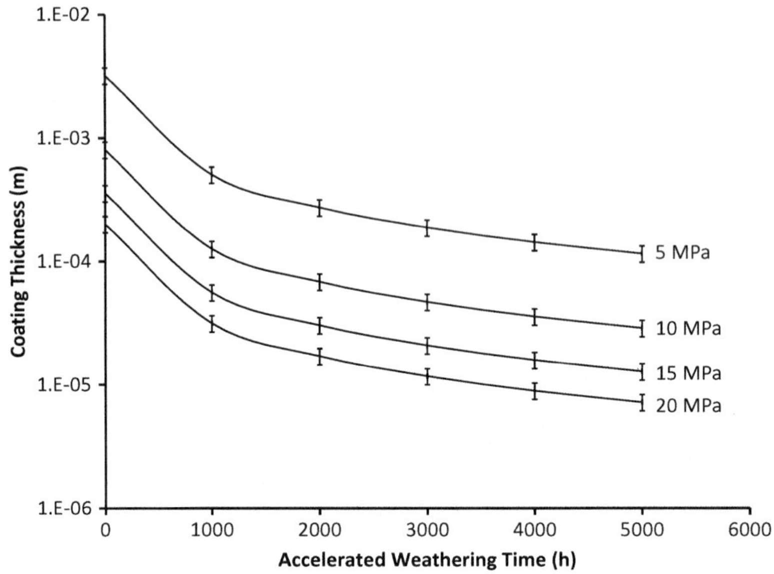

Fig. 20.4 Failure time envelope for an automotive clearcoat whose fracture energy as a function of time is shown in Fig. 20.3. Thicker coatings fail earlier than thinner coatings. Each curve represents a different tensile stress in the coating. Areas above each curve represent combinations of weathering time and coating thickness where cracking is predicted. Below each line, failure is not expected

the stress. For example, a 100 μm thick coating subjected to a 15 MPa stress would fail after less than 1000 h of accelerated weathering. However, a 20 μm coating would be predicted to last over 2500 h. One must, of course, keep in mind that other failure mechanisms need to be considered and that thinner coatings may lead to inferior opacity and other long-term durability issues. Also, while a slow buildup of stress over time is the norm, extreme cold excursion can induce large tensile stresses in most coating systems, and externally applied stresses, such as a stone impact on a vehicle, can induce failures rapidly.

Evolution of Cracking Patterns

The patterns seen in weathered industrial coatings are strikingly similar to those seen in centuries old works of art and in thin layers of dried mud in geological formations. These patterns are all characterized by cellular structures whose density increases as time progresses, which is shown schematically in Fig. 20.5.

The fractal nature of the cracking patterns is a result of the stresses that are generated in the coatings and the subsequent relief of those stresses once a crack propagates. As a material ages, the stress increases and the toughness decreases.

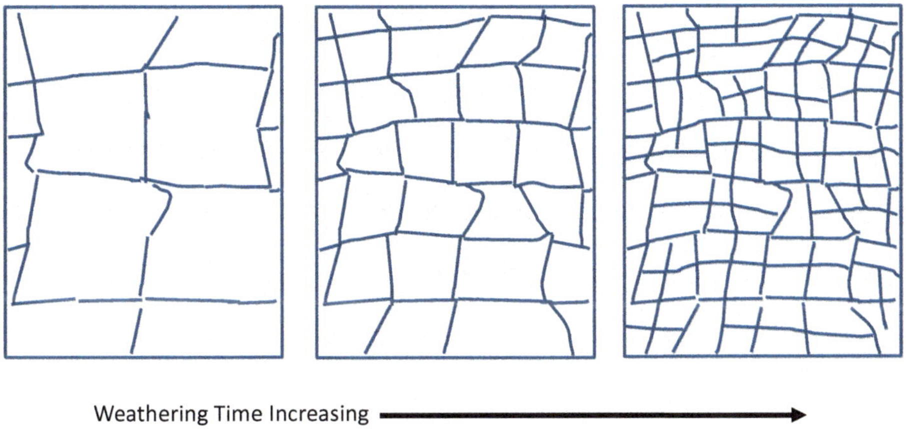

Fig. 20.5 Schematic of the cracking pattern in a coating as time progresses. Increased crack density can be due to either continued embrittlement of the coating and/or increasing stresses in the coating

Eventually, cracks will begin to form as the local driving force, G, exceeds the fracture energy of the coating, G_c. New cracks can form as long as the local driving force is sufficiently high. However, near previously formed cracks, the stresses relax and new cracks will not grow in their vicinity. Thus, an equilibrium crack density will be created, after which no new cracks will grow. For the case of uniaxial tension, the crack spacing is approximated by Eq. (20.2).

As the coating continues to age, however, the stresses will continue to increase and/or the coating will continue to embrittle. Eventually, the driving force between the already-present cracks will increase sufficiently such that again $G > G_c$ of the coating, and new cracks can form approximately at the midpoint between two previously existing cracks. If the coating does not continue to embrittle, the stress must double to allow the crack spacing to drop by ½. Alternatively, if the stress is constant, G_c must drop to ¼ of its original value for the crack spacing to be halved [17, 19, 32]. These relationships are strictly true for coatings stressed in uniaxial tension. When coatings are stressed in biaxial tension, solutions to the stress field around cracks are complex and are typically solved numerically [8, 18, 33]. However, similar trends have been found, and the uniaxial solutions can be used to guide the understanding of the more complex stress states.

Conclusions

The long-term mechanical performance of coatings is critical to the long-term performance of the substrates the coatings protect and beautify. Environmentally induced degradation of most coatings leads to embrittlement, which can be quantified via a variety of means. For failures dominated by cracking, higher stresses and thicker coatings lead to shorter lifetimes, due to the greater driving force for cracking that both variables produce. From a process control point of view, uniform application of coating film builds allows for a more precise definition of coating failure times. Examination of a coating's cracking pattern can provide insight into both the brittleness and stress state of a coating and help to understand the failure mechanisms and strategies to prevent failures in objects painted with industrial coatings as well as modern and historical works of art.

References

1. Chen, J.: On the determination of coating toughness during nanoindentation. Surf. Coat. Technol. **206**(13), 3064–3068 (2012)
2. Wredenberg, F., Larsson, P.-L.: On the stress distribution at scratching of thin film structures. J. Coat. Technol. Res. **7**(5), 623–635 (2010)
3. Boerman, A.E., Perera, D.Y.: Measurement of stress in multicoat systems. J. Coat. Technol. **70** (6), 69–75 (1998)
4. Nichols, M.E., Darr, C.A.: Effect of weathering on the stress distribution and mechanical performance of automotive paint systems. J. Coat. Technol. **70**(885), 141–149 (1998)
5. Perera, D.Y., Schutyser, P.: Effect of physical aging on thermal stress development in powder coatings. Prog. Org. Coat. **24**, 299–307 (1994)
6. Gerlock, J.L., Kucherov, A.V., Nichols, M.E.: On the combined use of UVA, HALS, photo-oxidation, and fracture energy measurements to anticipate the long-term weathering performance of clearcoat/basecoat automotive paint systems. J. Coat. Technol. **73**(918), 45–45 (2001)
7. Shenoy, V.B., Schwartzman, A.F., Freund, L.B.: Crack patterns in brittle thin films. Int. J. Fract. **103**, 1–17 (2000)
8. Liang, J., Huang, R., Prevost, J.H., Suo, Z.: Evolving crack patterns in thin films with the extended finite element method. Int. J. Solids Struct. **40**(10), 2343–2354 (2003)
9. Tuckwell, G., Lonergan, L., Jolly, R.: The control of stress history and flaw distribution on the evolution of polygonal fracture networks. J. Struct. Geol. **25**(8), 1241–1250 (2003)
10. Karpowicz, A.: A study on the development of cracks in paintings. J. Am. Inst. Conserv. **29**(2), 169–180 (1990)
11. Michalski, S.: Paintings—their response to temperature, relative humidity, shock, and vibration. In: Mecklengerg, M.F. (ed.) Art in Transit, pp. 223–248. National Gallery of Art, Washington, DC (1991)
12. Lonyuk, M., Bosma, M., Riemslag, A.C., Zuidema, J., Bakker, A., Janssen, M.: Stone-impact damage of automotive coatings: a laboratory single-impact tester. Prog. Org. Coat. **58**(2-3), 241–247 (2007)
13. Gu, X., et al.: Multiscale physical characterization of an outdoor-exposed polymeric coating system. J. Coat. Technol. Res. **6**(1), 67–79 (2008)
14. Croll, S.G.: Application and limitations of current understanding to model failure modes in coatings. J. Coat. Technol. Res. **10**(1), 15–27 (2012)
15. Nichols, M., Darr, C., Smith, C., Thouless, M., Fischer, E.: Fracture energy of automotive clearcoats—I. Experimental methods and mechanics. Polym. Degrad. Stab. **60**(2), 291–299 (1998)
16. Beuth Jr., J.L.: Cracking of thin bonded films in residual tension. Int. J. Solids Struct. **29**, 1657–1657 (1992)

17. Thouless, M.D., Olsson, E., Gupta, A.: Cracking of brittle films on elastic substrates. Acta Metall. **40**(6), 1287–1292 (1992)
18. Xia, Z.C., Hutchinson, J.W.: Crack patterns in thin films. J. Mech. Phys. Solids. **48**, 1107–1131 (2000)
19. Shorlin, K.A., de Bruyn, J.R., Graham, M., Morris, S.W.: Development and geometry of isotropic and directional shrinkage-crack patterns. Phys. Rev. E. **61**(6 Pt B), 6950–6957 (2000)
20. Perera, D., Tinh, N.: Hygroscopic stress and failure in coating/metal systems. In: Eurocoat Conference, Italy, 1996. Paint, Varnish, Ink and Adhesive Industry
21. Perera, D.Y., Vanden Eynde, D.: Moisture and temperature induced stresses (hygrothermal stresses) in organic coatings. J. Coat. Technol. **59**(748), 55–55 (1987)
22. Bailey, W.J., Endo, T.: Radical ring-opening polymerization and copolymerization with expansion in volume. In: Journal of Polymer Science: Polymer Symposia 1978, vol. 64, pp. 17–26. Wiley Online Library
23. Bailey, W.J.: Synthesis of monomers that expand on polymerization. J. Elast. Plast. **5**(3), 142–152 (1973)
24. Nichols, M.E., Darr, C.A.: The effect of weathering on the stress distribution and mechanical performance of automotive paint systems. In: Bauer, D.R., Martin, J.W. (eds.) Service Life Prediction Methodologies and Metrologies. ACS, Washington, DC (1999)
25. Corcoran, E.: Determining stresses in organic coatings using plate beam deflection. J. Paint Technol. **41**(538), 635 (1969)
26. Shiga, T., Narita, T., Ikawa, T., Okada, A.: Stress monitoring in thin polymer coatings using time resolved fluorescence. Polym. Eng. Sci. **38**(4), 693 (1998)
27. Chan, W.Y.F., Williams, J.G.: Determination of the fracture toughness of polymeric films by the essential work method. Polymer. **35**(8), 1666–1672 (1994)
28. Nichols, M.E.: Anticipating paint cracking : the application of fracture mechanics to the study of paint weathering. J. Coat. Technol. **74**(924), 39–46 (2002)
29. Gerlock, J.L., et al.: Testing accelerated weathering tests for appropriate weathering chemistry: ozone filtered xenon arc. J. Coat. Technol. **75**(936), 35–45 (2003)
30. Gerlock, J.et al.: A brief review of paint weathering research at Ford. In: Martin, J., Bauer, D.R. (ed.) Service Life Prediction Methodologies and Metrologies, vol. 805, pp. 212–249 (2001)
31. Nichols, M., Gerlock, J.: Rates of photooxidation induced crosslinking and chain scission in thermoset polymer coatings II. Effect of hindered amine light stabilizer and ultraviolet light absorber additives. Polym. Degrad. Stab. **69**(2), 197–207 (2000)
32. Thouless, M.D.: Crack spacing in brittle films on elastic substrates. J. Am. Ceram. Soc. **73**(7), 2144–2146 (1990)
33. Hutchinson, J.W., Suo, Z.: Mixed mode cracking in layered materials. Adv. Appl. Mech. **29**, 63–191 (1991)

Chapter 21
Automotive Paint Application

John R. Moore

Introduction

The automotive manufacturing industry has experienced tremendous transformation over the past several decades. Relentless global competition continues to raise expectations for ever-increasing quality without sacrificing price point. The competitive pressure reaches every aspect of the manufacturing process from reliability to fit and finish. As a consequence, the automotive paint industry has seen steadily rising pressure to improve corrosion resistance, appearance, and color glamor, while mounting regulation by environmental organizations has sought to minimize or eliminate solvent usage and waste. The entrance of emerging economies along with their lower labor cost has compounded the urgency to continue the aforementioned progress with higher efficiency. It is perhaps this final pressure that has had the most profound effect for the coating engineer. The drive for higher efficiency has fueled an evolution to progressively more sophisticated paint application systems. An automated paint line demands deeper understanding and control of key process variables than a manual line. Before automation, multiple manual paint operators shared the responsibility for paint quality; with automated systems, the process engineer bears a heavier responsibility. Therefore, it is incumbent on the engineer to use all available scientific understanding for designing and operating today's paint process.

This chapter begins with a brief overview of major adaptations of the automotive coating industry in response to market and regulatory forces and includes a description of the steps in a typical modern automotive paint production line. The second section deals with fundamental concepts and equations that should be readily understood and applied by process engineers tasked with the responsibilities of

J.R. Moore (✉)
Axalta Coating Systems, Coatings Technology Center, Wilmington, DE, USA
e-mail: john.r.moore@axaltacs.com

M. Wen, K. Dušek (eds.), *Protective Coatings*, DOI 10.1007/978-3-319-51627-1_21

designing or managing a paint process. The final portion of the chapter has been devoted to a technical review of process variables and equations related to pneumatic and high-speed electrostatic rotary atomization.

The Evolution of Automotive Coating Application from 1980 to Today

Over the last three decades, the technology of coating application has been driven to increasingly higher level of efficiency by two major market forces: environmental regulation [1] and global competition [2]. The National Environmental Policy Act was officially adopted by Congress in 1970, and by the end of the decade, the Clean Air Act required states to implement "reasonably available control technology" (RACT)[1] for all sources of emissions. At the same time, the US auto industry began feeling the pressure of Japan's increasing production with high-quality and lower cost offerings. Together these forces propelled much of the US auto industry's process improvements including coating efficiency.

The first reaction of US auto manufacturers to the regulatory pressure was to drive coating suppliers toward the technology containing fewer volatile organic compounds (VOCs). The paint industry responded with "high solids" solventborne paints and eventually waterborne alternatives. However, as the EPA standards continued the drive toward lower emissions, the need for other methods of reduction became vital.

The gap between the reasonable lower limit of paint VOC and the EPA mandate created an opportunity for equipment suppliers to offer innovative solutions. Two nearly universally accepted options focused on the beginning and end of paint process. At the exhaust end, VOC abatement equipment such as thermal oxidizers became standard equipment in new paint shops and available as a retrofit in existing plants. Thermal oxidizers destroy VOCs in solvent-laden air coming from the spray booths and ovens by passing it through a very high temperature (760–870 °C) before returning it to the atmosphere.

At the spray end, the technology for electrostatic painting, which had grown in acceptance after its introduction by Ransburg in the 1940s, came into the mainstream. Applying a charge to atomized paint droplets from a pneumatic spray gun increases the proportion of droplets that deposit on a grounded car body. The portion of coating that deposits on the substrate (car body) in relation to the total sprayed amount is considered transfer efficiency. For comparison, a conventional spray gun without electrostatics has a transfer efficiency of about 15–40%.

[1]The United States Environmental Protection Agency (EPA) defines RACT as "the lowest emission limitation that a particular source is capable of meeting by the application of control technology that is reasonably available considering technological and economic feasibility" 44 FR 53761 (Sept. 17, 1979).

Fig. 21.1 Robotic application of clearcoat by rotary atomizer [3]

However, when a charge is applied, the transfer efficiency can increase to 65%. Since less paint needs to be sprayed to achieve the desired film thickness, fewer VOCs are emitted from the process.

By the 1970s and 1980s, the industry began adopting an even more efficient electrostatic paint atomizer technology, known as the electrostatic rotary bell (ESRB) atomizer. It was an adaptation of electrostatic rotary disc technology for the automotive paint industry. Rather than using air jets to cause breakup, droplets are generated by feeding the paint onto the inner surface of a spinning electrically charged cup. As the paint becomes charged, it is propelled toward the grounded car body with the assistance of a ring of air jets that surround the cup edge. This technology is capable of delivering 75–85% of the paint [4] to the intended target under typical manufacturing conditions, leaving only the remainder as waste. Manufacturers quickly perceived the inherent benefit of this highly efficient mode of coating and began integrating bells into their paint lines. By the mid-1990s, ESRB technology had become common in auto manufacturer's paint shops for painting the exterior body panels, and today they are adapted to fit on robotic arms for automatic painting of both interior and exterior surfaces of the vehicle (Fig. 21.1).

The early part of the new millennium brought the coating suppliers back as the lead technology drivers with the development of "Harmonized[2] Coating Technologies®." These are advanced coatings designed to work together as a system

[2]Axalta Coating Systems defines Harmonized Coating Technologies as formulations or application systems designed to increase productivity, save energy, and reduce costs in vehicle manufacture.

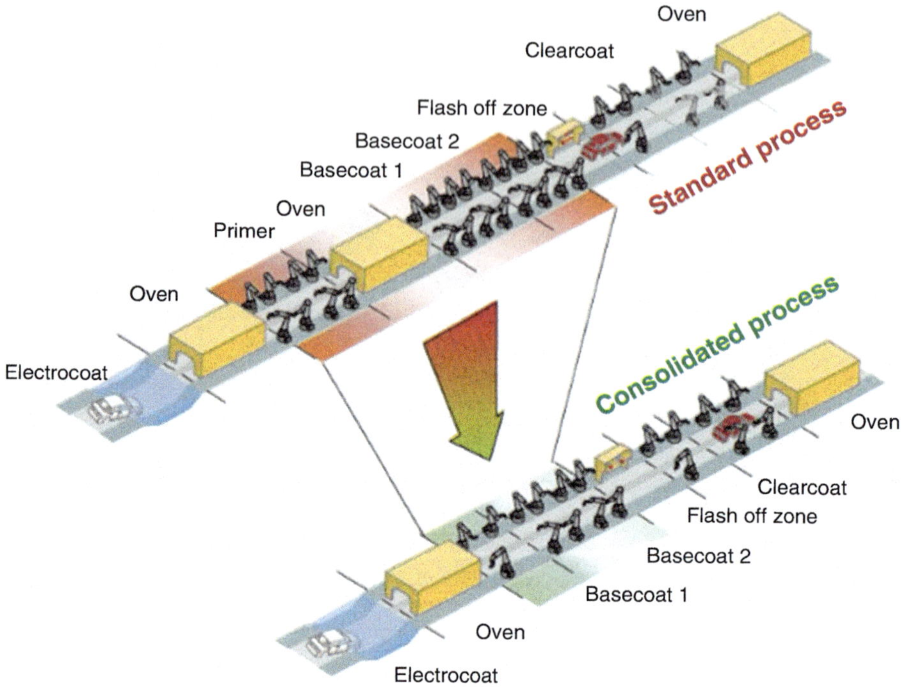

Fig. 21.2 An example of a consolidated paint system made possible by a Harmonized Coating Technology®

allowing primer, basecoat, and clearcoat to be deposited in a wet-on-wet fashion with a single bake. In comparison, a typical pre-2000 paint shop had a primer system consisting of a booth with rotary atomizers followed by a bake oven. After bake, the primed body passed to a second booth for color coat (basecoat) which was followed by clearcoat application and bake. By leveraging advancements in the science of coatings and rheology, the same layers can now be applied one after another without the requirement of baking in between. This development provides car manufacturers the option of eliminating the primer spray booth and oven, thereby leading to reduced energy consumption and CO_2 generation. Figure 21.2 shows an example of paint shop simplification made possible by one variety of Harmonized Coating Technology.

While US manufacturers were adjusting to the new processes required to meet EPA mandates, they were also facing pressure to improve quality as Japanese imports became known for their attention to detail. W. Edwards Deming is credited with indoctrinating Japanese manufacturers in the rigor of statistical process control as they rebuilt after World War II [3]. Manageable processes designed and controlled to produce minimal defects became a hallmark of the Japanese auto industry. Through the pressure of competition, Deming's teachings eventually made their way into the USA and European automotive coating processes. Up to the early

1980s, paint quality metrics consisted mainly of film thickness measurements and subjective visual appearance assessments. By the turn of the century, highly repeatable and sophisticated measurements of gloss, surface texture, and multi-angle color had become commonplace. Today, robots with millimeter scale accuracy for position and speed control have largely replaced humans as paint applicators. Similarly, automated vision systems are now capable of capturing appearance measurements from each car body panel for statistical process control (SPC). Over the last three decades, auto manufacturers, coating suppliers, and equipment providers have had a share in the transformation from the art of manual painting into the science of automated coating application.

Typical Automotive Coating Sequence

Automobile factories are typically designed with three major manufacturing areas. The first area is the body shop. In this section, body panels of steel and aluminum are welded and assembled together to form the complete body shell. The completed body is passed by conveyor to the paint shop where it is cleaned and coated with rust-inhibiting undercoats and attractive topcoats. A typical paint process sequence is shown in Fig. 21.3, where a car body goes through the conversion coating, electrocoat, primer, basecoat, and the clearcoat processes. The painted body is then sent to the final assembly area where it receives the drivetrain, interior, and all the details that make it a complete vehicle.

Conversion Coating

Car bodies enter the paint shop as bare metal shells after being stamped and welded together in the body shop. This shell is cleaned and rinsed in preparation for deposition of a phosphate coating, which forms as the body is immersed in a dilute solution of phosphoric acid and phosphate salts. The chemical reaction of the mild acid with metal causes the dissolved salts to fall out of solution and form a layer of

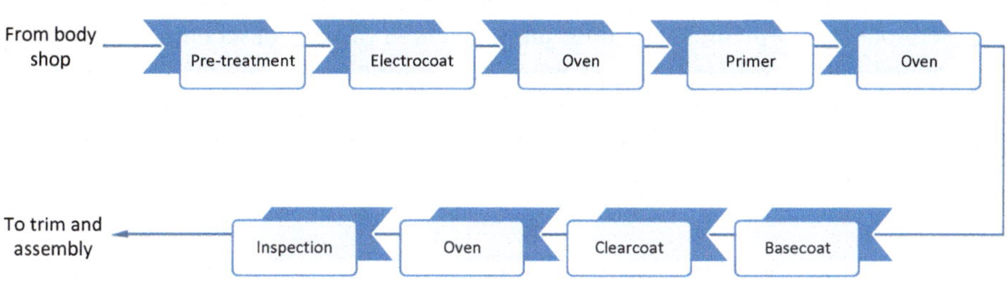

Fig. 21.3 A typical sequence of processes involved in painting a car body

insoluble crystalline phosphates on the metal surface. The phosphate layer becomes the foundation for adhesion of all subsequent layers of coating to the metal substrate. After further rinsing, the phosphated body receives a primer by immersion in a waterborne suspension known as electrocoat.

Electrocoat

In this process, electrically charged particles of paint are deposited from a water suspension onto a submerged body. Although electrocoating may be carried out when the part to be coated is used as the anode or cathode, the cathodic configuration provides superior corrosion resistance and has therefore become the industry's process of choice [5]. In cathodic deposition, the body is grounded and the bath receives a positive DC charge from electrodes within the electrocoat tank. Nichols and Tardiff explain that the positively charged paint particles migrate to the cathode (the car body) where water in the solution is reduced, resulting in hydrogen evolution and hydroxyl formation. The hydroxyl ions neutralize the charged paint particles, and they precipitate onto the surface of the car body [6]. According to Ohm's Law, electrical current will preferentially take paths of least resistance. Therefore, in this process, the coating is drawn to the grounded metal directly in front of the electrode first and then spread to cover all the other bare metal regions as the path to the ground is impeded by the deposited film. The ultimate coating thickness is regulated by the voltage to the electrodes. Once sufficient thickness of coagulated electrocoat deposits to form an insulating barrier, electrical deposition can no longer occur. In the final steps of the process, the body is removed from the bath, rinsed, and sent to a baking oven [8] (Fig. 21.4).

Primer, Basecoat, and Clearcoat

The immersion process for the conversion coating and electrocoat are intended to form films that cover both the inner and outer surfaces of the metal body parts in order to provide adhesion to the phosphated metal and corrosion protection. Once the electrocoat is deposited, all subsequent layers of primer, basecoat, and clearcoat are applied to the exterior only as sprays. These sprays are generated from either a pneumatic atomizer or electrostatic bell. High pressure spray equipment is not typically used due to the complexity and safety concerns surrounding paint systems pressurized to 8000 kPa (1200 psi).

Fig. 21.4 Electrocoated car bodies in inspection area [8]

Practical Application

This section is dedicated to practical considerations for process engineers when preparing for a new coating process. A grasp of these basic concepts and equations will help the engineers design for the proper number of atomizers (rotary or pneumatic guns) while thinking through process variables such as gun speed and index, which impact cycle time and required paint gun flow rate.

Uniform Film

The goal of every automotive spray coating process is to achieve the desired film thickness with minimum waste and imperfections while obtaining maximum uniformity. Variations in coating thickness may result in a number of appearance and flow-related defects. While most people are familiar with orange peel that may occur if film thickness is too low and sags if it is too high, the specifications for today's high-performance automotive finishes go far beyond these two properties. Along with chip resistance, scratch resistance, acid resistance, color fastness, adhesion, and flexibility, they are also required to achieve optical qualities rivaling that of a mirror. It is the emphasis on appearance quality and durability that requires precise control of both the uniformity of the deposited coating and the manner in which it is deposited.

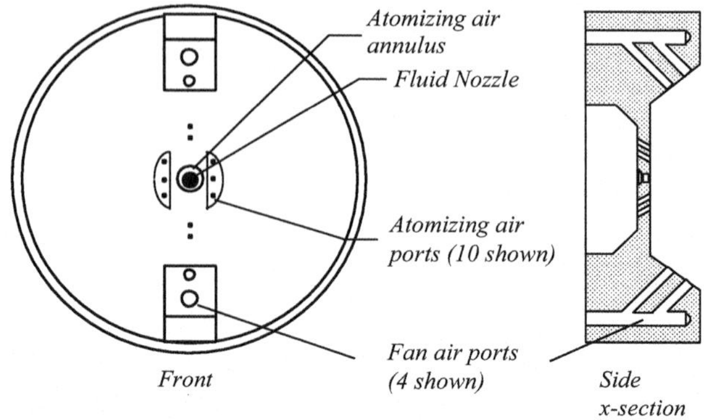

Fig. 21.5 Typical OEM air cap for pneumatic atomizer

In order to generate a uniform film of paint on a large surface, it is important to optimize the spray pattern emitted from the atomizer. Both pneumatic guns and rotary bells distribute the atomized paint into a pattern by the use of air jets. Guns use a combination of "atomizing" and "fan air," and bells use "shaping air" to adjust pattern diameter. The geometry of the spray jet becomes the "brush" that deposits the paint on the intended target. In the case of a pneumatic atomizer, the pattern is produced from an "air cap" (Fig. 21.5) located at the exit of the fluid nozzle. The air cap has an annulus encircling the fluid nozzle that creates an air jet parallel to the direction of liquid flow. The interaction of this high-speed air jet with the fluid column begins the work of atomization and directs the stream toward opposing fan air jets with the assistance of small atomizing air holes drilled on the face of the air cap. Air from the annulus and small drillings is referred to as atomizing air. As the partially atomized stream reaches the intersecting fan air jets, the liquid column is further atomized and flattened into a fan-shaped pattern. The shape of the resulting atomized paint fan pattern can be predictably adjusted by regulating the combination of paint flow rate, atomizing air, and fan air.

We mapped the two-dimensional air jet velocity from a typical OEM air cap by traversing an anemometer at a distance of six inches downstream from the nozzle (Fig. 21.6a). To make the measurement, the air cap was positioned with the fan air jets arranged vertically as shown in Fig. 21.5. The lower image is a "static fan pattern" which is a topographical map of dried paint film thickness deposited on a steel panel positioned 12 in. downstream of the nozzle. Graphing the measured film thickness provides information concerning the relative droplet density within the fan pattern.

Several observations can be made from these two images. In the upper image, the chosen settings for atomizing and fan air have resulted in an air jet with two regions of higher air velocity seen to the left and right of center with somewhat lower velocity at the centerline. The outer regions of the pattern show steadily decreasing velocity due to the transfer of momentum to the surrounding ambient

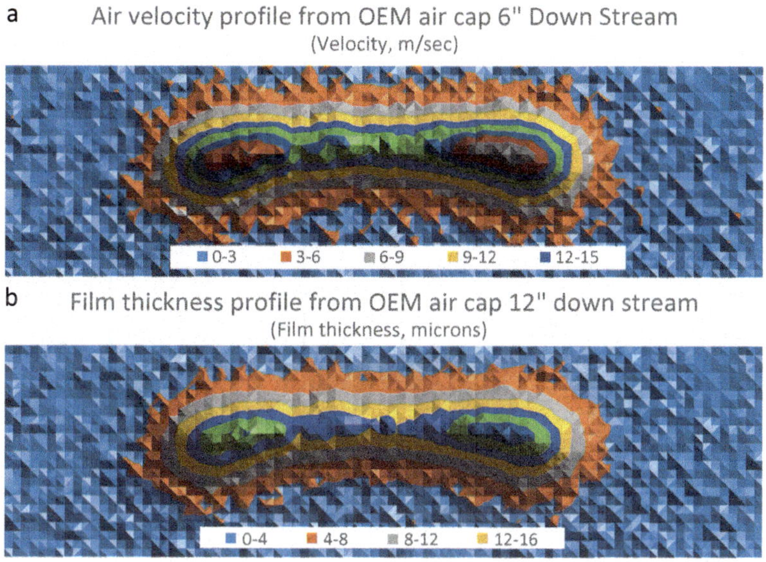

Fig. 21.6 Contour profile of air velocity (**a**) and static fan pattern (**b**)

air. It is evident from the lower image that the atomized paint droplets are not evenly distributed. Regions of greater paint thickness correspond to areas of higher velocity as seen in the upper chart. This phenomenon can be explained by the Bernoulli principle which states that higher speed of an inviscid flow of a nonconducting fluid occurs simultaneously with a decrease in pressure. Therefore, as the atomized droplets are propelled toward the target within the spray jet, they tend toward regions of lower pressure which correspond with regions of higher velocity air. A more uniform film deposition is possible by adjusting the atomizing air, fan air, or paint flow rate. In this case, a slight decrease in fan air volume will eliminate the fan splitting and result in a more uniform pattern. Although an example is beyond the scope of this chapter, the interaction of paint flow rate, atomizing air, and fan air to form a fan pattern can be statistically modeled though standard design of experiment (DOE) methodology.

With the static fan pattern obtained in Fig. 21.6b, we can determine the thickness of a sprayed coating applied by a process where the pattern moves across the panel to be painted with vertical strokes represented by the black lines of Fig. 21.7. At the end of each stroke, the spray gun is shifted or "indexed" by a fixed distance (as shown by the gray arrows) to prepare for the next stroke. The strokes continue until the pattern is off the opposite edge of the panel. This painting technique is referred to as the stroke and index process and is commonly used for robotic application. The strokes may go horizontal, vertical, or even diagonally across the part being painted, but the strokes should remain parallel to each other with a constant index distance between each one. The goal is to deposit a uniform film of paint from edge to edge by overlapping the pattern through a series of parallel

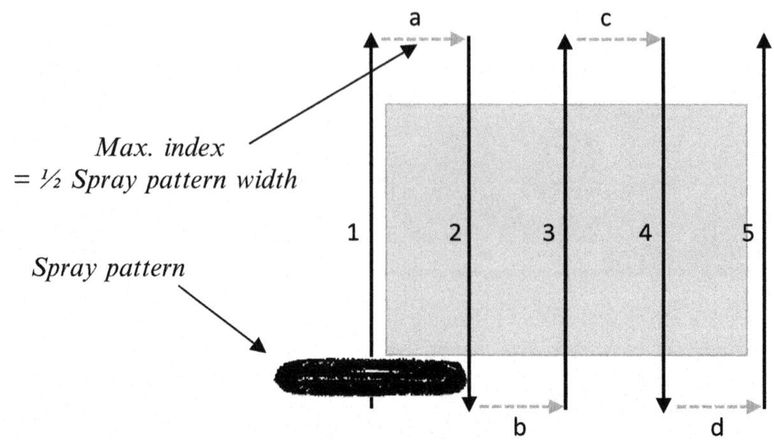

Fig. 21.7 Panel sprayed with vertical strokes (1–5). Each successive stroke is indexed to the right by a fixed distance (a–d) until reaching the end of the panel

strokes, which progress from one end of the part to the other. In order to supply sufficient overlap of the pattern, the maximum index distance should be no larger than half of the spray pattern width as shown above.

It is desirable to design a paint process with minimal film thickness variation and if possible to do it with minimal use of expensive auto parts for verification. The following "off-line" process can be used to optimize thickness uniformity. Final thickness uniformity over a flat painted panel is a function of the film thickness deposited from a single stroke of the spray gun and the amount of overlap between successive strokes [9]. Figure 21.8 is a graph of the film thickness profile deposited by a single stroke using the same air and paint flow rate of Fig. 21.6b. The profile thickness data were obtained by setting a pneumatic gun to the desired atomizing air, fan air, and flow rate settings. Then, the gun was triggered while moving across a test panel at a customer defined speed and distance. After baking the panel, film thickness measurements were taken in regular intervals across the dried coating to map the film thickness from edge to edge in order to create the film thickness vs. position profile. Figure 21.9a shows the film thickness profile of Fig. 21.8 indexed three times over four strokes (light gray lines). The dark black line above the profiles represents the final film thickness and is obtained by summing the thickness values of each individual stroke. Therefore, it can be used to provide an estimate of the uniformity that can be expected from the process. Since the total pattern width is 800 mm, the 400 mm index distance shown represents 50% overlap

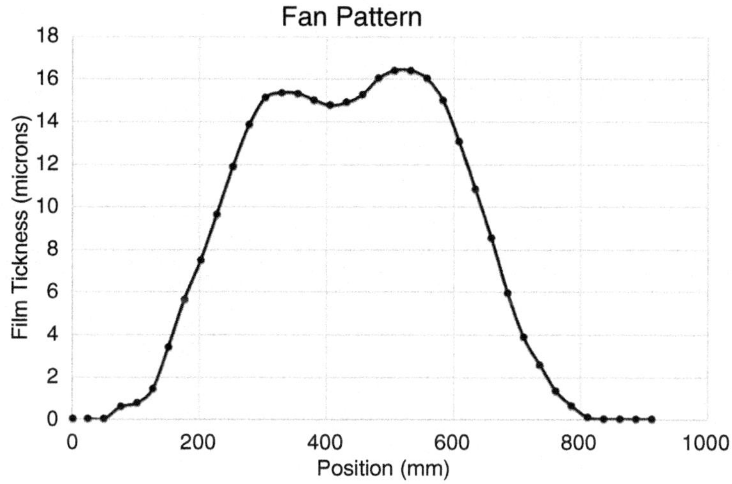

Fig. 21.8 Film thickness profile of a single stroke using the same air and paint settings as Fig.21.6b

of sprayed coatings between two adjacent spray strokes. With this fan profile and 50% overlap, one can expect a 5 μm variation from peak to trough. The same profile with 66% overlap gives a 12 μm variation (Fig. 21.9b). Here, the index is 266 mm or 1/3 of the pattern width. Figure 21.9c shows the variation expected for 75% overlap (200 mm index) drops to 2.5 μm. This example illustrates the impact of overlap on final film uniformity and describes an off-line method for optimizing uniformity of a spray process. These graphs also demonstrate how average film thickness (black line) is directly proportional to overlap.

The preceding example shows it is possible to optimize film thickness uniformity by optimizing the process index. More frequently, however, the process index is not the first variable to consider changing. Large robotic paint paths are time-consuming to create, and a change to the index requires a significant amount of work. For this reason, engineers will typically choose to optimize the fan profile in order to minimize variability. By decreasing the fan air setting used in Fig. 21.8, it is possible to generate the profile shown in Fig. 21.10. Notice how this profile lends itself to 50% overlap in Fig. 21.11.

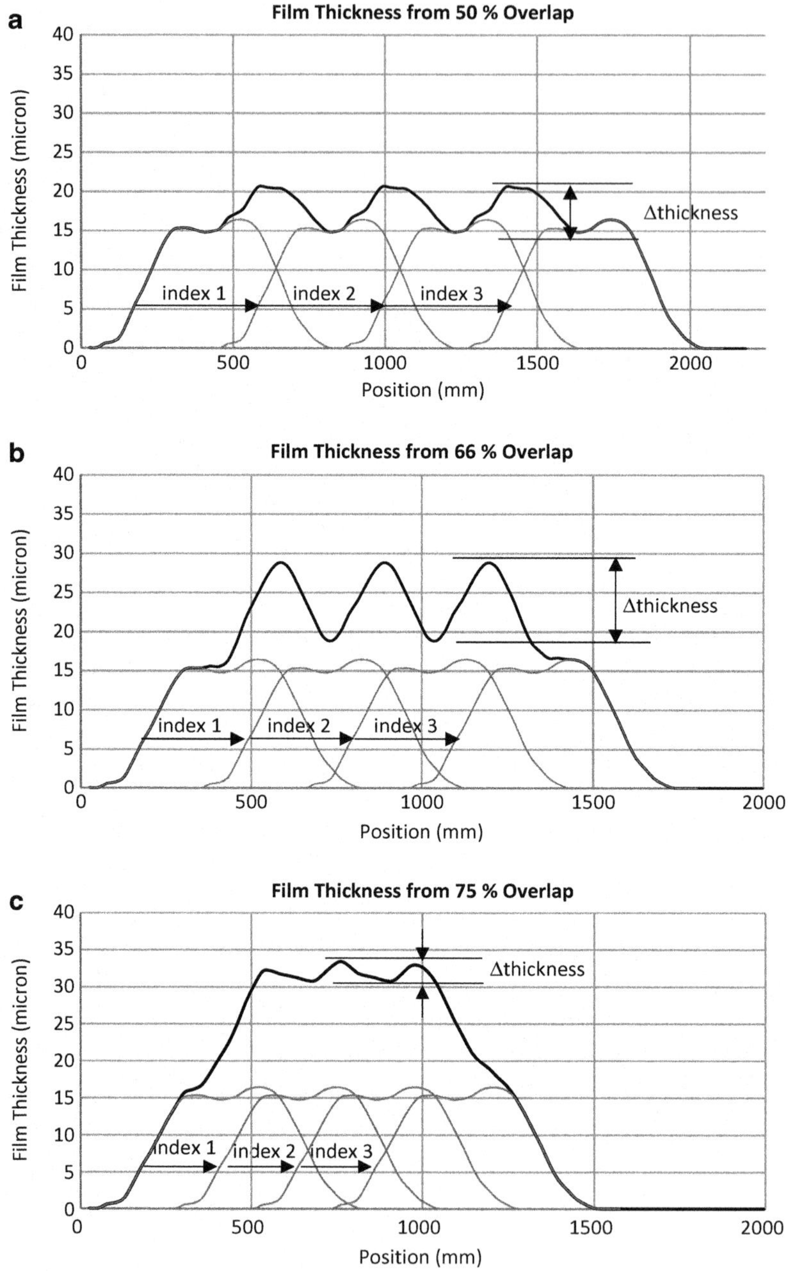

Fig. 21.9 Graphical depiction of resultant film thickness as a function of fan pattern overlap. (**a**) 50% overlap, (**b**) 66% overlap, (**c**) 75% overlap

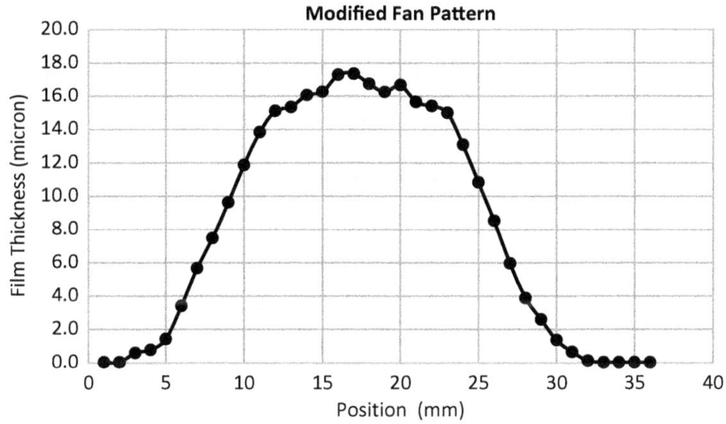

Fig. 21.10 Film thickness profile obtained by modified fan pattern

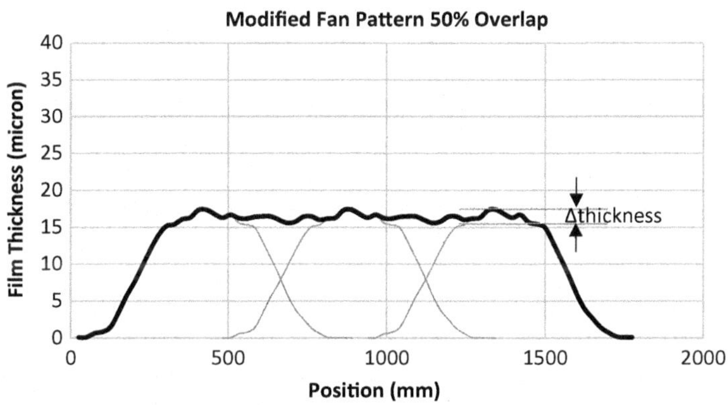

Fig. 21.11 Improved film thickness uniformity at 50% overlap

Predicting Deposited Film Thickness from Application Parameters

After an optimized fan profile and the resultant overlap profile are determined, a final process adjustment entails increasing or decreasing gun speed or paint flow rate in order to achieve the recommended film thickness. All colors in automotive basecoats have a recommended target film thickness based on the ability to achieve "hiding" or opacity. Hiding power is a measure of the capability of a coating to hide a colored background. Some colors like silver metallic or black will become opaque at low film thickness, while other colors like red or yellow will require thicker film

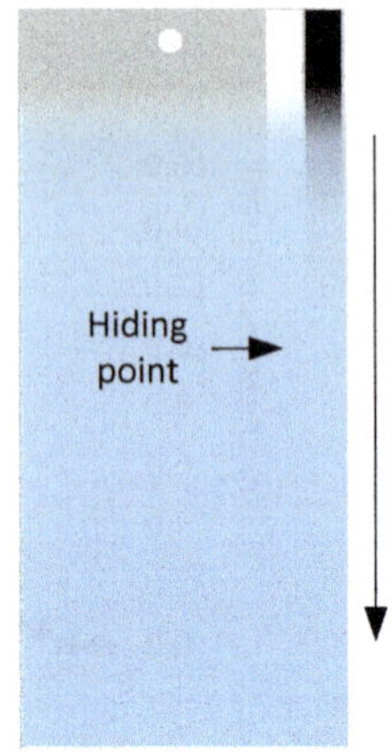

Fig. 21.12 A wedge paint applied over a black and white test strip to measure hiding film thickness

for hiding. The hiding film thickness for a specific color is quantified by spraying a film with a gradual change of thickness from thin to thick, called a wedge paint, over a black and white test strip. The dry thickness at which an observer can no longer see a color difference between the black and white side of the test strip is referred to as the hiding film thickness. For most colors, the hiding film thickness is also the recommended process film thickness. However, for colors that achieve hiding at very low film thickness, there will typically be a minimum specification to ensure optimum paint properties. The paint supplier can supply a film thickness requirement for their coating system. Once the hiding values are known, the engineer must consider how to tune the process flow rate in order to achieve the specified final film thickness for each color. The following example demonstrates a method for approximating process flow rate based on process and paint specifications (Fig. 21.12).

Flow Rate Calculation Example

Figure 21.13 shows a 150 mm wide paint roller that is coating a wall with no overlap between strokes. It is rolling on 100% solids paint with 100% transfer efficiency (no loss) at a speed of 800 mm/s and index of 150 mm. Assume there is no reaction during drying of the coating and no shrinkage caused by the pure binder due to physical aging. Calculate the flow rate needed to deposit 25 μm in two equal coats given:

- Roller width (pattern width) = 150 mm
- Roller (robot tip) speed = 800 mm/s
- Index (distance between successive strokes) = 150 mm

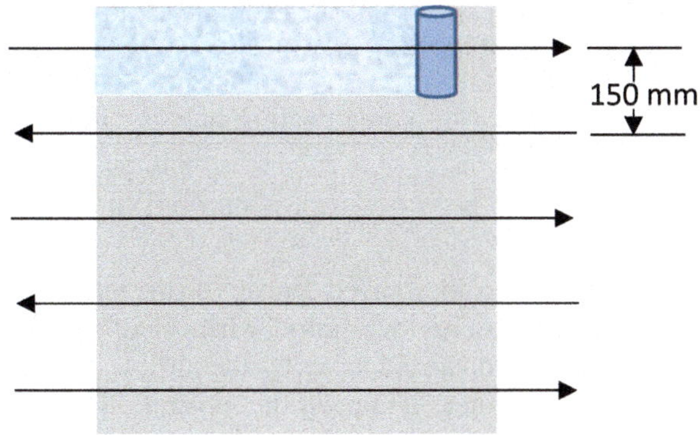

Fig. 21.13 Roller coating example

- 0% overlap
- Paint volume solids = 100%
- Transfer efficiency = 100%

Solution: Since the paint is 100% solids, there is no solvent evaporation to cause film shrinkage while drying, and therefore the volume of paint deposited equals the final film volume.

$$\text{flow rate} = \text{area coating rate} \times \text{film thickness per coat} \tag{21.1}$$

$$\text{Area coating rate} = \text{robot speed} \times \text{pattern width} = 120,000\ \text{mm}^2/\text{s} \tag{21.2}$$

$$\text{flow rate} = 120,000\ \text{mm}^2/\text{s} \times 0.0125\ \text{mm} = 90\ \text{cc/min}$$

In reality, most paints are not 100% solids. Typical automotive OEM basecoats are 15–50% volume solids as sprayed. This percentage represents the volume of the sprayed paint that remains to form the paint film after drying. The other 50–85% is volatile consisting mainly of solvent (or water in the case of waterborne paint) used to modify viscosity for optimal atomization, flow, and leveling. As volatiles evaporate from the film surface, the volume of paint film decreases accordingly resulting in decreased film thickness. Film shrinkage will continue until all volatiles are evaporated. For example, assuming there is no further shrinkage occurring from the cure reaction of the solid binder part, a 30 μm thick film at 50% volume solids will shrink to approximately 15 μm after all volatiles are removed. It should be noted that in reactive coating systems, shrinkage caused by the crosslinking reaction during cure may need to be considered. For instance, in acrylic-melamine clearcoat, the amount of alcohol released can be as high as 15% weight loss [10], and the corresponding volume shrinkage can be significant. However, for initial estimation of the flow rate required to achieve a specified film thickness, it is reasonable to ignore shrinkage from crosslinking.

– Assume the paint from our example is 50% volume solids and film shrinkage from crosslinking is negligible[3]:

$$\text{flow rate} = \frac{\text{area coating rate} \times \text{film thickness per coat}}{\text{volume solids}} \tag{21.3}$$

$$\text{flow rate} = 180\ \text{cc/min}$$

Not all of the paint sprayed deposits on the intended target (the car body). A fraction of the spray is lost as "overspray" when small droplets from the atomization process are entrained and carried away by the airflow generated by the deflected air stream from the spray gun as it impacts the substrate. The ratio of coating solids deposited to the amount of coating solids used during the coating process is called transfer efficiency. Note that coating solids may be measured by either volume or weight since the density of the solids remains constant. This equation uses the "as sprayed" volume solids which may be obtained from the coating supplier.

– Assume only 50% of the paint solids make it from the atomizer to deposit on the panel (50% transfer efficiency).

$$\begin{aligned}\text{flow rate} &= \frac{\text{area coating rate} \times \text{film thickness per coat}}{(\text{volume solids}) \times (\text{transfer efficiency})} \\ &= 360\ \text{cc/min}\end{aligned} \tag{21.4}$$

Finally, successive strokes from an atomizer typically have overlapping fan patterns. That is, the pattern may be 150 mm wide, and the index between coats may be 75 mm, yielding a 50% overlap. In this case, each portion of the target is essentially sprayed twice in a single coat.

– Using this final equation, it is possible to estimate the flow rate needed for any coating process. The area coating rate and percent overlap can be obtained from a process audit or design criteria.

$$\begin{aligned}\text{flow rate} &= \frac{\text{area coating rate} \times \text{film thickness per coat} \times (1 - \text{OL})}{(\text{volume solids}) \times (\text{transfer efficiency})} \\ &= 180\ \text{cc/min}\end{aligned} \tag{21.5}$$

where

$$\text{OL} = 1 - \left(\frac{\text{index}}{\text{fan width}}\right) \tag{21.6}$$

[3]Use the decimal form of all percentages in each of the following equations.

As previously mentioned, most paints contain some volume of solvent that evaporates to form the hardened film. For this reason, the calculations of transfer efficiency are based on the portion of the paint that remains behind (nonvolatiles). ASTM D5286 provides the standard method for measuring the transfer efficiency of a process based on the weight of paint solids deposited relative to the weight of paint solids sprayed. The calculation requires knowledge of the paint weight solids, measurement of the weight gain on a sprayed part after bake, and measurement of the weight of paint sprayed as shown in the following example.

Transfer Efficiency Example Calculation

A paint sample with weight solids of 0.5 (50%) is sprayed from a gravity feed spray gun with initial weight (paint plus gun) of 1000 g. All spray is directed toward a 100 g large flat panel as the gun traverses and indexes inside the borders of the panel. When spraying is complete, the panel is baked to remove all volatiles and weighed at 250 g. The spray gun with remaining paint is reweighed at 400 g. Calculate the transfer efficiency of the gun.

$$\text{total solids deposited} = (250 - 100) = 150\ \text{g} \tag{21.7}$$

$$\begin{aligned}\text{Total solids sprayed} &= \text{Total paint sprayed} \times wt\ \text{solids}\\ &= (1000 - 400) \times 0.5 = 300\ \text{g}\end{aligned} \tag{21.8}$$

$$\%\text{transfer efficiency} = \left(\frac{\text{total solids deposited}}{\text{total solids sprayed}}\right) \times 100 = 50\% \tag{21.9}$$

Figure 21.14 demonstrates the relative effect of two process variables, speed and target distance, on transfer efficiency. Panels were sprayed by a standard stroke and index process with a gun mounted on a machine programmed for three levels of traverse speed and three levels of gun to target distance. Transfer efficiency decreased linearly with distance, while gun speed demonstrated no clear trend

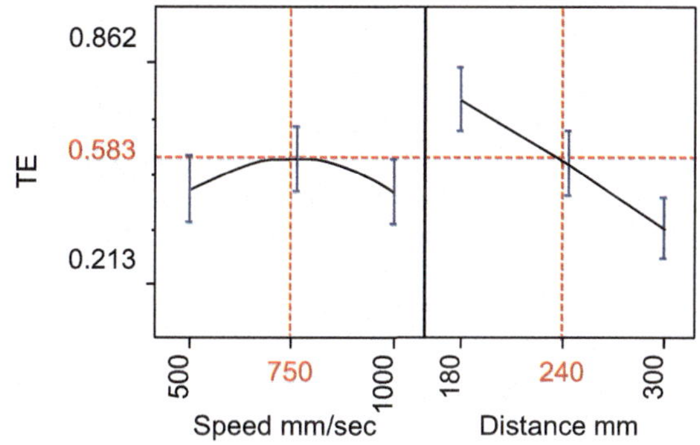

Fig. 21.14 Transfer efficiency vs. speed and distance

Table 21.1 Typical ranges of transfer efficiency for common automotive atomizers [11]

Atomizer	TE Range (%)	Initial estimate for panel painting (%)
Conventional air spray	15–40	30
HVLP/compliant	30–60	50
Electrostatic air spray	40–80	65
Electrostatic bell	70–95	85

over the tested range. Other variables such as air pressure for a gun, bell rotation speed for a rotary atomizer, or even booth air velocity may also influence the transfer efficiency of a process and can be quantified by a similar DOE. However, for simplicity, Table 21.1 contains approximate values for use as a starting point. These estimates are for "panel painting." It should be assumed that transfer efficiency will decrease when painting small or complex parts due to increased overspray.

Wet Film Nonvolatiles

It is often desirable to measure the attributes of an applied wet film from a coating operation for the purpose of problem solving or to enable comparisons to other processes. Consider the deposition of a multi-layer process: A primer coating gets applied by a rotary atomizer spinning at 60,000 rpm on the door of a car body to a wet film thickness on the order of 50–70 μm. This primer coating may be 50% weight solids as sprayed with the remaining 50% consisting of solvents and other volatile components. However, depending on the conditions of the application process, the film generated by the depositing primer droplets may measure 80% solids or more simply due to the evaporation of solvent from the droplets during the spray (atomization) process. In this case only 20% of the wet film is solvent available to assist with flow and leveling. On the other hand, if the bell rotational speed is reduced to 25,000 rpm, the droplets will become larger with proportionally less surface area available for evaporation[4], and the resulting film may be only 65% solids (35% volatile). This increase of solvent content in the film may result in smoother appearance due to better flow, but it may also lead to a sag or other process defect.

A similar effect can be produced by changing the distance between the atomizer and the intended target. In this case, the droplets will have different travel time to get to the target providing a varying window for solvent to evaporate on the way. Not surprisingly, booth environmental conditions and booth air velocity can also have an effect. For all these reasons, it is helpful to have a means to quantify the solids of the film. This metric is known as "wet film nonvolatiles" or WFNV.

[4]Large droplets have lower surface area to volume ratio than small droplets.

$$\%\text{solvent in the film} = \left(\frac{\Delta\text{weight}}{\text{wet film weight}}\right) \times 100 \tag{21.10}$$

$$\%\text{WFNV} = 100 - \left(\frac{\Delta\text{weight}}{\text{wet film weight}}\right) \times 100 \tag{21.11}$$

where Δ weight = wet film weight (after spray) – dry film weight (after bake).

The measurement is very similar to a weight solids measurement, except the paint is applied to a pre-cut and weighed foil which is attached to a part during the coating operation and removed immediately after deposition is complete. The wet foil is weighed and baked. The change in baked weight vs. initial wet film weight provides the weight percent of volatiles in the wet film. This parameter can be used as a fingerprint of the application process and is helpful for statistical process control or as a means to compare the deposited film of one application process to another.

Atomization

Atomization can be defined as any process whereby a bulk liquid is converted into small droplets. Figure 21.15 shows atomization as the result of a large droplet impacting a surface. The inertia of the drop impact overcomes viscous and surface tension forces to flatten the spherical droplet into an expanding sheet until a maximum diameter is reached. Beyond this point, the sheet begins to contract under the influence of surface tension forming cylindrical ligaments. These ligaments quickly stretch and thin due to inertia of the mass at the free end of the

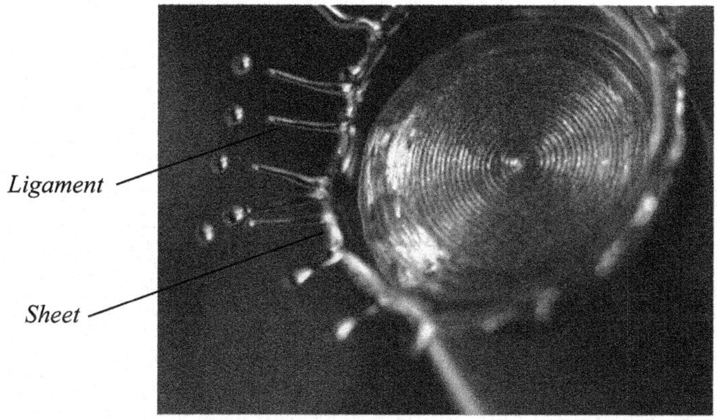

Fig. 21.15 A water droplet impacting a raised surface shows the formation of a liquid sheet, ligaments, and droplets, $T_R = 3$

Fig. 21.16 A paint droplet with $T_R \approx 2900$ impacting a raised surface. Ligaments persist much longer than a Newtonian liquid due to higher extensional viscosity brought on by the presence of polymer

ligament and capillary force seeking to minimize surface energy. Once the capillary force exceeds extensional force inside the ligament, the cylindrical sections collapse and droplets are produced. The magnitude of extensional force that a ligament can withstand before rupture and droplet pinch off is related to its extensional viscosity. Newtonian liquids like water or glycerol water solutions have extensional viscosity three times their shear viscosity. This relationship is known as the Trouton ratio (T_R). When a small amount of polymer is added to a Newtonian solution, T_R can increase by several orders of magnitude [12]. The result of increased extensional viscosity is evident in stretched ligaments and delayed droplet pinch off as seen in Fig. 21.16. For spatter coatings, incomplete atomization is desirable. However, for automotive coatings requiring high-quality appearance, the atomization process must result in discrete droplets that coalesce on the part to be painted in order to form a smooth film. For the sample in Fig. 21.16, a more energetic mode of atomization is required to produce droplets.

Pneumatic Atomization

Marmottant and Villermaux proposed a three-stage mechanism [13] for atomization of a low velocity cylindrical liquid jet surrounded by a high-velocity annular air jet (Figs. 21.17 and 21.18). When the liquid jet exits the nozzle, the liquid surface experiences high shear induced by the annular flow of air, thus generating wave crests on the surface of the jet known as Kelvin-Helmholtz instability. Then, as the wave crests grow, air gets pulled into the liquid column causing Raleigh-Taylor instability. Finally, ligaments are formed and stretched until they detach from the liquid core, eventually forming droplets as the result of Raleigh-Plateau instability. Literature refers to this type of nozzle as a pneumatic or co-flowing airblast atomizer [14]. For typical automotive paint spray guns, a similar "atomizing air" annulus is followed immediately by opposing fan air jets which flatten the accelerating paint cloud into a fan-shaped pattern while assisting the atomization process.

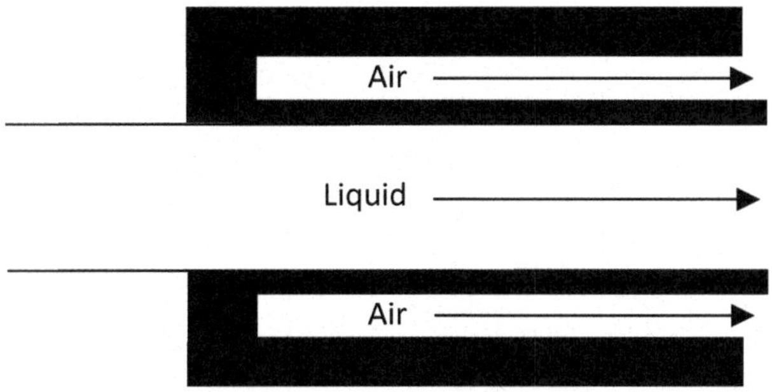

Fig. 21.17 Cross-section of an airblast atomizer showing a liquid jet with a high-velocity annular co-flowing air jet. This is similar to the atomizing air from a typical automotive air spray gun

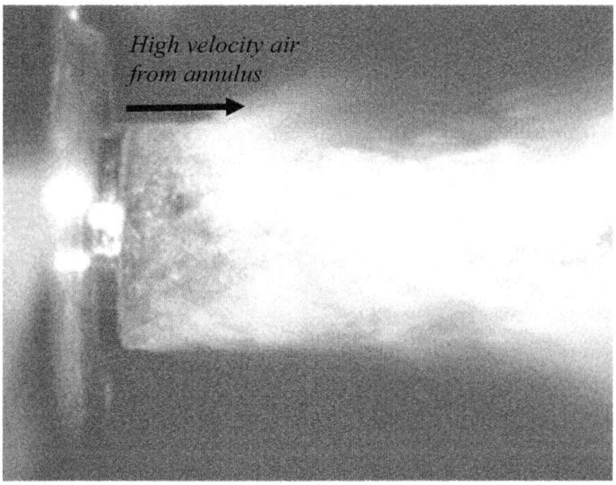

Fig. 21.18 Atomization from an airblast atomizer commonly used for paint spray. High-velocity air exits annulus surrounding the liquid jet shearing off droplets

A histogram of droplet sizes from a typical pneumatic atomizer for a waterborne automotive basecoat is shown in Fig. 21.19. The distribution plot shows a polydisperse spray with an arithmetic mean diameter of 21.5 μm and the median diameter of 18.9 μm. Keshavarz et al. indicate that the distribution of droplet diameters is best described by a gamma distribution [15].

The droplet sizing shown above was obtained by LaVision® ParticleMaster using a technique known as shadowgraphy [16]. This method employs a pulsed laser to backlight the spray while a synchronized camera takes images through a long-distance microscope. The bright light and short pulse of the laser freezes motion providing non-blurred images of individual droplets in flight. Figure 21.20 shows such an image with two exposures produced by a double pulse from the laser. Arrows indicate the distance traveled over the time between laser pulses. Thus, shadowgraphy provides information on both droplet sizes and velocity.

Since a spray consists of a range of droplet sizes, it is difficult to verbally convey all the information contained in the distribution plot. For this reason, it is common to simplify the entire series of droplet sizes to a single parameter such as the arithmetic mean or median diameter. However, neither of these parameters are

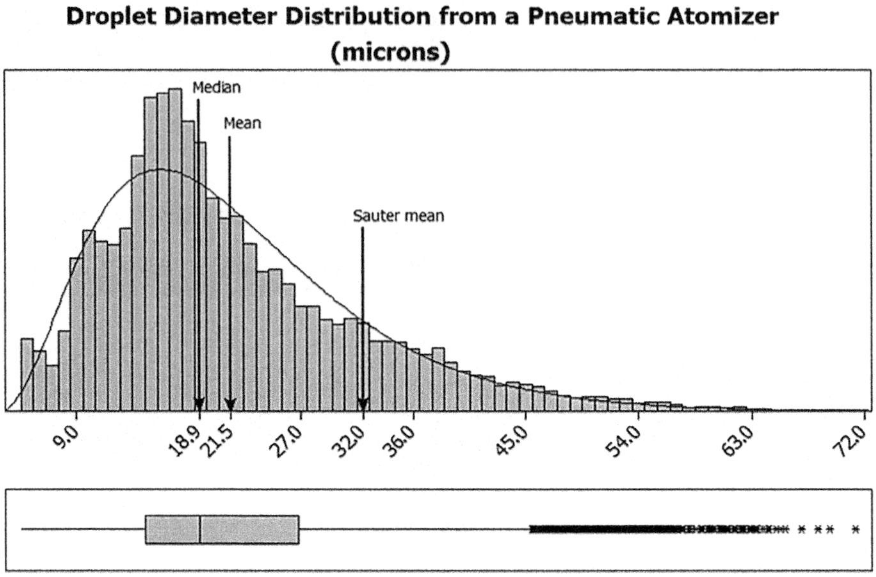

Fig. 21.19 Size distribution of droplets sprayed from a typical automotive pneumatic atomizer with automotive waterborne basecoat. Viscosity = 30 mPa·s. A gamma distribution is used to fit the data

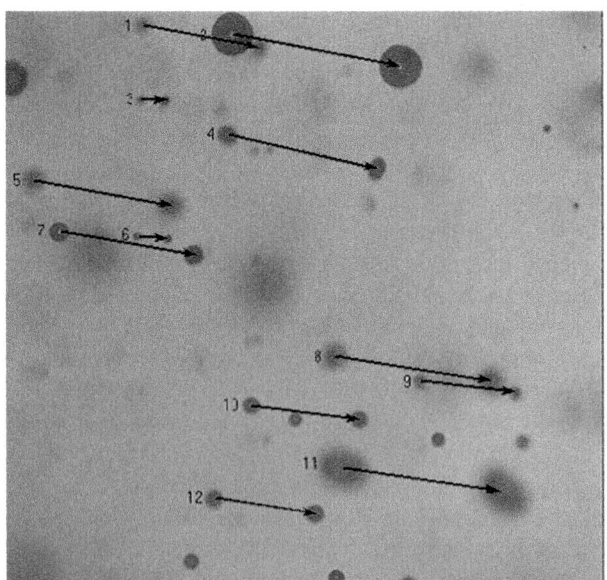

Fig. 21.20 Image of paint droplets 10 cm downstream from a typical pneumatic atomizer. Image taken by shadowgraphy with double-pulsed laser. *Arrows* indicate distance traveled in 10 ms

particularly sensitive to a small fraction of large droplets. Large droplets from paint atomizers are frequently the source of defects in automotive finishes. For this reason, it is important to include them into the characterization so they may be minimized. For example, an impact crater may be formed when a large droplet splashes onto a previous wet coating layer. The energy of the collision causes a visible depression and rim as the name implies. Alternately, large droplets deposited prior to a subsequent coating layer may result in a noticeable loss of appearance by introducing surface texture. The Sauter mean diameter (SMD) is a more helpful parameter to describe a paint spray because it is more sensitive to the presence of large droplets. It is defined as the diameter of a droplet that has the same ratio of volume to surface area as the entire spray. As such it is a more meaningful parameter for comparing the fineness of a spray than the mean or median diameter. SMD is calculated based on the following equation [14]:

$$\mathrm{SMD} = \frac{\sum N_i D_i^3}{\sum N_i D_i^2} \tag{21.12}$$

Where i denotes the ith diameter size range considered, N_i is the number of droplets in the ith size range, and D_i is the middle diameter of the ith size range.

Rizk and Lefebvre (1989) [14] developed an empirical equation for a "twin fluid air blast atomizer" helpful for providing insight to the factors involved in pneumatic atomization:

$$\begin{aligned} \mathrm{SMD} = {} & 0.48d\left(\frac{\sigma}{\rho_{\mathrm{gas}} U^2 d}\right)^{0.4}\left(1+\frac{1}{\mathrm{ALR}}\right)^{0.4} \\ & + 0.15\mathrm{d}\left(\frac{\mu_{\mathrm{liquid}}^2}{\sigma d \rho_{\mathrm{liquid}}}\right)^{0.5}\left(1+\frac{1}{\mathrm{ALR}}\right) \end{aligned} \tag{21.13}$$

where

d = nozzle diameter
σ = liquid surface tension
ρ_{gas} = density of air
ρ_{liquid} = density of liquid
U = air velocity
ALR = air/liquid mass ratio
μ_{liquid} = viscosity of liquid

This equation predicts that increases in surface tension or viscosity will result in higher SMD and thus a coarser spray. Conversely, increasing air to liquid mass ratio will decrease SMD. Figure 21.21 shows statistical output from an experiment investigating the relationship between air/liquid ratio and Sauter mean diameter of the spray for a common automotive spray gun. Our data has acceptable agreement with Eq. 21.13 confirming SMD is directly proportional to paint flow rate and inversely proportional to the volume of air used for atomization.

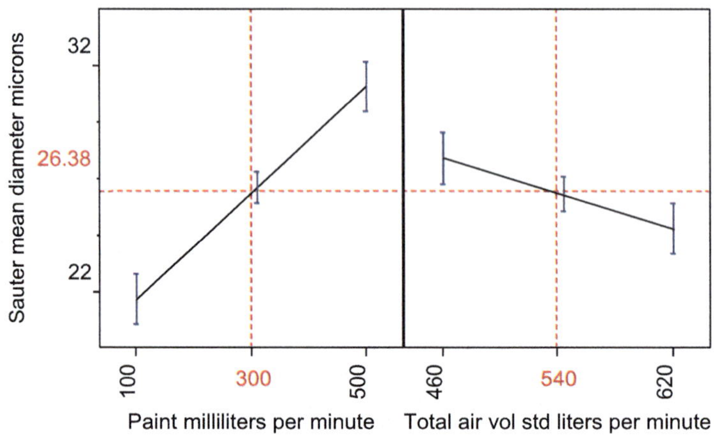

Fig. 21.21 Sauter mean diameter of droplets vs. paint flow rate and air volume

Rotary Atomization

For an electrostatic bell, paint is fed through the back of an electrically charged spinning cone called a "bell cup" (Fig. 21.22). As the paint contacts the inner surface of the bell cup, it spreads into a film and acquires a charge while being accelerated to the edge by centrifugal force. On contact with air, the film breaks down into ligaments and ultimately droplets. The momentum imparted to the ligaments causes them to spiral outward in an involute trajectory as they leave the edge. Within the ligament, capillary forces are balanced by elastic forces determined by the rheology of the coating. Once capillary forces dominate, surface tension works to break the ligament into multiple droplets. These droplets retain momentum due to their ejection from the bell edge causing them to spread radially outward.

If left undisturbed, the plume of the bell spray would have a very large diameter with little forward velocity as droplets sling from the cup. However, the atomizer is fitted with a ring (or rings) of holes behind the bell cup (Fig. 21.22). Shaping air flows from these jets and over the back of the spinning bell cup, thus directing the atomized liquid toward the target. Increasing the volume of shaping air acts to decrease the diameter of the pattern. The lower limit for pattern diameter is determined by the size of the bell cup and angle at which the shaping air deflects around it. Typical shaping air volume flow rate ranges from 100 to 500 slpm (standard liters per minute), and typical pattern diameters are 15 to 50 cm. Im et al. [17] examined the spatial distribution of droplets emanating from a bell and conclude that larger droplets stay on the outside of the fan pattern due to their inertia. Smaller droplets have greater tendency to become entrained into a recirculation zone that exists in front of the bell cup. von Karman and later Sohrab [18, 19] explained that rotation of a spinning disk results in an axial flow toward the disk to replace the fluid (air) being pumped radially outward by centrifugal forces (Fig. 21.23a).

Fig. 21.22 Bell cup and shaping air ring for rotary atomizer (*right*) bell cup with serrations (*left*)

Fig. 21.23 (**a**) An inward radial flow is induced by a rotating disk. (**b**) Small droplets become entrained into the recirculation zone in front of rotary atomizer caused by the combination of rotating cup and shaping airflow

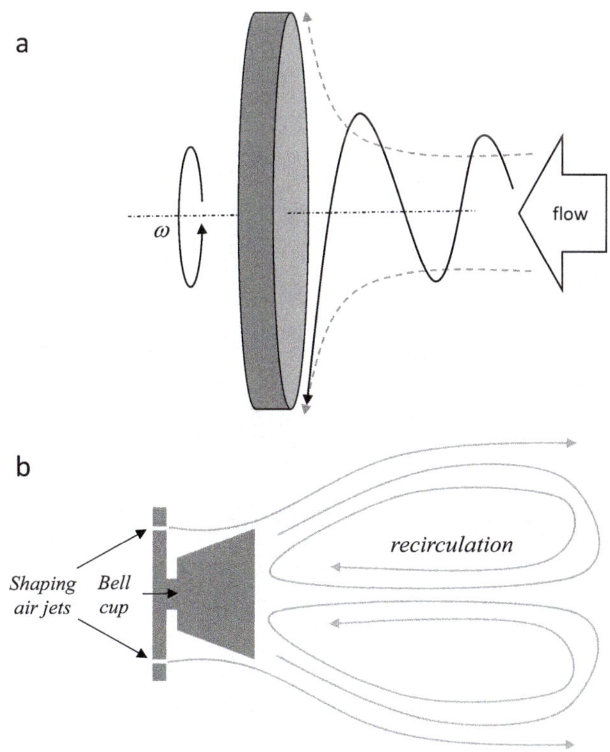

The rotation of the bell cup on an ESRB produces a similar flow. The combination of the inward radial flow and the flow of shaping air over the bell cup results in a recirculating flow (Fig. 21.23b).

The inner surface of a bell cup may be smooth or have machined lines called serrations that extend approximately one to 2 mm in from the outer edge of the cup. The purpose for the serrations is to intentionally form ligaments as paint leaves the bell edge. That is, the serrations are designed to prohibit paint from escaping the bell cup as a film. Under low flow rate conditions, there is little noticeable difference between the serrated and non-serrated photos, pictures A and D in Fig. 21.24. Both show relatively even spacing between ligaments, and ligament length appears similar. However, under high paint flow rate at the same bell rotational speed, the serrated bell cup retains regular spacing between the ligaments (picture B in Fig. 21.24), while the non-serrated bell cup (picture E in Fig. 21.24) shows more random spacing and irregular ligament length. Pictures C and F in Fig. 21.24 demonstrate the effect of high flow rate combined with increased viscosity. For both serrated and non-serrated cups, higher viscosity produces longer, more stable ligaments and postpones disintegration into droplets; however, the serrated cup retains ligament mode breakup (picture C in Fig. 21.24). Film mode is clearly evident for the non-serrated cup (picture F in Fig. 21.24).

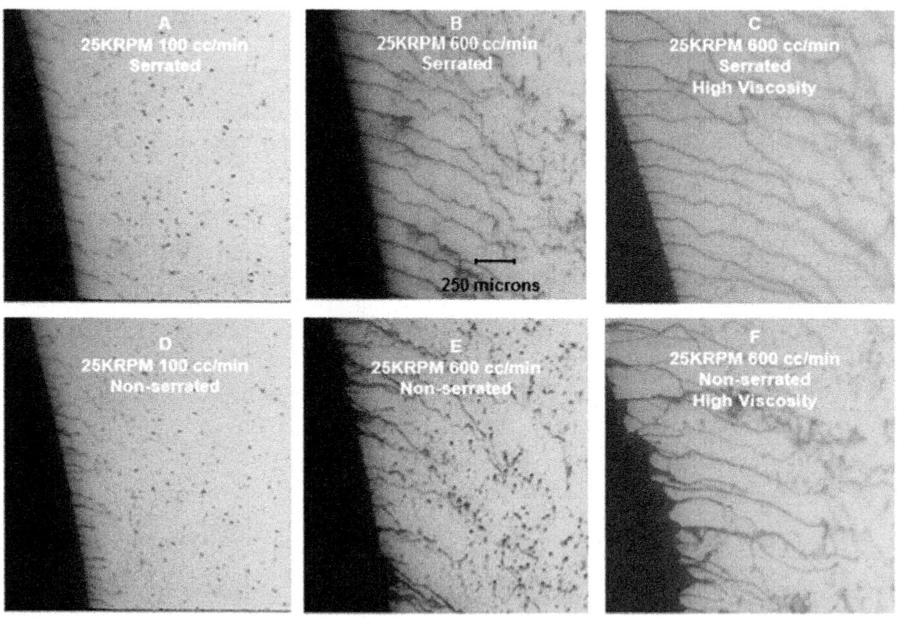

Fig. 21.24 Waterborne basecoat from serrated (**a–c**) and non-serrated bell cups (**d–f**) at low and high paint flow rates. (**a**, **b**, **d**, and **e**) basecoat viscosity = 30 mPa·s. (**c** and **f**) basecoat viscosity = 60 mPa·s

Waterborne and solventborne basecoats can be effectively deposited under the film mode of atomization; however, clearcoats will frequently fail from a defect known as "air entrapment." Domnick modeled the flow field between the stationary fluid tip and rotating center section of a bell cup and concluded that the chaotic flow in this region can possibly result in the formation of air bubbles in the paint issuing from the bell cup, which get deposited along with the paint droplets in the paint film [20]. Whether these bubbles remain in the deposited film or are able to migrate out through the air/liquid interface is dependent on the rheological properties of the paint. Hendrix et al. present a second possible mechanism for the inclusion of air in a deposited film. Aerodynamic drag on a large droplet in a spray generates a local increase in pressure on the leading edge of the droplet producing a concave deformation that results in an entrapped air bubble upon deposition [21]. In either case, the air becomes buried in the film and can potentially get trapped during cure. The ensuing defect appears as multitudes of visible spherical voids in the film on the order of 0.25–0.5 mm in diameter. The presence of serrations that extend to the edge of the bell cup so as to result in ligament mode breakup is frequently effective at mitigating this defect. It should be noted that some bell cups marketed as serrated may have serrations that stop short of the edge thus opening the possibility for film mode breakup.

Droplet size was measured near the outer edge of the fan pattern produced from a 65 mm non-serrated bell cup spinning at 20 Krpm with paint flow rate of 360 mL/min and shaping air volume flow rate of 465 slpm. Images were captured by LaVision® Shadowgraphy using an OEM solid white, waterborne basecoat. The distribution shown in Fig. 21.25 is monomodal with a mean diameter of 55.8 μm and Sauter mean diameter of 68.6 μm. Domnick similarly observed a monomodal distribution from a

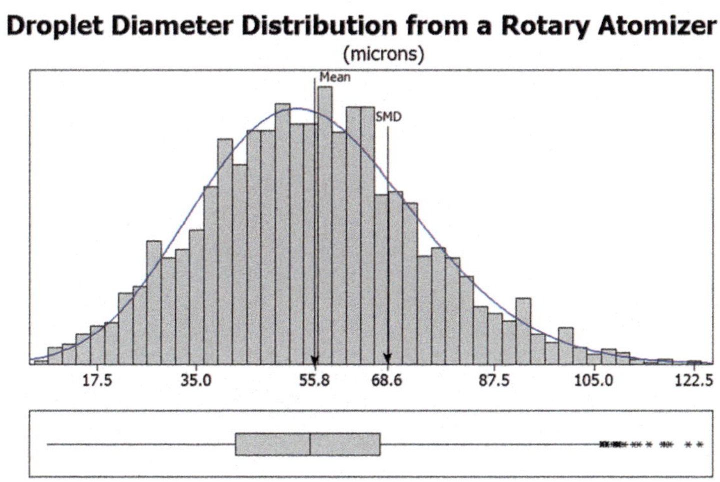

Fig. 21.25 Droplet size distribution from rotary atomizer spraying waterborne basecoat. Non-serrated bell cup, 360 mL/min, 20 Krpm, viscosity = 60 mPa·s

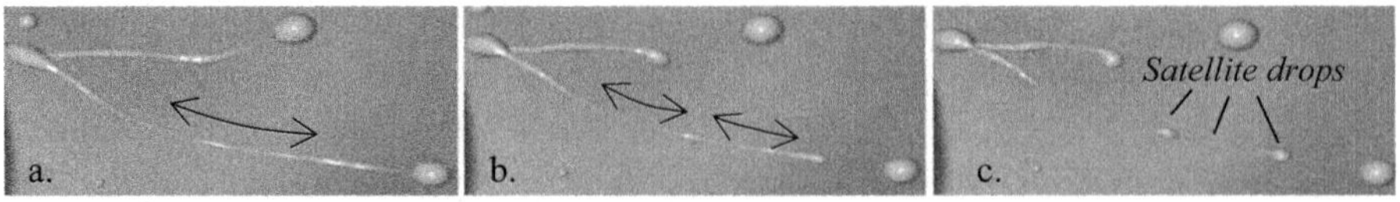

Fig 21.26 Ligament stretch resulting in the formation of satellite drops. *Arrows* indicate direction of capillary flow

non-serrated bell cup at both low and high rotational speeds. However, he described bimodal distributions from serrated bell cups as the rotational speed increased [22]. Basu et al. also reported a bimodal distribution from a serrated 65 mm bell cup using a typical OEM solventborne silver basecoat [23]. The bimodal distributions may be explained by the production of satellite drops. Bousfield et al. studied the necking behavior and pinch off of Newtonian and viscoelastic liquids and concluded satellite droplets are formed due to the nonlinear effects of a collapsing liquid filament [24]. Inertial effects in a stretching ligament prevent the liquid in the ligament from draining back to the main drop before pinch off occurs. The image series in Fig. 21.26 shows the production of satellite droplets. As the ligament stretches, capillary forces work to push the liquid back to the main drop on either side while extensional forces prevent droplet pinch off (Fig. 21.26a). When capillary forces exceed extensional forces, the ligament breaks (Fig. 21.26b). Remaining liquid fragments retract due to surface tension and form satellite drops (Fig. 21.26c).

The modal difference noted in the droplet size distributions from serrated and non-serrated bell cups may be an artifact of the uniform ligament formation imposed by the presence of serrations. The uniform ligaments from a serrated bell cup may regularly produce a distribution of main and satellite drops, thus the bimodal distribution. Non-serrated bell cups generate less uniform ligaments as seen in E and F in Fig. 21.24, leading to a more normal droplet size distribution.

Tanasawa et al. modeled the atomization from ligament and film mode and developed equations for each [25]:
Ligament mode:

$$\mathrm{SMD} = .119\,\frac{Q^{0.1}\sigma^{0.5}}{Nd^{0.5}\rho_{\mathrm{L}}^{0.4}\mu_{\mathrm{L}}^{0.1}} \tag{21.14}$$

Film mode:

$$\mathrm{SMD} = 15.6\frac{Q^{0.5}}{N}\left(\frac{\sigma}{d^2\rho_{\mathrm{L}}}\right)^{0.4} \tag{21.15}$$

where
d = bell diameter
N = rotational speed (rev/s)

σ = liquid surface tension
Q = volumetric flow rate
μ_L=liquid viscosity
ρ_L = liquid density

We substituted typical solventborne properties in Eqs. (21.1, 21.4, and 21.15) for the graph in Fig. 21.27 and waterborne basecoat parameters in the graph shown in Fig. 21.28. In both cases, Domnick's results from viscoelastic paints generated larger droplet size than what is predicted from Tanasawa's Newtonian model.

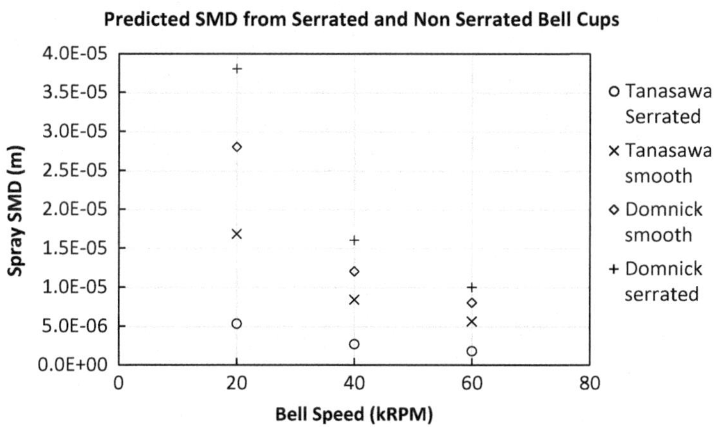

Fig. 21.27 SMD predictions from Tanasawa equations for serrated and non-serrated cups using typical solventborne basecoat property data. Values from Domnick added for comparison [22]

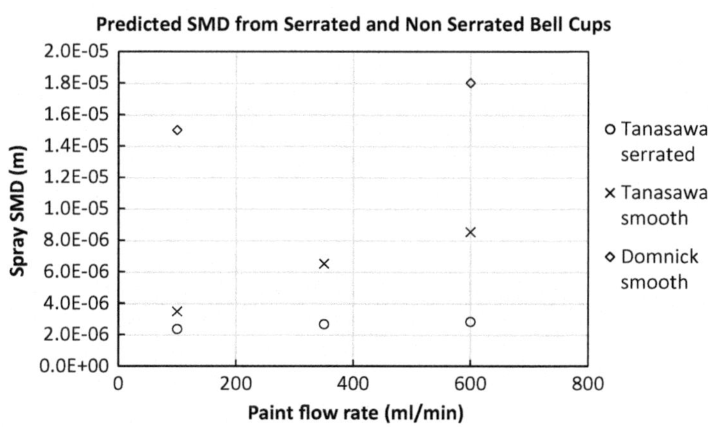

Fig. 21.28 SMD predictions from Tanasawa equations for serrated and non-serrated cups using typical waterborne basecoat property data at 40 Krpm. Values from Domnick added for comparison [26]

Nevertheless, both agree that bell rotational speed is the dominant factor for controlling droplet size. Domnick's data contradict Tanasawa's prediction that serrated bell cups would produce smaller droplets than the smooth ones. The reason for the discrepancy may lie in a difference in the mode of atomization as summarized by Ahmed and Youssef [26]. Tanasawa assumes sheet/film mode atomization from a smooth bell cup with Newtonian liquid. If the flow rate and bell speed in combination with the rheology of the clearcoat used in Domnick's study produced ligaments instead of a sheet, smaller droplets may be expected.

Tanasawa predicted ligament mode atomization to be very robust to flow rate changes while film mode was expected to produce larger droplets as flow rate increases. A more recent study by Domnick [27] examined the atomization of a waterborne basecoat by a rotary atomizer. Plotting his results for a smooth cup along with Tanasawa's predicted values shows the rate of increase in SMD with increasing flow rate is similar.

The topic of atomization continues to be an area of interest for industry and academia. For a more complete treatment of the dynamics of jet breakup, please refer to Eggers et al. [28].

Electrostatic Deposition by Rotary Atomizer

During the electrostatic rotating bell (ESRB) atomization process, the paint is given an electrostatic charge of 40,000–90,000 V. The charge effectively increases the fraction of paint that deposits on the target by creating an attractive force between the charged paint droplets and grounded car body. If the paint being sprayed has a relatively high electrical resistivity, which is typical in solvent-based paints, the charge may be applied directly to the bell cup. According to Dominick et al. [29], the direct charge method generates an electric field between the bell cup and the grounded car body, and the paint receives a charge through direct charge exchange as it flows over the edge of the cup. However, if the coating is too electrically conductive, e.g., waterborne paints, applying a direct charge on it will lead to an electrical short as the voltage is free to travel up the conductive paint in the supply tube to the nearest electrical ground. To avoid this, high-conductivity paints are charged by an indirect charge method [30]. With indirect charge, there is no direct contact between paint and the electrode. This method employs an external array of finger-like electrodes which are located symmetrically around and just behind the bell cup. High voltage at the electrodes produce corona discharge, and droplets from the bell cup become charged as they pass through the ion flow in the air between the electrodes and grounded body. There is general agreement that the lower threshold of paint resistivity for the direct charge method is approximately 500–800 Ω. Less resistive coatings will generally require the indirect charge method.

Summary

Increasing expectations from quality savvy car buyers combined with the drive to ever greater manufacturing efficiency continue to supply motivation for further automation in the automotive paint area. Nevertheless, the move to automation is incomplete unless accompanied by fundamental understanding of paint atomization and coating processes. The basic concepts of film uniformity and film thickness control have been discussed. These two principles are foundational for both design and continuous improvement efforts relating to a production line. The flow rate equation should be used in combination with known cycle time constraints to determine the number of atomizers required for a new paint operation. Additionally, coating application parameters such as required tip speed and index can be rapidly assessed by its use. Whether application is by bell or gun, film uniformity may be improved by optimizing the fan profile in conjunction with the programmed index. Taking the time to determine the relationship between paint flow rate and atomizer set points on fan profile can save countless hours attempting to correct problems at line side. Finally, our review of simple equations related to jet breakup from pneumatic and rotary atomizers has sought to convey the contribution of paint characteristics and physical applicator parameters on spray droplet size distributions. Taken together these concepts and learnings aim to improve comprehension of the fundamental relationships between paint and process for the purpose of advancing toward defect-free and attractive automotive finishes.

Acknowledgments The author recognizes technicians Scott P. Ryan, Doug Zachwieja, and Andrea Robinson for their contribution to the experimental work on spray characterization and high-speed photography. Thanks also to Axalta European colleagues Peter Minko and Sven Radek for background on consolidated coating system as well as North American colleagues Joseph Maglich, Mike Koerner, and Eric Houze for reviewing and correcting content. Very special thanks to Kevin R. Ellwood of Ford Research and Innovation Center in Dearborn Michigan for his review and feedback on this chapter.

References

1. Bonifant B.: Competitive implications of environmental regulation in the paint and coatings industry. The Management Institute of Environment and Business. https://yosemite.epa.gov/ee/epa/eerm.nsf/vwAN/EE-0045-01.pdf/$file/EE-0045-01.pdf. (1994). Accessed 15 October 2015
2. Holweg, M.: The evolution of competition in the automotive industry. In: Parry, G., Graves, A.P. (eds.) Build to Order: The Road to the 5-day Car. Springer-Verlag, London http://www.springer.com/cda/content/document/cda_downloaddocument/9781848002241-c1.pdf?SGWID=0-0-45-586609-p173810311 (2008) Accessed 15 October 2015
3. Robots painting car body Andrei Kholmov/Shutterstock.com
4. Im, K.S., Lai, M.C., Yoon, S.J.: Spray characteristics on the electrostatic rotating bell applicator. KSME Int. J. **17**, 2053–2065 (2003)
5. Streitberger, H.J., Dossel, K.F.: Automotive Paints and Coatings, 2nd edn. Wiley-VCH, Frankfurt (2008)

6. Nichols, M., Tardiff, J.: Automotive coatings. In: Hughes, A.E., Mol, J.M.C., Zheludkevich, M.L., Buchheit, R.G. (eds.) Active Protective Coatings: Automotive Coatings, Springer Series in Materials Science, vol. 233, pp. 373–384 (2016)
7. Electrocoat Association: Electrocoat Association Electrocoating: A Guidebook for Finishers. Electrocoat Association, Cincinnati (2002)
8. Electrocoated Body Nataliya Hora/Shutterstock.com
9. Connor, D.C., Atar, P.N., Rizzi, A.A., Howie, C.: Experimental verification of deposition models for automotive painting with electrostatic rotating bell atomizers. In: Siciliano, B., Dario, P. (eds.) Experimental Robotics VIII, pp. 136–145. Springer, Berlin (2003)
10. Lin, J., Shen, J., Zimmer, M.E.: Appearance of automotive coatings. In: Wen, M., Dusek, K. (eds.) Protective coatings: film formation and properties. Springer, New York (2017)
11. Tolbert, R.: Paint Technology Handbook. CRC Press, Boca Raton (2008)
12. Dinic, J., Zhang, Y., Jimenez, N.L., Sharma, V.: Extensional relaxation times of dilute, aqueous polymer solutions. ACS Macro Letters **4**(7), 804–808 (2015)
13. Marmottant, P., Villermaux, E.: On spray formation. J. Fluid Mech. **498**, 73–111 (2004). doi:10.1017/S0022112003006529
14. Lefebvre, A.H.: Atomization and Sprays. Hemisphere, New York (1989)
15. Keshavarz, B., Houze, E.C., Moore, J.R., Koerner, M.R., McKinley, G.H.: Ligament mediated fragmentation of viscoelastic liquids. Phys. Rev. Lett. **117**(15), 154502 (2015)
16. Berg, T., Deppe, J., Michaelis, D., Voges, H., Wissel, S.: Comparison of particle size and velocity investigations in sprays carried out by means of different measurement techniques. ICLASS Paper ID ICLASS06-151 (2006)
17. Im, K.-S., Lai, M.-C., Yoon, S.-J.: Spray characteristics on the electrostatic rotating bell applicator. KSME Int. J. **17**, 2053–2065 (2003)
18. von Kármán, T.: Über laminare und turbulente Reibung. Angew. Math. Mech 1, 233–252 (1921); NACA T M. 1092 (1946).
19. Sohrab, S.H.: A modified theory of laminar flow near a rotating disk. IASME Trans. **1**, 152–159 (2005)
20. Domnick, J., Yang, Z., Ye, Q.: Simulation of the Film Formation at a High-Speed Rotary Bell Atomizer Used in Automotive Spray Painting Process. ILASS, Como Lake (2008)
21. Hendrix, H.W., Bouwhuis, W., van der Meer, D., Lohse, D., Snoeijer, J.H.: Universal mechanism for air entrainment during liquid impact. J. Fluid Mech. **789**, 708–725 (2016)
22. Domnick, J.: Effect of bell geometry in high-speed rotary bell atomization. ILASS, 23rd annual conference on Liquid Atomization and Spray Systems, Brno, Czech Republic (2010)
23. Basu, S. K., Davis, R., Hochstetler, S., Williams, C., Moncier, J., Harding, A., McCreight, K.: Atomization to appearance in electrostatic rotating bell coating application. 15th International Coating Science and Technology Symposium, St. Paul (2010)
24. Bousfield, D.W., Keunings, R., Marrucci, G., Denn, M.M.: Nonlinear analysis of the surface tension driven breakup of viscoelastic filaments. J. Nonnewton Fluid Mech. **21**, 79–97 (1986)
25. Tanasawa, Y.: In Atomization and Sprays. Hemisphere, New York (1989)
26. Ahmed, M., Youssef, M.S.: Characteristics of mean droplet size produced by spinning disk atomizers. J. Fluids Eng. **134**, 071103.1–071103.9 (2012)
27. Domnick, J.: High-speed rotary bell atomization of Newtonian and non-Newtonian fluids. ICLASS, 12th Triennial International Conference on Liquid Atomization and Spray Systems, Heidelberg, Germany (2012)
28. Eggers, J., Villermaux, E.: Physics of liquid jets. Rep. Prog. Phys. **71**, (2008). doi:10.1088/0034-4885/71/3/036601
29. Dominick, J., Scheibe, A., Ye, Q.: The simulation of the electrostatic spray painting process with high-speed rotary bell atomizers part I direct charge. Part. Part. Syst. Char. **22**, 141–150 (2005)
30. Dominick, J., Scheibe, A., Ye, Q.: The simulation of electrostatic spray painting process with high-speed rotary bell atomizers part II indirect charge. Part. Part. Syst. Char. **23**, 408–416 (2006)

Index

M. Wen, K. Dušek (eds.), *Protective Coatings*, DOI 10.1007/978-3-319-51627-1

D

I

Printed by Printforce, the Netherlands